Molecular Biology of
Membrane-Bound Complexes in
Phototrophic Bacteria

FEDERATION OF EUROPEAN MICROBIOLOGICAL SOCIETIES SYMPOSIUM SERIES

A Continuation Order Plan is available for this series. A continuation order will bring delivery of each new volume immediately upon publication. Volumes are billed only upon actual shipment. For further information please contact the publisher.

Molecular Biology of Membrane-Bound Complexes in Phototrophic Bacteria

Edited by

Gerhart Drews

Albert-Ludwigs University
Freiburg, Federal Republic of Germany

and

Edwin A. Dawes

University of Hull
Hull, United Kingdom

PLENUM PRESS • NEW YORK AND LONDON

Library of Congress Cataloging-in-Publication Data

Molecular biology of membrane-bound complexes in phototrophic bacteria
/ edited by Gerhart Drews and Edwin A. Dawes.
      p.   cm. -- (FEMS symposium ; no. 53)
    Proceedings of a symposium held under the auspices of the
Federation of European Microbiological Societies and the Deutsche
Forschungsgemeinschaft, 8/2-5/89, Freiburg im Breisgau, Fed. Rep. of
Germany.
    Includes bibliographical references.
    ISBN 0-306-43515-2
    1. Bacteria, Photosynthetic--Congresses. 2. Molecular
microbiology--Congresses.   I. Drews, G. (Gerhart) II. Dawes, Edwin
A. (Edwin Alfred) III. Federation of European Microbiological
Societies.   IV. Deutsche Forschungsgemeinschaft. V. Series.
QR88.5.M65 1990
589.9'088--dc20   90-7068
                                                                  CIP

Proceedings of a symposium held under the auspices of the
Federation of European Microbiological Societies and the
Deutsche Forschungsgemeinschaft, August 2–5, 1989,
in Freiburg im Breisgau, Federal Republic of Germany

© 1990 Plenum Press, New York
A Division of Plenum Publishing Corporation
233 Spring Street, New York, N.Y. 10013

Printed in the United States of America

PREFACE

Cells of phototrophic bacteria are fitted out with a characteristic and sometimes species-specific membrane-system: the continuous but differently composed cytoplasmic-intracytoplasmic membrane systems in purple bacteria, the cytoplasmic membrane with the attached light-harvesting chlorosomes in green bacteria and the intracytoplasmic thylakoids with the attached light-harvesting phycobilisomes in cyanobacteria.

During the long-lasting evolutionary process phototrophic bacteria have been adapted to numerous ecological niches and on this way they have developed various types of light-harvesting antenna systems and pigments. The evolutionary pressure on the development of efficient energy-transducing systems resulted, on the other hand, in homologous structures with a high similarity of primary amino-acid sequences of membrane-bound pigment-binding polypeptides and very similar principles of organization, realized, for example, in the photochemical reaction center and the ubiquinone-cytochrome b/c_1 oxidoreductase of many evolutionary remote organisms.

The bacterial photosynthetic and respiratory apparatuses are much simpler in composition and organization than the corresponding structures of higher organisms. They are, therefore, excellent model systems to study correlations between structure and function and assembly of these highly organized membrane particles. Biophysicists, biochemists and molecular biologists have in close cooperation, but using different methodical approaches, reached a clear progress in this field.

From the 150 contributions to the Symposium on molecular biology of membrane-bound complexes in phototrophic bacteria (Freiburg, August 2-5, 1989) 56 representative papers have been selected and combined in this volume.

The first group of articles is dedicated to structure and regulation of genes coding for pigment-binding membrane proteins and enzymes for bacterio-chlorophyll, carotenoid and cytochrome synthesis under the control of oxygen and light gradients.

Articles of the second chapter deal with composition, structure, organization and function of membrane-bound protein complexes. The localization, orientation and binding of pigments to proteins, the structure of polypeptides in the membrane, the process of excitation energy transfer in antenna complexes, the charge separation and quinone reduction in reaction centers and the reconstitution of the complexes from their subunits are some of the major features.

The third chapter summarizes progress in the analysis of different electron transport systems, the formation of gradients of charges and ions across membranes, specific transport systems, chemotactic sensory transduction and excitation of the flagellar motor.

All those who are interested in structure, function and formation of membrane-bound complexes will find facts and stimulating hypotheses in this field.

I wish to thank the Deutsche Forschungsgemeinschaft and Federation of European Microbiological Societies for financial support and all friends and colleagues who contributed to the success of the meeting.

Gerhart Drews

CONTENTS

I. GENE STRUCTURE AND ORGANIZATION
REGULATION OF GENE EXPRESSION

MOLECULAR GENETICS STUDIES OF GENE EXPRESSION AND PROTEIN STRUCTURE/FUNCTION RELATIONSHIPS IN PHOTOSYNTHETIC BACTERIA

Barry L. Marrs

Central Research and Development Department
E.I. du Pont de Nemours & Company, Inc.
Wilmington, Delaware 19880-0173

This introductory chapter provides an overview of the topics currently studied in photosynthetic bacteria with the tools of molecular genetics, and comments on the appropriate matching of tools and tasks.

Photosynthetic bacteria are a diverse assemblage of microorganisms, distributed over a variety of different habitats and possessing a correspondingly wide range of metabolic activities. Historically, the primary motivation for the study of these bacteria has been to gain a better understanding of photosynthesis itself, but an important secondary theme, which has been explored in many of the photosynthetic bacteria, is the incredible metabolic diversity displayed in the energy metabolism of this group. These dual themes are reflected in the tools of molecular genetics that are being applied to photosynthetic bacteria today. On the one hand, site-directed mutagenesis is being used extensively to probe the structure/function relationships of photosynthetic reaction centers and other energy transduction systems, while studies of the regulation of gene expression, employing the full range of molecular genetic techniques, seek to understand how organisms adapt to changing environmental conditions.

MOLECULAR GENETICS IN STRUCTURE/FUNCTION STUDIES

The basic paradigm for structure/function analysis of reaction centers or antennae by site-directed mutagenesis was published by Youvan et al. in 1985. These authors deleted the chromosomal copies of the reaction center genes, and then supplied engineered copies of those genes on plasmids. In this way they could relatively quickly generate sets of amino acid substitution mutations at points of interest in the reaction center or antenna proteins, allow the altered proteins to be synthesized, and analyze the phenotype of each. They adopted a powerful way of displaying the data generated from these experiments (Bylina and Youvan, 1987). The standard set of twenty amino acids are graphed in two dimensions by plotting molar volume vs hydrophobicity for each, and then contours are drawn around groups of amino acids that may be substituted at a given position to give a particular phenotype (herbicide resistance, photosynthetic competence, etc.). These plots typically give simple contours, suggesting that in many cases these two parameters are sufficient to describe the important features of a residue that determine its function.

Molecular Biology of Membrane-Bound Complexes in Phototrophic Bacteria
Edited by G. Drews and E. A. Dawes
Plenum Press, New York, 1990

It will be recognized from the above example that three genetic capabilities are essential for experimentation using site-directed mutagenesis. First, the relevant genes must be cloned. This capability is generally not a barrier to work with any species, but it is easier to work with organisms for which gene banks are readily available. DNA is usually cloned into *Escherichia coli,* and site directed mutagenesis conducted there by any of a variety of available techniques. Second, it is usually necessary to return engineered DNA to the species of origin. For the deletion/complementation paradigm described above, this requires replicons that are more or less stably maintained in the species of origin as well as in *Escherichia coli,* and carry selectable genetic markers, usually antibiotic resistance-conferring genes. Furthermore, a mechanism must exist for depositing DNA in the cytoplasm of the original species. Here the biology of the individual species varies widely, and some organisms are much better genetic hosts than others. For example *Rhodopseudomonas viridis* has proven most intractable as a genetic research object. Another factor that distinguishes the suitability of an organism for genetic studies is not a biological distinction but an historical truism. Organisms that have been previously used for genetic studies are better objects for additional genetic studies than virgin organisms, largely because of the size of the genetic tool kit available for each species. Third, there must be a convenient means for precise manipulations of regions on the chromosome of the species of origin. For the deletion/complementation paradigm, this can be accomplished by returning to the species of origin an engineered DNA segment in which a region of DNA has been replaced by a selectable marker, and then identifying clones in which homologous recombination has lead to a replacement of the chromosomal region in question by the engineered DNA. This is most easily accomplished in *R. capsulatus,* using the gene transfer agent-mediated technique developed by Scolnik and Haselkorn (1984). A different strategy for chromosomal engineering is required for other species, which do not have a system for delivering linear DNA fragments like those carried by the gene transfer agent of *R. capsulatus.* It is possible to introduce chromosomal deletions plasmids that cannot replicate in the target organism, but the resulting clones must be carefully screened to determine their true genotype, since a variety of genetic events other than homologous replacement can occur. It is conceivable that site-directed mutations could be directly introduced onto the chromosome, avoiding the deletion/complementation paradigm, but tools to accomplish this easily and reproducibly do not currently exist. Ease and reproducibility are important requirements for such tools, because thorough site-directed studies require the construction of not just one new mutation, but many changes at each site of interest. Tools for this purpose are under development in *R. capsulatus.*

"O, the June Bug she has golden wings, the Lightning Bug has fame; the Weevil has no wings at all, but he gets there just the same" (from an American folk song). The superior genetic systems of *Rhodobacter capsulatus* allowed the first cloning, complete sequencing and site-directed changes of photosynthetic reaction center genes, but reaction center crystals were first made, and their structures solved, with material isolated from *Rhodopseudomonas viridis.* *Rhodobacter sphaeroides* placed second in both of those races, but it is currently the only organism for which we have both crystal structure data and genetic systems capable of site-directed mutagenesis, so an increasing amount of structure/function work is being done in that system. Although crystal structures for the reaction centers of *R. capsulatus* are not available, it has been possible to analyze the results of mutational changes in *R. capsulatus* reaction centers by assuming significant structural homology accompanies the obvious amino acid sequence homologies between *Rhodobacter sphaeroides* and *Rhodobacter capsulatus* (78% identity for the L subunits of the *Rhodobacter* species, 77% for their M subunits; Williams et al, 1984). The general validity of this assumption is attested to by the coherence of results from numerous site-

directed changes when interpreted under the assumption of homology.
Another way to get around the lack of both good genetics and crystal structure
in the same organism would be to move the genes for either *R. sphaeroides* or
Rhodopseudomonas viridis reaction centers into an *R. capsulatus* deletion strain.
There is some evidence that at least the *R. sphaeroides* reaction centers would
function properly in the *R. capsulatus* cell milieu (Daldal, personal
communication).   Less global changes, involving partial swaps of one peptide or
peptide segment for another, will test how homologous the proteins from the
various species really are, but will probably not simplify the interpretation of
functional changes in terms of precise structural data, unless whole new crystal
structures are generated.

Cyanobacteria, the blue-green algae with plant-like photosynthetic
systems, have more recently gained the minimal set of genetic tools with which
site-directed experiments can be conducted.   The primary organisms used are
Synechococcus sp. PCC 7002 and *Synechocystis* 6803.   Photosystem II is studied,
again making the assumption of homology to crystal structures of the *R.
sphaeroides* and *Rhodopseudomonas viridis* reaction centers.   The water
splitting capabilities of PSII are not found in the simpler purple photosynthetic
bacteria, so this is an ideal area of study in which to apply the blue-green
systems.   Phycobilisomes, the well-characterized antenna system unique to
cyanobacteria, are also objects of investigation by site-directed mutagenesis
techniques.

MOLECULAR GENETICS IN GENE EXPRESSION STUDIES

There are at least three dimensions that are relevant to a discussion of
currents studies of gene expression in photosynthetic bacteria: what
physiological systems are under study, in which organisms are they being
studied, and at what level of genetic organization are the studies focussed.
Nitrogen fixation, pigment biosynthesis, hydrogen metabolism, and electron
transport join antennae and reaction center studies as those physiological
systems most explored at the level of gene expression in photosynthetic
bacteria.   In the accompanying papers we find a sampling of a few more genera
(Chloroflexus, Calothrix, Pseudanabaena) than are used in the
structure/function studies discussed above,   but the majority of the work is
conducted with *Rhodobacter* species or *Synechocystis* or *Synechococcus*.   One
notes that the purple sulfur bacteria and their remarkable transformations of
sulfur compounds are not among the objects of genetics-assisted study.   Most
unexpected is the range of levels of gene organization over which these
questions have been explored, running the gamut from genomic mapping,
through superoperons, operons, genes and control regions, right down to
individual mutations.

The most widely studied system in the photosynthetic bacteria is the
regulation of expression of genes coding for the photosynthetic apparatus
itself.   The range of regulatory phenomena exhibited in the photosynthetic
apparatus of *Rhodobacter* species, for example, includes changes in level of
expression in response to oxygen tension, light intensity and carbon source,
and changes in the composition of the apparatus in response to these same
signals.   Throughout these changes, neither free pigments nor free proteins
are accumulated to a measurable extent, implying that small molecule and large
molecule syntheses are somehow coordinated.   Furthermore, assembly and
accumulation of one type of antenna complex (light-harvesting II) is normally
dependent upon the synthesis of carotenoids, implying a mechanism for
coordinating that particular pigment/protein pair.

The availability of genetic tools often determines what studies are carried
out in pursuit of understanding, and studies of how regulation of gene

expression occurs fit this pattern. Some tools, such as hybridization-based assays (Southern, northern and dot blotting, S1 mapping etc.) and sequence analysis require only cloning technologies, and thus are applicable to virtually any system. With these techniques an investigator could theoretically measure changes in expression of the level of transcripts from any given gene, and study which genes are co-regulated and co-transcribed with other genes. The difficulty in using this approach alone lies in the variability of the metabolism of mRNA's in bacteria. Some segments of mRNA have short half-lives compared to others, ranging from a matter of seconds to tens of minutes, so the amount of signal measured by hybridization depends on both synthesis and degradation rates. Furthermore, since transcription, translation and degradation are known to proceed simultaneously on the same transcript in bacteria, different genes from the same operon may seldom or never be found on the same molecule of mRNA. Finally, complete knowledge of any hybridization probe is essential. If a restriction fragment used as a probe for a particular gene also carries other genetic regions, misleading data can easily be generated. For these and other reasons, it is a great aid to the interpretation of hybridization data to also have data reflecting gene expression at the translational level in the organism of interest. This may be accomplished by fusion of a reporter enzyme sequence to the genetic region under investigation, a technique that requires methods for introducing engineered DNA into the organism of interest. This is a limitation for many systems, and time consuming for any system, but the data gathered can be critical for understanding.

A description of the *puf* operon provides illustrations of some of these points. This operon includes, in the following order, the *pufQ* gene, required for bacteriochlorophyll synthesis, *pufB* and A, light harvesting I antenna peptides, *pufL* and *M*, reaction center proteins, and *pufX* of unknown function. Early work in *R. capsulatus* (Belasco et al, 1985), based on sequence analyses and S1 nuclease 5' end mapping, suggested that the *puf* operon began just upstream of the *pufB* gene. Functional analysis of this region, using fusions of a reporter enzyme to proteins of the *puf* operon, later showed that the operon actually begins approximately 700 bp upstream from the *pufB* gene, and a new gene, *pufQ* was located in the intervening region (Bauer *et al*, 1988). Similar confusion seems to currently exist in the *R. sphaeroides* system (Kansy and Kaplan, 1989), where workers have attempted to locate the beginning of the *puf* operon by analyzing various types of biochemical data in the absence of appropriate genetic data. They have described a great number of RNA fragments, and they suggest that the *puf* operon begins just upstream of the *pufB* gene. While it remains a logical possibility that *R. capsulatus* and *R. sphaeroides* are fundamentally different in regulation of expression of the *puf* operon, it seems more likely that when the proper experiments are done, the *R. sphaeroides* puf promoter will resemble that of *R. capsulatus*.

REFERENCES

Bauer, C.E., Young, D.A., and Marrs, B.L. (1988) J. Biol. Chem. 263: 4820-4827.
Belasco, J.G., Beatty, J.T., Adams, C.W., von Gabian, A., and Cohen, S.N. (1985) Cell 40: 171-181.
Bylina, E.J., and Youvan, D.C. (1987) Z. Naturforsch. 42c: 769-774.
Kansy, J.W., and Kaplan, S. (1989) J. Biol. Chem. 264: 13751-13759.
Scolnik, P.A., and Haselkorn, R. (1984) Nature 307: 289-292.
Williams, J.C., Steiner, L.A., Feher, G., and Simon, M.I. (1984) Proc. Natl. Acad. Sci. USA 81: 7303-7307.
Youvan, D.C., Ismail, S., and Bylina, E.J. (1985) Gene 38: 19-30.

PHYSICAL MAPPING OF THE GENOME OF *RHODOBACTER CAPSULATUS*

J. C. Williams*†§, B. L. Marrs†, and S. Brenner*

*Molecular Genetics Unit, Medical Research Council, Hills Road, Cambridge, CB2 2QH, England; and †Central Research and Development Department, Experimental Station, E. I. du Pont de Nemours & Co., Wilmington, DE 19880-0173, U. S. A.

INTRODUCTION

Among the photosynthetic bacteria, *Rhodobacter capsulatus* has been particularly well characterized genetically (reviewed in Scolnik and Marrs, 1987). By a variety of fine structure genetic mapping techniques, it has been previously determined that most of the genes involved in the synthesis of the photosynthetic apparatus in this organism are clustered in a small region of the genome (Yen and Marrs, 1976), and that region has been cloned (Marrs, 1981; Taylor, *et al.*, 1983)) and some of the genes sequenced (Youvan, et al., 1984). Mapping by R plasmid-mediated conjugation has produced a circular linkage map of the chromosome (Willison, *et al.*, 1985). A detailed study of the overall organization of the chromosome of *R. capsulatus* should assist in the understanding of the global regulation of its genes, and the availability in cloned form of all parts of the genome will facilitate the identification and sequencing of additional genes. Towards these ends we are determining a physical map of the chromosome of *R. capsulatus*.

A physical genome map consists of an ordered set of DNA clones, and it may be constructed by finding the overlaps among a random library of genomic clones. A method for determining these overlaps by restriction fingerprinting has been developed and used in mapping the genome of the nematode, *Caenorhabditis elegans* (Coulson, *et al.*, 1986). An additional type of physical mapping consists of ordering the large chromosomal fragments that may be produced by the action of restriction enzymes at rare sites and separated by pulsed field gel electrophoresis. It has been shown previously that an *Xba*I digest of *R. capsulatus* DNA yields a small number of fragments (McClelland, *et al.*, 1987). We have used the restriction fingerprinting technique in conjunction with the ordering of large *Xba*I fragments to develop a physical genome map for *R. capsulatus*.

§Present address: Department of Chemistry and Center for the Study of Early Events in Photosynthesis, Arizona State Universtiy, Tempe, AZ 85287-1604, U.S.A.

Molecular Biology of Membrane-Bound Complexes in Phototrophic Bacteria
Edited by G. Drews and E. A. Dawes
Plenum Press, New York, 1990

5

EXPERIMENTAL PROCEDURES

Cloning_ *R. capsulatus* SB1003 was grown aerobically in 0.3% yeast extract, 0.3% peptone, 2 mM CaCl$_2$ and 2 mM MgSO$_4$. The cells were harvested and resuspended in 100 mM EDTA, 5 mM Tris-Cl pH 8, 0.5% sodium dodecylsulfate, and 50 µg/ml proteinase K, incubated at 50°C for 2 hr, then extracted with phenol at 4°C for 15 min. The DNA was precipitated in ethanol, dispersed in 10 mM Tris-Cl pH 7.4, 1 mM EDTA, and partially digested with *Sau*3A. Large fragments (20-40 kilobase pairs (kb)) were isolated by phenol extraction after electrophoresis through two 0.4% low melting temperature agarose gels. The fragments were ligated with the cosmid vector Lorist2 (Gibson, *et al.*, 1987) which had been digested with *Bam*H1 and treated with calf intestinal alkaline phosphatase. The ligated DNA was packaged into λ extracts and used to infect *Escherichia coli* 1046 as described in Maniatis, *et al.* (1982).

Fingerprinting._ The technique for comparison of clones was adapted from that of Coulson, *et al.* (1986). An aliquot of cosmid DNA isolated from each clone was digested with *Bam*H1, and the resulting ends labeled by treatment with avian myeloblastosis virus reverse transcriptase, dGTP, [α-^{35}S]dATP, and ddTTP. After inactivation by heat of the reverse transcriptase, the samples were digested with *Sau*3A. A 6% polyacrylamide gel was used for electrophoresis of the samples. The gel was fixed, dried and autoradiographed, and the resulting fingerprint data were digitized using a scanning densitometer of local design and manipulated using the programs described by Sulston, *et al.*(1988).

*Xba*I fragment analysis._ *R. capsulatus* cells were encapsulated in agarose beads (Jackson and Cook, 1985) and lysed as described (McClelland, *et al.*, 1987). The DNA was digested with 100 units of *Xba*I and electrophoresed on a 0.8% agarose gel in 45 mM Tris, 45 mM boric acid, and 0.5 mM EDTA using ramped field inversion (MJ Research PPI-200, program 5). The gels were dried and hybridized directly with labeled DNA (Purrello and Balazs, 1983). The fragments were also excised from low melting temperature agarose gels, labeled by random priming (Feinberg and Vogelstein, 1983), and hybridized with cosmid DNA.

A library containing *Xba*I linking fragments was created by digesting the *Sau*3A partial fragments with *Xba*I and ligating with the vector λ2001 (Karn, *et al.*, 1984) that had been digested with *Xba*I and with the right arm of λ2001 that had been digested with *Bam*H1. These were then packaged and used to infect *E. coli* Q358. Clones from this *Xba*I/*Sau*3A λ library were fingerprinted to determine the matching cosmid clones. The cosmids were subsequently digested with *Xba*I to confirm the presence of an *Xba*I site. These cosmid clones were hybridized with a genomic *Xba*I digest to determine which *Xba*I fragments were contiguous on the chromosome.

Location of genes._ A set of 200 clones, having a 50% overlap between neighbors, was chosen that covered the entire map. These clones were screened by making a dot blot grid with cosmid DNA that had been isolated from each clone. Approximately 1 µg of DNA from each clone was spotted on a 10 by 20 cm grid on a GeneScreen (DuPont, DE) filter. The DNA was denatured by laying the filter on a solution of 0.5 N NaOH and 1.0 M NaCl, and then neutralized by laying the filter on a solution of 0.5 M Tris-HCl, pH 7.4 and 0.3 M NaCl. The filters were hybridized against labeled probes as described in Maniatis *et al.* (1982). After autoradiography, the blots were stripped of bound probe by incubation in the denaturing and neutralizing solutions. The blots were reused approximately 10 times. To locate genes on mapped cosmids, clones of genes isolated previously from *R. capsulatus* were used as probes. These probes were either labeled plasmid DNA, or labeled fragments of the plasmids cut from agarose gels, in the cases where the plasmid contained DNA that would hybridize with the cosmid vector. The restriction patterns of the cosmid clones that hybridized with the cloned genes were used to verify that these mapped clones contained the corresponding genes.

RESULTS

Map construction. A physical map of cosmids containing genomic DNA fragments from *R. capsulatus* was constructed using restriction fingerprint data. The overlaps between 1000 clones from a cosmid bank were analyzed by *Bam*H1/*Sau*3A restriction fingerprinting. The clones were joined in contigs (groups of clones with contiguous nucleotide sequences) based on the fingerprint data. The number of contigs rose to a maximum of 60 when 574 clones had been fingerprinted, and then decreased. After this initial mapping, 11 contigs remained. Several thousand additional clones were screened by hybridization with fragments of clones from the ends of the 11 contigs. Fragments were chosen that were found in the clones at the ends of contigs, but that were not in the penultimate clone and did not contain the vector. Fingerprints of 450 clones chosen in this way were analyzed, and the additional data used to join 8 contigs.

Fourteen fragments were observed after separation of a genomic *Xba*I digest by pulsed field gel electrophoresis, in agreement with the results of McClelland *et al.* (1987). The fragments were numbered from the smallest (#1) to the largest (#12). On most of the pulsed field gels, only 12 bands were visible, however it was determined that the 10th and 11th bands were doublets; the two bands in each case were labeled a or b. In addition, when the gels were run in low melting temperature agarose, the 10th and 11th band doublets did not separate. The exact molecular weights of the fragments were not determined, however, the fragments ranged in size from approximately 50 to 500 kb.

The restriction fingerprinting gave an initial indication of how the clones fit together in the map. Many of the clones had unambiguous overlaps of over 50% of the fingerprint bands. Hybridization of the *Xba*I fragments to representatives of the mapped clones indicated where overlaps that had a higher probability of being random should be made. For the smallest *Xba*I fragments (#1 to #9) the clones could be clearly positioned within the fragments. The large *Xba*I fragments #10a, #10b, #11a and #11b were not well separated on gels, and thus clones hybridizing with these fragments could not be assigned to one of the four individual fragments. The map is currently being refined with additional data. A portion of the map is shown in Figure 1.

Location of genes. The location in the map of the plasmid pRPS404 (Marrs, 1981), containing a photosynthetic gene cluster, was determined from a restriction fingerprint of this plasmid. The position and orientation of this region, bounded by the *puhA* gene encoding the H subunit of the photosynthetic reaction center and the *pufM* gene encoding the M subunit of the reaction center, were confirmed by restriction digests. Genes previously cloned from *R. capsulatus* that have been placed in the map by hybridization of the cloned DNA and confirmed with restriction digests include the LHII or B800-850 antenna structural genes (*puc A* and *B*; Youvan and Ismail, 1985), the three structural genes of the bc_1 complex (*pet A-C*; Daldal, *et al.*, 1987), the cytochrome c_2 structural gene (*cycA*; Daldal, *et al.*, 1986), the gene encoding

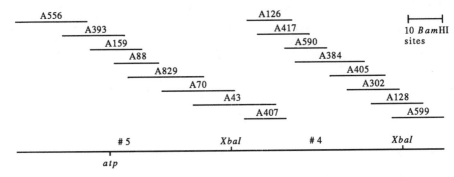

Fig. 1. Portion of physical map of *R. capsulatus* chromosome. Only the canonical clones are shown. The length of the line representing each clone is proportional to the number of bands in the fingerprint of that clone, and the degree of overlap between clones is proportional to the number of fingerprint bands shared between them. Each *Bam*HI site generates two fingerprint bands. The average cloned fragment was 35 kb. and yielded 15 fingerprint bands; thus 10 *Bam*HI sites are approximately equal to 50 kb. Below the clones are listed the positions of *Xba*I fragments #4 and #5, and the position of the clones hybridizing with a probe for the F_1 atpase structural genes from *Rhodopseudomonas blastica*.

glutamine synthetase (*glnA*; Scolnik, *et al.*, 1983), the nitrogenase structural genes (*nif H,D,K;* Avtges, *et al.*, 1983), and the 16S ribosomal RNA gene (*rrnA*; Schumann, *et al.*, 1986). Three copies of the 16S ribosomal RNA genes were found. Probes containing portions of the structural genes for the α and β subunits of the F_1 atpase (*atp*) from *R. blastica* (Tybulewicz, *et al.*, 1984) hybridized with a cosmid clone, however the identity of the corresponding *R. capsulatus* genes has not been confirmed.

DISCUSSION

The development of new mapping techniques has made it technically feasible to physically map the entire genetic content of an organism. The strategy used here to order the *R. capsulatus* clone banks is derived from that developed by Coulson, *et al.*, (1986) in mapping the genome of *C. elegans*, in which a modified type of restriction digest of the clones yields small fragments that are easily resolved and compared. Modifications of the technique were made to take advantage of the difference in base composition between *R. capsulatus* and *C. elegans*, including the use of *Bam*H1 rather than *Hind*III in the initial digest, and increasing the percentage of acrylamide in the gels for better separation of the relatively smaller size of the *Sau*3 A fragments. The information contained in the data set was not enough to completely order the map, and supplemental information in the form of large genomic restriction fragments was used to corroborate the overlaps among the clones. Further refinements of the fingerprinting technique should also provide more information from each clone in order to reduce the amount of overlap needed to form reliable joins.

The positions of the genes and gene clusters appear to be distributed evenly around the genome. The largest grouping of genes is the

8

superoperonic photosynthetic gene cluster. The genes for the B800-850 antenna proteins (*pucA* and *B*) were found to be far away from this cluster. This may reflect different evolutionary origins of the antenna complexes and the major photosynthesis gene cluster in the photosynthetic bacteria. The *crt, glnA* and *nif H,D,K* genes have been mapped genetically as well as physically (Willison, *et al.*, 1985), and the distances between these genes on the physical map corresponds approximately to their genetic linkage.

This ordered clone library is a resource that will facilitate future genetic studies on this organism. Screening for genes in the ordered library requires much less effort than is needed for screening random libraries, and the clone for any gene is available once its location on the map is known. With the physical arrangement of the genes on the chromosome known, the global organization of the genome can begin to be understood, and the extent to which the genes act in a concerted manner to effect a response to environmental stimuli can be probed.

Acknowledgements We thank J. Sulston and A. Coulson for assistance with biochemical techniques and computer analyses used in restriction fingerprinting; T. Allport for help with screening the banks and with fingerprinting; F. Daldal, P. Scolnik, J. Shively, J. Wall, J. Walker and D. Youvan for providing cloned DNA; and M. McClelland and R. Jones for sharing unpublished data.

REFERENCES

Avtges, P., Scolnik, P. A., and Haselkorn, R., 1983, Genetic and physical map of the structural genes (*nif*H,D,K) coding for the nitrogenase complex of *Rhodopseudomonas capsulata*, J. Bacteriol., 156:251.

Coulson, A., Sulston, J., Brenner, S., and Karn, J., 1986, Toward a physical map of the genome of the nematode *Caenorhabditis elegans*, Proc. Natl. Acad. Sci. USA, 83:7821.

Daldal, F., Cheng, S., Applebaum, J., Davidson, E. and Prince, R. C., 1986, Cytochrome c_2 is not essential for photosynthetic growth of *Rhodopseudomonas capsulata*, Proc. Natl. Acad. Sci. USA, 83:2012.

Daldal, F., Davidson, E., and Cheng, S., 1987, Isolation of the structural genes for the Rieske Fe-S protein, cytochrome *b* and cytochrome c_1, all components of the ubiquinol:cytochrome c_2 oxidoreductase complex of *Rhodopseudomonas capsulata*, J. Mol. Biol., 195:1.

Feinberg, A. P. and Vogelstein, B., 1983, A technique for radiolabeling DNA restriction endonuclease fragments to high specific activity, Anal. Biochem., 132:6.

Gibson, T. J., Coulson, A. R., Sulston, J. E., and Little, P. F. R., 1987, Lorist2, a cosmid with transcriptional terminators insulating vector genes from interference by promoters within the insert: effect of DNA yield and cloned insert frequency, Gene, 53:275.

Jackson, D. A. and Cook, P. R., 1985, A general method for preparing chromatin containing intact DNA, EMBO J., 4:913.

Karn, J., Matthes, H. W. D., Gait, M. J. and Brenner, S., 1984, A new selective phage cloning vector, λ2001, with sites for *Xba*I, *Bam*HI, *Hin*dIII, *Eco*RI, *Sst*I and *Xho*I, Gene, 32:217.

Maniatis, T., Fritsch, E. F., and Sambrook, J., 1982, "Molecular Cloning: A Laboratory Manual," Cold Spring Harbor Laboratory, Cold Spring Harbor, NY.

Marrs, B. L., 1981, Mobilization of the genes for photosynthesis from *Rhodopseudomonas capsulata* by a promiscuous plasmid, J. Bacteriol., 146:1003.

McClelland, M., Jones, R., Patel, Y., and Nelson, M., 1987, Restriction endonucleases for pulsed field mapping of bacterial genomes, Nucleic Acids Res., 15:5985.

Purrello, M. and Balazs, I., 1983, Direct hybridization of labeled DNA to DNA in agarose gels, Anal. Biochem., 128:393.

Schumann, J. P., Waitches, G. M., and Scolnik, P. A., 1986, A DNA fragment hybridizing to a *nif* probe in *Rhodobacter capsulatus* is homologous to a 16S rRNA gene, Gene, 48:81.

Scolnik, P. A. and Marrs, B. L., 1987, Genetic research with photosynthetic bacteria, Ann. Rev. Microbiol., 41:703.

Scolnik, P. A., Virosco, J., and Haselkorn, R., 1983, The wild-type gene for glutamine synthetase restores ammonia control of nitrogen fixation to Gln⁻ (*glnA*) mutants of *Rhodopseudomonas capsulata* , J. Bacteriol., 155:180.

Sulston, J., Mallett, F., Staden, R., Durbin, R., Horsnell, T., and Coulson, A., 1988, Software for genome mapping by fingerprinting techniques, CABIOS, 4:125.

Taylor, D. P., Cohen, S. N., Clark, W. G., and Marrs, B. L., 1983, Alignment of genetic and restriction maps of the photosynthesis region of the *Rhodopseudomonas capsulata* chromosome by a conjugation-mediated marker rescue technique, J. Bacteriol., 154:580.

Tybulewicz, V. L. J., Falk, G., and Walker, J. E., 1984, *Rhodopseudomonas blastica atp* operon, nucleotide sequence and transcription, J. Mol. Biol., 179:185.

Willison, J. C., Ahombo, G., Chabert, J., Magnin, J.-P., & Vignais, P. M., 1985, Genetic mapping of the *Rhodopseudomonas capsulata* chromosome shows non-clustering of genes involved in nitrogen fixation, J. Gen. Micro., 131:3001.

Yen, H-C., and Marrs, B. L., 1976, Map of genes for carotenoid and bacteriochlorophyll biosynthesis in *Rhodopseudomonas capsulata* , J. Bacteriol., 126:619.

Youvan, D. C., Bylina, E. J., Alberti, M., Begusch, H., and Hearst, J. E., 1984, Nucleotide and deduced polypeptide sequences of the photosynthetic reaction center, B870 antenna, and flanking polypeptides from R. capsulata, Cell, 37:949.

Youvan, D. C. and Ismail, S., 1985, Light-harvesting II (B800-850 complex) structural genes from *Rhodopseudomonas capsulata*, Proc. Natl. Acad. Sci. USA, 82:58.

PRELIMINARY STUDIES ON THE OPERON CODING FOR THE REACTION

CENTER POLYPEPTIDES IN *CHLOROFLEXUS AURANTIACUS*

Judith A. Shiozawa, Katalin Csiszár and
Reiner Feick

Max-Planck Institut für Biochemie
D-8033 Martinsried b. München BRD

INTRODUCTION

The isolated reaction center of the thermophilic, facultative photoheterotroph *Chloroflexus (C.) aurantiacus* is composed of two polypeptides (Shiozawa *et al.*, 1987). The genes encoding the two polypeptides have been cloned and sequenced (Shiozawa *et al.*, 1989 and Ovchinnikov *et al.*, 1988 a, b). The two genes are adjacent to one another; 17 bases separate the stop codon of the L-gene and the start codon of the M-gene. The deduced amino acid sequences are about 40% similar to the respective subunits of *Rhodobacter sphaeroides*. We have now undertaken studies to determine if the *puf* operon of this green nonsulfur bacteria is similar to that of Rhodospirillaceae.

MATERIALS AND METHODS

Sources of biochemicals, chemicals and isotopes have been described previously Shiozawa *et al.*, (1989).

Chloroflexus aurantiacus, strain J10-fl (Deutsche Sammlung von Mikroorganismen, Göttingen) were grown photoheterotrophically or aerobicaly in the dark at 55°C in Medium DG (Castenholz and Pierson, 1981).

Cellular RNA from *C.aurantiacus* was isolated essentially as described by Aiba *et al.* (1981). Glyoxal/DMSO denatured RNA were separated on 1% agarose gels and then transferred to GeneScreen™ (NEN Chemicals, Dreieich FRG) and hybridized as described in the GeneScreen™ instruction manual. Primer extension experiments were carried out following the procedure of Ausubel *et al.* (1988). The method of Hennighausen and Lubon (1987) was used for mobility shift assays. DNA sequencing of pUC5.2A and pUC4.1B was performed as described in Shiozawa *et al.* (1989).

Molecular Biology of Membrane-Bound Complexes in Phototrophic Bacteria
Edited by G. Drews and E. A. Dawes
Plenum Press, New York, 1990

RESULTS AND DISCUSSION

The transcript of the L and M genes of the *C. aurantiacus*
reaction center. Under steady state conditions, the predominant form of mRNA on which the L and M genes were found had a
relative mobility of about 2.3 kb (Fig. 1). Minor bands were
observed at about 3.1, 1.7 and 1.5 kb. The level of mRNA for
both genes is much lower in chemoheterotrophically grown cells
than in photoheterotrophically grown cells. It is likely that
both genes occur on the same transcript; this will be tested
with a DNA probe on which portions of both genes occur. In
preliminary primer extension experiments, the approximate
transcription start point was shown to be 130 bases 5'of the
ATG start codon of the L gene (Fig. 2).

Nucleotide sequencing upstream and downstream of the L
and M reaction center genes. Approximately 1 kb 5'of the L
gene and 350 bp 3'of the M gene have been sequenced (Figs. 3
and 4). A portion of the downstream sequence has been previously published (Ovchinnikov *et al.*, 1988b). Within this
overlapping region, a single base difference was found (base
number 2067 to 2071). The sequence we obtained reads ...
GGGGC... rather than GGGCA....Potential open reading frames
longer than 20 amino acids are shown on both of the figures.
None of the open reading frames found in the upstream region
showed any significant similarity (> 16%) to either the α and
β polypeptides of the B806-866 light-harvesting complex (Wechsler *et al.*, 1985 and 1987) or to *pufQ* (Bauer *et al.*, 1988 and
Adams *et al.*, 1989). Four potential ORFs were also found

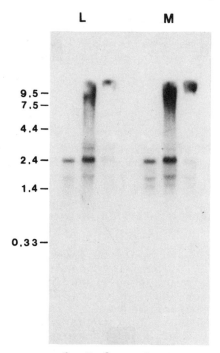

L　　**M**

9.5 —
7.5 —

4.4 —

2.4 —

1.4 —

0.33 —

p p a　　p p a

Fig. 1. Northern blots.
Cellular RNA from photohetero
trophically (p) and aerobically
(a) grown cells were hybridized
with nick-translated probes
specific for *pufL* (794 bp *Hinf*
I fragment) or for *pufM* (650 bp
*Xba*I-*Asp*718 fragment). The mobilities of RNA markers are in
dicated on the left. The
smearing in the 7.5 to 10 kb
region is due to DNA present in
the preparations.

downstream of the stop codon of the M gene. None of these
showed any similarity to the gene product of *pufX* (Youvan *et
al.*, 1984 and Farchaus, this volume). A computer search
through the combined MIPS protein data bank (Martinsried) did
not reveal any protein with significant sequence similarities
to the ORFs in the up- and downstream regions of the *C. aurantiacus puf* operon.

Since primer extension results indicated that the transcription of the L and M genes began about 130 bases upstream
of the start codon for the L gene, the nucleotide sequence in
this region was analyzed for consensus sequence homologies to
E. coli promoters. Between base pair numbers -142 and -116
(Fig. 3) were sequences showing very high similarity to the *E.
coli* promoter sequence for the σ_{70} subunit of RNA polymerase:

promoter sequence	-35	-10
E. coli	TTGACA	TATAAT
C. aurantiacus *puf* operon	TTGACA	TATTAT

Between the "-35" and "-10" regions of the *Chloroflexus* promoter sequence, the nucleotide sequence TATCA was found; this
probably corresponded to the weakly conserved -16 sequence
TATGT of *E. coli*. In *C. aurantiacus*, the sequence just 3'of
the Pribnow box (-10 sequence) was GC-rich which indicated
that this was probably a "weak" promoter. Slightly further
upstream in *C. aurantiacus* (particularly between -167 and -
156) were regions containing poly (dA)-(dT). These homopoly-
mer regions have been postulated to be areas of DNA contact
with the polymerase (Tavers, 1987).

Fig. 2.   Primer Extension.
Cellular RNA from photoheterotrophically grown *C. aurantiacus*
was hybridized with the 5'-labeled
oligonucleotide CGCTTTTGCTCTGCTCAT
which comprises the 5'terminus of
pufL and extended with AMV reverse
transcriptase. In this experiment,
RNA prepared by different methods
were subjected to primer extension.
The pattern in the fifth lane from
the left was obtained with 60 µg
RNA prepared by the method of Aiba
et al.(1981). The two lanes to the
left are PNK-labeled pBR322 digested
with *Hha*I.

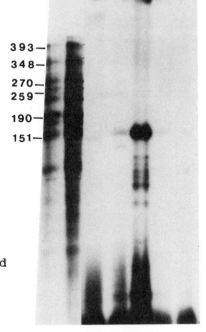

393—
348—
270—
259—
190—
151—

```
-945  TAATACGGTAAGCCCAATCAAGGTCTTCGCCATACATGAAGAACGCCTCATCGAGCAGAC  -886
      N  T  V  S  P  I  K  V  F  A  I  H  E  E  R  L  I  E  Q  T
                              M  K  N  A  S  S  R  P

-885  CGACTTCGCGCGACCACCGACGTGCGAACCATCATACACGCCCCGACCACCGCATCAACC  -826
      D  F  A  R  P  P  T  C  E  P  S  Y  T  P  R  P  P  H  Q  P
      T  S  R  D  H  R  R  A  N  H  H  T  R  P  D  H  R  I  N  L

-825  TCAGTCTCCACATCAGGATCGAGGTAAGTCAAATTATATGCAGCGAAGCGGCGGCTGCGC  -766
      Q  S  P  H  Q  D  R  G  K  S  N  Y  M  Q  R  S  G  G  C  A
      S  L  H  I  R  I  E  V  S  Q  I  I  C  S  E  A  A  A  A  R

-765  GGGAAGAGGCGGGCCAGCCCGAACAAGCGGTAAAAGGCAATTGCAGGAGTGGGGAAACTG  -706
      G  R  G  G  P  A  R  T  S  G  K  R  Q  L  Q  E  W  G  N  C
      E  E  A  G  Q  P  E  Q  A  V  K  G  N  C  R  S  G  E  T  A

-705  CGCCGACAGGCCAGATCGAGCGAACCATCGGGTAAGAGCAGTTTCGGTCCCACGACGCCG  -646
      A  D  R  P  D  R  A  N  H  R  V  R  A  V  S  V  P  R  R  R
      P  T  G  Q  I  E  R  T  I  G  *

-645  ACCGTTGGATGATGGTCGAGATAATCAACCATCCCATCGAGGGCACCGGGCGGTACTACG  -586
      P  L  D  D  G  R  D  N  Q  P  S  H  R  G  H  R  A  V  L  R
        M  M  V  E  I  I  N  H  P  I  E  G  T  G  R  Y  Y  G

-585  GTATCGTTATTGAGAAGCAGAATGTAGTCAGGTGGATCGGGTAAGGTAAGAATCTGCCGT  -526
      Y  R  Y  *
        I  V  I  E  K  Q  N  V  V  R  W  I  G  *

-525  AACGCCAGATTGTTACCGGCAGAAAAACCACCGTTGACCGGACTCTCGATCAGATGCACC  -466

-465  CACCCAAACCGCTCACGAACCATTGCGGCGCTACCATCAGTTGAGGCATTGTCAACCACC  -406

-405  CACACCGATAGATCGCAGCGGGTTGGCGAAGCAGCAATCGAGGCCAGACAATCAGCTAAC  -346

-345  AGATCGGCCCGGTTGTAGTTGAGGATAACAACGGCAAGACTACTCATTATCGACATGCCC  -286

-285  GGTCTGCACTGGTAGACAACCCTGGTTCAGTATACCACAGGTTATTCTTGTTTTATATGC  -226
                                                            M  Q

-225  AAGAAGCGAATGGTAATCGATTAGACGTTTGCAAAATCAATTTAAGGCTTCATAATGCAA  -166
      E  A  N  G  N  R  L  D  V  C  K  I  N  L  R  L  H  N  A  K

-165  AATTAAATATCGATACCTATAGCTTGACAATGTGTATCAGGCGTATTATACTGGCCGAAA  -106
      L  N  I  D  T  Y  S  L  T  M  C  I  R  R  I  I  L  A  E  K

-105  AGGCGGTAAAATTTTGCTTATCATTACACGTTATCCGAGTCAGGGCAACCGTAAGGCGCG  -46
      A  V  K  F  C  L  S  L  H  V  I  R  V  R  A  T  V  R  R  D

-45   ATGTGAGAACGGTCGCCGCGCCAACCGTCAGCAGAGGGAGTCGCTATGAGCAGAGCAAAA  14
      V  R  T  V  A  A  P  T  V  S  R  G  S  R  Y  E  Q  S  K  S
                                                M  S  R  A  K

15    GCGAAAGACCCCCGTTTTCCCGACTTTTCGTTCACCGTCGTTGAGGGTGCGCGGGCCACA  74
      E  R  P  P  F  S  R  L  F  V  H  R  R  *
      A  K  D  P  R  F  P  D  F  S  F  T  V  V  E  G  A  R  A  T

75    CGAGTACCGGGAGGGCGGACGATT  98
      R  V  P  G  G  R  T  I
```

Fig. 3. Nucleotide sequence of DNA region 5' of the L-gene of
         C. aurantiacus reaction center. Sequences of poten
         tial open reading frames longer than 20 amino acids.
         are shown. Shine-Dalgarno sites are indicated with
         a bar above the sequence. The two regions with bars
         above and below the sequence comprise the putative
         promoter sequence of the *C. aurantiacus puf* operon.
         The "A" in the start codon of the L gene equals 0.

```
1853   ACGCCACCGGTTTCACTGCCGTAGCGTTCGTGACACTTCTCTACCTTCCTCCCGGCCTCT   1912
       T  P  P  V  S  L  P  *

1913   CGCCCGTCATCAGACGGGCGAGAGGTTGTTTATTGCACGGGGAGAGGTGACTATCTGAAG   1972

1973   ACTGGTGTATCAAATCTGGCTTCATAACAGTCATTGCACGCCGTTTTCAGGTGTGCCACC   2032

2033   GGCGCGTCGGTGGCTTGTTGTCGGTGGGCCGGTAGGGGCACGGCATGCCGTGCCCCTACG   2092
                                                    M  P  C  P  Y  A

2093   CCTCCTGTCGTACCAGTAGAGACGACGCATGCGTCGCCCCTACCAGCGTCCAGCGGTGAA   2152
        S  C  R  T  S  R  D  D  A  C  V  A  P  T  S  V  Q  R  *
                             M  R  R  P  Y  Q  R  P  A  V  N

2153   TTGTGTAGGTGCTGCATCAGGGGGATCGTTGGCTGGCGTCTTCTGCCGGGTTTGATTCCA   2212
        C  V  G  A  A  S  G  G  S  L  A  G  V  F  C  R  V  *

2213   GGTGTTGATCAACGATGCGTATCGTTTTCAGGTGTGCGACCTGGCGAGGCGGAGGCTCGT   2272
         M  R  I  V  F  R  C  A  T  W  R  G  G  G  S  S

2273   CGTCGGCGGTCGTAGGGGCGACGCATGCGTAGCCCCTACCAGCGTCCAGCGCTGCATCGT   2332
        S  A  V  V  G  A  T  H  A  *
                             M  R  S  P  Y  Q  R  P  A  L  H  R

2333   GCAGGTGCTGCATCAGGGCGGTTGGTCAGGGTGTCCTTCCAGGTTTAA   2380
        A  G  A  A  S  G  R  L  V  R  V  S  F  Q  V  *
```

Fig. 4. Nucleotide sequence 3'of *pufM*. Potential open read
         ing frames longer than 20 amino acids are shown. The
         putative simple termination loop is underlined with a
         dotted line. Base numbering corresponds to that
         used in Shiozawa *et al.* (1989).

Mobility shift assays. The mobility shift or gel retar-
dation assay is one method by which regions of DNA containing
putative promoter sequences or thought to be involved in the
regulation of gene expression can be tested for protein binding activity. Since the upstream region between -165 to -105
showed such strong resemblance to an *E. coli* promoter sequence, we decided to test this region for protein binding activity.

     Figure 6A shows the results of a mobility shift experi-
ment using a 368 bp *Nci*I DNA fragment which extends from -286
bp upstream to 82 bp within the L gene. When the fragment was
incubated with 1.4 µg cellular protein a small portion of the
DNA fragment had a retarded mobility. The mobility retarda-
tion of the DNA fragment was practically 100% when 7 µg protein was added to the assay mix. As controls, a 351 bp *Nci*I
fragment from within the ampicillin gene on pUC19 (Fig. 6B)
and a 272 bp *Nci*I fragment (bp numbers 83 to 354, Fig. 3) from
the 5'end of the L gene (Fig. 6C) were tested. Neither of
these fragments showed a response.

     Since the 368 bp *Nci*I fragment contains the putative pro-
moter sequence, it is highly probable that RNA polymerase or
at least the σ subunit of the enzyme binds to the fragment
under the experimental conditions. Whether the different
mobility bands represent different forms of dissociated RNA
polymerase or whether other proteins are bound to this DNA
fragment is presently unknown.

15

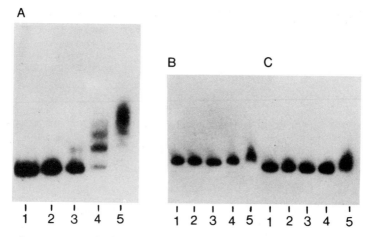

Fig. 5 Mobility shift assay. Three different
DNA fragments were incubated with 0,
0.14, 1.4, 7.0, and 14.0 µg cellular
protein from *C. aurantiacus*, lanes
1 - 5 respectively. A: 368bp *Nci*I
fragment just upstream of the L-gene.
B: 351bp *Nci*I fragment from the Amp[r]
gene in pUC 19. C: 272 bp *Nci*I fragment at the N-terminus of the L-gene.

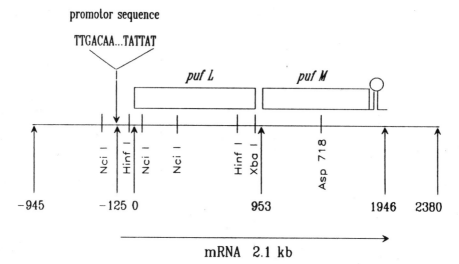

Fig. 6 Schematic representation of the *puf* operon of
C. aurantiacus. Also indicated on the map are
restriction sites of fragments used for North
ern blots (*Hinf*I-*Hinf*I and *Xba*I-*Asp*718) and
mobility shift assays (*Nci*I fragments).

SUMMARY

Based upon the data presented here, we believe that the *puf* operon of *C.aurantiacus* is composed of only two genes, the L and M genes of the reaction center (Fig. 6). Under steady state conditions, it appears that the message for the two genes is about 2.1 kb long and extends from 4-10 nucleotides 3'of a putative promoter sequence through a putative rho-independent terminator loop just beyond the M gene stop codon. The presence of nucleotide sequences 130 bases upstream of the L gene which strongly resemble the promoter sequence of *E. coli* RNA polymerase is also a unique occurrence among the *puf* operons examined thus far. A 368 bp DNA fragment from this region does exhibit protein binding ability. The most likely candidate for binding to this DNA fragment is RNA polymerase; through further studies we hope to determine the identity and function of the protein(s) which associate with this DNA fragment. Whether other, longer species of mRNA encoding the reaction center genes and other photosynthesis related genes occur and if trans-operon transcription is the normal mode of synthesis will also be assessed.

ACKNOWLEDGEMENTS

This research was supported by a grant from the *Deutsche Forschungsgemeinschaft* (SFB 143).

REFERENCES

Adams, C. W., Forrest, M. E., Cohen, S. N., and Beatty, J. T., 1989, Structural and functional analysis of transcriptional control of the *Rhodobacter capsulatus puf* operon J. Bacteriol. 171, 473.

Aiba, H., Adhya, S., and de Crombrugghe, B., 1981, Evidence for two functional *gal* promoters in intact *Escherichia coli* cells, J. Biol. Chem., 256:11905.

Ausubel, F. M., Brent, R., Kingston, R. E., Moore, D. D., Smith, Seidman, J. G,. and Struhl, K., 1988, "Current Protocols in Molecular Biology 1987-1988," John Wiley and Sons, New York.

Bauer, C. E., Young, D. A., anad Marrs, B. L., 1988, Analysis of the *Rhodobacter capsulatus puf* operon. Location of the oxygen-regulated promoter region and the identification of an additional *puf*-encoded gene, J. Biol. Chem. 263, 4820.

Castenholz, R. W. and Pierson, B. K., 1981, Isolation of members of the family Chloroflexaceae, in: "The Prokaryotes. A Handbook of Habitats, Isolation, and Identification of Bacteria," M. P. Starr, H. Stolp, H. G., Trüper, A. Balows, and H. G. Schlegel, eds., Springer-Verlag, Berlin.

Feick, R. G. and Fuller, R. C., 1984, Topography of the photosynthetic apparatus of *Chloroflexus aurantiacus*, Biochemistry 23:3693.

Hennighausen, L and Lubon, H., 1987, Interaction of protein witn DNA *in vitro*, in: "Methods in Enzymology," S. L. Berger and A. R. Kimmel, eds., Academic Press, Orlando Vol. 152.

Ovchinnikov, Yu. A., Abdulaev, N. G., Zolotarev, A. S., Shmukler, B. E., Zargarov, A. A., Kutuzov, M. A., Telezhinskaya, I. N., and Levina, N. B., 1988, Photosynthetic reaction centre of *Chloroflexus aurantiacus*. I. Primary structure of L-subunit, FEBS Lett., 231: 237.

Ovchinnikov, Yu. A., Abdulaev, N. G., Shmukler, B. E., Zargarov, A. A., Kutuzov, M. A., Telezhinskaya, I.N., Levina, N. B., and Zolotarev, A. S., 1988, Photosynthetic reaction centre of *Chloroflexus aurantiacus*. Primary structure of M-subunit, FEBS Lett., 232:364.

Shiozawa, J. A., Lottspeich, F., and Feick, R., 1987, The photochemical reaction center of *Chloroflexus aurantiacus* is composed of two structurally similar polypeptides, Eur. J. Biochem. 167:595.

Shiozawa, J. A., Lottspeich, F., Oesterhelt, D. and Feick, R., 1989, The primary structure of the *Chloroflexus aurantiacus* reaction center polypeptides, Eur. J. Biochem. 180:75.

Tavers, A. A., 1987, Structure and function of *E. coli* promoter DNA, CRC Critical Reviews in Biochemistry 22, 181.

Wechsler, T., Brunisholz, R., Suter, F., Fuller, R. C., and Zuber, H., 1985, The complete amino acid sequence of a bacteriochlorophyll *a* binding polypeptide isolated from the cytoplasmic membranes of the green photosynthetic bacterium *Chloroflexus aurantiacus*, FEBS Lett. 191:34.

Wechsler, T. D., Brunisholz, R. A. Frank, G., Suter, F., and Zuber, H., 1987, The complete amino acid sequence of the antenna polypeptide B806-866-β from the cytoplasmic membrane of the green bacterium *Chloroflexus aurantiacus*, FEBS Lett. 210:189.

Youvan, D. C., Bylina, E. J., Alberti, M., Begusch, H., and Hearst, J. E., 1984, Nucleotide and deduced polypeptide sequences of the photosynthetic reaction center, B870 antenna, and flanking polypeptides from *R. capsulata*, Cell 37, 949.

GENE STRUCTURE, ORGANISATION AND EXPRESSION OF THE LIGHT-HARVESTING B800-850 α AND β POLYPEPTIDES IN PHOTOSYNTHETIC BACTERIA

Monier Habib Tadros[1,2]

[1]Institut für Biologie 2, Mikrobiologie
Universität Freiburg, Schänzlerstr. 1
D-7800 Freiburg, F.R.G.

[2]European Molecular Biology Laboratory
Postfach 10.2209, Meyerhofstrasse 1
D-6900 Heidelberg, F.R.G.

INTRODUCTION

Members of the phototrophic non-sulfer purple bacteria produce ATP either by oxidative phosphorylation when growing aerobically in the dark or by photophosphorylation when grown anaerobically in the light. During the differentiation of the photosynthetic intracytoplasmic membrane, three pigment-protein complexes, besides other membrane bound complexes and cofactors, are synthesized. These are the photochemical reaction center (RC) and the light-harvesting antenna complexes B870 (LHI) and B800-850 (LHII). The light-harvesting complexes absorb light energy which is then directed to the photochemical reaction center with high efficiency.

In all analysed light-harvesting complexes of purple bacteria two bacteriochlorophyll-binding polypeptides, named α and β, have been found (Shiozawa et al., 1980; Brunisholz et al., 1981; Firsow and Drews, 1977; Tadros et al., 1982; Theiler et al., 1984). However, in the case of *Rhodopseudomonas (Rps.) palustris*, it has recently been shown that the LHII (B800-850) complex contains at least two major α-, two major β- and additional minor polypeptide species (Tadros and Waterkamp, 1989). Several workers have also observed that additional polypeptide species of the B800-850 complex are present in cells of *Rps. palustris* exposed to different light intensities (Hayashi et al., 1982a, b; Varga and Staehelin, 1985). It is possible that these polypeptide species also belong to the light-harvesting

Molecular Biology of Membrane-Bound Complexes in Phototrophic Bacteria
Edited by G. Drews and E. A. Dawes
Plenum Press, New York, 1990

19

complex since they migrate closely to the positions of α- and β- polypeptides on SDS-PAGE.

On the genetic level, it has previously been shown that in *Rhodobacter (Rb.) capsulatus* and *Rb. sphaeroides,* the B800-850 α- and β- polypeptides are encoded by two adjacent genes which form a single operon (named pucBA; Youvan and Ismail, 1985; Kiley and Kaplan, 1987). However, Tadros and Waterkamp (1989) have recently shown that the LHII α- and β- polypeptides in *Rps. palustris* are encoded by a multigene family.

In this paper, I will discuss new methods for the separation and characterisation of the polypeptide components belonging to a membrane complex under investigation. In addition, I will also describe the organisation and regulation of genes for the LHII complex. Although this article will mainly concentrate on *Rps. palustris*, similar processes may also occur in *Rb. capsulatus* and *Rb. sphaeroides.*

ISOLATION AND PURIFICATION OF THE POLYPEPTIDE COMPONENTS OF THE LHII COMPLEX

Previous studies on the light-harvesting complexes of *Rhodospirillaceae* have resulted in the isolation of only two bacteriochlorophyll binding polypeptides (α and β respectively). The methods used to separate these polypeptides has involved either gel filtration using Sephadex LH60 column (Shiozawa et al., 1980), or gel filtration followed by ion-exchange chromatography on DEAE columns (Brunisholz et al., 1981; Tadros et al., 1982; Theiler et al., 1984) or a combination of gel filtration, ion-exchange and FPLC (Wechsler et al., 1987). In addition, all these isolation procedures depended on extraction of the lypholized membrane or complex with organic solvents (chloroform/methanol 1.1 v/v containing 0.1 M ammonium acetate) since it was generally believed that all the light-harvesting polypeptides of the purple photosynthetic bacteria are soluble in organic solvents. However, it is possible that some of these polypeptides may be either less soluble or insoluble in organic solvents. We thus developed a strategy for separating the polypeptide components from the LHII complex in a water soluble system without employment of any detergents. For this, the native LHII complex from *Rps. palustris* was isolated from the intracytoplasmic membrane and its polypeptide components were separated directly by reverse phase HPLC (Tadros and Waterkamp, 1989). This step resolved at least two major α- (αa and αb), two major β- (βa and βb) and two additional minor polypeptide species. This strategy is the first example in which more than two bacteriochlorophyll-binding polypeptides from the LHI (B870) or LHII (B800-850) were purified. It is possible that the previous methods which utilized organic solvents for purification (Schiozawa et al., 1980; Brunisholz et al., 1981) failed to extract all the different polypeptide components belonging to the complex under

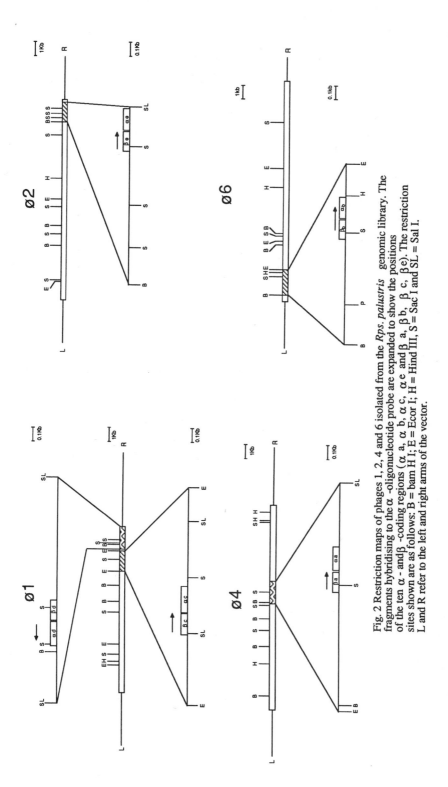

Fig. 2 Restriction maps of phages 1, 2, 4 and 6 isolated from the *Rps. palustris* genomic library. The fragments hybridising to the α -oligonucleotide probe are expanded to show the positions of the ten α - and β -coding regions (α a, α b, α c, α e and β a, β b, β c, β e). The restriction sites shown are as follows: B = bam H I; E = Ecor I; H = Hind III, S = Sac I and SL = Sal I. L and R refer to the left and right arms of the vector.

investigation. This problem was avoided by direct application of the isolated complex on HPLC. The polypeptides obtained were highly pure (Tadros and Waterkamp, 1989) and suitable for direct microsequence analysis. The additional advantage of this method is that dialysis is not required since the solvents used are volatile and can be easily removed by evaporation. In addition, only 70 min are required compared to several days necessary for the previous methods.

Theiler et al. (1984) have reported that the light-harvesting polypeptides of *Rb. sphaeroides* have a pronounced tendency to aggregate in the presence of water. However, in our isolation procedure, the freeze-dried LHII complex from *Rps. palustris* was dissolved in 50% formic acid and directly applied to a reverse-phase column. No aggregation was observed although the purified polypeptides were in contact with water all the time (Tadros and Waterkamp, 1989).

ANALYSIS AND IDENTIFICATION OF THE GENES CODING FOR THE LHII COMPLEX

In *Rb. capsulatus* and *Rb. sphaeroides,* only a single locus encoding both the alpha and beta polypeptides has been identified (Youvan and Ismail, 1985; Kiley and Kaplan, 1987). In order to investigate the organization of the genes encoding the LHII complex from

```
              T   A   C
5'  ATGAA CA GGN GNATTTGGAC    3'
          C   G   A

              T   A   C
5'  ATGAA CA GGN GNATCTGGAC    3'
          C   G   A

              T   A   C
5'  ATGAA CA GGN GNATATGGAC    3'
          C   G   A
```

Fig. 1 Oligonucleotide probes based on the amino-terminal sequence of one of the α polypeptides from the isolated B800-850 complex. N indicates any nucleotide.

22

Table 1

Sequences upstream of the ATG initiation codons contain purine-rich stretches (underlined), including nucleotides matching the predicted ribosome binding site (upper case).

Gen	5'		3'
LHII-ßa	ccatt<u>at</u><u>GAGG</u>tctcaaa		ATG
LHII-aa	ccc<u>aaaGGAG</u>tatagaat		ATG
LHII-ßb	tctc<u>aGGAGG</u>tttttaca		ATG
LHII-ab	cact<u>aaGGAG</u>tagatgca		ATG
LHII-ßc	caac<u>aaGAGG</u>gtcatcta		ATG
LHII-ac	cact<u>aaGGAG</u>ccactgag		ATG
LHII-ßd	tctctt<u>aGGAGG</u>ttatcc		ATG
LHII-ad	cact<u>aaGGAG</u>tatcgaag		ATG
LHII-ße	ctctc<u>aGGAGG</u>atttact		ATG
LHII-ae	cact<u>aaGGAG</u>tacgtaag		ATG

Rps. palustris, we constructed and screened a genomic library in λEMBL3 with degenerated oligonucleotides corresponding to the amino terminal sequence of one of the α polypeptides (Fig. 1).

The LHII α- and β- genes of *Rb. capsulatus*, *Rb. sphaeroides* and *Rps. palustris* share a high degree of homology at both the nucleotide and amino acid level (Table 2). Overall, at the amino acid level, the β- polypeptides are more conserved than the α. Whilst the nucleotide sequences of the α- polypeptide are more conserved than the β. However, this is not the case when the deduced amino-acid sequence from the coding regions of different α- and β-polypeptides of *Rps. palustris* are compared (Table 3).

Table 2

The percentage homology at the nucleotide and amino acid level of α and b between *Rb. sphaeroides* and *Rps. palustris* are compared using the sequence of *Rb. capsulatus* as a reference

Organism	Homology at amino-acid level		Homology at nucleotide level	
	β	α	β	α
Rb. sphaeroides	68.8	45.3	44.0	60.0
Rps. palustris				
αa - βa	52.2	44.4	44.2	63.9
αb - βb	53.1	46.3	41.3	63.9
αc - βc	51.0	44.4	46.0	57.4
αd - βd	49.0	46.3	44.0	62.2
αe - βe	53.1	48.1	46.0	64.8

Table 3

Homology at the nucleotide and deduced amino acid level between the different coding regions of α and β from *Rps. palustris* are compared using αa - βa (see figure 5) as a reference

Gene cluster	Homology at amino-acid level		Homology at nucleotide level	
	β	α	β	α
αb - βb	89.4	71.2	84.7	75.6
αc - βc	80.9	59.3	79.9	64.4
αd - βd	83.0	76.3	81.9	84.4
αe - βe	89.4	88.7	85.4	89.5

Six positively hybridising phages (named 1-6) were isolated and characterized (Fig.2).

The restriction maps of phage 1 and 3, and phage 5 and 6, overlap and hence for simplicity will be referred to as phage 1 and 5 respectively. Phage 1 contains two adjacent regions coding for the α- and β-polypeptides (named αc-βc and αd-βd) whilst all of the other phages contain only one region (αa-βa, αb-βb and αe-βe) (Fig 2). This suggests that *Rps. palustris* contains at least five coding regions for the α- and β-like polypeptides, respectively. This conclusion is supported by Southern blot analysis of genomic DNA (not shown).

DNA SEQUENCE ANALYSIS AND GENE ORGANIZATION OF THE LHII COMPLEX

The coding sequence of the α- and β- polypeptides in each of the phages were determined. The sequence of four independent coding regions are shown in Figure 3 and are compared with those from *Rb. capsulatus* and *Rb. sphaeroides*. The sequence of the fifth coding region which encodes for αe - βe is not shown (Tadros and Waterkamp, unpublished data).

There are several similarities in the organisation of the genes coding for the α- and β-polypeptides in *Rps. palustris*, *Rb. capsulatus* and *Rb. sphaeroides*. For example, the β-gene always precedes the α- gene. In addition, all the genes coding for the different β-polypeptides start with an ATG codon and terminate with a TAA codon. Furthermore, all coding regions are preceded by a polypurine-rich stretch of four to five nucleotides which is highly reminiscent of the Shine-Dalgarno ribosomal binding site that precedes the initiation codon of prokaryotic mRNAs (Shine and Dalgarno 1974) (Table 1).

There are, however, some differences. For example, the gap between the coding regions of the α- and β- polypeptide varies (Fig. 3).

ANALYSIS OF THE NON-CODING REGIONS OF THE α- AND β- GENES FROM *Rps palustris*

The non-coding sequence downstream of several α- genes contains a striking region of dyad symmetry (Fig. 4). These inverted repeats may form relatively stable stem-loop structures at the RNA level which could serve as transcriptional terminators. In several of these genes, the stem-loop structures are GC-rich and are followed by a run of thymidine residues, which is characteristic of rho-independent transcription termination in bacteria

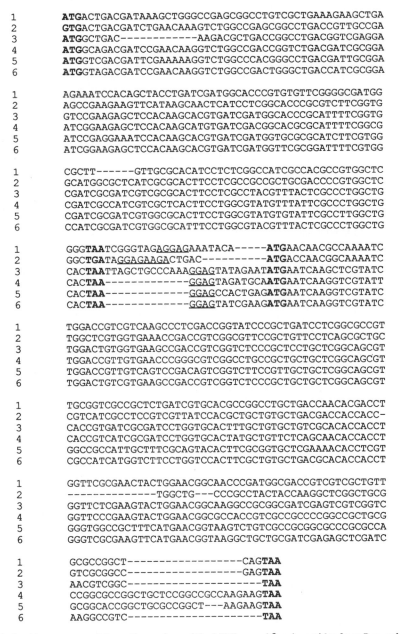

```
1    ATGACTGACGATAAAGCTGGGCCGAGCGGCCTGTCGCTGAAAGAAGCTGA
2    GTGACTGACGATCTGAACAAAGTCTGGCCGAGCGGCCTGACCGTTGCCGA
3    ATGGCTGAC-----------AAGACGCTGACCGGCCTGACGGTCGAGGA
4    ATGGCAGACGATCCGAACAAGGTCTGGCCGACCGGTCTGACGATCGCCGGA
5    ATGGTCGACGATTCGAAAAAGGTCTGGCCCACGGGCCTGACGATTGCGGA
6    ATGGTAGACGATCCGAACAAGGTCTGGCCGACTGGGCTGACCATCGCGGA

1    AGAAATCCACAGCTACCTGATCGATGGCACCCGTGTGTTCGGGGCGATGG
2    AGCCGAAGAAGTTCATAAGCAACTCATCCTCGGCACCCGCGTCTTCGGTG
3    GTCCGAAGAGCTCCACAAGCACGTGATCGATGGCACCCGCATTTTCGGTG
4    ATCGGAAGAGCTCCACAAGCATGTGATCGACGGCACGCGCATTTTCGGCG
5    ATCCGAGGAAATCCACAAGCACGTGATCGATGGTGCGCGCATCTTCGTGG
6    ATCGGAAGAGCTCCACAAGCACGTGATCGATGGTTCGCGGATTTTCGTGG

1    CGCTT------GTTGCGCACATCCTCTCGGCCATCGCCACGCCGTGGCTC
2    GCATGGCGCTCATCGCGCACTTCCTCGCCGCCGCTGCGACCCCGTGGCTC
3    CGATCGCGATCGTCGCGCACTTCCTCGCCTACGTTTACTCGCCCTGGCTG
4    CGATCGCCATCGTCGCTCACTTCCTGGCGTATGTTTATTCGCCCTGGCTG
5    CGATCGCGATCGTGGCGCACTTCCTGGCGTATGTGTATTCGCCTTGGCTG
6    CCATCGCGATCGTGGCGCATTTCCTGGCGTACGTTTACTCGCCCTGGCTG

1    GGGTAATCGGGTAGAGGAGAAATACA-----ATGAACAACGCCAAAATC
2    GGCTGATAGGAGAAGACTGAC----------ATGACCAACGGCAAAATC
3    CACTAATTAGCTGCCCAAAGGAGTATAGAATATGAATCAAGCTCGTATC
4    CACTAA-------------GGAGTAGATGCAATGAATCAAGGTCGTATT
5    CACTAA-------------GGAGCCACTGAGATGAATCAAGGTCGTATC
6    CACTAA-------------GGAGTATCGAAGATGAATCAAGGTCGTATC

1    TGGACCGTCGTCAAGCCCTCGACCGGTATCCCGCTGATCCTCGGCGCCGT
2    TGGCTCGTGGTGAAACCGACCGTCGGCGTTCCGCTGTTCCTCAGCGCTGC
3    TGGACTGTGGTGAAGCCGACCGTCGGTCTTCCCGCTCCTGCTCGGCAGCGT
4    TGGACGTTGTGAACCCGGGCGTCGGCCTGCCGCTGCTGCTCGGCAGCGT
5    TGGACCGTTGTCAGTCCGACAGTCGGTCTTCCGTTGCTGCTCGGCAGCGT
6    TGGACTGTCGTGAAGCCGACCGTCGGTCTCCCGCTGCTGCTCGGCAGCGT

1    TGCGGTCGCCGCTCTGATCGTGCACGCCGGCCTGCTGACCAACACGACCT
2    CGTCATCGCCTCCGTCGTTATCCACGCTGCTGTGCTGACGACCACCACC-
3    CACCGTGATCGCGATCCTGGTGCACTTTGCTGTGCTGTCGCACACCACCT
4    CACCGTCATCGCGATCCTGGTGCACTATGCTGTTCTCAGCAACACCACCT
5    GGCCGCCATTGCTTTCGCAGTACACTTCGCGGTGCTCGAAAACACCTCGT
6    CGCCATCATGGTCTTCCTGGTCCACTTCGCTGTGCTGACGCACACCACCT

1    GGTTCGCGAACTACTGGAACGGCAACCCGATGGCGACCGTCGTCGCTGTT
2    --------------TGGCTG---CCCGCCTACTACCAAGGCTCGGCTGCG
3    GGTTCTCGAAGTACTGGAACGGCAAGGCCGCGGCGATCGAGTCGTCGGTC
4    GGTTCCCGAAGTACTGGAACGGCGCCACCGTCGCCGCCCCGGCCGCTGCG
5    GGGTGGCCGCTTTCATGAACGGTAAGTCTGTCGCCGCGGCGCCCGCGCCA
6    GGGTCGCGAAGTTCATGAACGGTAAGGCTGCTGCGATCGAGAGCTCGATC

1    GCGCCGGCT----------------CAGTAA
2    GTCGCGGCC----------------GAGTAA
3    AACGTCGGC--------------------TAA
4    CCGGCGCCGGCTGCTCCGGCCGCCAAGAAGTAA
5    GCGGCACCGGCTGCGCCGGCT---AAGAAGTAA
6    AAGGCCGTC--------------------TAA
```

Fig. 3 Nucleotide sequences of the coding regions of the LH II α- and β-polypeptides from *Rps. palustris* (lines 3-6) are aligned with those from *Rb. capsulatus* (line 1; Yuovan and Ismail, 1985) and *Rb. sphaeroides* (line 2; Kiley and Kaplan, 1981). The sequences in lines 3-6 correspond to αa-βa, αb-βb, αc-βc and αd-βd respectively (see Fig. 2). The initiation codon (ATG or GTG) and the termination codon (TAA) are represented in bold letters.

(Platt, 1986). Possible rho-independent transcription termination signals have also been observed at the 3' ends of the *Rb. capsulatus puf* and *puc* operon mRNA (Zucconi and Beatty 1988; Chen et al. 1988 ; Tichy et al. 1989) and in the carotenoid biosynthetic gene cluster (Armstrong et al. 1989).

The 5' flanking sequence of several coding regions for the LHII complex in *Rps. palustris* also show significant homology (Tadros and Waterkamp, unpublished data). Interestingly, regions homologous to the δ70 consensus promoter, TTGACA N_{15-19} TATAAT (N, any nucleotide), used by the major RNA polymerase of *E. coli* (McClure 1985), are also present upstream of two of these gene clusters (Tadros and Waterkamp, unpublished data). Additional nucleotides in the spacer between the -35 and -10 regions are also present in the *Rps. palustris* sequences. Interestingly several genes from *Rb. capsulatus* also contain similar sequences (Armstrong et al., 1989) suggesting that such promoters may exist in photosynthetic bacteria.

Two of these phages (phage 2 and 6, Fig. 2) also contain highly conserved palindromic sequences which are repeated several times in the 5' flanking region (Tadros and Waterkamp, unpublished data). Palindromic sequences have also been observed 5' to the *puc* operon (Youvan and Ismail 1985) and several carotenoid biosynthetic genes in *Rb. capsulatus* (Armstrong et al., 1989). It is conceivable that these sequences may represent recognition sites for DNA-binding, transcriptional regulatory factors. This possibility is currently being investigated in my laboratory.

REGULATION OF EXPRESSION OF THE LIGHT HARVESTING α- AND β-GENES

The synthesis of the LHII complex is regulated by oxygen partial pressure and/or light intensity (Schumacher and Drews, 1978, 1979; Zhu and Hearst, 1986; Kiley and Kaplan, 1987; Zucconi and Beatty, 1988; Lee et al. 1989). The level of the mRNA for the LH-II polypeptides increases by lowering the O_2 concentration in cultures of both *Rb. capsulatus* and *Rb. sphaeroides* (Klug et al. 1985 Zhu et al., 1986; Zucconi and Beatty, 1988). The molecular mechanisms responsible for the regulation of the LHII genes by oxygen partial pressure are currently unclear. However, much is known about the regulation of the α- and β- polypeptides in response to different light intensities. For example, Zhu and Hearst (1986) have shown that the steady state level of B800-850 mRNA in *Rb. capsulatus* is higher in cells grown under high light intensity compared to low light grown cells. This is in contrast to the level of B800-850 complex in the cells suggesting that its regulation may be controlled by a posttranscriptional regulatory process. The authors suggest that the frequency of initiation of transcription of the B800-850 mRNAs is higher in cells grown under high light and that the relative amounts of the B800-850 complex under these conditions may be controlled by a translation or post translational event.

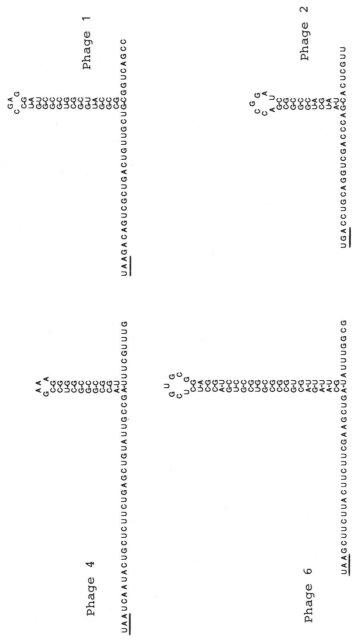

Fig. 4 Possible rho-independent transscriptional terminals. The termination codon (UAA) is underlined.

In *Rb. sphaeroides*, two B800-850 specific mRNAs have been identified (Lee et al., 1989). The small 0.5kb transcript is about 200 fold more abundant than the large (2.3kb) transcript. These two mRNAs share a common 5' terminus but differ in their 3' end. The use of mutant strains which are unable to express the B800-850 phenotype has shown that DNA region immediately downstream of *puc*BA encodes gene product(s) which may be essential for translational or posttranslational expression of the B800-850 α-and β- polypeptides. Interestingly, Tichy et al. 1989) have also proposed that coding region or DNA sequences downstream from the *puc* C in *Rb. capsulatus* are necessary for the expression of the *puc* operon.

In *Rps. palustris*, Northern blot analysis using different members of the multigene family has shown the presence of two *puc* mRNAs of 500 and 650 bp respectively (Waterkamp and Tadros, unpublished observations). Different members of the same family gene cluster, however, show a differential response in terms of both size and abundance of mRNA under different light intensities. The mechanisms responsible for this differential response are currently being investigated.

Acknowledgements

I wish to thank Riccardo Cortese, Dipak P. Ramji (Gene regulation and expression at EMBL) and Gerhart Drews (University of Freiburg) for faithful scientific discussions and critically reading the manuscript. I would also like to thank Helen Fry and Wendy Barker for typing the manuscript. The work performed in my laboratory was supported by Deutsche Forschungsgemeinschaft and the EMBL.

References

Armstrong, G.A., Alberti, M., Leach, F., and Hearst, J.E., 1989, Nucleotide sequence, organization, and nature of the protein products of the carotenoid biosynthesis gene cluster of *Rhodobacter capsulatus*, *Mol. Gen. Genet.* 216: 254-268.

Brunisholz, R.A., Cuendet, P.A., Theiler, R., and Zuber, H., 1981, The complete amino acid sequence of the single light harvesting protein from chromatophores of *Rhodospirillum rubrum* G-9, *FEBS Lett.* 129: 150-154.

Chen, C.Y., Beatty, J.T., Cohen, S.N., and Belasco, J.G., 1988, An intercistronic stem-loop structure functions as an mRNA decay terminator necessary but insufficient for *puf* mRNA stability, *Cell* 52: 609-629.

Firsow, N.N., and Drews, G., 1977, Differentiation of the intracytoplasmic membrane of *Rhodopseudomonas palustris* induced by variations of oxygen partial pressure or light intensity, *Arch. Microbiol.* 115: 299-306.

Hayashi, H., Nakano, M., and Morita, S., 1982a, Comparative studies of protein properties and bacteriochlorophyll contents of bacteriochlorophyll-protein complexes from spectrally different types of *Rhodopseudomonas palustris*, *J.Biochem.* 92: 1805-1811.

Hayashi, H., Miyao, M., and Morita, S., 1982b, Absorption and fluorescence spectra of light-harvesting bacteriochlorophyll-protein complexes from *Rhodopseudomonas palustris* in the near-infrared region, *J. Biochem.* 91: 1017-1027.

Kiley, P.J., and Kaplan, S., 1987, Cloning, DNA sequence, and expression of the *Rhodobacter sphaeroides* light-harvesting B800-850-α and B800-850-β genes, *J. Bacteriol.* 169: 3268-3275.

Klug, G.T., Kaufmann, N., and Drews, G., 1985, Gene expression of pigment-binding proteins of the bacterial photosynthetic apparatus: transcription and assembly in the membrane of *Rhodopseudomonas capsulata, Proc. Natl. Acad. Sci. USA* 82: 6485-6489.

Lee, J.K., Kiley, P.J., and Kaplan, S., 1989, Posttranscriptional control of *puc* operon expression of B800-850 light-harvesting complex formation in *Rhodobacter sphaeroides*, *J. Bact.* 171: 3391-3405.

McClure, W.R., 1985, Mechanism and control of transcription initiation in prokaryotes, *Ann. Rev. Biochem.* 54: 171-204.

Platt, T., 1986, Transcription termination and the regulation of gene expression, *Annu. Rev. Biochem.* 55: 339-372.

Schumacher, A., and Drews, G., 1978, The formation of bacteriochlorophyll-protein complexes of the photosynthetic apparatus of *Rhodopseudomonas capsulata* during early stages of development, *Biochim. Biophys. Acta* 501: 183-194.

Schumacher, A., and Drews, G., 1979, Effects of light intensity on membrane differentiation in *Rhodopseudomonas capsulata, Biochim. Biophys. Acta* 547: 417-428.

Shine, J., and Dalgarno, L., 1974, The 3'-terminal sequence of *Escherichia Coli* 16S ribosomal RNA: Complementary to nonsense triplets and ribosomes binding sites, *Proc. Natl. Acad. Sci. USA* 71: 1342-1346.

Shiozawa, J.A., Cuendet, P.A., Drews, G., and Zuber, H., 1980, Isolation and characterization of the polypeptide components from light-harvesting pigment-protein complex B800-850 of *Rhodopseudomonas capsulata, Eur.J.Biochem.* 111: 455-460.

Tadros, M.H., Zuber, F., and Drews, G., 1982, The polypeptide components from the light-harvesting pigment-protein complex II (B800-850) of *Rhodopseudomonas capsulata, Eur.J.Biochem.* 127: 315-318.

Tadros, M.H., and Waterkamp, K., 1989, Multiple copies of the coding regions for the light-harvesting B800-850 α and β polypeptides are present in the *Rhodopseudomonas palustris* genome, *EMBO J.* 5: 1303-1308.

Theiler, R., Suter, F., Wiemken, V., and Zuber, H., 1984, The light-harvesting polypeptides of *Rhodopseudomonas sphaeroides* R-26.1 I.Isolation, purification and sequence analysis, *Hoppe-Seyler's J. Physiol.Chem.* 365: 703-719.

Tichy, H.V., Oberlé, Béatrice, Stiehle, H., Schiltz, E., and Drews, G., 1989, Genes downstream from PUCB and PUCA are essential for the formation of the B800-850 complex of *Rhodobacter capsulatus, J. Bacteriol.*, in press.

Vagra, A.R., and Staehlin, L.A., 1985, Pigment-protein complexes from *Rhodopseudomonas palustris*: Isolation, characterization, and reconstitution into liposomes, *J. Bacteriol.* 161: 921-927.

Wechsler, T.D., Brunisholz, R.A., Frank, G., Suter, F., and Zuber, H., 1987, The complete amino acid sequence of the antenna polypeptide B806-866-β from the cytoplasmic membrane of the green bacterium *Chloroflexus aurantiacus*, *FEBS Lett.* 210: 189-194.

Whittenbury, R., and Mc Lee, A.G., 1967, *Rhodopseudomonas palustris* and *Rhodopseudomonas viridis*. Photosynthetic bacteria, *Arch. Mikrobiol.* 59: 324-334.

Youvan, D.C., and Ismail, S., 1985, Light-harvesting II (B800-850 complex) structural genes from *Rhodopseudomonas capsulata*, *Proc. Natl. Acad. Sci. USA* 82: 58-62.

Zhu, Y.S., Cook, D.N., Leach, F., Armstrong, G.A., Alberti, M., and Hearst, J.E., 1986, Oxygen-regulated mRNAs for light-harvesting and reaction center complexes and for bacteriochlorophyll and carotenoid biosynthesis in *Rhodobacter capsulatus* during the shift from anaerobic to aerobic growth, *J. Bacteriol.* 168: 1180-1188.

Zhu, Y.S., and Hearst, J.E., 1986, Regulation of the expression of the genes for light-harvesting antenna proteins LH-I and LH-II; reaction center polypeptides RC-L, RC-M, and RC-H; and enzymes of bacteriochlorophyll and carotenoid biosynthesis in *Rhodobacter capsulatus* by light and oxygen, *Proc. Natl. Acad. Sci. USA* 83: 7613-7617.

Zucconi, A.P., and Beatty, J.T., 1988, Post-transcriptional regulation by light of the steady-state levels of mature B800-850 light-harvesting complexes in *Rhodobacter capsulatus*, *J. Bacteriol.* 170: 877-882.

NEW GENETIC TOOLS FOR *Rhodobacter capsulatus* AND STRUCTURE

AND EXPRESSION OF GENES FOR CAROTENOID BIOSYNTHESIS

Glenn E. Bartley and Pablo A. Scolnik

E. I. DuPont De Nemours & Co., Inc.
Central Research and Development Department
Wilmington, DE 198800-0402, USA

INTRODUCTION

Photosynthetic bacteria are currently used as model systems for the study of genetic, biochemical and physical aspects of photosynthesis. In this article we briefly review genetic methods that we developed for use in *Rhodobacter capsulatus* and we present recent results on the molecular biology of carotenoid biosynthesis.

GENETIC TOOLS

For a description of the genetic tools available for *R. capsulatus* the reader is referred to the article by Scolnik and Marrs (1987), which described the principles and applications of capsduction, conjugation, transformation and transposon and interposon mutagenesis. After that review was published, we described a new set of expression vectors (Pollock et al, 1988). In these "tunable" expression vectors transcription of cloned sequences is controlled by a promoter for nitrogen fixation genes (*nif*), which can be activated at different levels by the nitrogen source used in the growth medium. The structure of one of these vectors and an example of regulated transcription are shown in Figs. 1 and 2.

In the first use of these vectors, Bauer et al. (1988) showed a correlation between the levels of expression of the PufQ protein and the synthesis of chlorophyll in *R. capsulatus* mutants. We have used the *nif* expression vectors to study the role of the CrtI protein in carotenoid biosynthesis (Bartley et al., 1989, see below).

Molecular Biology of Membrane-Bound Complexes in Phototrophic Bacteria
Edited by G. Drews and E. A. Dawes
Plenum Press, New York, 1990

33

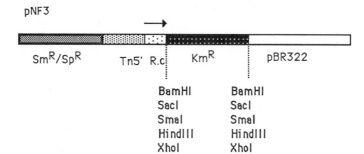

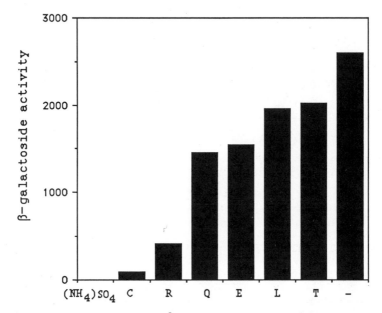

Fig 1.   Plasmid vector pNF3. A region of *R. capsulatus* DNA containing the *nif* promoter provides inducible transcription to genes cloned in the appropiate orientation. For a complete description see Pollock et al. (1988).

Fig 2. Expression of β-galactosidase driven by a *nif* with either (NH4)SO4 or the aminoacids shown in the one-letter code. For details see Pollock et al. (1988).

In many cases one wishes to express *R. capsulatus* genes in E. coli, either to study function or to produce antibodies. We designed an expression vector for E. coli that allows for high levels of expression in this organism and subsequent subcloning of the expressed gene into the *R. capsulatus* expression vector pNF3. The plasmid (pGBFT7) is shown in Fig.

3. This strategy for cloning requires that the direction of
transcription and at least a partial sequence around the start
point of translation be known. The gene is first directionally
cloned into pGBFT7 so the T7 promoter reads into the coding
region. Next, single-stranded DNA is produced and a *Nde*I site
is placed by site-directed mutagenesis at the translation start
codon. High levels of expression can be obtained in E. coli
strains bearing the gene for T7 polymerase (Bartley et al,
1989). Next, the gene is subcloned into pNF3 using the newly
created *Nde*I site and other restriction sites present in the
polylinker region. The resulting construct is introduced into
R. *capsulatus* by conjugation selecting for resistance to
spectinomycin.

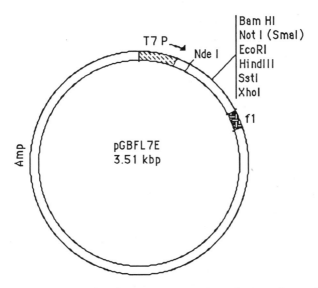

Figure 3. restriction map of pGBFL7E. T7P, T7
          promoter; f1, intergenic region of f1.

CAROTENOID BIOSYNTHESIS

The Carotenoid Gene Cluster

     We have studied the organization and expression of genes
for carotenoid biosynthesis. In our first paper on the subject
(Scolnik et al., 1980) we characterized six genes clustered in
one region of the genome and used a variety of spectroscopic
methods to characterize pigments accumulated by strains
containing point-mutation in those loci. Next, we located the
gene that codes for phytoene desaturase (*crt*I) by genetic
analysis of phytoene-accumulating mutants (Giuliano et al,
1986). In that same publication we also described an *in vitro*
system for the desaturation of phytoene. In addition to
providing a fast and sensitive method for studying the
synthesis of carotenoids *in vitro,* this technique provided the
first demonstration that ζ—carotene is an intermediate in the
synthesis of colored carotenoids. With the advent of

interposon mutagenesis (Scolnik and Haselkorn, 1984) it was possible to study the organization of the cluster in more detail. We constructed fifteen interposon insertions into the carotenoid gene cluster (Fig 4; Giuliano et al. 1988). These insertions define seven genes, whereas an eight one (*crtK*) was found by sequencing the cluster (Armstrong et al., 1989).

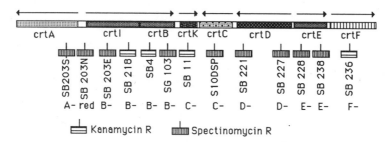

Figure 4. genetic map of the carotenoid gene cluster according to Giuliano et al. (1988) and revised according to Armstrong et al (1989). Interposon insertion mutants strains and the resulting phenotypes are shown.

With information about the location of genes and the direction of transcription, we used S1 protection analysis to study the response of four selected genes (*crtA, crtI, crtC* and *crtE*) to anaerobic conditions. The results indicated that anaerobiosis induces the expression of *crtA, crtC* and *crtE* but it has little effect on the expression of *crtI* (Giuliano et al., 1988). Interestingly, a new promoter for *crtE* is activated under anaerobic conditions.

In our analysis of the gene products we focused first on *crtI*, the gene that codes for phytoene desaturase. We sequenced this gene and characterized an open reading frame coding for a protein of Mr 57,997 (Bartley and Scolnik, 1989). Using the T7 and *nif* expression systems we overexpressed the CrtI protein in both E. coli and *R. capsulatus*. Protein from the E. coli overexpression was used to raise anti-CrtI antibodies in rabbits. In wild-type *R. capsulatus* the resulting antibody recognized a 60 kD protein which could not be extracted from the membrane with 2 M NaCl. This protein was not present in *crtI* mutant strains.generated by interposon mutagenesis, indicating that we have identified the correct open reading frame. In the interposon insertion strain SB 203N (Fig 4) little CrtI protein was detected. However, this strain contains wild-type levels of carotenoids, indicating that the cellular concentration of CrtI is not a limiting factor in carotenoid biosynthesis. Interestingly, this strain forms little light harvesting II (LH-II) antenna complex. Overexpression of CrtI leads to an increased accumulation of LH-II respect to wild-type. Taken together, these results suggest that CrtI influences the formation of the LH-II antenna system. Analysis of the deduced CrtI polypeptide sequence showed that the bacteriochlorophyll-binding motif A/GXXXH is present three times, opening the possibility that this protein also binds chlorophyll.

Expression of a *Neurospora crassa* cDNA in *R. capsulatus*

The early steps of carotenoid biosynthesis are common to both prokaryotes and eukaryotes. In the fungus *Neurospora crassa* the *alb-1* gene codes for phytoene desaturase. This gene has been cloned by T. Schmidhauser (C. Yanofky's lab), who is currently collaborating with us on the characterization of this gene. We have been able to complement *R. capsulatus crtI* mutants with an *alb-1* cDNA cloned into pNF3, indicating that *alb-1* and *crtI* are functionally equivalent. Comparison of the primary structure of the two proteins revealed extensive homology (data not shown) indicating that these two genes are evolutionary related.

BIBLIOGRAPHY

Armstrong, G.A, Alberti, M. Leach, F., and Hearst, J. E., 1989, Nucleotide Sequence, Organization, and Nature of the Protein Products of the Carotenoid Biosynthesis Gene Cluster of *Rhodobacter capsulatus*, Mol. Gen. Genet., 216:254-268.
Bartley, G. E., and Scolnik, P. A., 1989, Carotenoid Biosynthesis in Photosynthetic Bacteria, J. Biol. Chem., in press.
Giuliano, G., Pollock, D., Stapp, H., and Scolnik, P. A., 1988, A Genetic-Physical Map of the *Rhodobacter capsulatus* Carotenoid Biosynthesis Gene Cluster, Mol. Gen. Genet., 213:78-83.
Pollock, D., Bauer, C. E., and Scolnik, P. A., 1988, Transcription of the *Rhodobacter capsulatus nifHDK* Operon is Modulated by the Nitrogen Source. Construction of Plasmid Expression Vectors Based on the *nifHDK* Promoter, Gene., 65:259-275.
Scolnik, P. A., and Marrs, B.L., 1987, Genetic Research with Photosynthetic Bacteria, Ann. Rev.Microbiol., 41:703-726.
Scolnik, P. A., Walker, M. A., and Marrs, B. L.. 1980, Biosynthesis of Carotenoids Derived from Neurosporene in *Rhodopseudomonas capsulata*, J. Biol.Chem., 255:2427-2432.

ORGANIZATION OF THE *RHODOBACTER CAPSULATUS* CAROTENOID BIOSYNTHESIS GENE CLUSTER

[1,2]Gregory A. Armstrong, [2]Marie Alberti, [2]Francesca Leach and [1,2]John E. Hearst

[1]Department of Chemistry, University of California and
[2]Division of Chemical Biodynamics, Lawrence Berkeley Laboratory
Berkeley, CA 94720, USA

INTRODUCTION

Carotenoids are a major class of pigment molecules found in all photosynthetic organisms, and some nonphotosynthetic bacteria, fungi and yeasts (reviewed in Goodwin, 1980). In photosynthetic organisms carotenoids are not only essential physical quenchers of excited state triplet chlorophyll and bacteriochlorophyll (Bchl) and of singlet oxygen generated by these species, but also serve as accessory light-harvesting pigments (reviewed in Cogdell and Frank, 1987). The isolation of the R-prime plasmid pRPS404, containing a 46 kb region from the *Rhodobacter capsulatus* chromosome which complemented all known point mutation defects in photosynthesis, suggested that the genes encoding structural photosynthetic polypeptides and the enzymes of carotenoid and bacteriochlorophyll biosynthesis were clustered (Marrs, 1981). The genes encoding the reaction center and light-harvesting I polypeptides, flanking the pigment biosynthesis genes, were subsequently located and sequenced (Youvan et al., 1984a), as were the unlinked genes encoding the light-harvesting II antenna polypeptides (Youvan and Ismail, 1985). No DNA sequences were previously available for the genes encoding carotenoid biosynthetic enzymes from any carotenogenic organism. Thus, the determination of the nucleotide sequence and the organization of the *crt* genes from *R. capsulatus* is essential both to further studies of the gene products and of gene regulation. We have focused our attention on the characterization of the subcluster of *crt* genes within the photosynthesis gene cluster (for a description of the carotenoid biosynthesis pathway see Armstrong et al., 1989). Seven of the eight previously identified *R. capsulatus crt* genes were known to be clustered on the *Bam*HI-H, -G, -M, and -J fragments of pRPS404 in the order *crtA, I, B, C, D, E, F* from left to right on the genetic-physical map (Fig. 1) (Taylor et al., 1983; Zsebo and Hearst, 1984; Giuliano et al., 1988). These studies established that mutations causing Bchl⁻ phenotypes map within these four *Bam*HI fragments, flanking both ends of the *crt* gene cluster. We have determined the nucleotide sequence of an 11039 bp region encompassing the *Bam*HI-H, -G, -M, and -J fragments of pRPS404 (Armstrong et al., 1989). The nucleotide sequence reveals the presence of a new gene, *crtK*, not described in previous studies. We present here a comprehensive analysis of the DNA sequence and the gene organization, and discuss nucleotide sequences potentially involved in the initiation, regulation and termination of transcription within this region.

RESULTS

Alignment of the Nucleotide Sequence with Genetic-Physical Maps Identifies a New Gene, *crtK*

Sequencing across the *Bam*HI sites (Fig. 1) demonstrated that the *Bam*HI-J, -M, -G and -H

Molecular Biology of Membrane-Bound Complexes in Phototrophic Bacteria
Edited by G. Drews and E. A. Dawes
Plenum Press, New York, 1990

39

fragments are indeed contiguous. Fig. 1 shows the genes located within the 11039 base pair (bp) sequenced region (Armstrong et al., 1989). Because pRPS404 carries the *crtD223* point mutation (Marrs, 1981), the nucleotide and deduced polypeptide sequences determined reflect this deviation from the *R. capsulatus* wildtype sequences. The sequenced region contains three additional ORFs, designated *crtK*, ORF H and ORF J, distinct from any of the previously described *crt* genes. Interposon mutations introduced at *Sal*I (bp 5583) and *Nru*I (bp 6723) sites (Fig. 1) have both been proposed to lie within *crtC* because they result in the accumulation of neurosporene, a CrtC⁻ phenotype (Giuliano et al., 1988). Based on the DNA sequence, however, the interposons interrupt two distinct genes, which cannot be cotranscribed because of their convergent transcriptional orientations. Genetic-physical maps (Taylor et al., 1983; Zsebo and Hearst, 1984) have shown *crtC* to be bounded by *crtB* and *crtD*, with a gap left between crtB and *crtC*. On the basis of these studies, we designate the previously undetected gene found in this gap as *crtK* (Fig. 1).

Ribosome Binding Sites and Start Codons

The proposed amino acid sequences of the *crt* gene products and the ORFs (data not shown) correspond to the longest possible translations of ORFs possessing ribosome binding sites and typical *R. capsulatus* codon usage, located in the appropriate regions of the *crt* gene cluster. Translation of the nucleotide sequence in any of the alternative forward or reverse reading frames, with respect to a given gene, results in the frequent appearance of stop codons and the few alternative ORFs which do have ribosome binding sites show atypical codon usage (data not shown). ATA, CTA and TTA are never found among the 3038 predicted *crt* codons, while GTA (Val) appears only once, within *crtK* (Armstrong et al., 1989). All of the start codons proposed for the *crt* genes are preceded by purine-rich stretches containing possible ribosome binding sites (Fig. 2A) showing complementarity to the 3' end of the *R. capsulatus* 16 S rRNA (Fig. 2B) (Youvan et al., 1984b). An ATG start preceded by a ribosome binding site was not observed for ORFs in the region genetically mapped to *crtF* although a possible GTG start was found (Fig. 2A). We therefore propose that the coding

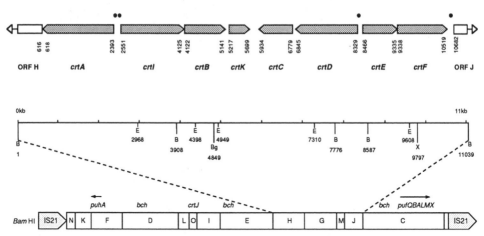

Fig. 1. Organization of the carotenoid biosynthetic gene cluster. Polarities of the *crt* genes (shaded) and ORFs (unshaded) are shown and putative transcriptional regulatory sites (Fig. 5) are indicated by (•). Numbers below the genes show the putative nucleotide positions of translational starts and stops. ORF H and ORF J extend beyond the region here, indicated by the detached arrowheads. A new start site has been assumed for *crtI* (Fig. 2A), replacing our previous proposal (Armstrong et al., 1989). Restriction sites referred to here or in previous genetic-physical mapping studies are indicated below the genes. B, *Bam*HI; E, *Eco*RI; Bg, *Bgl*II; X, *Xho*I. Boxes containing a letter indicate specific restriction fragments from the photosynthesis gene cluster of pRPS404 (Zsebo and Hearst, 1984), while the IS21 elements derived from the vector are indicated to the left and right. The locations of photosynthesis genes outside the *crt* gene cluster are shown above the boxes. The *pufB, A, L and M* genes encode the LH-I β, α and the RC-L, RC-M polypeptides, respectively, while the *puhA* gene encodes the RC-H polypeptide (Youvan et al. 1984a). Regions containing BchI biosynthetic genes are indicated by *bch*. The structure of *crtJ*, identified by a Tn5.7 insertion (Zsebo and Hearst, 1984) and separated from the other *crt* genes by about 12 kb (Fig. 1), is currently under study.

region of *crtF* begins with a GTG start codon. GTG start codons are used in about 8% of *Escherichia coli* genes (Stormo, 1986), and both the *fbcF* gene from *R. capsulatus* (Gabellini and Sebald, 1986), and the *pucB* gene of *Rhodobacter sphaeroides* have GTG starts. We originally proposed that the 5' end of the *crtI* coding region was a GTG codon found at bp 2650 (Armstrong et al., 1989), but more recent evidence suggests that the translation start corresponds to one of two upstream ATG codons located at bp 2551 and bp 2572, respectively (Armstrong G. A., and Hearst, J. E., unpublished data). We have assumed a *crtI* start at bp 2551 throughout the text and figures. Absolute confirmation of the deduced amino acid sequences will require the isolation of the gene products.

Organization of the Carotenoid Gene Cluster

The *crt* genes must form at least four distinct operons because of the inversions of transcriptional orientation which occur between *crtA-crtI*, *crtK-crtC*, and *crtD-crtE* (Fig. 1). *crtA* cannot be cotranscribed with the other *crt* genes because of its divergent orientation at one end of the gene cluster. An interposon insertion at an *Eco*RV site (bp 1303) and a transposon insertion (between bp 999-1244) cause Bchl⁻ phenotypes, although both of these mutations lie within the 3' end of the *crtA* gene (Giuliano et al., 1988; Armstrong, G. A. and Hearst, J. E., unpublished data). Mutations at the 5' end of *crtA* cause a CrtA⁻ phenotype but do not affect Bchl synthesis. The most likely explanation for these effects is the polar inactivation of ORF H or a downstream gene in the same operon (Fig. 1) in the 3' insertion mutants. This suggests that *crtA* is not cotranscribed with ORF H, although the promoter(s) for ORF H may overlap *crtA*. ORF H may thus be part of an operon required for Bchl biosynthesis. Groups of genes which could also form operons are *crtIBK*, *crtDC* and *crtEF*. A mutant bearing an interposon insertion at an *Apa*I site (bp 10713) within ORF J exhibits a Bchl⁻ phenotype, suggesting that this ORF may also belong to an operon required for Bchl biosynthesis (Giuliano et al., 1988). ORF J, located downstream from *crtF* (Fig. 1), does not appear to be transcribed as part of an operon including *crt* genes (Armstrong, G. A. and Hearst, J. E., unpublished data). The coding regions of the *crt* genes are closely spaced. In the most extreme case, the TGA stop codon of *crtI* overlaps the putative ATG start of *crtB*, reminiscent of the overlap between the coding regions of the *R. capsulatus pufL* and *pufM* genes (Youvan et al., 1984a).

A

Gene		5'	3'	Residues	MW
crtA		tcacaggGGAGGactgag	ATG	591	64761
crtB		ccgggccAAGGcGGcgca	ATG	339	37299
crtC		ggcgaAAAGGccttctcg	ATG	281	31855
crtD		tgcgtgcgGGAGcgagcg	ATG	494	52309
crtE		gcagcGGAGGgctctgtc	ATG	289	30004
crtF		cgccgaGAGGgGctgact	GTG	393	43004
crtI	(I)	gaaactaccgaAGAAAcc	ATG	524	57974
crtI	(II)	tccaAGAAcacAGAAggt	ATG	517	57226
crtK		ccacaaccGGAGGccatg	ATG	160	17607

B   5' AGAAAGGAGGTGAT..3'

          3' ₕₒUCUUUCCUCCACUA..5'

Fig. 2 A, B. Ribosome binding sites, start codons and the predicted gene products. (A) Sequences to the left of the ATG/GTG start codons contain purine-rich stretches (underlined), including nucleotides matching the predicted ribosome binding site (uppercase). The length of the gene product in amino acids and its calculated molecular weight are given to the right. The ribosome binding sites preceding two possible *crtI* start codons are shown, based on a revision of our original proposal for the 5' end of the *crtI* gene (Armstrong, G. A., Alberti, M., Leach, F. and Hearst, J. E., unpublished data). (B) shows the predicted ribosome binding site (above) as the DNA complement of the 3' end of the *R. capsulatus* 16 S rRNA (below) (Youvan et al. 1984b).

5' Non-coding Regions are A + T-rich

Fig. 3 illustrates the extreme asymmetry in % A + T content within the region encoding the *crt* gene cluster. Although the entire genome of *R. capsulatus* has an average A + T content of 34 %, the 5' flanking regions of *crtA-crtI*, *crtD-crtE* and ORF J are unusually A + T-rich, ranging up to 53 % in A + T content averaged over a 151 bp window. These A + T-rich regions contain DNA sequences which may bind transcription factors or serve as *E. coli*-like σ^{70} promoters (see below). The presence of A + T-rich islands in the 5' control regions of genes from an organism with a low average A + T content suggests a compelling selective pressure for the preservation of the nucleotide bias. Non-coding regions surrounding prokaryotic transcription initiation and termination points are A + T rich compared to the coding regions, as determined using a data base composed predominantly of *E. coli* genes (Nussinov et al., 1987). Specific structural features in A + T-rich regions of the chromosome may alert DNA-binding proteins to the presence of potential sites of action (Nussinov et al., 1987).

E. coli-like σ^{70} Promoter Sequences

We have located three sequences closely resembling the σ^{70} consensus promoter, TTGACA N_{15-19} TATAAT (N = any nucleotide) used by the major RNA polymerase of *E. coli* (McClure, 1985). The *R. capsulatus* sequences, found 5' to *crtI*, *crtD* and ORF J, are compared to the canonical *E. coli* promoter in Fig. 4. An optimal spacing of 17 bp is observed between the -35 and -10 regions in *E. coli* (McClure, 1985). The putative *crtD* and *crtI* promoters show spacings of 16 bp, while the putative ORF J promoter has a spacing of 17 bp.

No other sequences with a total of nine or more nucleotide matches to the *E. coli* σ^{70} consensus promoter, including five of the six most conserved nucleotides (Fig. 4), were found within the *crt* gene cluster (Fig. 1). We allowed a variable spacing of N_{15-19} between the -35 and -10 regions for these homology searches.

A Conserved Palindromic Motif is Related to a Recognition Site for DNA-binding Regulatory Proteins

We have identified a highly conserved palindromic nucleotide sequence, found four times in 5' flanking regions within the *crt* gene cluster. This motif occurs twice in the *crtA-crtI* 5' flanking region, and once each in the *crtD-crtE* and ORF J 5' flanking regions (Fig. 1). A search among other published *R. capsulatus* nucleotide sequences also revealed the presence of

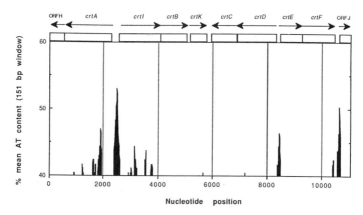

Fig. 3. Average A + T nucleotide content within the carotenoid gene cluster. Analysis of percentage A + T content throughout the DNA sequence was also performed using programs described by Pustell and Kafatos (1982). Locations and polarities of the *crt* genes are indicated at the top. ORF H and ORF J extend beyond the boundaries of the region shown. Nucleotide positions are as in Fig. 1. Percentage A + T content was calculated by averaging over a 151 bp window at 10 bp intervals, and values exceeding 40 % were plotted (average genomic percentage A + T content is 34 %). Note the unusually high mean A + T content in 5' flanking regions of genes.

this palindrome 5' to the coding region of the *puc* operon (Youvan and Ismail, 1985). Based on these five examples, the consensus sequence is TGTAART N_3 A N_2 TTACAC (R = purine) (Fig. 5B). The palindromes are centered anywhere from 162 bp (*pucB*) to 29 bp (*crtA*) from the start codon of the nearest gene (Fig. 5A). Each of the three putative *E. coli*-like σ^{70} promoters located within the *crt* gene cluster overlaps one of the palindromes (compare Figs. 4, 5). No additional palindromes were found when we required matches to each absolutely conserved nucleotide in the consensus (Fig. 5B) in a search of the coding and flanking regions of other published *R. capsulatus* nucleotide sequences encoding proteins (Armstrong et al., 1989), as well as from the 5' end of ORF J to the 5' end of the *pufQ* gene (Fig. 1) (M. Alberti, unpublished data). The *R. capsulatus* consensus palindrome shows strong similarity to a consensus sequence, TGTGT N_{6-10} ACACA, derived from the recognition sites of a collection of prokaryotic transcription factors containing examples of both positive and negative regulators (Fig. 5C) (Gicquel-Sanzey and Cossart, 1982; Buck et al., 1986). The *R. capsulatus* consensus palindrome is, in fact, very similar to the *E. coli* TyrR protein consensus recognition sequence TGTAAA N_6 TTACA (Yang and Pittard, 1987). TyrR is known to be a transcriptional regulator of genes required for aromatic amino acid metabolism. Based on these sequence similarities to the sites of action of known DNA-binding proteins, we propose that the *R. capsulatus* palindromic motif represents the binding site for a transcription factor.

<u>Rho-independent Transcription Termination Signals</u>

The region shown in Fig. 1 was searched for regions of dyad symmetry with the potential to form stem-loop structures in RNA. Possible secondary structures found between *crtK* and *crtC* include two GC-rich stem-loops, one 3' to *crtK* and the other 3' to *crtC*, each followed by a run of three thymidines (Armstrong et al., 1989). Single regions of dyad symmetry followed by thymidine-rich stretches were found 3' to the *crtI*, *crtB* and *crtF* genes, respectively. The combination of a GC-rich dyad symmetrical region, followed by several thymidine residues is characteristic of rho-independent transcriptional terminators in bacteria (Platt, 1986). Possible rho-independent termination signals were previously noted close to the 3' ends of the *R. capsulatus puf* and *puc* operon mRNAs as mapped by nuclease protection experiments (Zucconi and Beatty, 1988; Chen et al., 1988).

DISCUSSION

The minimum four operons in the *crt* gene cluster are *crtA*, *crtIBK*, *crtEF* and *crtDC* based on the polarities of the genes, although the latter operon seems unlikely because of the phenotypes of polar Ω interposon insertions within *crtD* (Giuliano et al., 1988). Possible rho-independent transcriptional terminators have been found 3' to the *crtI*, *B*, *K*, *C* and *F* genes (Armstrong et al., 1989), suggesting that the former three genes could form separate operons. The *R. capsulatus crt* gene cluster (Fig. 1) is bounded by genetic loci required for Bchl

```
             -35                          -10
  2489    TTGtaA atcggaattgac-gacc TATcAT.. 34bp..crtI
  8434    TTGgcA ttcgcacctacctgtg- TAaAcT...77bp..crtD
 10599    TTGACA gtcgggcgtgtaagttc aATgAT...54bp..ORF J
          ***  *  ***      *        *   *
          TTGACA      N15-19         TATAAT
```

Fig. 4. Comparison of sequences found 5' to *crtI*, *crtD* and ORF J with the *E. coli* σ^{70} consensus promoter. The numbers at left indicate the position of the 5' nucleotide of each sequence (Fig. 1). The distance in bp from each putative promoter to the start codon of the 3' gene is shown at the right. The consensus *E. coli* promoter is shown below with the six most highly conserved nucleotides underlined (McClure, 1985). Putative -35 and -10 regions (above) are indicated in boldface, and uppercase letters show matches to the *E. coli* consensus. Gaps (-) were placed between these regions to maximize the nucleotide alignment. Nucleotides absolutely conserved in all three *R. capsulatus* sequences are indicated by (*).

biosynthesis (Taylor et al., 1983; Zsebo and Hearst, 1984; Giuliano et al., 1988). The correspondence between these loci and specific ORFs has not yet been established, although mutations within ORF J and 5' to ORF H, in the 3' end of *crtA*, cause Bchl⁻ phenotypes (Giuliano et al., 1988; Armstrong, G. A. and Hearst, J. E., unpublished data). We therefore propose that ORF J and ORF H are part of two operons which include genes required for Bchl but not carotenoid biosynthesis, and which are transcribed outwards away from the *crt* gene cluster. *crtK* was not identified previously by interposon mutagenesis, presumably because of the similarity of CrtC⁻ and CrtK⁻ phenotypes and the fact that *crtC* and *crtK* are adjacent (Giuliano et al., 1988).

E. coli-like σ⁷⁰ promoters have never been observed in *Rhodobacter* (Kiley and Kaplan, 1988), nor have detailed data on any *R. capsulatus* promoters been available until recently. We have, however, found possible *E. coli*-like σ⁷⁰ promoters (McClure, 1985) 5' to *crtI*, *crtD* and ORF J (Fig. 4). Within the constraints of our homology search (see Results), no other *E. coli*-like σ⁷⁰ promoter sequences were found within the *crt* gene cluster, although the gene

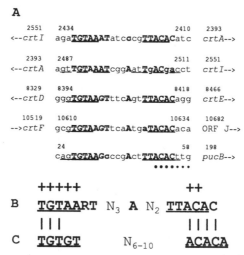

Fig. 5 A-C. Comparison of a palindromic motif found 5' to photosynthesis genes with a consensus regulatory protein binding site. (A) The genes flanking each palindrome are indicated to the left and right, respectively. Arrows show the directions of transcription. Numbers above each sequence show the nucleotide positions (as in (Fig. 1), except for *pucB* (see Youvan and Ismail 1985)) of the 5' or 3' ends of the flanking genes with respect to the location of the palindrome. Possible *puc* operon transcription initiation signals (Zucconi and Beatty, 1988) are indicated by (•). Complementary nucleotides in the two halves of the palindromes are underlined. (B) Nucleotides which match the *R. capsulatus* consensus are given in uppercase, while those that occur in positions defined by the consensus are shown in boldface. (+) indicates an absolutely conserved nucleotide in the palindrome. (C) The *R. capsulatus* consensus sequence is compared to a consensus derived from the recognition sites of the transcription factors NifA, AraC, CAP, LacI, GalR, LexA, TnpR, LysR and λ cII. Nucleotides conserved between the two consensuses are indicated by (|) between the sequences.

organization suggests that there must be promoters 5' to both *crtA* and *crtE* (Fig. 1). Whether these promoters have a weaker match to the σ^{70} consensus or perhaps have an entirely different structure remains to be determined. Neither Bchl nor carotenoids accumulate in *E. coli* strains harboring the *R. capsulatus* photosynthesis gene cluster carried on pRPS404 (Marrs, 1981). Our observation that the *R. capsulatus crtD* and *crtI* genes may have *E. coli*-like σ^{70} promoters, thus, was not anticipated. *E. coli* may fail to recognize at least one *R. capsulatus crt* promoter or lack the proper transcription factors required for *crt* gene expression. In addition, post-transcriptional regulation could also differ between the two species.

We have found five examples of a conserved nucleotide motif (Fig. 5A) in the 5' flanking regions of *R. capsulatus* photosynthesis genes. One example of the *R. capsulatus* palindromic motif occurs 5' to the *puc* operon (Fig. 5A), which encodes the LH-II antenna polypeptides. Zucconi and Beatty (1988) mapped the 5' triphosphate-containing ends of *puc* operon mRNAs and have suggested that a direct repeat of ACACTTG, located 5' to each of the two mapped mRNA start sites, may be involved in transcription initiation. The palindrome 5' to the *puc* operon overlaps the upstream ACACTTG sequence (Fig. 5A). and is located ~35 and ~50 nucleotides, respectively, upstream from the two 5' ends of the *puc* mRNAs.

Three other examples of the palindrome overlap the putative *E. coli*-like σ^{70} promoter sequences found 5' to *crtI*, *crtD* and ORF J (compare Figs. 4, 5). We propose a role for the palindromes in transcriptional regulation because of the extraordinary conservation of the motif and its sequence similarity to binding sites of known transcription factors, and because of its overlap with three putative *E. coli*-like σ^{70} promoters in the *crt* gene cluster. Overlap of the *R. capsulatus* palindromes with promoter sequences could be consistent with either positive or negative gene regulation. The regulatory sites may also be widely separated from the promoters with which they interact. Further experiments are in progress to define the interaction between the putative regulatory palindromes and sequences involved in transcription initiation.

The *puc* operon is highly regulated at the transcriptional level in response to oxygen tension (Klug et al., 1985). We have recently shown that expression of several *crt* genes is strongly induced during a shift from aerobic to photosynthetic growth (Armstrong, G. A. and Hearst, J. E., unpublished data), while Giuliano et al. (1988) have found an increase in the steady-state levels of 5' ends from *crtA*, *C* and *E* mRNAs in anaerobic versus aerobic cultures. The common feature of anaerobic gene induction would seem a reasonable explanation for the unexpected presence of identical transcriptional regulatory signals 5' to both the *puc* and *crt* operons. On the other hand, these regulatory sequences are not found close to the *puf* and *puh* operons (Fig. 1), whose expression is also induced by reduction of the oxygen tension (Clark et al., 1984; Klug et al., 1985). The palindromes (Fig. 5A) may thus bind a transcription factor involved in the regulation of a subset of the *R. capsulatus* photosynthesis genes. Whether a linkage exists between the expression of the *puc* operon and the regulated *crt* genes remains to be tested. We have determined the first nucleotide and deduced amino acid sequences of genes and genes products involved in carotenoid biosynthesis, and have also identified possible promoter, terminator and transcriptional regulatory signals which govern *crt* gene expression. Previous studies of *crt* gene regulation in *R. capsulatus* have been hampered by the lack of gene-specific probes (Clark et al., 1984; Klug et al., 1985; Zhu and Hearst, 1986; Zhu et al., 1986). The work presented here will facilitate an examination of the regulation of individual *crt* genes.

References

Armstrong, G. A., Alberti, M., Leach, F., and Hearst, J. E. (1989) Nucleotide sequence, organization, and nature of the protein products of the carotenoid biosynthesis gene cluster of *Rhodobacter capsulatus*, Mol. Gen. Genet., 216:254-268.
Buck, M., Miller, S., Drummond, M., and Dixon, R. (1986) Upstream activator sequences are present in the promoters of nitrogen fixation genes, Nature, 320:374-378.
Chen, C. Y., Beatty, J. T., Cohen, S. N., and Belasco, J. G. (1988) An intercistronic stem-loop structure functions as an mRNA decay terminator necessary but insufficient for *puf* mRNA stability, Cell, 52:609-619.

Clark, W. G., Davidson, E., and Marrs, B. L. (1984) Variation of levels of mRNA coding for antenna and reaction center polypeptides in *Rhodopseudomonas capsulata* in response to changes in oxygen concentration, J. Bacteriol., 157:945-948.

Cogdell, R. J., and Frank, H. A. (1987) How carotenoids function in photosynthetic bacteria, Biochim. Biophys. Acta, 895:63-79.

Gabellini, N., and Sebald, W. (1986) Nucleotide sequence and transcription of the *fbc* operon from *Rhodopseudomonas sphaeroides*, Eur. J. Biochem., 154:569-579.

Gicquel-Sanzey, B., and Cossart, P. (1982) Homologies between different procaryotic DNA-binding regulatory proteins and between their sites of action, EMBO J., 1:591-595.

Giuliano, G., Pollock, D., Stapp, H., and Scolnik, P. A. (1988) A genetic-physical map of the *Rhodobacter capsulatus* carotenoid biosynthesis gene cluster, Mol. Gen .Genet., 213:78-83.

Goodwin, T. W. (1980) The Biochemistry of the Carotenoids: Plants, Chapman and Hall, Ltd., New York, New York.

Kiley, P. J., and Kaplan, S. (1988) Molecular genetics of photosynthetic membrane biosynthesis in *Rhodobacter sphaeroides*. Microbiol. Rev., 52:50-69.

Klug, G., Kaufmann, N., and Drews, G. (1985) Gene expression of pigment-binding proteins of the bacterial photosynthetic apparatus: transcription and assembly in the membrane of *Rhodopseudomonas capsulata*, Proc. Nat. Acad. Sci., USA, 82:6485-6489.

Marrs, B. (1981) Mobilization of the genes for photosynthesis from *Rhodopseudomonas capsulata* by a promiscuous plasmid, J. Bacteriol., 146:1003-1012.

McClure, W. R. (1985) Mechanism and control of transcription initiation in prokaryotes. Annu. Rev. Biochem., 54:171-204.

Nussinov, R., Barber, A., and Maizel, J. V. (1987) The distributions of nucleotides near bacterial transcription initiation and termination sites show distinct signals that may affect DNA geometry, J. Mol. Evol., 26:187-197.

Platt, T. (1986) Transcription termination and the regulation of gene expression, Annu. Rev. Biochem., 55:339-372.

Pustell, J., and Kafatos, F. (1982) A convenient and adaptable package of DNA sequence analysis programs for microcomputers, Nucl. Acids. Res., 10:51-59.

Stormo, G. D. (1986) Translation initiation, in: "Maximizing Gene Expression," W. Reznikoff W and L. Gold, eds., Butterworths, Stoneham, Massachusetts.

Taylor, D. P., Cohen, S. N., Clark, W. G., and Marrs, B. L. (1983) Alignment of the genetic and restriction maps of the photosynthesis region of the *Rhodopseudomonas capsulata* chromosome by a conjugation-mediated marker rescue technique, J. Bacteriol., 154:580-590.

Yang J., and Pittard, J. (1987) Molecular analysis of the regulatory region of the *Escherichia coli* K-12 *tyrB* gene, J. Bacteriol., 169:4710-4715.

Youvan, D. C., Bylina, E.J., Alberti, M., Begusch, H., and Hearst, J. E. (1984a) Nucleotide and deduced polypeptide sequences of the photosynthetic reaction-center, B870 antenna, and flanking polypeptides from *R. capsulata*, Cell, 37:949-957.

Youvan, D. C., Alberti, M., Begusch, H., Bylina, E. J., and Hearst, J. E. (1984b) Reaction center and light-harvesting genes from *Rhodopseudomonas capsulata*, Proc. Nat. Acad. Sci., USA, 81:189-192.

Youvan, D. C., and Ismail, S. (1985) Light-harvesting II (B800-B850 complex) structural genes from *Rhodopseudomonas capsulata*, Proc. Nat. Acad. Sci., USA, 82:58-62.

Zhu, Y. S., and Hearst, J. E. (1986) Regulation of the expression of the genes for light-harvesting antenna proteins LH-I and LH-II; reaction center polypeptides RC-L, RC-M, and RC-H; and enzymes of bacteriochlorophyll and carotenoid biosynthesis in *Rhodobacter capsulatus* by light and oxygen, Proc. Nat. Acad. Sci., USA, 83:7613-7617.

Zhu, Y. S., Cook, D. N., Leach, F., Armstrong, G. A., Alberti, M., and Hearst, J. E. (1986) Oxygen-regulated mRNAs for light-harvesting and reaction center complexes and for bacteriochlorophyll and carotenoid biosynthesis in *Rhodobacter capsulatus* during the shift from anaerobic to aerobic growth, J Bacteriol., 168:1180-1188.

Zsebo, K. M., and Hearst, J. E. (1984) Genetic-physical mapping of a photosynthetic gene cluster from *R. capsulata*, Cell, 37:937-947.

Zucconi, A. P., and Beatty, J. T. (1988) Posttranscriptional regulation by light of the steady-state levels of mature B800-850 light-harvesting complexes in *Rhodobacter capsulatus*, J. Bacteriol., 170:877-882.

THE REGULATION OF BACTERIOCHLOROPHYLL SYNTHESIS IN RHODOBACTER

CAPSULATUS

Barry L. Marrs, Debra A. Young, Carl E. Bauer, and
JoAnn C. Williams

E. I. du Pont de Nemours & Co.
CR&D, Experimental Station, P.O. Box 80173
Wilmington, Delaware 19880-0173

INTRODUCTION

As members of the genus Rhodobacter respond to shifting environmental conditions, the amount and composition of the photosynthetic apparatus changes. The photosynthetic apparatus is composed of specific complexes of proteins and photosynthetic pigments, and the changes in the photosynthetic apparatus are achieved in a coordinated fashion so that neither pigment nor protein is accumulated in excess. It is especially important that neither bacteriochlorophyll (BChl) nor its precursors accumulate in excess, since these compounds are sensitizers to photodynamic killing processes. We have discovered a protein, the product of the pufQ gene, that mediates the coordination between pigment and protein synthesis. In the course of studying how this coordination of syntheses is achieved, we have found an unusual superoperonal arrangement of the genes coding for various elements of the photosynthetic apparatus.

The significance of previous examples of superoperonal gene clusters has been interpreted in terms of lateral gene transfer, stability and regulation. We have evidence that the superoperon for photosynthesis from Rhodobacter capsulatus (for which we have proposed the name BPScaps) exhibits transcriptional read-through, thus supporting the notion that the significance of the cluster is at least in part regulatory.

A full account of part of this work is in press in Molecular and General Genetics.

RESULTS AND DISCUSSION

The pufQ gene is the first gene in the puf operon, which includes the genes for the pigment-binding reaction center and light-harvesting proteins. Transcription of the puf operon is regulated in response to environmental signals, such as oxygen tension, and thus corrdination in synthesis between BChl and BChl-binding proteins is achieved in part by this genetic arrangement. The control region for the puf operon is located in a region of DNA that also serves to code for a structural gene involved in catalysis of a particular reaction in BChl synthesis, the bchA gene. This was demonstrated by analyzing the phenotypes of mutants generated by interposon mutagenesis in the region around the puf

Molecular Biology of Membrane-Bound Complexes in Phototrophic Bacteria
Edited by G. Drews and E. A. Dawes
Plenum Press, New York, 1990

promoters, P_{puf1} and P_{puf2}. Mutations downstream from these promoters clearly affect the bchA gene product and result in the accumulation of the characteristic Bchl precursor, P670. To demonstrate the accumulation of P670, an active copy of the pufQ gene must be included in the cells, since without this gene product little or no Mg-containing Bchl precursors of any type are accumulated. Codon preference plots of this region lead to the same conclusion, namely that the open reading frame for the bchA gene extends to within 4 bases of the open reading frame for pufQ.

The bchA gene and the bchC gene comprise a two gene operon that is transcribed in the same direction as the puf operon. Thus, an RNA polymerase molecule that transcribes the complete bchCA operon would find itself past the initiation point for transcription of the puf operon, and the operons may be said to overlap. We have also shown that transcriptional read-through occurs from the crtEF operon into the bchCA operon. This was done by constructing a series of protein fusions of beta galactosidase from Escherichia coli to the amino terminus of the bchC gene product and measuring the amount of beta galactosidase activity in the presence of different upstream genetic constructs. The key observation here is that interposons that interrupt the crtF gene decrease transcription of the downstream bchC gene.

Thus, there is a superoperon of three adjacent groups of genes, the first involved in carotenoid synthesis, the next in BChl synthesis, and the third coding for both pigment-binding proteins and BChl-regulating proteins. We speculate that this genetic arrangement allows for the production of low levels of photosynthetic apparatus under a variety of environmental conditions, which would be advantageous to the organism in rapidly changing environments. For example, during a shift from highly aerated to strictly anaerobic conditions, we have noticed that different species of bacteria require different periods of time to commence photosynthetic growth. This might be attributed to different amounts of reaction center synthesis maintained by different amounts of read through from promoters that are not shut down by the presence of oxygen.

The compact gene arrangement of these and other genes for photosynthesis might equally well reflect a situation occasioned by lateral gene transfer. These two hypotheses are not mutually exclusive.

THE CLONING AND ORGANISATION OF GENES FOR BACTERIOCHLOROPHYLL

AND CAROTENOID BIOSYNTHESIS IN RHODOBACTER SPHAEROIDES

Shirley A. Coomber[*], Maliha Chaudri and C.Neil Hunter

Department of Molecular Biology and Biotechnology
University of Sheffield, Sheffield S10 2TN
[*]Present address; Dupont Experimental Station,
P.O.Box 80402 Wilmington, Delaware 19880-0402, U.S.A.

INTRODUCTION

   Photosynthetic bacteria such as Rhodobacter sphaeroides and
Rb.capsulatus are capable of chemoheterotrophic growth in the dark, and
under conditions where oxygen is not limiting can repress the synthesis of
the photosynthetic apparatus almost completely. The removal of oxygen
initiates the coordinated synthesis of the pigments, proteins and lipids of
the photosynthetic membrane which grows as invaginations of the cytoplasmic
membrane (Niederman et al., 1976; Chory et al., 1984). By the time this
process is complete, light harvesting (LH) domains of several thousand
bacteriochlorophyll (bchl) and carotenoid (crt) molecules have been
assembled, consisting of LH2 units which surround and interconnect cores
containing LH1 and the photochemical reaction centre (Hunter et al., 1985.
Vos et al., 1988). Nevertheless, such cells still contain small amounts of
non-pigmented cytoplasmic membrane (Parks and Niederman, 1978) and membrane
regions enriched in newly synthesised pigment protein complexes.
(Niederman et al., 1979)

   The onset of photosynthetic membrane assembly relies upon the
coordinated expression of a large number of genes. Starting with the work
of Yen and Marrs in 1976, it was increasingly clear that many of these
genes are physically linked on a small stretch of the chromosome. In 1981,
Marrs isolated an R-prime plasmid which was able to complement an array of
lesions in the biosynthesis of photosynthetic pigments and proteins. A
genetic and physical map of the 46 kb cluster borne on this R-prime plasmid
emerged two years later. (Taylor et al., 1983). Further characterisation
by localised mutagenesis with transposon Tn 5.7 identified several new
putative genes for bacteriochlorophyll and carotenoid synthesis (Zsebo &
Hearst 1984). Interposon mapping and DNA sequencing has defined the crt
cluster in detail (Giuliano et al., 1988; Armstrong et al., 1989); the
regions around and including puh and puf genes for reaction centre and
light harvesting complexes have also been sequenced (Youvan et al., 1984).

Molecular Biology of Membrane-Bound Complexes in Phototrophic Bacteria
Edited by G. Drews and E. A. Dawes
Plenum Press, New York, 1990

In Rb.sphaeroides there were a few indications that genes for photosynthetic membrane assembly may be linked in a cluster similar to that found in Rb.capsulatus. Sistrom and colleagues (1984) suggested that genes for light-harvesting and reaction centre polypeptides and for pigment biosynthesis may be linked on an unspecified plasmid, pWS2. Rb.sphaeroides puh, puf and puc genes have been cloned and sequenced (Williams et al., 1983; Williams et al., 1984; Ashby et al., 1987; Kiley and Kaplan, 1987; Donohue et al., 1986; Williams et al., 1986; Kiley et al., 1987). Furthermore, Pemberton and Harding (1986) isolated two cosmids which between them encompass a 60 kb region of the Rb.sphaeroides chromosome, and which bear a 15 kb cluster of carotenoid biosynthesis genes. Subsequently, three clones pSCN5-1, pSCN6-1 and pSCN22-1 were isolated from a Rb.sphaeroides gene library cloned in E.coli using the mobilisable vector pSUP202 (Hunter and Coomber, 1988). Transfer of these clones restores wild type phenotype to mutants N5, N6 and N22 respectively, each of which is blocked at a different stage of bacteriochlorophyll synthesis. These clones were characterised by restriction endonuclease mapping using EcoR1, HindIII, PstI and BamH1. pSCN6-1 and pSCN5-1 were found to overlap when their restriction maps were compared.

More recently, several more overlapping clones were isolated from this Rb.sphaeroides gene bank. Following conjugative gene transfer from Escherichia coli these clones restored a wild type phenotype to several mutants unable to synthesise bacteriochlorophyll (Coomber and Hunter, 1989). The insert DNA was analysed by restriction mapping, and taken together these clones form the basis of the first restriction map, 45 kb in length, of the photosynthetic gene cluster of Rb.sphaeroides. This cluster is defined on one side by puh A encoding the reaction centre H polypeptide and on the other by the puf operon encoding reaction centre L and M apoproteins and light harvesting LH1 α and β polypeptides.

The availability of these clones in the vector pSUP202 facilitates the examination of cloned inserts by localised transposon Tn5 mutagenesis (Hunter, 1988). In this paper we report a physical map of this cluster obtained using this technique.

The position of each Tn5 insertion has been mapped and the mutant phenotype identified in each case. Analysis of these insertions provides a physical map of the photosynthesis cluster in Rb.sphaeroides.

MATERIALS AND METHODS

The isolation and restriction mapping of plasmids pSCN5H-1, pSCN5-1, pSCN6-1, pSCN6-20, pSCN22-1 and pSCN22-15 are described in Hunter and Coomber (1989). They are recombinants of the vector pSUP202 (Simon et al., 1983). The conditions for growth of Rb.sphaeroides on plates and in liquid culture are described by Hunter and Turner (1988). Localised Tn5 mutagenesis was carried out as in Hunter (1988) with one modification, described below.

Six clones in vector pSUP202, pSCN5H-1, pSCN5-1, pSCN6-1, pSCN6-20, pSCN22-1 and pSCN22-15 (Coomber and Hunter, 1989) were subjected to Tn5 mutagenesis as described in Hunter (1988). E. coli colonies which showed resistance to neomycin (Tn5), chloramphenicol and ampicillin (pSUP202) and which were assumed to bear different Tn5 insertions were chosen for each parental clone. Although Hunter (1988) used a triparental system to transfer pSUP202::Tn5 derivatives from E.coli to Rb.sphaeroides this was found to be unreliable.

Instead, 'miniprep' plasmid DNA was made from each of the colonies and transformed into E.coli strain 17-1 which contains the transfer functions of RP1 immobilized in the bacterial genome (Simon et al., 1983). Transfer of the various Tn5 bearing clones from E.coli 17-1 to Rb.sphaeroides was performed as described in Hunter and Turner (1988). After 5 days in the dark an average of 200 neomycin resistant colonies were observed on each plate . The plates were transferred to an anaerobic jar and illuminated for a further 3-4 days in order to see if any colonies had lost the ability to photosynthesise.

RESULTS

Characterisation of Tn5 mutant phenotypes

        Following transfer from E.coli a number of Tn5 bearing plasmids produced a variety of bchl⁻, crt⁻ and PS⁻ phenotypes in Rb.sphaeroides. Around 70% of the Tn5 bearing plasmids from each clone produced colonies of wild type phenotype only. These clones were assumed to have Tn5 inserted in either the vector itself, in a non-essential region of the Rb.sphaeroides insert DNA, or in a region close to the edge of the insert. Indeed one of the 'hot spots' for Tn5 insertion is the promoter region of the tetracycline resistance gene found on pSUP202 (Berg et al., 1983). It was noted that for any given transfer, the number of colonies displaying a single mutant phenotype varied between 5% and 40%. Hunter (1988) found that 1 in 10 Nm^R colonies produced by site directed Tn5 mutagenesis of the puf operon of Rb.sphaeroides showed a mutant phenotype and also lacked pSUP202, an observation which is consistent with double reciprocal crossover events. It should be noted that in experiments conducted with Rhizobium the proportion of kanamycin resistant transconjugants produced by double reciprocal crossover events in localised mutagenesis has been found to vary according to the length of flanking DNA. For example, when Tn5 was placed in the middle of a 4 kb DNA insert and transferred to Rhizobium, 1 out of 250 km^R transconjugants were the result of a double reciprocal crossover. In contrast, when Tn5 was located in the middle of a 13 kb DNA insert 1 out of 12 of the km^R transconjugants were the result of a double reciprocal crossover event (Noti et al., 1987).

        One typical mutant Rb.sphaeroides colony was chosen from each plate and restreaked. Each Rb.sphaeroides Tn5 mutant was grown in semiaerobic culture and subjected to spectral analysis. In general, mutants lacking bacteriochlorophyll were identified solely according to data obtained from absorption spectra. Rb.sphaeroides Tn5 mutants with lesions in carotenoid biosynthesis were identified partially from absorption spectra. To identify the mixtures of carotenoids present in each mutant a number were analysed further using TLC and HPLC techniques. The TLC work was performed by Professor Richard Cogdell (University of Glasgow, U.K.) and the HPLC work by Ann Connor and Dr. George Britton (University of Liverpool, U.K.). Using these results each carotenoid mutant phenotype was assigned to a carotenoid gene using nomenclature already established for Rb.capsulatus (Scolnik et al., 1980; Guiliano et al., 1988).

Mapping of transposon Tn5 insertions

        Each E.coli donor clone that produced a neomycin resistant mutant phenotype in Rb.sphaeroides was used as a source of plasmid DNA which was mapped using the restriction endonucleases PstI and

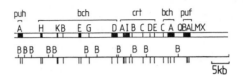

Fig.1   Map of the Rb.sphaeroides photosynthesis cluster.   Each
vertical line represents a Tn5 insertion.   Positions of
bch, crt, puf and puh genes are indicated.   B = Bam H1.

BamH1, as these enzymes cut Rb.sphaeroides DNA relatively frequently.
Although bands with elevated molecular weights resulting from
insertion of Tn5 were fairly easy to spot on agarose gels, DNA was also
transferred to nitrocellulose and probed with a radiolabelled 3.5kb HindIII
fragment of Tn5. (not shown).   To check that the positions mapped for each
Tn5 insertion were correct and to ensure that no large rearrangement of
flanking DNA had occurred, genomic DNA was prepared from a number of
Rb.sphaeroides Tn5 mutants.   The position of the Tn5 insertion in each case
was mapped using PstI and BamH1 digests, in conjunction with Southern
hybridization in which the blots were probed with fragments from the
Rb.sphaeroides photosynthetic cluster. (not shown).   Figure 1 shows all
the transposon Tn5 insertions mapped onto the photosynthetic cluster of
Rb.sphaeroides.

DISCUSSION

A large number of transposon Tn5 insertions which produce mutant
Rb.sphaeroides phenotypes have been mapped within the photosynthetic gene
cluster.   Each bchl⁻, crt⁻ or PS⁻ phenotype has been marked by at least one
Tn5 insertion except in the case of bchG.   Although the error in mapping
Tn5 insertions varies from 100 - 300 bp depending on fragment size, the
number and extent of Tn5 insertions which produce each phenotype help
overcome the inaccuracies of mapping.   We hope to add more Tn5 insertions
to this map in later work.

The crt genes

In this work a crt cluster of similar size and composition to that of
Rb.capsulatus has been shown in Rb.sphaeroides.   Although Tn5 insertions
which show a crtF mutant phenotype have not been isolated, an appropriate
gap exists in the overall Rb.sphaeroides map which could accommodate this
gene.   Interposon mutagenesis techniques will be used to identify the crtF
mutant phenotype.   The carotenoidless or blue-green mutant phenotypes of
crtB and crtE cannot be phenotypically distinguished and occupy separate
positions in the crt gene cluster of Rb.sphaeroides, as found in
Rb.capsulatus.   Two Tn5 insertions were found to accumulate the carotenoid
phytoene attributed by Zsebo and Hearst (1984) and Guiliano et al., (1986)
to lesions in the crtI gene.

crtC and crtD occupy a position in Rb.sphaeroides similar to that
found in Rb.capsulatus.   Two transposon Tn5 insertions which produce a crtA

mutant phenotype were positioned at the end of the crt cluster next to crtI and B in Rb.sphaeroides, in a position similar to that found in Rb.capsulatus. (Zsebo and Hearst, 1984; Giuliano et al., 1988). Overall, this work is in good agreement with that previously published for Rb.sphaeroides (Pemberton and Harding, 1986) and with data obtained from Rb.capsulatus.

The bch genes

In Rb.capsulatus Zsebo and Hearst (1984) identified three new bch genes, bchI, bchJ and bchK within the photosynthetic cluster. None of the Tn5 mutants studied here showed a bchJ or bchI phenotype. Moreover, Guiliano et al., (1988) showed that an interposon positioned 0.9 kb away from the crtA gene produced a mutant phenotype consistent with a lesion in bchD, not bchI. It should be noted that the discrimination between bchI and bchD phenotypes depends on the presence of a trace of pigment absorbing at 590nm (Zsebo and Hearst, 1984) so confusion is likely to occur. Nevertheless, the Tn5 insertions reported here are ascribed to bchD; more insertions are necessary to establish the size of this gene.

In Rb.capsulatus mutant phenotypes similar to that found for bchD also result from lesions in two further genes, bchH (Taylor et al., 1983; Zsebo and Hearst, 1984) and bchK (Zsebo and Hearst, 1984). In Rb.sphaeroides two pairs of Tn5 insertions both showing bchD-like mutant phenotypes appear in positions similar to those in Rb.capsulatus and so they have been named bchH and bchK. The bchD and bchH gene products are thought to be involved in the magnesium chelation and methylation steps preceding Mg protoporphyrin monomethyl ester formation. However since it is not currently possible to distinguish between bchD, bchH and bchK phenotypes on the basis of absorption spectra, further characterisation of these mutants is necessary.

Two Tn5 insertions producing a phenotype attributed in Rb.capsulatus to lesions in the bchE gene (Taylor et al; 1983, Biel and Marrs, 1983) were mapped in a similar position in the Rb.sphaeroides gene cluster (Fig. 1). A single Tn5 insertion which produced a phenotype attributed to bchG in Rb.capsulatus (Taylor et al., 1983; Biel and Marrs, 1983) mapped in a comparable position in the Rb.sphaeroides photosynthetic cluster (Fig 2).

One Tn5 insertion showed a phenotype attributed in earlier work to mutant N5 (Coomber et al., 1987, Hunter and Coomber, 1988) and to bchB mutants in Rb.capsulatus (Biel and Marrs, 1983). Zsebo and Hearst (1983) reported a Tn5.7 insertion that they attributed to a lesion in bchF which led to the accumulation of a pigment absorbing at 630nm previously used by Biel and Marrs (1983) as an indication of a lesion in bchB. Here we have identified a similar pigment and we therefore revert to the earlier nomenclature of Biel and Marrs (1983) and designate this bchB. The bchF mutant phenotype accumulates intermediates of bacteriochlorophyll synthesis absorbing at 730nm and 665nm (Taylor et al, 1983); no similar mutant phenotype was found in this work.

In Rb.capsulatus lesions in the gene designated bchL are phenotypically similar to those in bchB although these genes are over 7 kb apart (Zsebo and Hearst, 1984). So far no insertions have been mapped which correspond to this gene.

Two further bch genes, bchA and bchC have been located between the crt cluster and puf operon in Rb.capsulatus (Taylor et al., 1983). Tn5 insertions producing phenotypes typical of lesions in bchC and bchA map in a similar position in Rb.sphaeroides. This agrees with the position of these genes obtained from complementation data (Coomber and Hunter 1989).

A cluster of Tn5 insertions which produce a high fluorescence phenotype mapped within the puhA gene which has been fully characterised in Rb.sphaeroides (Williams et al., 1986; Donohue et al., 1986). In Rb.capsulatus, mutants of similar phenotype have been attributed to lesions in the puhA gene (Youvan et al. 1983; Youvan et al., 1984).

Finally, we include a cluster of Tn5 insertions described previously by Hunter (1988) in and around puf genes encoding reaction centre L and M subunits and α and β polypeptides.

It is clear that a great deal of work must be carried out, particularly with regard to the putative bch genes, in order to establish firmly their precise function in the biogenesis of the bacterial photosynthetic apparatus.

ACKNOWLEDGMENTS

This work was supported by a research grant to C.N.H. from the Science and Engineering Research Council (SERC). Maliha Chaudri is the recipient of an SERC Studentship.

REFERENCES

Armstrong, G.A., Alberti, M., Leach, F. and Hearst, J.E., 1989, Nucleotide sequence, organisation, and nature of the protein products of the carotenoid biosynthesis gene cluster of Rhodobacter capsulatus, Mol.Gen.Genet., 216: 254-268.

Ashby, M.K., Coomber, S.A. and Hunter, C.N., 1987, Cloning, nucleotide sequence and transfer of genes for the B800-850 light harvesting complex of Rhodobacter sphaeroides, FEBS Letters, 213: 245-248.

Berg, D.E., Schmandt, M.A. and Lowe, J.B., 1983, Specificity of transposon Tn5 insertion, Genetics, 105: 813-828.

Biel, A.J. and Marrs, B.L., 1983, Transcriptional regulation of several genes for bacteriochlorophyll biosynthesis in Rhodopseudomonas capsulata in response to oxygen, J.Bacteriol., 156: 686-694.

Chory, J., Donohue, T.J., Varga, A.R., Staehelin, L.A. and Kaplan, S., 1984, Induction of the photosynthetic membranes of Rhodopseudomonas sphaeroides: biochemical and morphological studies, J.Bacteriol, 159: 540-554.

Coomber, S.A., Ashby, M.K. and Hunter, C.N., 1987, Cloning and oxygen regulated expression of genes for the bacteriochlorophyll biosynthetic pathway in Rhodopseudomonas sphaeroides, in: "Progress in Photosynthesis Research", vol.4, pp. 737-740. J. Biggins, ed., Martinus Nijhoff, Dordrecht:

Coomber, S.A. and Hunter, C.N., 1989, Construction of a physical map of the 45kb photosynthetic gene cluster of Rhodobacter sphaeroides, Arch. Microbiol., 151: 454-458.

Donohue, T.J., McEwan, A.G. and Kaplan, S., 1986, Cloning and expression of the Rhodobacter sphaeroides reaction centre H gene, J. Bacteriol., 168: 962-972.

Giuliano, G., Pollock, D. and Scolnik, P.A., 1986, The crt I gene mediates the conversion of phytoene into coloured carotenoids in Rhodopseudomonas capsulata, J.Biol.Chem., 261: 12925-12929.

Giuliano, G., Pollock, D., Stapp, H. and Scolnik, P.A., 1988, A genetic-physical map of the Rhodobacter capsulatus carotenoid biosynthesis gene cluster, Mol. Gen. Genet., 213: 78-83.

Hunter, C.N., Kramer, H.J.M and van Grondelle, R., 1985, Linear dichroism and fluorescence emission of antenna complexes during photosynthetic unit assembly in Rhodopseudomonas sphaeroides, Biochim. Biophys. Acta, 807: 44-51.

Hunter, C.N., 1988, Transposon Tn5 mutagenesis of genes encoding reaction centre and light-harvesting LH1 polypeptides of Rhodobacter sphaeroides, J. Gen. Microbiol., 134: 1481-1489.

Hunter, C.N. and Coomber, S.A., 1988, Cloning and oxygen-regulated expression of the bacteriochlorophyll biosynthesis genes bch E, B, A, and C of Rhodobacter sphaeroides. J.Gen. Microbiol., 134: 1491-1497.

Hunter, C.N. and Turner, G., 1988, Transfer of genes coding for apoproteins of reaction centre and light harvesting LH1 complexes to Rhodobacter sphaeroides. J. Gen. Microbiol., 134: 1471-1480.

Kiley, P.J, and Kaplan, S., 1987, Cloning, DNA sequence and expression of the Rhodobacter sphaeroides light harvesting B800-850 α and B800-850 β genes, J.Bacteriol., 169: 742-750.

Kiley, P.J., Donohue, T.J., Havelka, W.A. and Kaplan, S., 1987, DNA sequence and in vitro expression of the B875 light harvesting polypeptides of Rhodobacter sphaeroides, J. Bacteriol., 169: 742-750.

Marrs, B., 1981, Mobilization of the genes for photosynthesis from Rhodopseudomonas capsulata by a promiscuous plasmid, J.Bacteriol., 146: 1003-1012.

Niederman, R.A., Mallon, D.E. and Langan, J.J., 1976, Membranes of Rhodopseudomonas sphaeroides. IV. Assembly of chromatophores in low-aeration cell suspensions, Biochim. Biophys. Acta., 440: 429-447.

Niederman, R.A., Mallon, D.E. and Parks, L.C., 1979, Membranes of Rhodopseudomonas sphaeroides VI. Isolation of a fraction enriched in newly synthesised bacteriochlorophyll a-protein Complexes, Biochim. Biophys. Acta, 555: 210-220.

Noti, J.D., Jadadish, M.N. and Szalay, A.A., 1987, Site directed Tn5 and transplacement mutagenesis : methods to identify symbiotic nitrogen fixation genes in slow-growing Rhizobium. Methods in Enzymology, 154: 197-217.

Parks, L.C. and Niederman, R.A., 1978, Membranes of Rhodopseudomonas sphaeroides. V. Identification of bacteriochlorophyll a - depleted cytoplasmic membrane in phototrophically grown cells, Biochim. Biophys. Acta., 511: 70-82.

Pemberton, J.M., Harding, C.M., 1986, Cloning of carotenoid biosynthesis genes from Rhodopseudomonas sphaeroides, Curr. Microbiol., 14: 25-29.

Scolnik, P.A., Walker, M.A. and Marrs, B.L., 1980, Biosynthesis of carotenoids derived from neurosporene in Rhodopseudomonas capsulata, J.Biol. Chem., 225: 2427-2432.

Simon, R., Priefer, U. and Puhler, A., 1983, A broad host range mobilization system for in vivo genetic engineering: transposon mutagenesis in Gram negative bacteria, Biotechnology, 1: 784-791.

Sistrom, W.R., Macalusa, A. and Pledger, R., 1984, Mutants of Rhodopseudomonas sphaeroides useful in genetic analysis, Arch Microbiol., 138: 161-165.

Taylor, D.P., Cohen, S.N., Clark, W.G. and Marrs, B.L., 1983, Alignment of genetic and restriction maps of the photosynthesis region of the Rhodopseudomonas capsulata chromosome by a conjugation-mediated marker rescue technique, J. Bacteriol.. 154: 580-590.

Vos, M., van Dorssen, R.J., Amesz, J., van Grondelle, R. and Hunter, C.N., 1988, The organisation of the photosynthetic apparatus of Rhodobacter sphaeroides: studies of antenna mutants using singlet-singlet quenching, Biochim. Biophys. Acta., 933: 132-140.

Williams, J.C., Steiner, L.A., Ogden, R.C., Simon, M.I. and Feher, G., 1983, Primary structure of the M subunit of the reaction center from Rhodopseudomonas sphaeroides, Proc. Natl. Acad. Sci. USA, 80: 6505-6509.

Williams, J.C., Steiner, L.A., Feher, G. and Simon, M.I., 1984, Primary structure of the L subunit of the reaction center of Rhodopseudomonas sphaeroides, Proc. Natl. acad. Sci. USA, 81: 7303-7308.

Williams, J.C., Steiner, L.A. and Feher, G., 1986, Primary structure of the reaction centre from Rhodopseudomonas sphaeroides, PROTEINS: Structure, Function Gen., 1: 312-325.

Yen, H.C. and Marrs, B., 1976, Map of the genes of carotenoid and bacteriochlorophyll biosynthesis in Rhodopseudomonas capsulata, J. Bacteriol., 126: 619-629.

Youvan, D.C., Hearst, J.E. and Marrs, B.L., 1983, Isolation and characterisation of enhanced fluorescence mutants of Rhodopseudomonas capsulata, J. Bacteriol., 154: 748-755.

Youvan, D.C., Bylina, E.J., Alberti, M.H., Begusch, H. and Hearst, J.E., 1984, Nucleotide and deduced polypeptide sequences of the photosynthetic reaction-center, B870 antenna, and flanking polypeptides from R.capsulata, Cell, 37: 949-957.

Zsebo, K.M. and Hearst, J.E., 1984, Genetic-physical mapping of a photosynthetic gene cluster from R.capsulata, Cell, 37: 937-947.

STRUCTURAL ANALYSIS OF THE *Rhodobacter capsulatus bchC* BACTERIOCHLOROPHYLL

BIOSYNTHESIS GENE

Cheryl L. Wellington [1] and J. Thomas Beatty [1, 2]

Departments of Microbiology [1], and Medical Genetics [2], University of British Columbia, Room 300, 6174 University Boulevard, Vancouver, British Columbia, Canada, V6T 1W5.

INTRODUCTION

The tetrapyrrole pigment bacteriochlorophyll (bchl) *a* is an essential photopigment in *Rhodobacter capsulatus*, being necessary for transfer of light energy between the three pigment-protein complexes in the photosynthetic apparatus and for charge separation in the photosynthetic reaction center (RC) (for reviews see Drews, 1985; Kiley and Kaplan, 1988). Because mutants blocked in bchl biosynthesis fail to accumulate the structural polypeptides of the RC and light-harvesting (LH) complexes when grown under conditions that would normally induce the formation of the photosynthetic apparatus (Dierstein, 1983; Klug et al., 1986) and are also impaired in carotenoid accumulation under low oxygen conditions (Biel and Marrs, 1985), it is accepted that the production of bchl is required for formation of many of the components of the photosynthetic apparatus. Therefore, an improved understanding of the mechanism and regulation of bchl synthesis would enhance our understanding of the biogenesis of the photosynthetic apparatus.

Bchl *a* is synthesized from a series of reactions that can be conveniently divided into two halves, the early steps of which are common to the biosynthetic pathways of other tetrapyrroles (such as hemes and corrinoids), and begin with the formation of δ-aminolevulinic acid (δ-ALA) (Jones, 1978). Eight molecules of δ-ALA are condensed and modified in a series of reactions to form the tetrapyrrole intermediate protoporphyrin IX. The pathways of bchl and heme biosynthesis diverge at this point, by the insertion of either Mg^{2+} or Fe^{2+} into the tetrapyrrole macrocycle of protoporphyrin IX, whereas the vitamin B_{12} biosynthetic branch splits from the common pathway before protoporphyrin IX is synthesized. The work of Lascelles and colleagues has been instrumental in elucidating many of these early steps in tetrapyrrole biosynthesis, and also in developing purification strategies and assays for most of these early enzymes (Jones, 1978; Lascelles, 1978). In contrast, the enzymes that convert protoporphyrin IX to bchl *a* have been much less amenable to experimental analysis. These enzymes are encoded by the *bch* genes, of which eight are known in *R. capsulatus* (Biel and Marrs, 1983). A mutation in any of the *bch* genes blocks a biosynthetic reaction and results in the accumulation of a colored intermediate that is usually identifiable on the basis of its absorption or fluorescence emission spectra. Through the analysis of many *bch* mutant strains, a "metabolic grid" has been proposed for ordering the various reactions that comprise the late steps in bchl *a* biosynthesis (Pudek and Richards, 1975), and the locations of *bch* genes have been mapped on the *R. capsulatus* chromosome (Yen and Marrs, 1976; Biel and Marrs, 1983; Taylor et al., 1983).

Because assays have not yet been developed for most of the enzymes of the magnesium branch, purification of these enzymes and direct detection of cloned *bch* genes has been difficult. Although many of the *R. capsulatus bch* genes have been roughly localized to cloned DNA fragments, the starts, ends, and sequences of these genes cannot be determined without additional information. We chose to identify a *bch* gene by creating an in-frame fusion with the *Escherichia coli lac'Z* gene,

Molecular Biology of Membrane-Bound Complexes in Phototrophic Bacteria
Edited by G. Drews and E. A. Dawes
Plenum Press, New York, 1990

57

and using the DNA sequence of this fusion to identify the reading frame and start site of the *R. capsulatus* gene. Subsequent *in vitro* mutation and replacement of the chromosomal allele with the mutated copy allowed the identification of the cloned open reading frame as the *bchC* gene, which encodes an enzyme that catalyzes the penultimate step in the biosynthesis of bchl *a* (Biel and Marrs, 1983). In this report we present a structural analysis of this *bchC* gene.

MATERIALS AND METHODS

Bacterial strains and plasmids

R. capsulatus strain B10 (Marrs, 1974; Weaver et al., 1975), and the *bchC* mutant CW100 derived from B10 (Wellington and Beatty, 1989) have been described. The structures of the promoter fusion vector pTB931 and of the *bchC-lac'Z* fusion plasmid pCW1 have both been described (Wellington and Beatty, 1989).

DNA sequence analysis

Fragments of DNA to be sequenced were subcloned into M13 mp18 and mp19 vectors (Yanisch-Perron et al., 1985). Dideoxy chain-termination reactions were carried out essentially as described (Smith, 1980) except that c^7-dGTP was used instead of dGTP (Mizusawa et al., 1986; Barr et al., 1986). Computer assisted analyses of the nucleotide and proposed amino acid sequences were performed using algorithms as cited in the figure legends.

RESULTS AND DISCUSSION

The products of a *Sau*3A 1 partial digest of the 5.5 kb *Eco*RI H fragment of pRPS404 (Taylor et al., 1983) were randomly inserted into the unique *Bam*HI site of the promoter fusion vector pTB931 in order to create a fusion bank. In this study we chose to restrict the source of *R. capsulatus* DNA to the *Eco*RI H fragment which was known to contain segments of at least two *bch* genes, *bchC* and *bchA* (Taylor et al., 1983), which biased the fusion bank toward the isolation of a *bch* gene. One resultant fusion plasmid, named pCW1, contained an *R. capsulatus* insert of approximately 450 bp which directed oxygen-regulated expression of β-galactosidase when pCW1 was mobilized into *R. capsulatus*, suggesting that the *R. capsulatus* insert encoded the 5' regulatory region and amino terminus of an *R. capsulatus* photosynthesis gene. Subsequent mutagenesis and complementation studies identified this amino terminus as the 5' coding region of the *bchC* gene (Wellington and Beatty, 1989), which encodes 2-desacetyl-2-hydroxyethyl bacteriochlorophyllide *a* dehydrogenase.

The DNA sequence of the *bchC* gene was determined and is presented in Fig. 1. Because the amino terminus of the *bchC* gene was originally isolated as an in-frame fusion to the *lac'Z* gene, the correct reading frame of the cloned *bchC* amino terminus could be deduced by alignment of the *lac'Z* reading frame across the junction with the *R. capsulatus* insert. The sequence of the remainder of the *bchC* gene was obtained by sequencing overlapping DNA segments subcloned from the *Eco*RI H fragment. The coding region of the *bchC* gene is proposed to begin at an ATG located 10 nt downstream from a potential ribosome binding site (GGAG), and to extend to two tandem termination codons located 942 nt downstream of the proposed ATG intiation codon. The *bchC* gene would therefore encode a protein of 314 amino acids which is predicted to have a molecular weight of 36,006 Da.

It is noteworthy that another open reading frame was found to begin at base position 1129 in Fig. 1, such that it overlaps the tandem termination codons TAA TGA of the *bchC* gene. Because the ATG start codon of this second open reading frame was preceded by a possible ribosome-binding site (GGAAAG), and contains typical *R. capsulatus* codons (see below), it may encode the amino terminus of the *bchA* gene.

In order to check for possible frameshifts in the sequence shown in Fig. 1, a codon preference plot (Gribskov et al., 1984) was generated for all three reading frames of the sequenced region using a codon preference statistic compiled from 24 *R. capsulatus* genes (Armstrong et al., 1989; Daldal et al., 1986; Davidson and Daldal, 1987; Masepohl et al., 1988; Schumann et al., 1986; Youvan et al.,

```
          10        20        30        40        50        60        70        80
    i  e  a  e  r  g  *
GATCGAGGCCGAACGCGGCTGACACGGCTGCGTTCGGACCCGGCTTTGACCCGGGGGTCAGAAAGTCGCACATCCGTCTG
Sau 3A 1
          90       100       110       120       130       140       150       160
TCGCAAAAGTGTCTAATCAAATTGACAGTCGGGCGTGTAAGTTCAATGATACACACAGGCGTGATCAGCCCGACTCTCCG

         170       180       190       200       210       220       230       240
                       m  e  t  q  v  v  i  m  s  g  p  k  a  i  s  t  g  i  a
GCCCGATCATACCGGGAGCAAGAAATGGAAACGCAAGTCGTCATAATGTCCGGGCCCAAGGCCATCTCGACGGGCATCGC
               SD
         250       260       270       280       290       300       310       320
  g  l  t  d  p  g  p  g  d  l  v  v  d  i  a  y  s  g  i  s  t  g  t  e  k  l  f
CGGTCTGACCGACCCCGGGCCGGGGGACCTCGTCGTGGATATCGCCTATTCCGGCATTTCGACTGGCACCGAGAAATTGT

         330       340       350       360       370       380       390       400
  w  l  g  t  m  p  p  f  p  g  m  g  y  p  l  v  p  g  y  e  s  f  g  e  v  v
TCTGGCTCGGCACCATGCCACCCTTCCCGGGCATGGGATATCCGCTTGTTCCCGGCTACGAAAGCTTCGGAGAGGTCGTT

         410       420       430       440       450       460       470       480
  q  a  a  p  d  t  g  f  r  p  g  d  h  v  f  i  p  g  a  n  c  f  t  g  g  l  r
CAAGCCGCCCCCGACACCGGCTTCCGACCGGGCGATCACGTCTTCATTCCCGGCGCCAACTGCTTCACCGGCGGGTTGCG
                              Sau 3A I
         490       500       510       520       530       540       550       560
  g  l  f  g  g  a  s  k  r  l  v  t  a  a  s  r  v  c  r  l  d  p  a  i  g  p  e
CGGGCTGTTCGGCGGGGCGTCGAAGCGCCTTGTCACGGCCGCCTCGCGCGTTTGTCGGCTGGATCCCGCCATCGGCCCCG

         570       580       590       600       610       620       630       640
  g  a  l  l  a  l  a  a  t  a  r  h  a  l  a  g  f  d  n  a  l  p  d  l  i  v
AGGGCGCGCTTCTGGCCCTTGCCGCCACCGCGCGGCATGCGCTGGCCGGGTTTGACAATGCTCTGCCGGATCTGATCGTC

         650       660       670       680       690       700       710       720
  g  h  g  t  l  g  r  l  l  a  r  l  t  l  a  a  g  g  k  p  p  m  v  w  e  t  n
GGCCACGGCACCCTCGGGCGCCTTCTGGCCCGTCTGACCCTGGCTGCCGGTGGCAAGCCGCCGATGGTCTGGGAAACCAA

         730       740       750       760       770       780       790       800
  p  a  r  r  t  g  a  v  g  y  e  v  l  d  p  e  a  d  p  r  r  d  y  k  a  i  y
TCCTGCCCGTCGCACGGGCGCGGTCGGCTACGAGGTTCTGGACCCCGAAGCCGATCCCCGGCGCGACTACAAGGCCATCT

         810       820       830       840       850       860       870       880
  d  a  s  g  a  p  g  l  i  d  q  l  v  g  r  l  g  k  g  g  e  l  v  l  c  g
ATGACGCCTCGGGCGCGCCCGGTCTGATCGACCAGCTCGTCGGGCGTCTGGGCAAGGGCGGGGAACTGGTGCTGTGCGGC

         890       900       910       920       930       940       950       960
  f  y  t  v  p  v  s  f  a  f  v  p  a  f  m  k  e  m  r  l  r  i  a  a  e  w  q
TTCTATACGGTGCCGGTCAGCTTCGCCTTTGTTCCCGCCTTCATGAAGGAAATGCGCCTGCGCATCGCCGCCGAATGGCA

         970       980       990      1000      1010      1020      1030      1040
  p  a  d  l  s  a  t  r  a  l  i  e  s  g  a  l  s  l  d  g  l  i  t  h  r  r  p
GCCGGCCGACCTTTCGGCCACGCGCGCGCTGATCGAAAGCGGGGCGCTCTCGCTGGATGGTCTCATCACGCATCGTCGCC

        1050      1060      1070      1080      1090      1100      1110      1120
  a  a  e  a  a  e  a  y  q  t  a  f  e  d  p  d  c  l  k  m  i  l  d  w  k  d
CCGCGGCGGAGGCGGCCGAGGCCTATCAGACCGCTTTCGAAGACCCTGACTGCCTGAAGATGATCCTTGACTGGAAAGAT
                                                                         SD
        1130      1140      1150      1160      1170      1180      1190
  a  k  *  m  t  d  a  p  n  l  k  g  f  d  a  r  l  r  e  e  a  a  e  e  p
GCAAAATAATGACTGACGCACCCAACCTGAAGGGATTTGACGCCCGTCTGCGGGAAGAAGCCGCCGAAGAGCCC
```

Fig. 1. Nucleotide and deduced amino acid sequences of the C-terminus of the *crtF* gene, the entire *bchC* gene, and the potential N-terminus of the *bchA* gene. Proposed coding sequences (in single letter code) extend from nucleotides 1 to 19 for the *crtF* C-terminus, nucleotides 185 to 1126 for the *bchC* gene, and nucleotides 1129 to 1194 for the *bchA* N-terminus. The two *Sau*3A I sites used to construct the fusion between the *bchC* and *lac'Z* genes in pCW1 are indicated, and potential ribosome binding sites (SD) are underlined. The *crtF* gene has been independently sequenced by Armstrong et al. (1989), and Young et al. (1989) have independently determined the DNA sequence of the *crtF* C-terminus and the *bchC* N-terminus.

1984; Youvan and Ismail, 1985), for a total of 7017 codons. This plot (Fig. 2) shows that codon usage in the proposed reading frame corresponds well with codon bias observed in other sequenced *R. capsulatus* proteins (Fig. 2A, plot 2), whereas there is very poor correspondence in each of the other two reading frames (Fig. 2A, plots 1 and 3). Therefore, although confirmation of the proposed amino acid sequence of the *bchC* gene requires receipt of the true amino acid sequence of the BchC protein itself, it seems likely that the reading frame that we propose is correct. Examination of codon usage for the *bchC* gene (Table 1) shows that, like all other *R. capsulatus* genes sequenced to date, the *bchC* gene shows a strong bias against codons ending in A or T, and an extreme bias against the codons TTA, CTA, GTA, and ATA. Of the 7331 codons now sequenced in *R. capsulatus*, only one GTA codon has been found in the *crtK* gene (Armstrong et al., 1989), and the codon ATA has been found for the first time within the *bchC* gene.

Examination of the hydrophobicity plot (Fig. 3A) generated by the method of Kyte and Doolittle (1982) shows that the BchC protein is predicted to be slightly hydrophobic overall and

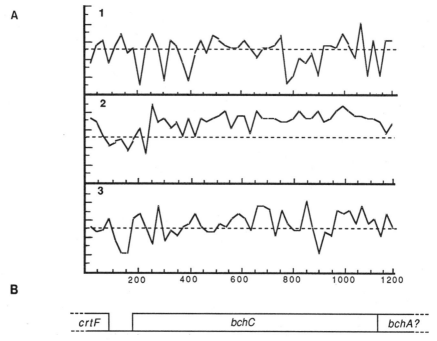

Fig. 2. Codon preference plot generated for the DNA sequence in Fig. 1 using a program written by Gribskov et al. (1984). Codon preference values calculated for this sequence using a window of 24 nucleotides were plotted using the computer program *TS Graph II*. Each of the three plots in panel A correspond to translation of the nucleotide sequence in one of three reading frames: plot 1 represents reading frame 1 (translation begins with base position 1); plot 2 represents reading frame 2 (translation begins with base position 2); and plot 3 represents reading frame 3 (translation begins with base position 3). Nucleotide positions are numbered along the horizontal axis, and relative codon preferences, measured as distance from a baseline (indicated by the dashed lines across each of the plots) are given along the vertical axis. Peaks above the dashed line indicate more frequently used codons, while peaks below the dashed line indicate less frequently used codons, according to a codon frequency table compiled from nucleotide sequences of 24 independently sequenced *R. capsulatus* genes (see text). The open reading frames of the *crtF*, *bchC*, and *bchA* genes are shown in panel B and are proposed to be translated in reading frame 2 for the *crtF* C-terminus and the *bchC* gene, and reading frame 1 for the *bchA* N-terminus.

Table 1. Codon usage of the *R. capsulatus bchC* gene, showing the number of times each codon was found in the 316 codons (including the two tandem termination codons) of the *bchC* gene.

TTT	phe	F	2	TCT	ser	S	0	TAT	tyr	Y	5	TGT	cys	C	1
TTC	phe	F	11	TCC	ser	S	2	TAC	tyr	Y	3	TGC	cys	C	3
TTA	leu	L	0	TCA	ser	S	0	TAA	OCH	Z	1	TGA	OPA	Z	1
TTG	leu	L	2	TCG	ser	S	7	TAG	AMB	Z	0	TGG	trp	W	4
CTT	leu	L	7	CCT	pro	P	2	CAT	his	H	2	CGT	arg	R	4
CTC	leu	L	6	CCC	pro	P	13	CAC	his	H	2	CAC	arg	R	10
CTA	leu	L	0	CCA	pro	P	1	CAA	gln	Q	2	CGA	arg	R	1
CTG	leu	L	19	CCG	pro	P	9	CAG	gln	Q	3	CGG	arg	R	3
ATT	ile	I	2	ACT	thr	T	1	AAT	asn	N	2	AGT	ser	S	0
ATC	ile	I	11	ACC	thr	T	10	AAC	asn	N	1	AGC	ser	S	3
ATA	ile	I	1	ACA	thr	T	0	AAA	lys	K	3	AGA	arg	R	0
ATG	met	M	8	ACG	thr	T	7	AAG	lys	K	7	AGG	arg	R	0
GTT	val	V	5	GCT	ala	A	3	GAT	asp	D	7	GGT	gly	G	4
GTC	val	V	11	GCC	ala	A	27	GAC	asp	D	12	GGC	gly	G	22
GTA	val	V	0	GCA	ala	A	1	GAA	glu	E	9	GGA	gly	G	2
GTG	val	V	3	GCG	ala	A	11	GAG	gly	E	6	GGG	gly	G	11

that it has four more pronounced hydrophobic regions that may interact with the cell membrane: from Ala-122 to Ala-139; from Gly-142 to Gly-169; from Val-219 to Met-250; and from Ala-265 to Leu-275. Secondary structure analysis of the proposed *bchC* sequence predicts several regions that are likely to form α-helices (Fig. 3B): Arg-108 to Pro-121; Ala-128 to Ala-146; Arg-159 to Gly-169; Val-187 to Asp-193; Ala-241 to Pro-260; Ala-265 to Leu-275; His-283 to Pro-301. It is interesting to note that although three of these predicted helices (Ala-128 to Ala-146; Arg-159 to Gly-169; and Ala-265 to Leu-275) are relatively hydrophobic, only the helix extending from Ala-128 to Ala-146 is of sufficient length to span the membrane.

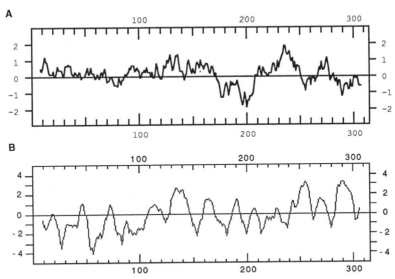

Fig. 3. Secondary structure analysis of BchC. Amino acids are numbered along the horizontal axis. Panel A: Hydropathy plot of BchC generated by the computer program *DNA Strider* (Marck, 1988) using the algorithm of Kyte and Doolittle (1982) with a window of 15 amino acids. Positive values indicate hydrophobic regions, and negative values indicate hydrophilic regions. Panel B: Prediction of α-helix formation in BchC using the algorithm of Garnier et al. (1978) with a window of 15 amino acids. Positive values indicate regions most likely to form an α-helix, and negative values indicate a low probability of α-helix formation.

Because the BchC enzyme has a bacteriochlorin tetrapyrrole as a substrate, we inspected the proposed amino acid sequence for regions that resemble bacteriochlorin binding sites in other pigment-binding proteins. Although ligands to bchls in both RC and LH polypeptides include conserved histidine residues found within a hydrophobic region, the sequence of amino acids flanking these conserved histidine residues have different features. Youvan and Bylina (1985) have noted that when the amino acid sequences of purple bacterial LH polypeptides are aligned along these conserved histidine residues, there appears a conserved alanine residue four amino acids prior to the histidine. Although a similar alignment of some sequenced RC polypeptides (Williams et al., 1986) reveals no such conserved alanine residue four amino acids prior to the histidine ligand of the special pair, Brunisholz and Zuber (1988) have observed that in both RC and LH polypeptides, there are several clusters of aromatic amino acids that flank the histidine ligand, and analysis of the three dimensional structure of crystallized reaction centers from *R. viridis* indicates that several of these aromatic amino acids are in close proximity to the histidine ligand of the special pair (Michel et al., 1986). The proposed BchC sequence contains four histidine residues at amino acids 85, 138, 154, and 283 (see Fig. 1). His-138 resembles a bchl binding site most strongly, being the only histidine found within a hydrophobic α-helix, and the only histidine preceeded by an alanine residue (Ala-143) four amino acids prior to the histidine residue. Interestingly, only His-85 is near four aromatic (phenylalanine) residues that are found in approximately the positions proposed by Bruinisholz and Zuber (1988). Because it is currently impossible to unambiguously predict a bacteriochlorin tetrapyrrole binding site from a primary sequence it will be necessary to experimentally test the BchC protein and mutant derivatives to correlate which, if any, of these structural features have functional significance for enzymatic activity.

The analysis presented here is the first structural investigation of any chlorophyll biosynthesis gene from any organism to be reported. It is hoped that information obtained by isolation and analysis of this and other chlorophyll biosynthesis genes may be used to overcome some of the difficulties encountered in the study of the late steps of chlorophyll biosynthesis, especially in the purification of these enzymes and their study by such techniques as site-directed mutagenesis. Such investigations will hopefully elucidate some of the mechanisms by which biosynthesis of the photosynthetic pigments and polypeptides is coordinately regulated, and will undoubtedly enhance our understanding of the biogenesis of the photosynthetic apparatus.

ACKNOWLEDGEMENTS

We thank D.A. Young and B.L. Marrs for valuable discussions and for sharing their unpublished data, W.R. Richards and B.R. Green for helpful comments, and D. Trimbur for the generous use of the *TS Graph II* software. C.L.W. was supported by a Canadian N.S.E.R.C. postgraduate fellowship. This research was supported by Canadian N.S.E.R.C. operating grant A-2796 to J.T.B.

REFERENCES

Armstrong, G.A., Alberti, M., Leach, F., and Hearst, J.E., 1989, Nucleotide sequence, organization, and nature of the protein products of the carotenoide biosynthesis gene cluster of *Rhodobacter capsulatus*, Mol. Gen. Genet., 216:254.
Barr, P.J., Thayer, R.M., Laybourn, P., Najarian, R.C., Seela, F., and Tolan, D.R., 1986, 7-Deaza-2'-deoxyguanosine-5'-triphosphate: Enhanced Resolution in M13 Dideoxy Sequencing, BioTechniques, 4:428.
Bauer, C.E., Young, D.A., and Marrs, B.L., 1988, Analysis of the *Rhodobacter capsulatus puf* operon, J. Biol. Chem., 263:4820.
Biel, A.J., 1986, Control of bacteriochlorophyll accumulation by light in *Rhodobacter capsulatus*, J. Bacteriol., 168:655.
Biel, A.J., and Marrs, B.L., 1983, Transcriptional regulation of several genes for bacteriochlorophyll biosynthesis in *Rhodopseudomonas capsulata* in response to oxygen, J. Bacteriol., 156:686.
Biel, A.J., and Marrs, B.L., 1985, Oxygen does not directly regulate carotenoid biosynthesis in *Rhodopseudomonas capsulata*, J. Bacteriol., 162:1320.
Brunisholz, R.A., and Zuber, H., 1988, Primary structure analyses of bacterial antenna

polypeptides: Correlation of aromatic amino acids with spectral properties; Structural similarities with reaction center polypeptides, in "Photosynthetic Light-Harvesting Systems," H. Scheer and S. Schneider, eds., Walter de Gruyter & Co., Berlin -New York.

Daldal, F., Chen, S., Applebaum, J., Davidson, E., and Prince, R.C., 1986, Cytochrome c_2 is not essential for photosynthetic growth of *Rhodopseudomonas capsulata*, Proc. Natl. Acad. Sci. USA, 83:2012.

Davidson, E., and Daldal, F., 1987, *fbc* operon, encoding the Rieske Fe-S protein, cytochrome *b*, and cytochrome c_1 apoproteins previously described from *Rhodopseudomonas sphaeroides*, is from *Rhodopseudomonas capsulata*, J. Mol. Biol., 195:25.

Dierstein, R., 1983, Biosynthesis of pigment-protein complex polypeptides in bacteriochlorophyll-less mutant cells of *Rhodopseudomonas capsulata* YS, FEBS. Lett., 160:281.

Drews, G., 1985, Structure and functional organization of light-harvesting complexes and photochemical reaction centers in membranes of phototrophic bacteria, Microbiol. Rev. 49:59.

Garnier, J., Osguthorpe, D.J., and Robson, B., 1978, Analysis of the accuracy and implications of simple methods for predicting the secondary structure of globular proteins, J. Mol. Biol., 120:97.

Gribskov, M., Devereux, J., and Burgess, R.R., 1984, The codon preference plot: graphic analysis of protein coding sequences and prediction of gene expression, Nucleic Acids Res., 12:539.

Jones, O.T.G., 1978, Biosynthesis of porphyrins, hemes, and chlorophylls, in "The Photosynthetic Bacteria," R.K. Clayton, R.K., and W.R. Sistrom, eds., Plenum Press, N. Y., N., Y.

Kiley, P.J., and Kaplan, S., 1988, Molecular genetics of photosynthetic membrane biosynthesis in *Rhodopseudomonas sphaeroides*, Miobiol. Rev. 52:50.

Klug, G., Liebetanz, R., and Drews, G., 1986, The influence of bacteriochlorophyll biosynthesis on formation of pigment-binding proteins and assembly of pigment protein complexes in *Rhodopseudomonas capsulata*, Arch. Microbiol., 146:284.

Kyte, J., and Doolittle, R.F., 1982, A simple method for displaying the hydropathic character of a protein, J. Mol. Biol., 157:105.

Lascelles, J., 1978, Regulation of Pyrrole Synthesis, in "The Photosynthetic Bacteria," R.K. Clayton and W.R. Sistrom, eds., Plenum Press, N. Y., N. Y.

Marck, C., 1988, 'DNA Strider': a 'C' program for the fast analysis of DNA and protein sequences on the Apple Macintosh family of computers, Nucleic Acids Res., 16:1829.

Marrs, B., 1974, Genetic recombination in *Rhodopseudomonas capsulata*, Proc. Natl. Acad. Sci. USA,71:971.

Masepohl, B., Klipp, W., and Pühler, A., 1988, Genetic characterization and sequence analysis of the duplicated *nifA/nifB* region of *R. capsulatus*, Mol. Gen. Genet., 212:27.

Michel, H., Epp, O., and Deisenhofer, J., 1986, Pigment-protein interactions in the photosynthetic reaction centre from *Rhodopseudomonas viridis*, The EMBO Journal, 5:2445.

Mizusawa, S., Nishimura, S., and Seela, F., 1986, Improvement of the dideoxy chain termination method of DNA sequencing by use of deoxy-7-deazaguanosine triphosphate in place of dGTP, Nucleic Acids Res., 14 :1319.

Pudek, M.R., and Richards, W.R., 1975, A possible alternate pathway of bacteriochlorophyll biosynthesis in a mutant of *Rhodopseudomonas sphaeroides*, Biochemistry, 14:3132.

Schumann, J., Waitches, G., and Scolnik, P., A DNA fragment hybridizing to a *nif* probe in *Rhodobacter capsulatus* is homologous to a 16 S rRNA gene, Gene, 48:79.

Smith, A.H.J., 1980, DNA sequence analysis by primed synthesis, Methods Enzymol., 166:560.

Taylor, D.P., Cohen, S.N., Clark, W.G., and Marrs, B.L., 1983, Alignment of genetic and restriction maps of the photosynthesis region of the *Rhodopseudomonas capsulata* chromosome by a conjugation-mediated marker rescue technique, J. Bacteriol., 154:580.

Weaver, P.F., Wall, J.D., and Gest, H., 1975, Characterization of *Rhodopseudomonas capsulata*, Arch. Microbiol., 105:207.

Wellington, C.L., and Beatty, J.T., 1989, Promoter mapping and nucleotide sequence of the *bchC* bacteriochlorophyll biosynthesis gene from *Rhodobacter capsulatus*, Gene, in press.

Williams, J.C., Steiner, L.A., and Feher, G., 1986, Primary structure of the reaction center from *Rhodopseudomonas sphaeroides*, Proteins: Structure, Function, and Genetics,1:312.

Yanisch-Perron, C., Vieira, J., and Messing, J., 1985, Improved M13 phage cloning vectors and host strains: nucleotide sequences of M13 and pUC vectors, Gene,33:103.

Yen, H.-C., and Marrs B., 1976, Map of genes for carotenoid and bacteriochlorophyll biosynthesis in *Rhodopseudomonas capsulata*, J. Bacteriol., 126:619.

Youvan, D.C., Bylina, E.J., Alberti, M., Begusch, H., and Hearst, J.E., 1984, Nucleotide and deduced polypeptide sequences of the photosynthetic reaction-center, B870 antenna, and flanking polypeptides from *R. capsulata*, Cell 37:949.

Youvan, D.C., and Ismail, S., 1985, Light harvesting II (B800-850 complex) structural genes from
Rhodopseudomonas capsulata, Gene, 38:19.

Young, D.A., Bauer, C.E., Williams, J.C., and Marrs, B.L., 1989, Genetic evidence for superoperonal
organization of genes for photosynthetic pigments and pigment-binding proteins in
Rhodobacter capsulatus, Mol. Gen. Genet., in press.

THE puf B,A,L,M GENES ARE NOT SUFFICIENT TO RESTORE THE

PHOTOSYNTHETIC PLUS PHENOTYPE TO A puf L,M,X DELETION STRAIN

J.W. Farchaus, H. Gruenberg, K.A. Gray,
J. Wachtveitl, B. DeHoff[*] , S. Kaplan[°] and
D. Oesterhelt

Dept. of Membrane Biochemistry, Max-Planck
Institut fuer Biochemie, Martinsried, FRG.
[*]Eli Lilly and Co. Lilly Corporate Center,
Indianapolis, IN.
[°]Dept. of Microbiology, Univ. of Illinois,
Urbana,IL

Introduction

The purple non-sulfur bacterium *Rhodobacter sphaeroides*
has the capacity to thrive as an anaerobic phototroph. The
ability to convert light energy into chemical energy within
the cell is driven by a photosynthetic unit consisting of two
light-harvesting antennae complexes and a reaction center
(RC). The light-harvesting complexes have been labelled LHI
and LHII based on their respective single (875 nm) and double
(800-850 nm)infrared absorption maxima. The RC responsible for
the primary light driven electron transfer is composed of
three protein subunits H, M and L and contains 4 molecules of
bacteriochlorophyll a (Bchl a), 2 molecules of bacteriopheophytin a, one non-heme iron and two molecules of ubiquinone. Two of the Bchl a molecules are monomeric and are
referred to as the voyeur Bchl while the other two form a
dimer, or so-called special pair (1). The RC special pair
undergoes a reversible photobleaching that can be measured at
860 nm or 600 nm. These two maxima have been assigned to the
Q_y and Q_x transition bands of the special pair, respectively
(2).

The genes for all components of the photosynthetic unit
have been cloned and sequenced. The genes encoding the RC L
and M subunits and those for LHI are located adjacent to one
another on the genome and are transcribed as a single polycistronic message indicating that they form a single operon,
now termed the *puf* operon. The *puf* operon in *Rhodobacter
capsulatus* is known to be composed of six genes *puf* Q (BChl a
synthesis), *puf* B (LHI ß-polypeptide), *puf* A (LHI - α
polypeptide), *puf* L (RC L-subunit), *puf* M (RC M-subunit) and
puf X (unknown function)(3).

A great deal of effort has recently been invested in
determining the exact location of the oxygen sensitive

Molecular Biology of Membrane-Bound Complexes in Phototrophic Bacteria
Edited by G. Drews and E. A. Dawes
Plenum Press, New York, 1990

promoter region responsible for the regulation of transcription of this operon. It has been established that the
promoter region lies some 700 bp 5' of the *puf* B gene (4,5).
It has been suggested that the initial transcript of this
operon is 3400 nt long and includes all six *puf* genes
mentioned above. This transcript is then quickly processed by
endo- and exonucleolytic ribonucleases to yield a five gene
(B,A,L,M and X) transcript of 2700 nt, which is in turn
further degraded to yield a longer lived message of 500 nt
coding for only the B and A genes (5).

Using interposon mutagenesis a kanamycin resistance gene
was specifically inserted into the genomic DNA of *R.
sphaeroides* simultaneously deleting the *puf* L,M and X genes
(6). The resulting photosynthetic minus (PS⁻) RC deletion
strain could be complemented in *trans* with a 5.3 kb DNA
fragment containing the *puf* B,A,L,M,X genes, an additional
1580 bp upstream of the *puf* B gene and 1200 bp downstream of
puf X (6). The fact that this fragment could restore the PS⁺
phenotype in *trans* indicated that the promoter region and
other possible *cis*-acting elements necessary for the transcription and translation of this operon were present.
However, several questions remained unanswered regarding the
complicated regulation of the *puf* operon and in particular as
to what role is played by the region downstream of *puf* M. It
was then of interest to determine the minimal fragment size
required to restore the PS⁺ phenotype to the deletion strain
and to examine the function of the *puf* X gene and the region
3' of the *puf* operon.

We describe here experiments that address the question of
function for the X gene and downstream region(s) and provide
the nucleotide sequence of the *R. sphaeroides puf* X gene. It
is concluded that the X gene is necessary for normal photosynthetic growth.

Materials and Methods

Bacterial Strains and Growth Condidtions

The *R. sphaeroides* deletion strain PUFΔLMX21, *E. coli*
strains and growth conditions were described in more detail
elsewhere (6). The pO_2 of the cultures at mid/late log phase
was 3-4 mm Hg as measured with a Trioxmatic EO 200 Clark-type
oxygen electrode connected to an Oxymatic 2000 microprocessor
oximeter from WTW (Weilheim,FRG). Photoheterotrophic growth
curves were measured using semi-aerobically grown cells as
inoculum and illumination (60 W/m^2) was with far-red light
(>680 nm) defined by two plexiglass cut-on filters (Plexiglass
Röhm, Darmstadt,FRG). The *E. coli* strain S17-1 was used to
mobilize plasmids into *R. sphaeroides* using the diparental
filter mating procedure described by Davis *et al.*(7).
Kanamycin (25 μg/ml)was added to deletion strain cultures as
was tetracycline (2 μg/ml) when the strain harbored the
plasmid pRK404 (7) or derivatives thereof.

Recombinant DNA techniques

The 5.3 kb *puf* operon shuttle fragment described
previously was inserted into the phasmid pMa/c (pBR322
derivative,colE1 replicon, F1 ori, Apʳ, Cmʳ; obtained from Dr.
H.J.Fritz, MPI,Martinsried FRG) and used for site-directed

mutagenesis to add an additional *Hind* III endonuclease restriction site three bp after the stop codon for the *puf* M gene. The 3.8 kb *Bam* HI-*Hind* III fragment was subcloned into the broad host range plasmid pRK404 to create pRKXmut2. The 60 bp DNA sequence for the transcription terminator stem-loop structure was cloned into pRK404 using four oligo-nucleotides, two for each strand with an overlap of six bases. The oligo-nucleotides were mixed in equimolar ratios, dried in a speed-vac and annealed in 10 mM TRIS pH8, 50 mM NaCl, 5 mM $MgCl_2$ by heating to 90° C for 3 min followed by gradual cooling to room temperature over 30 min. The resulting double stranded DNA had an intact *Pst* I site at the 5'end with a *Hind* III site 13 bp 3'of it. The 3'end was also complementary to *Hind* III but had a single base exchange that inactivated this restriction site, leaving a single *Hind* III site 5' of the terminator stem-loop. This oligonucleotide construct was then ligated into pRK404 and subsequently digested with *Pst* I and *Hind* III. This was digested again with *Hind* III and *Bam* HI and the 3.8 kb *Bam* HI-*Hind* III *puf* operon construct was inserted yielding pRKXmut2T.

For RNA isolations cell growth was monitored until the cells reached mid-late log phase at which point the cultures were placed on ice, harvested immmediately at 0° C and frozen in liquid N_2. RNA was isolated as described in Ref.8 with the exception that an additional RNase free DNase I (Boehringer Mannheim, FRG) digestion was carried out after precipitation of the RNA and followed by another phenol extraction and re-precipitation of the RNA. Purified RNA was denatured by glyoxalation prior to agarose electrophoresis. RNA standards (BRL Gibco) to determine message size were treated in an identical manner. RNA was transferred to Gene Screen (DuPont de Nemoirs, NEN division) by electroblotting. Hybridization of the Northern blots was carried out as described by the manufacturer. The final two washes after hybridization were carried out in 0.1 x SSC at 68° C for 30 min each.

Spectroscopy

Reaction center photobleaching in intact cells was measured using a photodiode array spectrophotometer (9) with the modification that the measuring beam intensity was reduced with a 1%T neutral density filter and defined by a Schott RG 665 filter. The cells were suspended in medium containing 25% w/v Ficoll to reduce light scattering and subsequently exposed to the measuring beam, which was strong enough to be weakly actinic. The RC bleaching was then measured as the difference between an intitial spectrum and a second spectrum 63 ms after the initial data points had been collected.

Results and Discussion

In previous work it was shown that the 5.3 kb *puf* operon *Bam* HI-*Hind* III fragment shown in Fig. 1A (see below) was sufficient to restore the PS[+] phenotype to the PS[−] *puf* L,M and X deletion strain PUF△LMX21 (6). The absence of the genomic copies of the L and M genes in PUF△LMX21 and presence and specific insertion of the kanamycin resistance in the place of the *puf* genes were confirmed using southern blots. The absence of the X gene in this strain was shown using dot blots (data not shown). Since the 5.3 kb *puf* construct restored the PS[+] phenotype contained an additional 1200 bp downstream of the *puf* operon, 500 bp of which had also been deleted in the genomic DNA of the deletion strain PUF△LMX21, it was then of interest to determine what, if any, role was played by the DNA region downstream of *puf* M.

Using the site-directed mutagenesis phasmid pMa/c containing the 5.3 kb construct, a second *Hind* III restriction site was introduced three bp after the stop codon for *puf* M. This left the M gene and all DNA upstream intact, but allowed the removal of the 1600 bp including the X gene, the transcription terminating stem-loop structure(s) at the 3' end of the large *puf* transcript and an additional 1100 bp downstream of the *puf* operon.

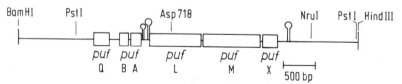

Fig. 1. Restriction map of the 5.3 kb *puf* operon construct used in this study. The figure includes all genes of the *puf* operon and the stem-loop structures located at the 3' end of the 2700 nt (3' of *puf* X) and 500 nt (3' of *puf* A) transcripts. The DNA between the *Asp* 718 and *Nru* I sites was deleted in the genome of PUF△LMX21.

The shortened *puf* operon construct Xmut2 was then partially sequenced to insure that no changes had occurred in the RC genes or upstream regions. The unique restriction sites shown upstream as well as in structural genes were identified and confirmed by sequence analysis.

The shortened *puf* operon construct Xmut2 was subcloned into the low copy number, mobilizable, broad host range plasmid pRK404 to create pRKXmut2. This was introduced conjugally into the *puf* L,M and X deletion strain PUF△LMX21. The resulting exconjugants showed comparable growth rates to exconjugates of PUF△LMX21 containing the 5.3 kb construct in the same plasmid (pRK5.3) under chemoheterotrophic, low pO_2 conditions known to cause gratuitous expression of the *puf* operon. Further tests under photoheterotrophic conditions revealed that pRKXmut2 did not restore the PS$^+$ phenotype.

It was initially hypothesized that the deletion of the stem-loop structure at the 3' end of the *puf* operon could cause the PS$^-$ phenotype by drastically reducing the stability of the transcript coding for the RC polypeptides. The result would be the inability to translate sufficient RC to support photoheterotrophic growth. To test this the stem-loop structure found at the 3' end of the wild-type large transcript was inserted into pRKXmut2 immediately 3' of the *puf* M gene using oligonucleotides (see Materials and Methods). The resulting construct termed pRKXmut2T was sequenced to confirm the presence of the stem-loop and to confirm that the M gene was unaltered.

RNA was isolated from the chemoheterotrophically low pO_2(3-4 mm Hg) grown wild-type, deletion strain PUF△LMX21, or PUF△LMX21 complemented in *trans* with pRK5.3, pRKXmut2 or pRKXmut2T. The Northern blot shown in Fig. 2 was probed with the fragment shown in Fig.1B after random priming. The wild-type RNA (lane 1) showed the expected two messages for the *puf* operon at 2700 and 500 nt coding for the *puf* operon B,A,L,M and X genes and B and A genes, respectively. The RNA from PUF△

LMX21 (lane 2) contained only the 500 nt message due to the absence of the L,M and X genes. This confirmed what had been seen at the DNA level, that only the three structural genes had been deleted and that the upstream promoter region remained intact. Complementation of the PUF△LMX21 with pRK5.3 (lane 3) restored a 2700 nt transcript indicating that the oxygen sensitive promoter region was present and functional in the plasmid-born *puf* construct. The large increase in the amount of message seen in the pRK5.3 case indicated a copy number effect on the transcription of the plasmid-born construct under the low pO_2 conditions used here. PUF△ LMX21 complemented with pRKXmut2 (lane 4) also showed a similar copy number effect, but only a smear was visible in the region

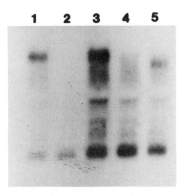

Fig. 2. Northern blot analysis of total RNA isolated from semi-aerobically grown cells. Agarose gels were run with 5 μg of total RNA per probe. Lane 1 wild-type, lane 2 PUF△LMX21, lane 3 PUF△ LMX21(pRK5.3), lane 4 PUF△LMX21(pRKXmut2) and lane 5 PUF△LMX21 (pRKXmut2T).

corresponding to the large transcript. The insertion of the stem-loop structure in pRKXmut2T restored a discrete band in the region of the large transcript (lane 5) comparable to that seen with the wild-type (lane 1) making it unlikely that message instability was the basis for the observed PS⁻ phenotype.

It was then of interest to examine the same strains grown under the same conditions at the translational level. PUF△ LMX21(pRK5.3) and PUF△ LMX21(pRKXmut2T) cultures, grown under the same conditions used to isolate RNA, were used to examine the amount of RC present in intact cells based on the bleaching of the Q_y region of the RC absorption spectrum. Figure 3 shows the bleaching spectra of PUF△ LMX21(pRK5.3) and PUF△ LMX21(pRKXmut2T) based on the same total number of cells are nearly identical. The spectra show the expected RC special pair Q_y bleaching maximum at 865 nm and the derivative-like feature due to the electrochromic shift of the RC voyeur Bchl a absorption (780-810 nm) caused by the charge separation within the RC upon excitation (9). Further analysis indicated that both strains showed a light dependent RC driven membrane potential resulting in an electrochromic shift of the carotenoid in the light-harvesting complexes. This can be taken as an indication that the RCs were inserted into the membrane in both cases. Further examination of the membranes using SDS-polyacrylamide gel electrophoresis indicated that there where no differences in total RC protein (Data not shown).

Having established that transcription was not inhibited and that RC was present in PUF△ LMX21(pRKXmut2T) further experiments were conducted to test the phenotype. The initial inoculum for the growth curves shown in Fig. 4 were chemo-heterotrophic low pO_2 cultures such as those used to isolate

69

RNA or to measure RC photobleaching. In this way it was certain that RC was present in the inoculum and transcription was unaffected. In addition to PUF△LMX21(pRK5.3) and PUF△ LMX21(pRKXmut2T) two other strains confirmed by sequence analysis were included to determine whether it was the upstream, structural or the deleted downstream 1600 bp that were responsible for the observed PS⁻ phenotype.

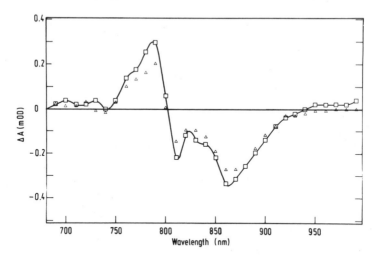

Fig. 3. RC photobleaching in intact semi-aerobically grown cells. Photobleaching assays were carried out as desribed in Materials and Methods. Cells were used for measurement when the cultures had reached the late-log phase. The cultures were diluted in the cuvette to a final count of 3.0×10^5 cells/ml. PUF△LMX21(pRK5.3) (□) and PUF△ LMX21(pRKXmut2T) (△).

The first strain labelled K6 was also PUF△LMX21 complemented in *trans* with pRKXmut2T, while K6/1600 was PUF△LMX21 complemented with the same plasmid pRKXmut2T after the 1600 bp fragment 3' of the *puf* M gene was re-inserted. It can readily be seen that the PUF△LMX21 strains complemented with pRKXmut2T showed no growth under photoheterotrophic conditions for 4 days at which point growth of presumed suppressor strains was observed. The period prior to growth varied between 4 and 6 days, while on the other hand the PUF△ LMX21(pRK5.3) or the control K6/1600 routinely exhibited growth under these conditions after 15-20 hrs. This experiment makes it clear that the lack of photosynthetic growth can be mapped to the region of 1600 bp beginning after the *puf* M gene.

It was necessary to determine whether the putative suppressor strains were phenotypically distinct or whether the inital lack of photosynthetic growth may have been due to the fact that the cells simply had lost the ability to adapt from semi-aerobic to photosynthetic conditions. To challenge this hypothesis the PS⁺ strains of PUF△LMX21(pRKXmut2T) generated under photosynthetic pressure were cultured three times semi-aerobically before being used as inoculum for photosynthetic growth. The photosynthetic cultures after repeated sub-culturing semi-aerobically did not exhibit a lag prior to the onset of growth, indicating they were phenotypically distinct from the inoculum used to test photoheterotrophic growth intitially. To further test this a semi-aerobic culture was diluted serially and plated under photosynthetic and chemo-heterotrophic conditions with the expectation that the cell count under both conditions would be identical if the long lag was simply an adaptation phenomenon. After continued growth for more than a week under photosynthetic conditions, single colonies able to suppress the PS⁻ phenotype were observed at a frequency of 10^{-6} to 10^{-7}.

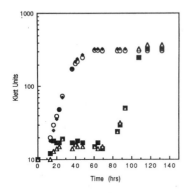

Fig. 4. Photoheterotrophic growth curves. Residual tetracycline from the pre-cultures was 0.07 µg/ml. PUF△LMX21 (pRK5.3) (O), PUF△ LMX21(pRKXmut2T) (■), K6 (△) and K6/1600 (◆).

Initial tests to determine if there was a toxic effect of illumination indicated that there was no decrease in CFU/ml as a function of illumination time.

A second interpretation of the PS⁻ phenotype could be that the shortened *puf* construct is somehow initially much less efficient at trapping and transferring light energy to the RC. Room temperature and low temperature (77K) spectro-scopic analysis of the light-harvesting complexes in intact PUF△LMX21(pRK5.3) and PUF△LMX21(pRKXmut2T) cells grown under the semi-aerobic conditions described above were examined. It was immediately established that the plasmid-born *puf* operon construct that was known to cause increased transcription also resulted in increased amounts of spectrally intact LHI measured relative to the wild-type cells. It was further seen that this was a copy number effect of the plasmid-born LHI genes when the *R. sphaeroides* wild-type showed the same increase in LHI after the introduction of the same plasmid constructs. This was further verified by comparing a *puf* construct missing the promoter region cloned into the same plasmid. This construct introduced into the wild-type had no effect on the total amount or ratio of the two light-harvesting complexes.

Given a copy number effect existed it was necessary to compare PUF△LMX21(pRKXmut2T) with PUF△LMX21(pRK5.3) and not with the original wild-type. Whole cell spectra from semi-aerobic cultures of the two strains were compared from early log phase to stationary growth phase. An approximate 20% increase in the amount of LHI was observed in PUF△LMX21 (pRKXmut2T) relative to PUF△LMX21(pRK5.3) and a 20-30% decrease in LHII. The changes in the ratio of the two light-harvesting complexes in PUF△LMX21(pRKXmut2T) could be reversed by re-inserting the 1600 bp fragment removed down-stream of *puf* M. This demonstrated that the region responsible for this effect also mapped to the 1600 bp downstream of *puf* M.

The ORF X (initially termed C2397) was identified in *R. capsulatus* and found to code for a polypeptide of 78 amino acids (11). A similar ORF has also been identified in *R. sphaeroides* and termed *puf* X (3). The DNA sequence of the *R. sphaeroides* X gene confirmed by two groups independently from two different wild-type strains of *R. sphaeroides* (2.4.1 and ATCC 17023) shown in Fig. 5.

```
stop M              M   A   D   K   T   I   F   N   D   H   L   N   T   N   P
5'TGAGGAGCGATCACAATGGCTGACAAGACCATCTTCAACGATCACCTCAACACCAATCCG
        10        20        30        40        50        60
3'ACTCCTCGCTAGTGTTACCGACTGTTCTGGTAGAAGTTGCTAGTGGAGTTGTGGTTAGGC


   K   T   N   L   R   L   W   V   A   F   Q   M   M   K   G   A   G   W   A   G
5'AAGACCAACCTTCGCCTCTGGGTCGCTTTCCAGATGATGAAGGGTGCGGGCTGGGCTGGC
        70        80        90       100       110       120
3'TTCTGGTTGGAAGCGGAGACCCAGCGAAAGGTCTACTACTTCCCACGCCCGACCCGACCG


   G   V   F   F   G   T   L   L   L   I   G   F   F   R   V   V   G   R   M   L
5'GGCGTGTTCTTCGGGACGCTCCTTCTCATCGGGTTCTTCCGGGTGGTCGGGCGGATGCTT
       130       140       150       160       170       180
3'CCGCACAAGAAGCCCTGCGAGGAAGAGTAGCCCAAGAAGGCCCACCAGCCCGCCTACGAA


   P   I   Q   E   N   Q   A   P   A   P   N   I   T   G   A   L   E   T   G   I
5'CCGATCCAGGAGAACCAGGCTCCGGCGCCGAACATCACCGGCGCTCTGGAGACCGGGATC
       190       200       210       220       230       240
3'GGCTAGGTCCTCTTGGTCCGAGGCCGCGGCTTGTAGTGGCCGCGAGACCTCTGGCCCTAG


   E   L   I   K   H   L   V   *
5'GAGCTGATCAAGCATCTCGTCTGA
       250       260
3'CTCGACTAGTTCGTAGAGCAGACT
```

Fig. 5. DNA and derived peptide sequence of *puf* X gene from *R. sphaeroides* ATCC 17023. A possible Shine-Dalgarno sequence is underlined. The phasmid pMa containing the 5.3 kb construct was packaged as a pseudo-phage using the helper phage M13K07. The single-stranded DNA was sequenced using a series of oligonucleotide primers, the first of which corresponded to the 3' end of the *puf* M gene.

The stop codon for *puf* M is followed immediately by the Shine-Dalgarno for the *puf* X gene and a single ORF encoding a polypeptide of 82 amino acids.

Hydropathy plots of the X genes from *R.sphaeroides* and *R. capsulatus* were performed to determine if there is structural homology at the level of secondary structure and to determine whether the polypeptide might be membrane-associated. The hydropathy plots shown in Fig. 6 indicated that the overall hydrophobicity of the two polypeptides were very similar. They also appear to have a very similar overall structural arrangement in that they probably have one membrane spanning helix located in the middle of the sequence.

The structural design of one membrane spanning helix surrounded on both sides by hydrophilic domains was also found to be the case for the *puf* Q gene from *R. sphaeroides* (Data not shown) and *R. capsulatus* (4). The *psb*H gene from chloroplasts was then chosen as a control to check that the observed hydropathy plots were not simply an artefact of the computer algorithm when short polypeptides are examined. The *psb*H gene product has 73 amino acids, a similar overall hydropathy and is thought to be associated with photosystem II (12). The hydropathy analysis of this gene also indicated a single helix, but positioned at the COOH-terminus allowing us to rule out a computer artefact as the explanation for the similar structures of the two X genes and their similarity to *puf* Q.

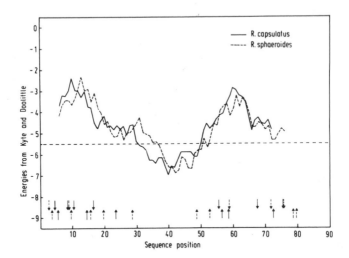

Fig. 6. The *R. sphaeroides* (dotted line) and *R. capsulatus* (solid line) X protein hydropathy plots were generated using the GENMON (Genetic Monitor) program supplied by the GBF-Gesellschaft fuer Biotechnologische Forschung GmbH, Braunschwieg, FRG. The plot was generated using a mean of 11 amino acids. Positively and negatively charged residues are indicated by arrows pointing upwards and downwards, respectively. The charged amino acids for the *R. sphaeroides* X peptide are indicated by the dotted arrows and those for *R. capsulatus* X are indicated by the solid arrows. The dotted line across the figure indicates the average energy for each polypeptide, all parts below this line are considered hydrophobic and everything above it hydrophilic.

Sequence comparison of the X gene with the other components of the *puf* operon indicated that there is no obvious homology and a search of the EMBL protein data bank failed to identify meaningful homologous sequences. A comparison of the *R. sphaeroides* X sequence to that from *R. capsulatus* (Fig. 7) using the program of Needleman and Wunsch (13) under conditions allowing the insertion of up to two gaps in the protein sequence to allow maximum alignment indicated that the two sequences showed at best 55 percent similarity (allowing for conservative changes) and 29 percent identity (conserved residues). The conserved residues were not trivial, but obvious functional residues such as histidines were not conserved. There is also no apparent pattern to the distribution of positively or negatively charged amino acids in the predicted hydrophilic portions of either polypeptides near the N- or C-termini (see Fig. 6). This could indicate that it is the overall secondary structure which has been conserved and that individual residues as such are less important for the required function. An indication of this was given by the fact that the *R. sphaeroides* sequence has two arginines at the COOH-terminal end of the possible membrane spanning region that are absent in *R. capsulatus*, making the possible membrane spanning region shorter in *R. sphaeroides* but not eliminating it.

```
      1 ..MADKTIFNDHLNTNPKTNLRLWVAFQMMKGAGWAGGVFFGTLLLIGFF 48
        | ||.:     ..:...| :: :|:: ||  || ::: .|: .| |:
      1 MSMFDKPF...DYENGSKFEMGIWIGRQMAYGAFLGSIPFLLGLGLVLGS 47

     49 RVVGRMLPIQENQAPAPNITG.ALETGIELIKHLV 82
        .:| |||  ..:|||.| .|: .:: :.|::
     48 YGLGLMLPERAHQAPSPYTTEVVVQHATEVV*... 79
```

Fig. 7. Sequence comparison of *puf* X genes from *R. capsulatus* and *R. sphaeroides*. The top row is always the *R. sphaeroides* sequence data. The introduction of two gaps shown here yielded the highest percent similarity 55% (29% identity) with one gap indicating 49% similarity (27% identity) and no gap indicating 43% similarity (23% identity).

The rather low homology between the X genes of *R. sphaeroides* and *R. capsulatus* was surprising as similar analysis of the *puf* B, A, L and M genes revealed 85-90 percent similarity and 75-80 percent identity between the two species when the same gap restrictions were chosen. This rather low homology could reflect inherent differences in the requirement for a functional X gene between the species. Other differences between the closely related species in the requirement for LHI for RC assembly (14) and on the effect of deletion of the cytochrome c_2 (15) have been noted previously

Other workers have reported studies using an X minus *puf* operon construct (16). They indicate however, that complementation of a *puf* minus deletion strain with this construct restored the PS[+] phenotype. The difference between this study and that reported earlier cannot be easily explained, but one difference was in the type of deletion strain used for the studies. In this case 500 bp of DNA downstream of the *puf* X gene was deleted along with the *puf* L,M, X genes. It is not known if there are ORF(s) in this region in *R. sphaeroides*, but other workers (10) have identified an ORF encoding a 641 amino acid polypeptide (or two smaller ORFs) in this region in *R. capsulatus*. It is unlikely that a 641 amino acid polypeptide is responsible for the observed differences here as the region 3' of the 5.3 kb *puf* operon construct known to restore the PS[+] phenotype is not large enough to code for such a large protein. This points rather directly at a smaller ORF downstream of the *puf* operon and/or the X gene as the cause of the observed PS[-] phenotype of the shortened *puf* operon construct pRKXmut2T.

Further experiments were conducted just prior to the publication deadline of this work to determine the phenotype of a *puf* construct ending with X and the wild-type stem-loop terminators. The initial results indicated that the re-insertion of the X gene was sufficient to restore the PS[+] phenotype in the absence of the 500 bp 3' of the *puf* operon.

Further experiments are underway to determine whether this region is transcribed. The function of the X gene and the manner in which the cells so readily suppress the PS⁻ phenotype is still unexplained, but it is intriguing to postulate that these gene(s) may play a structural role in the correct assembly of a functional bioenergetic system necessary to convert light energy to chemical energy.

References

1. Feher, G., J.P. Allen, M.Y. Okamura and D.C. Rees. 1989.Structure and function of bacterial photosynthetic reaction centers. Nature 339:111-116.

2. Warshel, A. and W.W. Parson. 1987.Spectroscopic properties of photosynthetic reaction centers 1. Theory. J.Am.Chem.Soc. 109:6143-6152.

3. Donohue, T.J., P.J. Kiley and S. Kaplan. 1988.The *puf* operon region of *Rhodobacter sphaeroides*. Photosyn.Res. 19:39-61.

4. Bauer, C.E., D.A. Young and B.L. Marrs. 1988.Analysis of the *Rhodobacter capsulatus puf* operon location of the oxygen regulated promoter region and the identification of an additional *puf* encoded gene. J.Biol.Chem. 263:4820-4827..

5. Adams, C.W., M.E. Forrest, S.N. Cohen and J.T. Beatty. 1989.Structural and functional analysis of transcriptional control of the *Rhodobacter capsulatus puf* operon. J. Bacteriol. 171:473-482.

6. Farchaus, J.W. and D. Oesterhelt. 1989.A *Rhodobacter sphaeroides puf* L, M and X deletion mutant and its complementation in *trans* with a 5.3 kb *puf* operon shuttle fragment. EMBO J. 8:47-54.

7. Davis, J., T.J. Donohue and S. Kaplan. 1988.Construction characterization and complementation of a *puf* negative mutant of *Rhodobacter sphaeroides*. J.Bacteriol. 170:320-329..

8. Aiba, H., S. Adhya and B. de Crombrugghe. 1981.Evidence for two functional Gal promoters in intact *E.coli*. J.Biol.Chem. 256:11905-11910.

9.Uhl,R.,B. Meyer and H. Desel. 1985. A polychromic flash photolysis apparatus (PFPA). J.Biochem.Biophys.Methods 10:35-48.

10.Breton, J. 1988. in:"The Photosynthetic Bacterial Reaction Center:Structure and Dynamics." eds. Breton,J. & Vermiglio,A. (Plenum,New York) pp 59-69.

11.Youvan, D.C., E.J. Bylina, M. Alberti, H. Begush and J.E. Hearst. 1984. Nucleotide and deduced polypeptide sequences of the photosynthetic reaction center B-870 antenna and flanking polypeptides from *Rhodopseudomonas capsulata*. Cell 37:949-958.

12.Westhoff,P.,J.W. Farchaus and R.G. Herrmann 1986. The gene for the M_r 10,000 phosphoprotein associated with photosystem

II is part of the *psb* B operon of the spinach plastid chromosome. Curr.Genet.11:165-169.

13.Needleman,S.B. and Wunsch,C.D. 1970. J.Mol.Biol.,48:443-453.

14.Jackson, W.J., P.J. Kiley, C.E. Haith, S. Kaplan and R. Prince 1987. On the role of the light-harvesting B880 in the correct insertion of the reaction center of *Rhodobacter capsulatus* and *Rhodobacter sphaeroides*. FEBS Lett. 215:171-174.

15.Donohue,T.J., A.G. McEwan, S. Van Doren, A.R. Crofts, and S. Kaplan. 1988. Phenotypic and genetic characterization of cytochrome c_2 deficient mutants of *Rhodobacter sphaeroides*. Biochem.27:1918-1925.

16.Klug, G. and S.N. Cohen. 1988.Pleiotropic effects of localized *Rhodobacter capsulatus puf* operon deletions on production of light-absorbing pigment-protein complexes. J.Bacteriol. 170:5814-5821.

REGULATION OF FORMATION OF PHOTOSYNTHETIC LIGHT-HARVESTING

COMPLEXES IN *RHODOBACTER CAPSULATUS*

Béatrice Oberlé, Hans-Volker Tichy,
Ulrike Hornberger, Gerhart Drews

Institute of Biology 2, Microbiology
Albert-Ludwigs-University, Schänzlestrasse 1
7800 Freiburg, Fed. Rep. Germany

INTRODUCTION

Facultative phototrophic bacteria such as *Rhodobacter* (*Rb.*) *capsulatus* can adapt to different modes of energy metabolism. Under anaerobic growth conditions light energy is captured by the antenna system of the photosynthetic apparatus and transduced into an electrochemical proton gradient across the membrane, which can be used for ATP production. In dark cultures proton motive force is generated on the expense of substrate oxidation by the respiratory chain under aerobic conditions with oxygen as electron acceptor and under anaerobic conditions with dimethyl sulfoxide, trimethylamine N-oxide, nitrate or nitrous oxide as alternative electron acceptors (Ferguson et al. 1987).

The respiratory electron transport system is present under all growth conditions, although in different concentrations. The photosynthetic apparatus, however, is synthesized under low oxygen tensions only (Kaufmann et al. 1982). Variations of light intensity modify the amount and composition of the photosynthetic apparatus, especially the amount of pigment-protein complexes and the amount of intracytoplasmic membranes per cell (Reidl et al. 1985).

The light energy is harvested by the antenna pigment complexes LHI (B870) and LHII (B800-850), which form a continuous network for transfer of excitation energy. The excitons (mobile excited states) can migrate over the antenna system of many photosynthetic units of one membrane vesicle before being trapped by a reaction center, where the charge separation takes place (Hunter et al. 1989).

Each antenna complex consists of two different small pigment-binding polypeptides, which span the membrane once by a hydrophobic, α-helical domain, two or three mol of bacteriochlorophyll (Bchl) and one or two mol of carotenoids. A third, non-pigment-binding polypeptide is in the complex LHII (Drews 1988). The LHI complex, the immediate donor of exitation energy for the reaction center, surrounds and assembles with the reaction center in stoichiometric amounts forming the core complex.

The genes for the pigment-binding polypeptides of LHI and reaction center complexes are clustered in the *puf* operon (Youvan et al. 1984) and expressed together (Klug et al. 1985). The assembling of a functional reaction center seems to be dependent on the simultaneous formation of LHI, and vice versa, for a full developed LHI complex expres-

Molecular Biology of Membrane-Bound Complexes in Phototrophic Bacteria
Edited by G. Drews and E. A. Dawes
Plenum Press, New York, 1990

77

sion of the reaction center is a prerequisite in *Rb. capsulatus* but not in *Rb. sphaeroides* (Jackson et al. 1987, Youvan et al. 1985, Bylina et al. 1986, Klug and Cohen 1988).

The LHII complex interconnects the core complexes and is formed in variable amounts depending on light intensity and other factors (Drews 1985, 1988, Hunter et al. 1989). The pigment-binding polypeptides of LHII are encoded by the *puc* operon localized oppositely to the photosynthetic gene clusters on the chromosome (Williams 1989, this book). The genes of the *puc* operon are expressed independently from the *puf* operon and with a different kinetic (Klug et al. 1985). Mutants defective in formation of reaction center and LHI are able to assemble stable LHII complexes (Feick and Drews 1978).

The stable formation of all pigment-protein complexes is completely dependent on the synthesis and incorporation of Bchl into the complexes (Bylina et al. 1986, Klug et al. 1986). In this communication it will be shown that the formation of the LHII (B800-850) complex in *Rb. capsulatus* is regulated by loci localized in different regions of the genome. A third *puc* gene has been identified.

RESULTS

The pigment-binding polypeptides of the LHII antenna complex (Mr α 7,322 and ß 4,579; Tadros et al. 1983, 1985) are encoded by the *puc* operon, which has been sequenced (Youvan and Ismail 1985). A third polypeptide (, Mr 14,000) has no influence on the absorption spectrum and does not bind Bchl (Feick and Drews 1978, 1979), but seems to be essential for the formation of the complex (Kaufmann et al. 1984): Two Tn5 mutants defective in formation of the B800-850 complex (Kaufmann et al. 1984) have been studied more thoroughly in order to learn more about the regulation of LHII formation. In the mutant NK3 the transposon Tn5 was found to be inserted 526 bp downstream from the *pucA* termination codon (Fig. 1).

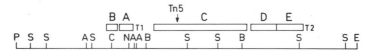

Fig. 1. Physical and genetic map of the 4.5 kbp *PstI-EcoRI* fragment of *Rhodobacter capsulatus* 37b4 DNA containig the genes *pucB*, *pucA*, and *pucE*, and ORFC and ORFD (open boxes marked B, A, C, D, and E). The terminator like sequences are designated T1 and T2. The arrow indicates the insertion of the transposon Tn5 in the mutant NK3. Restriction sites: A, *ApaI*; B, *BalI*; C, *ClaI*; E, *EcoRI*; N, *NruI*; P, *PstI*; S, *SmaI*.

A region of about 3,000 bp including the *pucB* and *pucA* genes and DNA downstream from *pucA* was sequenced. Three open reading frames (ORF's C, D and E) were found (Fig. 1 and Tichy et al. 1989). The deduced amino acid sequence of ORF E was found to be identical with the N terminus and two tryptic peptides of the alkaline-solubilized Mr-14,000 subunit of the isolated LHII complex. Therefore ORF E was named *pucE* and thought to be the gene coding for the -subunit of the LHII complex (Tichy et al. 1989).

Why did the transposon insertion downstream from *pucBA* operon inhibit the formation of the LHII complex (Kaufmann et al. 1984)? Northern experiments using an 1,065 bp *SmaI* fragment containing sequences of the ORF's C, D and E as a probe, showed a strong increase of hybridizing 1.2 kb RNA in wild-type cells after a shift from aerobic to semiaerobic conditions (Fig. 2).

The levels of the *pucAB* and *pucDE* mRNA increased after induction with the same extent and time course. The maximal level was observed after 45 to 60 min. In the mutant NK3 *pucDE* mRNA was not detectable in aerobic or semiaerobic cultures (Fig. 2, 3).

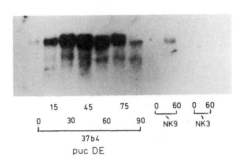

15   45   75   0   60   0   60

0   30   60   90   NK9   NK3

37b4

puc DE

Fig. 2. Autoradiogram of a hybridization of total RNA extracted from wild-type strain
37b4 and the derivative mutant strains NK3 and NK9, separated on form-
aldehyde-agarose gels, with the ^{32}P-labeled DNA probe *pucDE* (see text). The
RNA was isolated at different times after lowering of oxygen tension in the
culture. Fifteen μg total RNA were applied to each slot.

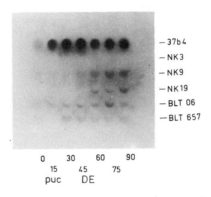

    — 37b4
    — NK3
    — NK9
    — NK19
    — BLT 06
    — BLT 657

0   30   60   90
   15   45   75
puc   DE

Fig. 3. Dot blot analysis. Hybridization of 10 μg total RNA with the ^{32}P-labeled DNA
SmaI fragment (ORF CDE). The mutants used for hybridization were described in the text.

After 60 min. of induction the level of *pucBA* mRNA was about eight-fold lower
than in the wild-type strain (Fig. 4).

The maximal level of *pucBA* mRNA was observed after 75 to 90 min. In accordance with the mRNA data small amounts of   and ß polypeptides were detectable in
membranes of induced NK3 cells, but no   polypeptide. It is hypothesized that one of the
genes downstream of *pucA* has a regulatory function which effects the expression of the
pucBA genes.

The mutant strains NK3, NK9 and NK19 are phenotypically different. All are
LHII negative, but NK3 synthesizes colored carotenoids, while NK9 and NK19 do not.
Mutants NK3, NK9 and NK19 synthesize, after induction, small amount of   and ß poly-
peptides and insert them into the membrane, but no   polypeptide was detectable
(Kaufmann et al. 1984). Mutants with similar phenotypes have been described by Zsebo

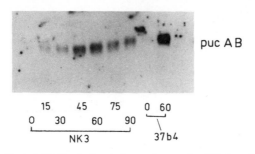

puc A B

```
        15      45     75    0  60
    0      30      60     90    ⌐‾|
    └────────────────────┘    37b4
              NK3
```

Fig. 4. Hybridization of total RNA, isolated from the strains 37b4 and NK3 after different times of semiaerobic incubation (min). The DNA probe was *pucBA*.

and Hearst (1984). The mutants NK9 and NK19 were reconstituted to wild-type by a conjugative transfer of the plasmide pRBH203 (H. Boos, unpublished) containing the *BamHI* H fragment of pRPS404 (Yen and Marrs 1976). On this fragment the *crtI* gene besides others, is localized (Yen and Marrs 1976, Zsebo and Hearst 1984, Giuliano et al. 1988). In the mutant NK9 the transposon Tn5 was found to be inserted in the *crtI* gene (Fig. 5).

The *crtI* gene product catalyzes the synthesis of an early component of the carotenoid pathways (Armstrong et al. 1989). The insertion of the transposon in this gene explains the inhibition of carotenoid synthesis in the strain NK9. Like Bchl, carotenoid may be an essential component of the LHII complex and the inhibition of its synthesis may block the assembly of LHII. However, a mutant strain GK2 was found, which forms the LHII complex including the   polypeptide in absence of colored carotenoids (Dörge et al. 1987). The insertion of the transposon Tn5 in the *crtI* gene has a pleiotropic effect. Besides the inhibition of carotenoid synthesis and a shift of the absorption peak from 874 nm (NK3) to 871 nm (NK9) the transposon insertion in NK9 inhibits the formation of the 14,000 polypeptide (Kaufmann et al. 1984).

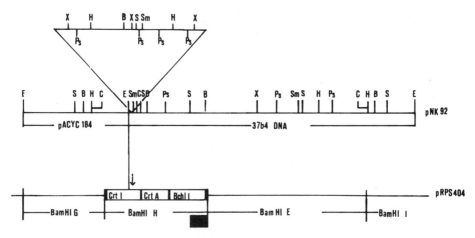

Fig. 5. Physical map of pNK92. The plasmid contains about 10.5 kbp *Rhodobacter capsulatus* NK9 DNA in the vector pACYC 184. The Tn5 insertion is indicated by the triangle. The hybridization probe used to determine the homologous regions on pNK92 and the *BamHI-H* fragment from pRPS404, is black boxed. The regions on the map to the right side of the vertical line are corresponding DNA regions. The arrow points to the Tn5 insertion and indicates that the insertion is in the crtI gene. Restriction enzymes: B, *BamHI*; C, *ClaI*; E, *EcoRI*; H, *HindIII*; Ps, *PstI*; S, *SalI*; Sm, *SmaI*; X, *XhoI*.

The mRNA formation was studied with the same probes (*pucDE* and *pucBA*) as used for the mutant NK3. The amount of *pucDE* mRNA in the induced NK9 cells was tenfold lower than in the respective wild-type cells after 60 min. of induction (Fig. 2, 3). The formation of *pucBA* mRNA was even stronger inhibited. The amount was 25-fold lower compared with the respective wild-type mRNA (Fig. 2, 6).

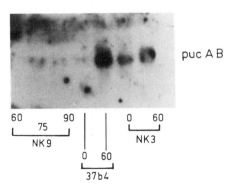

Fig. 6. Hybridization of total RNA, isolated from the strains 37b4, NK3, and NK9, against *pucBA* DNA.

So we concluded that in both mutant strains, NK3 and NK9, the transposon insertion effects a down regulation of *pucBA* mRNA formation and of transcription of genes localized downstream from *pucA* (ORF DE). Although the amounts of mRNA's found in the mutants NK3 and NK9 were different, the assembly of functional B800-850 light-harvesting complexes was completely inhibited in both strains. We have at present no experimental data explaining why the transcription of both DNA sequences, *pucBA* and *pucDE*, was affected by the transposon insertion.
Recent results gave some ideas how the mutation could affect transcription. Conserved palindromic nucleotide motifs have been found in the 5' flanking regions of *Rb. capsulatus crtA*, *crtI*, *crtE*, and *pucB* genes (Youvan and Ismail 1985, Zucconi and Beatty 1988, Armstrong et al. 1989, Tichy et al. 1989). They show strong sequence homology to consensus binding sites of DNA-binding proteins, which function as repressor or activator of transcription. The two halves of the consensus palindrome, located upstream from the *pucBA* mRNA (Zucconi and Beatty 1988), are separated by 12 nucleotides center to center (Armstrong et al. 1989). They may be involved in transcription initiation of the photosynthetic genes. It is also speculated that the transposon insertion affects directly or indirectly the expression of regulatory genes and their products, activator proteins or gene products which are essential for protein import into membranes.

Recently we have shown that in two Bchl-less mutants no mRNA for pigment-proteins was formed (Klug et al. 1986). To check the hypothesis that an inhibition of Bchl synthesis affects the transcription of LHII-specific genes we used Northern hybridization with probes from *pucDE* to detect mRNA in the two Bchl-free transposon mutants BLT 657 and BLT 06.

The mutant BLT 06 synthesizes carotenoids but no Bchl or colored precursors and seems to be defective in at least the synthesis of the α and ß polypeptides of both LH complexes. The strain BLT 657 showed the same defects in its protein and pigment synthesis as BLT 06 but excretes a tetrapyrrole derivative absorbing at 663 nm (in vivo) and 657 nm (in aceton-methanol). The semiaerobic culture has a yellow-green color, the cell sediment is orange-brown colored. In the strain BLT 06 the transposon was found to be inserted in a sequence corresponding to the *BamHI*-E fragment of the R' plasmid pRPS404 (Yen and Marrs 1976). In both strains the level of mRNA hybridizing with *pucDE* probes increased only very weakly after a shift from aerobic to semi-aerobic cultur conditions (Fig. 3).

Using a *SmaI* fragment from pVK1 (1266 nt, *pucBA* and part of ORF C) as a probe an increase of hybridizing mRNA after induction was observed with a maximum at about 80 min. The level of LHII specific mRNA in the strain BLT 657, as compared with the wild-type strain was very low. The experiments confirm earlier results that mutants defective in Bchl synthesis are impaired in transcription of genes coding for LH polypeptides.

In all of the Bchl-less strains the low molecular polypeptides of the antenna complexes are very weakly expressed (Fig. 7).

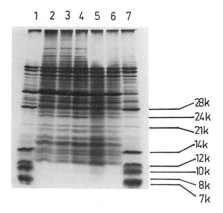

Fig. 7. SDS polyacrylamide (11.5 to 16.5 %) gelelectrophoresis of membrane proteins, isolated from the strains 37b4 (wild type), lanes 1 and 7, and the Bchl-less strains BLT 746 lane 2, BLT 657 lane 3, BLT 767 lane 4, BLT 06 lane 5, BLT 03 lane 6. The membranes were isolated form semiaerobically grown cells. The apparent Mr's of the bands. stained with coomassie blue, are indicated on the right side: 14,000; 10.000 and 7000 LHII; 12,000; 8000 LHI.

A pulse-chase experiment (labeling of proteins with 35 S-Met) indicated a weak synthesis of LH polypeptides. The label on ß polypeptides disappeared during chase (M. Yang, diploma thesis). These results confirm published observations that in absence of Bchl no stable LH complexes are formed and that the ß polypeptides generally are less stably incorporated than α polypeptides.

ACKNOWLEDGEMENT

The research work was supported by the grants Dr 29/31-3 A and G of Deutsche Forschungsgemeinschaft.

REFERENCES

Armstrong, G.A., Alberti, M., Leach, F., and Hearst, J.E., 1989, Nucleotide sequence, organization, and nuture of the protein products of the carotenoid biosynthesis gene cluster of *Rhodobacter capsulatus*. Mol. Gen. Genet. 216: 254.

Bylina, E.J., Ismail, S., and Youvan, D.C., 1986, Site-specific mutagenesis of bacteriochlorophyll-binding sites affecting biogenesis of the photosynthetic apparatus, in: "Microbial Energy Transduction" pp. 63-70, D.C. Youvan and F. Daldal, eds., Cold Spring Harbor Lab.

Dörge, B., Klug, G., and Drews, G., 1987, Formation of the B800-850 antenna pigment-protein complex in the strain GK2 of *Rhodobacter capsulatus* defective in caroteinoid synthesis. Biochim. Biophys. Acta 892:68.

Drews, G., 1985, Structure and functional organization of light-harvesting complexes and photochemical reaction centers in membranes of phototrophic bacteria, Microbiol. Rev. 49:59.

Drews, G., 1988, Organization and assembly of bacterial antenna complexes, in: "Photosynthetic Light-Harvesting Systems, Organization and Function, H. Scheer and S. Schneider, eds., pp 233-246, Walter de Gruyter Berlin, New York.

Feick, R., and Drews, G., 1978, Isolation and characterization of light-harvesting bacteriochlorophyll-protein complexes from *Rhodopseudomonas capsulata*, Biochim. Biophys. Acta 501:499.

Feick, R., and Drews, G., 1979, Protein subunits of bacteriochlorophylls B802 and B855 of the light-harvesting complex II of *Rhodopseudomonas capsulata*, Z. Naturforschg. 34c:196.

Ferguson, S.J., Jackson, J.B., and McEwan, A.G., 1987, Anaerobic respiration in the Rhodospirillaceae. FEMS Microbiol. Rev. 46:117.

Giuliano, G., Pollock, D., Stapp, H., and Scolnik, R.A., 1988, A genetic physical map of the *Rhodobacter capsulatus* carotenoid biosynthesis gene cluster. Mol. Gen. Genet. 213:78.

Hunter, C.N., van Grondelle, R., and Olsen, J.D., 1989, Photosynthetic antenna proteins: 100 ps before photochemistry starts, Trends Biochem. Sci. 14:72.

Jackson, W.J., Kiley, P.J., Haith, C.E., Kaplan, S., and Prince, R.C., 1987, On the role or light-harvesting B880 in the correct insertion of the reaction center of *Rhodobacter capsulatus* and *Rb. sphaeroides*, FEBS Lett. 215:171.

Kaufmann, N., Hüdig, H., and Drews, G., 1984, Transposon Tn5 mutagenesis of genes for the photosynthetic apparatus in *Rhodopseudomonas capsulata*, Mol. Gen. Genet. 198:153.

Kaufmann, N., Reidl, H.-H., Golecki, J.R., Garcia, A.F., and Drews, G., 1982, Differentiation of the membrane system in cells of *Rhodopseudomonas capsulata* induced to synthesize the photosynthetic apparatus. Arch. Microbiol. 131:313.

Klug, G., and Cohen, S.N., 1988, Pleiotropic effects of localized *Rhodobacter capsulatus* *puf* operon deletion on production of light-absorbing pigment-protein complexes, J. Bacteriol. 170:5814.

Klug, G., Kaufmann, N., and Drews, G., 1985, Gene expression of pigment-binding proteins of the bacterial photosynthetic apparatus, transcription and assembly in the membrane of *Rhodopseudomonas capsulata*, Natl. Acad. Sci. USA 82:6485.

Klug, G., Liebetanz, R., and Drews, G., 1986, The influence of bacteriochlorophyll biosynthesis on formation of pigment-binding proteins and assembly of pigment protein complexes in *Rhodopseudomonas capsulata*, Arch. Microbiol. 146:284.

Reidl, H., Golecki, J.R., and Drews, G., 1985, Composition and activity of the photosynthetic system of *Rhodobacter capsulatus*. The physiological role of the B800-850 light-harvesting complex, Biochim. Biophys. Acta 808:328.

Tadros, M.H., Suter, F., Drews, G., and Zuber, H., 1983, The complete amino acid sequence of the large bacteriochlorophyll-binding polypeptide from the light-harvesting B800-850 of *Rhodopseudomonas capsulata*, Eur. J. Biochem. 129:533.

Tadros, M.H., Frank, R., and Drews, G., 1985, The complete amino acid sequence of the small bacteriochlorophyll-binding polypeptide B800-850 ß from light-harvesting complex B800-850 of *Rhodopseudomonas capsulata*, FEBS Lett. 183:91.

Tichy, H.-V., Oberlé, B., Stiehle, H., Schiltz, E., and Drews, G., 1989, Genes downstream from *pucB* and *pucA* are essential for formation of the B800-850 complex of *Rb. capsulatus*. J. Bacteriol. 171: in press.

Youvan, D.C., Bylina, E.J., Alberti. M., Begusch, H., and Hearst, J.E., 1984, Nucleotide and deduced polypeptide sequences of the photosynthetic reaction center, B870 antenna and flanking polypeptides from *R. capsulatus*. Cell 37:949.

Youvan, D.C., Ismail, S., and Bylina, E.J., 1985, Chromosomal delition and plasmide complementation of the photosynthetic reaction center and light-harvesting genes from *Rb. capsulatus*, Gene 38:19.

GENETICS OF CYTOCHROMES C_2 and BC_1 COMPLEX OF PHOTOSYNTHETIC BACTERIA

Fevzi Daldal

Department of Biology, Plant Science Institute
University of Pennsylvania
Philadelphia, PA 19104

INTRODUCTION

Photosynthetic bacteria are endowed with a multitude of b- and c-type cytochromes that are important components of cellular energy transduction pathways operational during the metabolism of various substances[1]. They are often soluble and located in the periplasmic space, or are integral membrane proteins spanning the lipid bilayer. Soluble species may also associate, under certain conditions, with various membrane-embedded complexes. The presence or absence of a given set of cytochromes, as well as their relative amounts, although barely studied in many instances, appear well regulated. Different growth conditions induce, or preclude, the synthesis of various cytochromes in function of the cellular needs[2]. Expectedly, the metabolic pathways and the components involved are complex and numerous, but their knowledge is essential for a complete understanding of the biology of the cell. In the past, biochemical studies of cytochromes have been extremely fruitful, allowing the determination of their structural and functional properties[1,2]. In more recent years, these studies have also been complemented by molecular biological approaches directed at the analysis of the corresponding DNA structures which contain the information responsible for their structure and their regulated synthesis.

Among the many interesting proteins participating to microbial energy transduction we have chosen to study, by genetic approaches, the cytochromes c_2 and bc_1 complex as a model system of interacting soluble and membrane-bound redox-protein pair. Cytochrome c_2 is a small protein found in the periplasm of bacteria and is a diffusable electron carrier between the photochemical reaction center, the ubiquinol:cytochrome c_2 oxidoreductase (or the cyt bc_1 complex) and the cytochrome oxidase, respectively. It may also convey electrons to some other soluble and membrane-bound components, and very recently, its ability to donate electrons to the nitrous oxide reductase of R. $capsulatus$ has been recognized (B. Jackson, personal communication). Cyt bc_1 complex is an integral membrane complex essential only for photosynthetic, but not for respiratory, growth of R. $capsulatus$ because of the presence of an alternate oxidase[3]. It contains three prosthetic-group carrying catalytic subunits, the cytochromes b and c_1 and the Rieske FeS protein, catalyzes the reoxidation of quinol to quinone and concomittently, translocates protons and separates charges across the cellular membrane. In what follows our recent

Molecular Biology of Membrane-Bound Complexes in Phototrophic Bacteria
Edited by G. Drews and E. A. Dawes
Plenum Press, New York, 1990

85

molecular genetic studies related to the cytochromes c_2 and bc_1 complex will be summarized. The topics of our current focus are: 1- What is the molecular nature of the cyt c_2-independent photosynthetic electron pathway of *Rhodobacter capsulatus*? 2- What are the structural determinants of the quinone processing sites of the cyt bc_1 complex?

CYTOCHROME C_2

Recent experiments have shown that cyt c_2 is not universally obligatory for photosynthetic electron flow[4,5]. Deletion of the gene encoding the cyt c_2 apoprotein from the photosynthetic bacterium *Rhodobacter capsulatus* has shown that, in this species, photosynthetic growth occurs without cyt c_2, and that the cyt bc_1 complex can donate electrons to the photosynthetic reaction center via a cyt c_2-independent electron pathway[6]. Similar experiments performed on the closely related bacterium *Rhodobacter sphaeroides* have demonstrated that, in contrast, this species is unable to grow photosynthetically in the absence of cyt c_2 [7,8]. However, it was possible to isolate from a cyt c_2-less mutant of *R. sphaeroides* "suppressor" strains that can grow photosynthetically and that contain elevated amounts of another "cyt c_2-like" soluble cytochrome[9]. Thus, although a cyt c_2-independent photosynthetic growth can be revealed in both of these species under specific conditions, the nature of the electron pathways involved appears different: First, the cyt c_2-independent growth of *R. capsulatus* does not require, or accompany, the overproduction of any other "cyt c_2-like" soluble cytochrome[4,5]. Second, in the absence of the cyt bc_1 complex none of the other soluble, or membrane-bound, cytochromes detectable in *R. capsulatus* act as an electron donor to the photo-oxidized reaction center[5]. Finally, in sharp contrast with *R. sphaeroides*[8], a cyt c_2-independent electron pathway between the reaction center and the cyt bc_1 complex can also be demonstrated in sphaeroplasts prepared from wild type strains of *R. capsulatus*[5,6].

The cyclic photosynthetic electron pathway may be divided into three distinct components: An electron donor (cyt bc_1 complex), an electron acceptor (reaction center) and an electron carrier (cyt c_2) between these two membrane-bound complexes. In principle, each of these components, or a combination of them, may constitute the molecular basis of the difference between these two species in their ability to conduct cyt c_2-independent electron flow. Considering that these two bacterial species are closely related various possibilities could be tested genetically *in vivo* by exchanging between them some of the components involved and by scoring the ability of the hybrids obtained to perform the cyt c_2-independent photosynthetic growth. To test whether the cyt bc_1 complex constituted the structural basis of this difference we have used the structural genes of the three catalytic subunits of the cyt bc_1 complex (Rieske iron-sulfur protein, cytochrome b and cytochrome c_1) that we have previously cloned from both *R. capsulatus* and *R. sphaeroides*[10]. Earlier, using interposon mutagenesis the chromosomal copies of the cyt bc_1 complex and the cyt c_2 structural genes of *R. capsulatus* were eliminated and isogenic pairs of cyt bc_1^-, c_2^+ and cyt bc_1^-, c_2^- single and double mutants were isolated[5]. The cyt bc_1 genes of *R. sphaeroides* were introduced *in trans* into these mutants using a plasmid-borne copy (pGAB291) along an expressed copy of *R. capsulatus* cyt bc_1 genes (pR14A) used as a control[10]. The merodiploid strains thus obtained were tested for their photosynthetic growth ability and the data obtained are presented in Table I. That the growth observed was indeed photoheterotrophic was confirmed by testing its dependence on anaerobiosis+light and its sensitivity to atrazine and myxothiazol, specific inhibitors of the photochemical reaction center and the cyt bc_1 complex[11]. The overall growth properties of the trans-species hybrids were found complicated, and especially, the *R. sphaeroides* cyt bc_1 complex was noticed to be unable to complement for photosynthetic growth a cyt bc_1^-, cyt c_2^- double mutant of *R. capsulatus* on minimal RCV medium containing malate as the sole carbon source, although it can do so either when malate was replaced with fructose or in rich MPYE media containing yeast extract and peptone[12]. Why this merodiploid cannot grow on C4-dicarboxylic acids as the sole source of carbon is not yet clear. Nonetheless, even its conditional ability to restore photosynthetic growth of a cyt bc_1-less mutant of *R. capsulatus* in the presence, and in the absence, of cyt c_2 clearly indicated that the cyt bc_1 complexes could be interchanged between these two species without the loss of the cyt c_2-independent photosynthetic growth ability. This finding suggested that the molecular basis of this ability cannot rely solely on the properties of *R. capsulatus* cyt bc_1 complex[12].

Table 1. Growth characteristics of *R. capsulatus* merodiploid strains harboring the *R. capsulatus* (pR14A) and *R. sphaeroides* (pGAB291) cyt bc_1 complexes on rich (MPYE) and on minimal media (RCV) containing malate*/fructose** as the sole carbon source under photosynthetic and respiratory growth conditions.

Strains : relevant phenotypes	Photosynthetic Growth MPYE	RCV
MT-CBC1 : cyt bc_1^-	-	-
MT-GS18 : cyt bc_1^-, cyt c_2^-	-	-
pR14A/MT-CBC1 : R. cap. cyt bc_1/cyt bc_1^-	+	+*/+**
pR14A/MT-GS18 : R. cap. cyt bc_1/cyt bc_1^-, cyt c_2^-	+	+*/+**
pGAB291/MT-CBC1 : R. sph. cyt bc_1/cyt bc_1^-	+	+*/+**
pGAB291/MT-GS18 : R. sph. cyt bc_1/cyt bc_1^-, cyt c_2^-	+	-*/+**

The assessment of the relationship between the c_2-independent photosynthetic growth of *R. capsulatus* and the nature of the acceptor site (i.e., the photochemical reaction center and its accompanying light harvesting complexes (LH) I and II) is more complicated. Unlike the cyt bc_1 complex subunits, the structural genes of the LH I and II complexes and the photochemical reaction center subunits are not all organized as contiguous operons clustered together. The two subunits of the LH I complex and the L and M subunits of the reaction center are part of the *puf* operon. The H subunit of the reaction center and the two subunits of the LH II complex constitute the *puh* and *puc* loci, respectively, and are distant from each other on the chromosome. Fortunately, an *R. capsulatus* mutant containing a deletion covering the two subunits of the light harvesting complex I and the L and M subunits, U 15g, as well as a double mutant, U43, carrying in addition a mutation eliminating the light harvesting II complex were described previously by Bylina et al.[13]. Furthermore, a plasmid, pCT1, containing the two subunits of the LH I complex and the L and M subunits of the reaction center of *R. sphaeroides* was shown to complement for photosynthetic growth the *R. capsulatus* mutant U43 by Zilsel et al.[14], and these materials were kindly provided to us by Drs. D. C. Youvan and T.

Beatty. To test the effect of the exchange of *R. capsulatus* reaction center and the light harvesting complex I with those of *R. sphaeroides* in the presence and absence of the light harvesting complex II we have constructed the cyt c_2^- derivatives of strains U15g and U43, and used the plasmid pCT1 as the source of *R. sphaeroides* components. In these experiments the plasmid pU2922 containing the *puf* operon of *R. capsulatus* was used as a control[15]. The merodiploid strains obtained were then tested for their photosynthetic growth abilities on MPYE rich media and the results obtained are presented in Table 2.

Under our conditions the photosynthetic growth observed with the merodiploid strains pCT1/U43 and pCT1/U15g containing hybrid *R. sphaeroides/R. capsulatus* photosynthetic complexes was weaker than that seen with pU2922/U15g and pU2922/U43 containing only *R. capsulatus* subunits. However, in each case, cyt c_2^- derivatives of these merodiploids showed photosynthetic growth abilities similar to their respective parents. These experiments suggested that the cyt c_2-independent photosynthetic growth can still operate when the light harvesting complex I and the L and M subunits of the reaction center of *R. capsulatus* were replaced with the corresponding components from *R. sphaeroides*. Moreover, comparison of the growth data obtained using strains U15g and U43 further indicated that the presence of the photosynthetic growth was also unaffected by the presence or absence of light harvesting complex II.

Table 2. Growth characteristics on MPYE rich medium of merodiploid *R. capsulatus* strains carrying functional *R. capsulatus/R. sphaeroides* hybrid photosynthetic complexes in the presence and in the absence of cyt c_2 of *R. capsulatus*.

Strain	Relevant Characteristics			Growth
	LH II	[LH I ; Rxc]	Cyt c_2	
U15g*	+	[- ; -]	+	Ps$^-$
U15g-G4/S4	+	[- ; -]	-	Ps$^-$
pU2922*/U15g	+	[LH I ; L/M] $_{cap}$	+	Ps$^+$
pU2922/U15g-G4/S4	+	[LH I ; L/M] $_{cap}$	-	Ps$^+$
pCT1*/U15g	+	[LH I ; L/M] $_{sph}$	+	Ps$^+$
pCT1/U15g-G4/S4	+	[LH I ; L/M] $_{sph}$	-	Ps$^+$
U43*	-	[- ; -]	+	Ps$^-$
U43-G4/S4	-	[- ; -]	-	Ps$^-$
pU2922/U43	-	[LH I ; L/M] $_{cap}$	+	Ps$^+$
pU2922/U43-G4/S4	-	[LH I ; L/M] $_{cap}$	-	Ps$^+$
pCT1/U43	-	[LH I + L/M] $_{sph}$	+	Ps$^+$
pCT1/U43-G4/S4	-	[LH I + L/M] $_{sph}$	-	Ps$^+$

*The strains and plasmids U15g, U43 and pU2922 and pCT1 were kindly provided by Drs. D. C. Youvan[15] and T. Beatty[14], respectively.

In summary, although the molecular basis of the cyt c_2-independent electron pathway of *R. capsulatus* operating between the photochemical reaction center and the cyt bc_1 complex is not yet clearly understood it appears that it is not an intrinsic property of either the cyt bc_1 complex or the reaction center and its accessory light harvesting components. It may be argued that it involves a soluble carrier similar to the iso-cyt c_2 of *R. sphaeroides*, perhaps present only at very low levels and in a state tightly bound to the membrane so that its detection is difficult and that it is still present in sphaeroplasts. The reconstituted systems obtained by the incorporation of purified reaction center and the cyt bc_1 complexes of *R. capsulatus* into ptoteoliposomes have indicated that electron transfer between these complexes is dependent on the presence of cyt c_2[16], suggesting that under these conditions the two complexes may not interact directly for electron transfer. On the other hand, this pathway may be basically different from that operating in *R. sphaeroides* and may not involve a cyt c_2-like soluble electron carrier[6]. If so, then the electron transfer from the cyt bc_1 complex to the reaction center may involve either the cyt bc_1 complex only or in addition to it a yet to be discovered membrane-bound redox protein operating between these two latter complexes[12]. If such a component exists then one approach to uncover its presence may be the isolation and analysis of non photosynthetic derivatives of a cyt c_2^- mutant of *R. capsulatus*.

CYTOCHROME BC_1 COMPLEX

A better understanding of how the cyt bc_1 complex functions may be achieved by better defining the structure of its components and their interactions with the redox-active prosthetic groups. Although the subunit composition of cyt *bc₁* complexes depends on their origins[17] they always contain two *b*-type cytochromes, of different spectroscopic and thermodynamic properties, carried by a single polypeptide of approximately 40 kDa, a *c*-type cytochrome of about 30 kDa, and a 2Fe2S cluster containing protein of about 20 kDa. The primary structure of these subunits have been determined by the isolation of the corresponding structural genes and the determination of their nucleotide sequences in several bacterial species including *Rhodobacter capsulatus*[10,18], *Rhodobacter sphaeroides*[10], *Paracoccus denitrificans*[19], *Bradyrhizobium japonicum*[20] and the cyanobacterium *Nostoc*[21]. Currently, there is no three dimensional structure for either the cyt bc_1 or its chloroplast analog the cyt b_6f complex. In its absence it is our belief that genetic analyses may further our understanding of the structure and the function of this complex. In what follows the general features of a useful genetic system that we have developed over the past few years to analyze spontaneous- and induced-mutations affecting the cyt bc_1 complex will be described[22]. In this system a silent inteposon located closely to the cyt bc_1 gene cluster was used to map among various spontaneous mutations those that are linked to the cyt bc_1 genes. The tagging interposon can be used as a selectable handle to clone the chromosomal mutations without having any need for a selectable phenotype. The location on the cloned bc_1 gene cluster of a mutation may be determined fairly precisely by estimating the distance between the tagging interposon and the mutation under study based on recombination frequency between the mutant and wild-type alleles. This region can then be sequenced directly to determine the molecular nature of this mutation. That indeed the mutation thus defined constitute the molecular basis of the phenotype(s) originally observed can be further demonstrated by the use of a specifically constructed plasmid, pPET1. This plasmid contains an expressed copy of the cyt bc_1 genes and several unique restriction sites so that a restriction fragment swapping experiment can exchange the sequenced fragment that carry the mutation with its wild-type counterpart on pPET1. The newly reconstructed cyt bc_1 cluster may then be introduced into a cyt bc_1 deletion background (MT-RBC1) to further establish by complementation the phenotype of the mutation studied.

This general system is also useful for isolating site-directed mutants in various subunits of the cyt bc_1 complex. The mutations generated on phage M13 derivatives containing fragments of the cyt bc_1 genes may be transferred easily to pPET1 and then shuttled into an appropriate background for their study. An additional property of our sytem is the overproduction of wild-type, or mutant, cyt bc_1 complexes carried by pPET1[22]. Obviously, this is an advantage for *in vitro* studies as it has been shown recently by the easier purification of this complex (D. Robertson, personal communication). Further, this overproduction is also useful for screening mutants with various properties such as lack of assembly or increased resistance to various inhibitors[22]. However, proper physiological studies often need the production of mutant complexes from their natural expression regions, and this requires the integration of the mutations made *in vitro* into their natural loci on the chromosome. In this genetic system allele replacement can be achieved by the introduction of the allele to be transfered into a GTA (gene transfer agent, a transducing phage-like particule specific to *R. capsulatus*)-overproducer stain deleted for the cyt bc_1 genes. The GTA thus obtained may be used to exchange, and repair if possible, an antibiotic-marked deletion of the cyt bc_1 cluster. In such a cross the antibiotic-sensitive transductants can only be generated by the replacement of the resident chromosomal deletion with the incoming mutant allele. This system has now been used succesfully for the analysis of over 100 spontaneous inhibitor-resistant mutants of *R. capsulatus*[23] and over 50 site-directed mutations, and it is hoped that it will also be useful for the analysis of the second-site revertants of the nonfunctional cyt bc_1 mutants of *R. capsulatus*.

The availibility of several lines of approaches including interspecies primary sequence comparison, physicochemical analyses of the properties of the prosthetic groups in unliganded state in model compounds and the extension of this knowledge to the liganded state in the protein, direct probing of the predicted amino acid residues by site-directed mutagenesis now enables one to locate with some confidence the liganding residues of the subunits of the cyt bc_1 complex. However, our basic knowledge about the structure of the active sites of the cyt bc_1 complex is still very poor. Mechanistic considerations suggests that a cyt bc_1 complex contains two distinct catalytic domains on each side of the membrane[24]. The quinol oxidation (Q_z) site is on the outer positive side, binds a quinol molecule and converts it to a quinone by transferring an electron to the Rieske FeS center and another to the lower potential cyt b heme (b_L). The electron accepted by the cyt b_L is subsequently transferred to the cyt b_H and this high potential heme then reduces a quinone trapped at the quinone reduction (Q_c) site located in the vicinity of the inner negative face of the membrane[24]. In the past, although it has been possible to probe these sites separately using specific inhibitors it has been difficult to localize them unambiguously by chemical modifications using labelled qinone or quinol derivatives. Among the specific inhibitors myxothiazol, mucidin and stigmatellin[11] interfere with the electron transfer between ubiquinol, Rieske FeS protein and cyt b_L at the Q_z site. A second class of inhibitors such as UHDBT and UHNQ acts on the electron flow further downstream from the Q_z site, between the FeS protein and the cyt c_1. Finally, the inhibitors like antimycin, funiculosin and HQNO affect the electron flow from cyt b_H to quinone at the Q_c site. If indeed these chemicals act at defined sites to inhibit the function of the cyt bc_1 complex then the study of a collection of inhibitor resistant (Inh[R]) mutants may reveal information about the binding of the inhibitors to, and the catalysis of quinone by, this complex. To date, mutants resistant to different classes of inhibitors have been isolated in both the budding[25] and the fission[26] yeast, and in mouse[27] mitochondria. We have also reported the isolation of similar mutants from the photosynthetic bacterium *Rhodobacter capsulatus*[28]. Many of the Qz-inhibitor mutants that we have isolated earlier have now been analyzed using the genetic system described above[22] and the data obtained are summarized in Table 3. Further, they are also compared to those known in yeast[29] and mouse[30] mitochondrial cyt b.

Table 3. Bacterial and mitochondrial cyt *b* mutations conferring resistance to $Q_{z(o)}$-inhibitors.

Position R.c./yeast	Organism	Amino acid substitution	Phenotype	cyt *b* domain
140/125	Bacteria	M -> I	Myx^R, Stg^R	$Q_{z(o)}$ I
144/129	Bacteria	F -> L	Myx^R, Stg^R	$Q_{z(o)}$ I
	Yeast	F -> L	Myx^R	
	Bacteria	F -> S	Myx^R	
152/137	Bacteria	G -> S	Myx^R, Muc^R	
	$Q_{z(o)}$ I			
	Yeast	G -> R	Myx^R, Muc^R	
158/143	Bacteria	G -> D	Ps^-	$Q_{z(o)}$ I
	Mouse	G -> A	Myx^R	
162/147	Yeast	I -> F	Stg^R	$Q_{z(o)}$ I
163/148	Bacteria	T -> A	Stg^R	$Q_{z(o)}$ I
	Mouse	T -> M	Stg^R	
279/256	Yeast	N -> Y	Myx^R, Muc^R	$Q_{z(o)}$ II
298/275	Yeast	L -> S, F, T	Myx^R, Muc^R	$Q_{z(o)}$ II
333/292	Bacteria	V -> A	Stg^R	$Q_{z(o)}$ II
336/295	Mouse	L -> F	Stg^R	$Q_{z(o)}$ II

Data related to yeast and mouse mitochondrial cyt *b* mutations are from di Rago et al.,[29] and Howell and Gilbert[30], respectively. Position numbers for cyt *b* residues are given for both *R. capsulatus* (R.c) and *S. cerevisiae* (S. c) and separated by /.

Using myxothiazol, mucidin and stigmatellin affecting the quinol oxidation site of the cyt bc_1 complex we have isolated eight different classes of mutations of which seven were located in the cyt b subunit, in two distinct areas called Q_zI and Q_zII[33]. The first of these regions is limited by the amino acid residues 140 to 163 and contains the mutations M140I, F144S, F144L, G152S and T163A providing resistance to myxothiazol, stigmatellin, mucidin and their combinations (Table 3). Further, the G158D mutation found in the strain R126 with a non functional cyt bc_1 complex was also located in this region, suggesting that QzI may define a domain involved both in catalysis of quinone and the binding of inhibitors. Of the remaining two mutations the V333A was found in the QzII region extending from the amino acid residues 279 to 336 of the cyt b, toward its COOH-end, where several other mutations in yeast[29] and mouse[30] mitochondrial cyt b have also been observed. Interestingly, this mutation conferred increased sensitivity to myxothiazol while providing resistance to stigmatellin, perhaps suggesting the presence of interaction between the QzI and QzII regions. Finally the last mutation, L106P, was located in a region far away from both of QzI and QzII domains raising the possibility that secondary inhibitor-binding domains remote from the quinol binding site may exit in cyt b and that the cyt bc_1 complex may accomodate the quinol as well as some of the Qz-inhibitors without mutual exclusion[31]. The overall data accumulated using inhibitor-resistance mutations now indicate that the nature and the degree of inhibitor-resistance of the cyt bc_1 complex depends on the position of the residue mutated, the nature of the substitution at this position and the general context provided by the entire structure of the cyt b subunit. For example in *R.capsulatus* while the F144S substitution yields resistance to myxothiazol only that of F144L confers resistance to both myxothiazol and stigmatellin.

Table 4. Mitochondrial cyt *b* mutations conferring resistance to $Q_{i(c)}$-inhibitors and hypothetical bacterial cyt *b* mutations predicted to convey inhibitor-sensitivity.

Position R.c/S.c	Organisms		
	S.cerevisiae	Mouse	*R.capsulatus*
33/17	I->F : Diu^R	**[F->I : Diu^S]**[a]	I
46/31	N->K : Diu^R	N	**[I->N : Diu^S]**
52/37	G->V : Ana^R	G ->V : Ana^R	**[A->G: Ana^S]**
248/225	F->L,S : Diu^R	Y	F
251/228	K->I : Ana^R	K	K
255/232	**[T->G : HQNO^S]**	G->D: HQNO^R	**[A->G: HQNO^S]**

[a]Predicted mutations and their corresponding phenotypes are shown between brackets in bold characters. Data related to *S. cerevisiae* and mouse mitochondrial cyt *b* mutations are from di Rago and Colson[32] and Howell and Gilbert[30], respectively. Position numbers for cyt *b* residues are given for both *R. capsulatus* (R.c) and *S. cerevisiae* (S. c) and are separated by /.

Although the inhibitors affecting specifically the quinone reduction site of the cyt bc_1 complex is without effect on photosynthetic bacteria *in vivo* they are very potent inhibitors of respiration in eukaryotes. In yeast[32] and mouse[30] mitochondria mutations conferring resistence to antimycin, diuron, funiculosin and HQNO were also located in two distinct regions of cyt b, QcI and QcII, far away from the QzI and QzII domains. A comparison of these regions between the mitochondrial and bacterial cyt b suggests that the bacterial cyt bc_1 complex may be naturally resistant to the Q_c inhibitors diuron, antimycin, funiculosin and HQNO because of the natural substitutions I46, A52, I213 and A255 already present in *R. capsulatus* cyt b (Table 4). The validity of this interesting observation now needs to be tested directly by isolation of appropriate reverse mutations using site-directed mutagenesis.

ACKNOWLEDGEMENT

The work in this laboratory was supported by grants from National Institute of Health (GM 38237) and University of Pennsylvania Research Foundation. I acknowledge the participation of my coworkers, Mariko K. Tokito and Emmanuel Atta-Asafo-Adjei, to various facets of this research.

REFERENCES

1. Bartsch, R. G. 1978. In "The photosynthetic bacteria" (Clayton, R. and Sistrom, W., Eds) pp.249-279. Plenum Press, New York.
2. Meyer, T. E., and M. A. Cusanovich. 1989. Biochim. Biophys. Acta. **975**:1-28.
3. La Monica, R. F. and B. L. Marrs. 1976. Biochim. Biophys. Acta. **423**:431-439.
4. Daldal, F., S. Cheng, J. Applebaum, E. Davidson and R. C. Prince. 1986. Proc. Natl. Acad. Sci. **83**:2012-2016.
5. Prince, R. C. and F. Daldal. 1987. Biochim. Biophys. Acta. **894**:370-378.
6. Prince, R. C., E. Davidson, C. E. Haith and F. Daldal. 1986. Biochemistry **25**:5208-5214.
7. Donohue, T. J., A. G. McEwan and S. Kaplan. 1986. J. Bacteriol. **168**:962-972.
8. Donohue, T. J., A. G. McEwan, S. van Doren, A. R. Crofts and S. Kaplan. 1988. Biochemistry **27**:1918-1925.
9. Fitch, J., V. Cannac, T. E. Meyer, M. A. Cusanovich, G. Tollin, J. van Beumen, M. A. Rott and T. J. Donohue. 1989. Arch. Biochem. Biophys. **271**:502-507.
10 Daldal, F., E. Davidson and S. Cheng. 1987. J. Mol. Biol. 195:1-24.
11. von Jagow, G. and T. A. Link. 1986. In "Methods in Enzymology" (Fleischer, S. and Fleischer, B., Eds) vol. **126**, pp.253-271. Academic Press.
12. Davidson, E., R. C. Prince, C. E. Haith and F. Daldal. 1989. Submitted.
13. Youvan, D. C, S. Ismail and E. J. Bylina. 1985. Gene **38**:19-30.
14. Zilsel, J., T. G. Lilburn and J. T. Beatty. 1989. FEBS Lett., in press.
15. Bylina, E. J., R. V. M. Jovine and D. C. Youvan. 1989. Bio/Technology, **7**:69-74.
16. Crielaard, W., N. Gabellini, K.J. Hellingwerf and W. N. Konings. 1989. Biochim. Biophys. Acta, **974**:211-218.
17. Ljungdahl, P. O., J./ D. Pennoyer, D. E. Robertson and B. L. Trumpower. 1987. Biochim. Biophys. Acta, **891**:227-241.
18. Gabellini, N. and W. Sebald. 1986. Eur. J. Biochem. **154**:569-579.
19 Kurowski, B. and B. Ludwig. 1987. J. Biol. Chem. **262**:13805-13811.
20. Thony-Meyer, L., D. Stax and H. Hennecke. 1989. Cell, **57**:683-697.
21. Kallas, T., S. Spiller and R. Malkin. 1988. Proc. Natl. Acad. Sci. **85**:5794-5798.
22. Daldal, F. 1989. In preparation.
23. Daldal, F., M. K. Tokito, E. Davidson and M. Faham. 1989. Submitted.
24. Robertson, D. E. and P. L. Dutton. 1988. Biochim. Biophys. Acta, **935**:273-291.
25. Subik, J. 1975. FEBS Lett. **237**:31-34.
26. Lang, B., G. Burger, K. Wolf, W. Banlow and F. Kaudewitz. 1975. Mol. Gen. Genet. **137**:353-363.
27. Howell, N., A. Bantel and P. Huang. 1983. Somatic Cell Genetics, **9**:721-743.
28. Daldal, F., E. Davidson, S. Cheng, B. Naiman and S. Rook. 1986. In "Microbila Energy Transduction" (Youvan, D. C. and Daldal, F., eds). pp. 113-119. Cold Spring Harbor Press, Cold Spring Harbor.
29. di Rago, J. P., J-Y Coppee and A-M Colson. 1989. J. Biol. Chem. in press.
30. Howell, N. and K. Gilbert. 1988. J. Mol. Biol. **203**:607-618.
31. Brandt, U., H. Schagger and G. von Jagow. 1988. Eur. J. Biochem. **173**:499-506.
32. di Rago, J. P. and A-M Colson. 1988. J. Biol. Chem. 263:12564-12570.

SOLUBLE CYTOCHROME SYNTHESIS IN *RHODOBACTER SPHAEROIDES*

Timothy J. Donohue, Janine P. Brandner, Janice E. Flory,
Barbara J. MacGregor, Marc A. Rott and Brenda A. Schilke

Bacteriology Department
University of Wisconsin-Madison
1550 Linden Drive
Madison, WI 53706

INTRODUCTION

Purple non-sulfur photosynthetic bacteria such as *Rhodobacter sphaeroides* can grow by aerobic respiration, by photosynthesis under anaerobic conditions in the light, or by anaerobic respiration in the dark if electron acceptors such as dimethylsulfoxide (DMSO), trimethylamine-N-oxide or nitrous oxide are present (1) (*R. sphaeroides* sp. *denitrificans* can also use nitrate or nitrite; 2,3). Given this metabolic and energetic versatility, it is not surprising that this Gram-negative bacterium contains many cytochromes whose synthesis can be environmentally regulated.

We are interested in how environmental factors regulate cytochrome synthesis. *R. sphaeroides* is particularly well suited to this analysis because of its well characterized cytochrome complement, and its versatility in adapting energy metabolism to its environment. This review summarizes the regulation of soluble cytochrome synthesis. It focuses on cyt c_2 (which is required for photosynthetic growth of wild type cells), cyt c_{554} (whose synthesis is controlled by oxygen and other factors), and isocyt c_2 (a cytochrome discovered in mutants that grow photosynthetically in the absence of cyt c_2).

ENERGY METABOLISM IN *R. SPHAEROIDES*

The aerobic respiratory chain(s)

The *R. sphaeroides* aerobic respiratory system contains both cyt c_2-dependent and cyt c_2-independent branches (4). Cyt c_2 is a diffusible periplasmic electron carrier which, in the presence of oxygen, transfers electrons between the cyt b/c_1 and cyt a/a_3 complexes.

Members of the "cyt c_2" family (cyt c_2, mitochondrial cyt c, and possibly isocyt c_2; see below) are highly homologous, and they have similar surface and electrostatic properties (5). Cyt c_2 and mitochondrial cyt c can substitute for each other in electron transport reactions *in vitro* (6), and the redox chains in which they participate *in vivo* contain similar catalytic centers, polypeptide subunits and mechanisms of proton pumping (7).

Molecular Biology of Membrane-Bound Complexes in Phototrophic Bacteria
Edited by G. Drews and E. A. Dawes
Plenum Press, New York, 1990

The *R. sphaeroides* cyt c_2-independent aerobic electron transport chain (ETC) is not well characterized. A b-type terminal oxidase which can accept electrons from reduced soluble c-type cytochromes has been purified from membranes (8). However, the affinity of this "oxidase" for c-type cytochromes is low. It is not known if this is a physiologically significant reaction *in vivo*, or whether other redox proteins function in the cyt c_2-independent aerobic ETC.

Photosynthetic electron transport

Anaerobiosis induces cytoplasmic membrane differentiation and synthesis of the intracytoplasmic membrane (ICM) that houses the photosynthetic apparatus. The molecular mechanisms regulating synthesis of the ICM and its associated proteins are the subject of intense investigation (1).

The *R. sphaeroides* photosynthetic ETC is the best studied system when one considers structure, function, assembly and regulation (1). Light-induced oxidation of bacteriochlorophyll molecules within the reaction center (RC) ultimately leads to quinone reduction. Reduced quinone diffuses from the RC to the cyt b/c_1 complex where its step-wise oxidation results in proton pumping across the bilayer (9). The essential role of *R. sphaeroides* cyt c_2 in wild type electron transfer between the cyt b/c_1 complex and light-oxidized RC complexes has been demonstrated both biochemically and genetically (10).

Quinone oxidation and proton pumping within the cyt b/c_1 complex is best understood in *R. sphaeroides*. The organization of pigments, polypeptides and the transbilayer distribution of amino acids in the *R. sphaeroides* RC are available from the X-ray crystal structure (11). *R. sphaeroides* cyt c_2 has also been crystallized (12) and cyt c_2 residues which interact with membrane bound redox complexes have been identified (13). Thus, *R. sphaeroides* is a logical choice for studying how soluble redox proteins (i. e., cyt c_2 in wild type cells, or isocyt c_2 in *spd* mutants; see below) interact with membrane-bound redox partners.

Other redox proteins reduce light-oxidized RC complexes in some systems. Cyt c_2-deficient mutants of the closely related bacterium *Rhodobacter capsulatus* are able to grow photosynthetically (14). In contrast to cyt c_2-independent photosynthesis in *R. sphaeroides* (see below), this does not require a periplasmic protein in *R. capsulatus* (15). The soluble copper protein plastocyanin transfers electrons from the chloroplast cyt b_6/f complex to the photosystem I RC (16), however *Chlamydomonas reinhardtii* replaces plastocyanin with a soluble cytochrome of the "cyt c_2" family if environmental copper is limiting (17).

Anaerobic respiratory pathways

Soluble cytochromes have been implicated in anaerobic respiratory pathways of both *R. sphaeroides* and *R. capsulatus* (18-20). Some anaerobic respiratory pathways share components (cyt b/c_1 complex or cyt c_2) with the aerobic and photosynthetic chains (3, 18). However, electron flow through other anaerobic respiratory pathways is resistant to cyt b/c_1 complex inhibitors, suggesting that they are independent of these redox complexes (18).

ROLE OF SOLUBLE CYTOCHROMES

Five soluble cytochromes (cyt c_2, cyt c_{554}, cyt c', cyt c_3, and SHP) have been purified from wild type cells by Bartsch, Meyer, Cusanovich and coworkers (Table 1; 21, 22).

The periplasmic location of cyt c_2 and cyt c' was demonstrated by their fractionation when spheroplasts were prepared, or by their entrapment within sealed ICM vesicles (23). Cyt c_{554} and isocyt c_2 are also periplasmic proteins

(24). The subcellular location of the other soluble redox proteins has not been determined.

There are ~0.5 moles of cyt c_2 per mole of RC complex in the ICM (1), and spectroscopic and genetic experiments confirmed the role of cyt c_2 in the *R. sphaeroides* photosynthetic ETC (10). However, the function of the other soluble cytochromes is unknown. When purified, reduced cyt c_{554} or cyt c' are unable to donate electrons to light-oxidized RC complexes (22). Cyt c_{552} is believed to be involved in nitrate respiration in *R. sphaeroides* sp. *denitrificans*, since it has only been purified from cells grown photosynthetically in the presence of nitrate (2). Isocyt c_2 has not been detected in wild type *R. sphaeroides* (24, 25); it was discovered in cells containing mutations which suppress the photosynthesis-deficient phenotype of strains lacking cyt c_2 (*spd* mutants, see below).

TABLE 1

Rhodobacter sphaeroides Soluble C-Type Cytochromes

Component	λ MAX[a]	MW(kD)[b]	pI	$E_M.7$(mV)[c]	Photo[d]	Aero[d]	Ref
cyt c_2	550	14.0	5.5	352	15	16	21,22
cyt c_{554}	554	14.0	4.1	203	1	6	21,22
cyt c'	549	12.5	4.9	30	14	ND[f]	21,22
cyt c_3	546	21.0	4.1	-22	4	ND	21,22
SHP[e]	551.5	13.0	4.4	-254	2	ND	21,22
cyt c_{552}	552	13.5	4.2	252	NA[f]	NA	2
isocyt c_2[g]	552	13.5	4.5	295	3	NA	25

[a]Wavelength of alpha band maximum in reduced minus oxidized spectrum.
[b]Determined by SDS-PAGE.
[c]Midpoint potential in millivolts at pH 7.0.
[d]μmol heme/kg wet weight cells grown either photosynthetically (Photo) or in stationary phase aerobic cells (Aero).
[e]Sphaeroides heme protein.
[f]NA=Data not available; ND=None detectable (<0.5 μmoles/kg).
[g]Data are from analysis of cytochrome levels in *R. sphaeroides spd* mutants (see text).

The alpha maxima of *R. sphaeroides* soluble cytochromes are very similar (Table 1), so it is difficult to either measure the level, or ascribe a physiological function to a single species solely by spectroscopic analysis. For example, analysis of *R. sphaeroides* cyt c_2-deficient mutants suggested that cyt c_{554} (and not cyt c_2 as generally accepted) was the major soluble c-type cytochrome under aerobic conditions (26). Bartsch and coworkers confirmed the aerobic induction of cyt c_{554} in wild type cells (22). Despite the fact that cytochromes with characteristics distinct from cyt c_2 and similar to cyt c_{554} were purified from aerobic cells (27), a role for cyt c_{554} in *R. sphaeroides* aerobic respiration has not been considered.

Synthesis of several soluble cytochromes in *R. sphaeroides* is controlled by environmental factors. Table 1 shows that in the one comprehensive study, the levels of cyt c_{554} were highest under aerobic conditions, while cyt c', cyt c_3, and SHP levels were increased under photosynthetic conditions. Although the cyt c_2 levels in Table 1 are the same under aerobic and photosynthetic conditions, the authors suggest that these values were probably compromised by the use of stationary phase cells (22). Experiments in our laboratory using exponential phase aerobic cells have shown that cyt c_2 levels are ~7-10 fold higher under anaerobic conditions than in aerobically grown cells (26).

Photosynthetic growth in the presence of nitrate induces synthesis of cyt c_{552} and reduces the level of cyt c´ and cyt c_3 in R. sphaeroides sp. denitrificans (2).

REGULATION OF CYT C_2 SYNTHESIS

Studies on soluble cytochrome synthesis initially focused on cyt c_2, since information was available on its redox properties and redox functions. Consequently, more information is available on how expression of the cyt c_2 structural gene (cycA) is regulated.

Organization and transcription of the cycA gene

CycA transcription is ~2-fold higher under anaerobic conditions than in the presence of oxygen, but it is unaffected by changes in light intensity under photosynthetic conditions or the presence of alternate electrons acceptors (28). We initially reported 2 major cycA-specific transcripts of ~740 and 920 nt in wild type cells (28); however higher resolution Northern blot analysis has identified additional species in both wild type and cycA⁻ cells complemented with plasmids containing at least 475 bp of upstream DNA (26). We are currently determining whether these additional species are produced by unique promoters or are processed forms of a larger (~920 nt) primary transcript.

We have identified at least one promoter within ~90 bp of cycA. This region directs sufficient cycA transcription to complement a cycA⁻ strain. Transcription from this region appears oxygen-independent when levels of the ~740 nt mRNA are measured by Northern blot analysis or when cycA reporter molecules, either operon fusions to Escherichia coli β-galactosidase (lacZ) or protein fusions to alkaline phosphatase (phoA), containing ~90 bp of upstream DNA are used.

Northern blot analysis indicates that the ~920 nt cycA-specific mRNA is ~5-7 fold higher under anaerobic conditions than in cells grown in the presence of oxygen. β-galactosidase levels in cells harboring a cycA:lacZ operon fusion under control of this upstream region are increased ~7-fold under anaerobic conditions. Thus at least one cycA promoter is highly oxygen-regulated.

Given this pattern of cycA transcription, an obvious question is raised: how is activity of individual cycA promoters regulated? Transcription of the yeast cyt c structural gene is regulated by a heme-dependent activator protein (29). Transcription of the R. sphaeroides cycA gene is unaffected by exogenous heme precursors or analogs (30) which activate yeast cyt c transcription. We have isolated trans-acting mutations which affect cycA transcription in order to identify such regulatory factors.

A related question is why cycA contains multiple promoters? The DNA sequence of the region upstream of cycA suggests that the ~920 nt transcript could encode gene products not present on the ~740 nt species. We are currently determining whether additional gene products exist, and if they function in regulating cyt c_2 synthesis.

Is cyt c_2 synthesis post-transcriptionally regulated?

The ~2-fold increase in total cycA-specific mRNA under anaerobic conditions is accompanied by an ~7-10-fold increase in cyt c_2 levels (26). This suggests that cyt c_2 synthesis is post-transcriptionally regulated by oxygen. Post-transcriptional control could occur at synthesis of the cyt c_2 precursor polypeptide, processing of the precursor into an apoprotein, heme attachment to the apoprotein, or export to the periplasm. It is also possible that gene products encoded by the largest cycA transcript function to facilitate post-transcriptional regulation of cyt c_2 synthesis.

Post-transcriptional control over cyt c_2 biogenesis might be coordinated with increased tetrapyrrole biosynthesis which occurs under anaerobic conditions (1). If so, this regulatory system would appear somewhat specific, since levels of cyt c_{554} are lower under anaerobic conditions where levels of cyt c_2 and other soluble species are increased.

Post-transcriptional control over *R. sphaeroides* cyt c_2 levels also exists in *E. coli* (31). In this heterologous host, cyt c_2 antigen is detected only under anaerobic respiratory conditions when heme is attached to CycA (31). *E. coli* only synthesizes c-type heme under anaerobic respiratory conditions (32), so we assume that heme is attached to CycA by the host system for c-type heme synthesis. In contrast, CycA::PhoA protein fusions are present in the *E. coli* periplasm under aerobic growth conditions (33). Thus, CycA is either not exported or is unstable in the *E. coli* periplasm in the absence of heme attachment.

Sequences within CycA involved in function

Heme is covalently attached at Cys15 & Cys18 of mature cyt c_2; His19 and Met100 are non-covalent heme ligands. We have constructed a series of CycA::PhoA protein fusions at different positions within CycA using the TnphoA system of Beckwith and coworkers (34) for analysis of exported proteins. Heme is attached to at least one hybrid protein with a fusion junction downstream of the heme attachment site but before the methionine ligand. Thus, all the non-covalent heme ligands of cyt c_2 are not required for heme attachment in *R. sphaeroides*.

Lysine residues in either the amino (residue 10) or carboxy (residues 95-106) terminus of cyt c_2 (124 total residues) interact with membrane bound redox complexes (13). A CycA::PhoA protein fused at Gln119 complements a $cycA^-$ mutant for photosynthetic growth. Thus, this hybrid protein is at least partially functional in both heme attachment and association with membrane bound redox complexes. These hybrid proteins may also be useful to map CycA domains which interact with redox partners.

MUTANTS IN CYTOCHROME SYNTHESIS

This section summarizes how genetic studies have been utilized to test whether individual redox proteins function in a given ETC. These studies demonstrate that specific cytochrome mutants can provide new insights into cytochrome function or the regulation of cytochrome synthesis.

Mutations affecting soluble cytochrome synthesis

Molecular genetics has significantly changed models for electron transport in photosynthetic bacteria. The phenotypes of $cycA^-$ derivatives of *R. sphaeroides* (10) and *R. capsulatus* (14) provided genetic evidence for differences in photosynthetic electron transport, despite the fact that spectroscopic analysis was interpreted as indicating that cyt c_2 was an obligate intermediate in both species (23). The $cycA^-$ mutants also confirmed the existence of cyt c_2-independent aerobic and anaerobic respiratory pathways in both organisms.

Future analysis of electron transport pathways should be facilitated by using existing strains with specific lesions in one branch (cyt b/c_1^-, cyt c_2^-, cyt a/a_3^-, etc.). Such strains are logical parents in selections or screens for mutations which abolish activity of an alternative branch. *R. capsulatus* mutants lacking several c-type cytochromes have defined pathways which do not require any of these proteins (35-37); assuming that these strains contain single mutations, they should also provide new information on c-type cytochrome biogenesis once their genetic lesions are identified.

Analysis of the cyt c_2-independent aerobic ETC in *R. sphaeroides* may be aided by the conditional aerobic phenotype of *cycA⁻* strains. *cycA⁻* strains grow normally in a succinate-based minimal medium (10, 26); however they cannot grow aerobically in complex media which allows growth of wild type strains. Heme peroxidase assays indicate that growth media affects levels of soluble and membrane-bound cytochromes.

Does cyt c_{554} participate in aerobic respiration?

Recent evidence suggests that we should consider whether cyt c_{554} participates in aerobic respiration. Brandner et al., noted significant quantities of cyt c_{554} in aerobically grown cultures of *R. sphaeroides cycA⁻* strains (26). From this analysis, they suggested that cyt c_{554} was the major soluble cytochrome under aerobic conditions in wild type cells (26). However their studies could not preclude the possibility that cyt c_{554} synthesis was increased in cells lacking cyt c_2. We now know that cyt c_{554} is a major soluble cytochrome in aerobically grown wild type cells, and that cyt c_{554} levels are not significantly altered in *cycA⁻* cells.

Media which reduces cyt c_{554} levels in *cycA⁺* cells does not support aerobic growth of *cycA⁻* strains. Therefore, cyt c_{554} synthesis can be correlated with the ability to grow under aerobic conditions in the absence of cyt c_2. We have identified sequences homologous to cyt c_{554} structural gene probes in order to directly test whether this protein functions in aerobic respiration.

Cyt c_{554} levels are ~3-fold higher than cyt c_2 under aerobic conditions in succinate-based minimal media; under anaerobic conditions in the same media cyt c_2 is ~15-fold more abundant than cyt c_{554} (10, 26). The decrease in cyt c_{554} levels under anaerobic conditions is also consistent with a role for this protein in aerobic respiration. Cyt c_{554} is the only known *R. sphaeroides* redox carrier present at a higher level in aerobic cells. Therefore, we wish to compare the environmental and genetic signals regulating cyt c_{554} synthesis to those controlling synthesis of anaerobically-induced components of the photosynthetic ETC such as cyt c_2, isocyt c_2, and bacteriochlorophyll-binding proteins. Such studies should shed light on how cells control synthesis of individual ETC components to adapt energy metabolism to environmental changes.

Mutations suppressing the photosynthesis-deficient phenotype of *R. sphaeroides* cycA⁻ strains

The loss of rapid RC reduction and the lack of electron flow through the cyt b/c_1 complex both confirmed spectroscopic models for photosynthetic electron flow, and provided a biochemical explanation for why *R. sphaeroides cycA⁻* mutants did not grow photosynthetically (10). During this analysis, spontaneous mutations which suppressed the photosynthesis-deficient phenotype of *cycA⁻* strains (*spd* mutations) were observed in ~1-10 in 10^7 cells (10, 24). All spontaneous *spd* mutants analyzed grow photosynthetically with wild type doubling times, and they contain a new periplasmic cytochrome which we have named isocyt c_2 (24, 25). The redox properties of isocyt c_2 (Table 1) are similar to cyt c_2. There is amino acid sequence homology between isocyt c_2 and cyt c_2, particularly at the amino terminus and in residues believed to interact with membrane-bound redox complexes (25). Thus the properties of isocyt c_2 are consistent with its proposed role in RC reduction.

Our experiments indicate that RC reduction is ~1000-fold slower in *spd* mutants than in *cycA⁺* cells, but that electron flow through the cyt b/c_1 complex is only ~5-7 fold slower (10). RC reduction is dependent on a periplasmic protein in *spd* mutants, since both electron flow through the cyt b/c_1 complex and cytochrome c-dependent RC reduction are abolished in spheroplasts. This analysis is preliminary, since we must determine how these kinetics are influenced by the concentration of isocyt c_2, or by differences in the affinity of isocyt c_2 for membrane redox partners.

Isocyt c_2 is detectable in all spontaneous *spd* mutants grown photosynthetically and isocyt c_2 levels are increased ~2-fold during growth via DMSO respiration (24). This suggests that isocyt c_2 synthesis can respond to environmental factors in *spd* mutants. The level of isocyt c_2 under photosynthetic conditions in many *spd* mutants is ~20% the level of cyt c_2 found in $cycA^+$ cells. We have not yet detected isocyt c_2 in $cycA^+$ cells and we estimate that isocyt c_2 is at least 100-fold higher in *spd* mutants than in $cycA^+$ cells (25).

Several spontaneous *spd* mutants also contain isocyt c_2 under aerobic conditions (24). Allele-specific synthesis of isocyt c_2 under aerobic conditions is consistent with a model where *spd* mutants contain regulatory mutations which allow increased isocyt c_2 synthesis. The frequency at which spontaneous *spd* mutants occur suggests that they contain a single mutation, and this frequency is increased by the frame-shift mutagen ICR-191 (24). We are currently testing whether *spd* mutations inactivate a regulatory system which represses isocyt c_2 synthesis in $cycA^+$ cells.

Spontaneous *spd* mutants have additional phenotypes. Heme peroxidase assays demonstrate that *spd* mutants contain at least one other soluble cytochrome in addition to isocyt c_2 that is not found in $cycA^+$ cells (24). The identity of this additional species is not known, but *spd* mutations appear to inactivate a system which represses synthesis of more than one soluble cytochrome.

Levels of B800-850 complexes are derepressed ~25% in *spd* mutants (24). Many mutations can affect synthesis of B800-850 complexes (1), so future experiments need to address whether this phenotype is a direct consequence of the *spd* mutation, or if it results from the reduced rate of photosynthetic electron transport. If the latter were true, it would suggest that cells can respond to a reduced rate of RC reduction by increasing synthesis of B800-850 complexes.

SUMMARY AND FUTURE PERSPECTIVES

The study of cytochromes has increased our understanding of biology. Analysis of cytochromes in photosynthetic bacteria began with the purification of hematin-containing compounds by Kamen and co-workers over 30 years ago (38). A systematic study of cytochromes in the "cyt c_2" family allowed some of the first evolutionary trees to be constructed (39, 40). The similarity between ETC components of eubacteria, mitochondria and chloroplasts is well documented. Partly because of this, it is believed that procaryotic and eucaryotic redox chains evolved from an ETC similar to that in photosynthetic bacteria.

Analysis of light-driven redox chains which utilized cyt c_2 was the basis for some of the earliest models of energy generation. The chemiosmotic theory gained experimental support when light-induced proton pumping across a bilayer by either an intact ETC or by a purified membrane-bound redox complex was demonstrated (16, 41). The recent structural analysis of the photosynthetic RC allows detailed molecular models to be formulated for how light is converted into an energized membrane (11, 42).

Future studies of how photosynthetic bacteria control cytochrome synthesis need to explain how synthesis of individual proteins is regulated by oxygen, other electron acceptors, or reducing power. Facultative bacteria

101

(43) and yeast (30) activate transcription of cytochrome genes in response to stimuli such as anaerobiosis, heme availability, or alternate electron donors or acceptors. Although transcriptional control is clearly involved, we have not detected heme-stimulated transcription of R. sphaeroides cyt c_2, nor have sequences homologous to activators of cytochrome genes in other microorganisms been found. Classical genetic approaches to isolating mutations in "global" regulators of cytochrome synthesis (spd mutants ?), or lesions affecting a specific step in cytochrome synthesis (transcription, export, or heme attachment) will be particularly useful in this regard.

Under photosynthetic conditions, these bacteria contain other redox chains which could limit electron flow to the RC by competing for a common pool of electrons. Cells restrict electron flow to oxygen under low oxygen photosynthetic conditions, to alternate electron acceptors under anaerobic conditions, or to CO_2 fixation when a strongly reduced carbon source is lacking (18, 44). Future studies need to determine if and how cells monitor their redox state and transduce this information into signals which control either activity or synthesis of individual ETC components. A thorough understanding of this redox balancing and its control is required to determine precisely which intracellular signals regulate cytochrome synthesis. Such an analysis will undoubtedly require combined biochemical, energetic and genetic approaches.

ACKNOWLEDGEMENTS

Recent work in our laboratory has been supported by a grant from the National Institutes of Health (GM37590), or by a grant-in-aid from the Wisconsin Alumni Research Foundation. J. P. B., B. J. M., and M. A. R. have received support from a National Institutes of Health Cellular and Molecular Biology Training Grant (GM07215), while B. A. S. is the recipient of a National Science Foundation Predoctoral Fellowship.

BIBLIOGRAPHY

1. Kiley, P. J., and S. Kaplan. 1988. Molecular genetics of photosynthetic membrane biosynthesis in Rhodobacter sphaeroides. Microbiol. Rev. 52:50.
2. Michalski, W., D. J. Miller, and D. J. D. Nicholas. 1986. Changes in the cytochrome composition of Rhodopseudomonas sphaeroides sp. denitrificans grown under denitrifying conditions. Biochim. Biophys. Acta 849:304.
3. Itoh, M., S. Mizukami, K. Matsuura, and T. Satoh. 1989. Involvement of the cyt b/c_1 complex and cytochrome c_2 in the electron transfer pathway for NO reduction in a photodenitrifier Rhodobacter sphaeroides sp. denitrificans. FEBS Lett. 244:81.
4. Zannoni, D., and A. Baccarini-Melandri. 1980. Respiratory electron flow in facultative photosynthetic bacteria, p. 183. In D. Knowles (ed.), Diversity of bacterial respiratory systems. CRC Press, Inc., Boca Raton, Fla.
5. Meyer, T. E., and M. D. Kamen. 1982. New Perspectives on c-type cytochromes. Adv. Prot. Chem. 35:105.
6. Tiede, D. M. 1987. Cytochrome c orientation in electron-transfer complexes with photosynthetic reaction centers of Rhodopseudomonas sphaeroides and when bound to the surface of negatively charged membranes: Characterization by optical linear dichroism. Biochemistry 26:397.
7. Crofts, A. R., and C. A. Wraight. 1983. The electrochemical domain of photosynthesis. Biochim. Biophys. Acta 726:149.
8. Takamiya, K. 1983. Properties of the cytochrome c oxidase activity of cytochrome b_{561} from photoanaerobically grown Rhodopseudomonas sphaeroides. Plant & Cell Physiol. 24:1457.
9. Crofts, A. R., S. W. Meinhardt, and J. R. Bowyer. 1982. The electron transport chain of Rhodopseudomonas sphaeroides, p. 477. In B. L. Trumpower (ed.) Functions of quinones in energy conserving systems. Academic Press Inc., New York.
10. Donohue, T. J., A. G. McEwan, S. Van Doren, A. R. Crofts and S. Kaplan.

1988. Phenotypic and genetic characterization of cytochrome c_2-deficient mutants of *Rhodobacter sphaeroides*. Biochemistry 27:1918.

11. Feher, G., J. P. Allen, M. Y. Okamura, and D. C. Rees. 1989. Structure and function of bacterial reaction centres. Nature 339:111.

12. Allen, J. P. 1988. Crystallization and preliminary X-ray diffraction analysis of cytochrome c_2 from *Rhodobacter sphaeroides* J. Mol. Biol. 204:495.

13. Tiede, D. M., D. E. Budil, J. Tang, O. El-Kabbani, J. R. Norris, C.-H. Chang, and M. Schiffer. 1988. Symmetry breaking structures involved in the docking of cytochrome c and primary electron transfer in reaction centers of *Rhodobacter sphaeroides*, p. 13. *In* J. Breton and A. Vermeglio (eds.). The photosynthetic bacterial reaction center. Plenum Publishing Corp. New York.

14. Daldal, F., S. Cheng, J. Applebaum, E. Davidson, and R. C. Prince. 1986. Cytochrome c_2 is not essential for photosynthetic growth of *Rhodopseudomonas capsulatus*. Proc. Natl. Acad. Sci. USA 83:2012.

15. Prince, R. C., E. Davidson, C. E. Haith, and F. Daldal. 1986. Photosynthetic electron transfer in the absence of cytochrome c_2 in *Rhodopseudomonas capsulata*: cytochrome c_2 is not essential for electron flow from the cytochrome b/c_1 complex to the photochemical reaction center. Biochemistry 25:5208.

16. Crofts, A. R., and P. M. Wood. 1978. Photosynthetic electron transport chains of plants and bacteria and their role as proton pumps. Curr. Top. Bioenerg. 7:175.

17. Merchant, S., and L. Bogorad. 1987. The Cu(II)-repressible plastidic cytochrome c: Cloning and sequence of a complementary DNA for the pre-apoprotein. J. Biol. Chem. 262:9062.

18. Ferguson, S. J., J. B. Jackson, and A. G. McEwan. 1987. Anaerobic respiration in the *Rhodospirillaceae*: characterisation of pathways and evaluation of role in redox balancing during photosynthesis. FEMS Microbiol. Rev. 46:117.

19. McEwan, A. G., A. J. Greenfield, H. G. Wetzstein, J. B. Jackson, and S. J. Ferguson. 1985. Nitrous oxide reduction by members of the family *Rhodospirillaceae* and the nitrous oxide reductase of *Rhodopseudomonas capsulata*. J. Bacteriol. 164:823.

20. McEwan, A. G., D. J. Richardson, H. Hudig, S. J. Ferguson, and J. B. Jackson. 1989. Identification of cytochromes involved in electron transport to trimethylamine-N-oxide/dimethylsulphoxide reductase in *Rhodobacter capsulatus*. Biochim. Biophys. Acta 973:308.

21. Meyer, T. E., and M. A. Cusanovich. 1985. Soluble cytochrome composition of the purple phototrophic bacterium *Rhodopseudomonas sphaeroides* ATCC 17023. Biochim. Biophys. Acta 807:308.

22. Bartsch, R. G., R. P. Ambler, T. E. Meyer, and M. A. Cusanovich. 1989. Effect of aerobic growth conditions on the soluble cytochrome content of the purple phototrophic bacterium *Rhodobacter sphaeroides*: Induction of cytochrome c_{554}. Arch. Biochem. Biophys. 271:433.

23. Prince, R. C., A. Baccarini-Melandri, G. A. Hauska, B. A. Melandri, and A. R. Crofts. 1975. Asymmetry of an energy transducing membrane: the location of cytochrome c_2 in *Rhodopseudomonas sphaeroides* and *Rhodopseudomonas capsulata*. Biochim. Biophys. Acta 387:212.

24. Rott, M. A., and T. J. Donohue. *Rhodobacter sphaeroides spd* mutations allow cytochrome c_2-independent photosynthetic growth. J. Bacteriol. (In press).

25. Fitch, J. V. Cannac, T. E. Meyer, M. A. Cusanovich, G. Tollin, J. Van Beeumen, M. A. Rott, and T. J. Donohue. 1989. Expression of a cytochrome c_2 isozyme restores photosynthetic growth of *Rhodobacter sphaeroides* mutants lacking the wild type cytochrome c_2 gene. Arch. Biochem. Biophys. 271:502.

26. Brandner, J. P., A. G. McEwan, S. Kaplan, and T. J. Donohue. 1989. Expression of the *Rhodobacter sphaeroides* cytochrome c_2 structural gene. J. Bacteriol. 171:360.

27. Orlando, J. A. 1962. *Rhodopseudomonas sphaeroides* cytochrome c-553. Biochim. Biophys. Acta 57:373.

28. Donohue, T. J., A. G. McEwan, and S. Kaplan. 1986. Cloning, DNA sequence and expression of the *Rhodobacter sphaeroides* cytochrome c_2 structural gene. J. Bacteriol. 168:962.

29. Pfeifer, K., B. Arcangioli, and L. Guarente. 1987. Yeast HAP1 activator competes with the factor RC2 for binding to the upstream activation site UAS1 of the CYC1 gene. Cell <u>49</u>:9.

30. Guarente, L., and T. Mason. 1983. Heme regulates transcription of the CYC1 gene of *S. cerevisiae* via an upstream activation site. Cell <u>32</u>:1279-1286.

31. McEwan, A. G., S. Kaplan, and T. J. Donohue. 1989. Synthesis of the *Rhodobacter sphaeroides* cytochrome c_2 in *Escherichia coli*. FEMS Microbiol. Lett. <u>59</u>:253.

32. Cole, J. A. 1968. Cytochrome c_{552} and nitrite reduction in *Escherichia coli*. Biochim. Biophys. Acta <u>162</u>:356.

33. Varga, A., and S. Kaplan. Construction, expression and localization of a CycA::PhoA fusion protein in *Rhodobacter sphaeroides* and *Escherichia coli*. J. Bacteriol. (In press).

34. Manoil, C., and J. Beckwith. 1985. *TnphoA*: A transposon probe for protein export signals. Proc. Natl. Acad. Sci. USA <u>82</u>:8129.

35. Hudig, H. N. Kaufmann, and G. Drews. 1986. Respiratory deficient mutants of *Rhodopseudomonas capsulata*. Arch. Microbiol. <u>145</u>:378.

36. Davidson, E., R. C. Prince, F. Daldal, G. Hauska, and B. L. Marrs. 1987. *Rhodobacter capsulatus* MT113: a single mutation results in the absence of c-type cytochromes and in the absence of the cyt b/c_1 complex. Biochim. Biophys. Acta <u>890</u>:292.

37. Kranz, R. G. 1989. Isolation of mutants and genes involved in cytochromes c biosynthesis in *Rhodobacter capsulatus*. J. Bacteriol. <u>171</u>:456.

38. Vernon, L. P., and M. D. Kamen. 1954. Hematin compounds in photosynthetic bacteria. J. Biol. Chem. <u>211</u>:643.

39. Dickerson, R. E., R. Timkovich, and R. J. Almassy. 1976. The cytochrome fold and the evolution of bacterial energy metabolism. J. Mol. Biol. <u>100</u>:473.

40. Meyer, T. E., M. A. Cusanovich, and M. D. Kamen. 1986. Evidence against use of bacterial amino acid sequence data for the construction of all-inclusive phylogentic trees. Proc. Acad. Natl. Sci. USA <u>83</u>:217.

41. Clark, A. J., N. P. J. Cotton, and J. B. Jackson. 1983. The relation between membrane ionic current and ATP synthesis in chromatophores from *Rhodopseudomonas capsulata*. Biochim. Biophys. Acta <u>723</u>:440.

42. Deisenhofer, J., O. Epp, K. Miki, R. Huber, and H. Michel. 1984. X-ray structure analysis of a membrane protein complex. J. Mol. Biol. <u>180</u>:385.

43. Stewart, V., 1988. Nitrate respiration in relation to facultative metabolism in *Enterobacteria*. Microbiol. Rev. <u>52</u>:190.

44. Richardson, D. J., G. F. King, D. J. Kelly, A. G. McEwan, S. J. Ferguson, and J. B. Jackson. 1988. The role of auxiliary oxidants in maintaining redox balance during phototrophic growth of *Rhodobacter capsulatus* on propionate or butyrate. Arch. Microbiol. <u>150</u>:131.

Post-transcriptional Control of the Expression of Photosynthetic Complex

Formation in Rhodobacter sphaeroides

Samuel Kaplan

Department of Microbiology
The University of Texas Medical School at Houston
Houston, TX   77225

INTRODUCTION

Rhodobacter sphaeroides is a facultative photoheterotroph, able to grow on a variety of reduced organic carbon sources either in the presence or absence of O_2, in either the light or dark. When grown anaerobically in the light, the organism is a photoheterotroph utilizing light energy captured and transferred via a series of functionally interrelated spectral complexes. Energy ultimately transferred to the reaction center complex (RC) is the source of oxidation of one of a special pair of bacteriochlorophyll (Bchl) molecules (Sauer, 1986)[1].

The most abundant of these spectral complexes is the B800-850 complex composed of two polypeptides designated β and α with molecular sizes of 5448 and 5599 Da, respectively (Kiley and Kaplan, 1987)[2]. Liganded to the polypeptides are three molecules of Bchl and one molecule of carotenoid (Car). The amount of the B800-850 varies inversely with respect to light intensity and is obligately required for growth under low light conditions (Meinhardt, et al., 1985)[3]. The B875 spectral complex is comprised of a β and an α polypeptide of molecular sizes 5457 and 6809 Da, respectively, which are associated with two molecules of Bchl and two of Car (Kiley, et al., 1987)[4]. The B875 complex is in fixed stoichiometry, approximately 15:1, to the RC (DeHoff, et al., 1988)[5]. The RC is composed of three polypeptides designated H, M, and L, the latter two of which are associated with the Bchl, bacteriopheophytin, non-heme Fe, and quinones all involved in the reversible oxidation/ reduction of the RC (Feher and Okamura, 1975)[6]. Light energy trapped by the B800-850 is obligately passed to the B875 as exciton energy which is in turn passed to the RC where one of a special pair of Bchl's is oxidized. One very useful physiological response of Rb. sphaeroides is that under anaerobic, dark conditions a fully functional photosynthetic membrane is produced although when in the presence of an external electron acceptor, the functional photosynthetic membrane is not required for anaerobic dark growth (Yen and Marrs, 1977)[7].

All of the spectral complexes as well as the components of the electron transport chain and numerous other oxidoreductases are found in a specialized photosynthetic membrane system (ICM). The ICM is only present under semiaerobic or anaerobic growth conditions and it arises as numerous regularized invaginations, comprising a functionally separable

Molecular Biology of Membrane-Bound Complexes in Phototrophic Bacteria
Edited by G. Drews and E. A. Dawes
Plenum Press, New York, 1990

domain from the cell membrane, which is present under all growth conditions. Although the cellular level of ICM varies inversely to light intensity this variation is approximately 3-fold going from high to low light intensities whereas the variation in B800-850 is approximately 9-12 fold under similar conditions (Kiley and Kaplan, 1988)[8]. However, the number of RC's per unit ICM area is independent of light intensity (Yen, et al., 1984)[9].

Although there are many important and significant questions which need to be addressed regarding this experimental system, one very important question relates to the "assembly" of these three spectral complexes, namely: What factors are required to ensure that each complex is appropriately assembled and present as a functional complex within the ICM? For the moment, we will not concern ourselves with the stoichiometry existing between the various complexes or the regulation at the genetic level, which serves to give rise to each of the complexes in their appropriate ratios and amounts.

We could imagine that the process of spectral complex assembly is passive, i.e. polypeptides, Bchl and Car are made, and they spontaneously find each other after having arrived at the ICM and, there, undergo autogenous assembly. The results described below will demonstrate that this is not the case.

RESULTS

The first observation which indicated the existence of unassembled polypeptides not present in a photosynthetic complex, despite normal photosynthetic growth, stemmed from observations of mutant strain RS103 (Meinhardt, et al., 1985)[3]. This mutant contains a point mutation which results in the accumulation of approximately 30% of the W.T. level of the β and α polypeptides of the B875 complex, yet possesses no B875 complexes (Kiley, et al., 1988)[10] and (DeHoff, et al., 1988)[5]. We were able to demonstrate that the structural genes, pufBA, encoding these polypeptides are normal and, thus, we were not dealing with a structural gene mutation in either pufB or pufA. We will return to RS103 later. In the second instance, a deletion of the transcription terminator between pufA and pufL resulted in a mutant strain of Rb. sphaeroides, although containing approximately 7% of the W.T. level of B875 complexes, possessed nearly 55% of the W.T. level of the β and α polypeptides normally found in the B875 complex (DeHoff, et al., 1988)[5]. Again, we concluded that the apopolypeptides of the B875 spectral complex could exist within the photosynthetic membranes in a form other than in an active complex. In both of the examples cited above, we were dealing with mutant strains. In the case of RS103, the mutation was outside the puf region and in the case of the terminator deletion, the mutation was internal to the puf operon.

The significance of these earlier findings became apparent when we constructed a deletion of the puhA gene (Sockett, et al., 1989)[11] which encodes the H polypeptide and maps 31-kb from the puf operon. This deletion extends several hundred bp upstream of the puhA structural gene in the resulting mutant strain. Although not possessing any H-polypeptide or H-specific mRNA and not growing photosynthetically, this strain does possess a very weak RC signal as shown in Figure 1. You will note that from a comparison of the top panel (mutant) and the bottom panel (wild type), the mutant signal is over 100-fold weaker than the signal obtained from the wild type. Further, this signal is significantly less stable than the wild type signal. In previous experiments we had shown that the H polypeptide was unique when compared

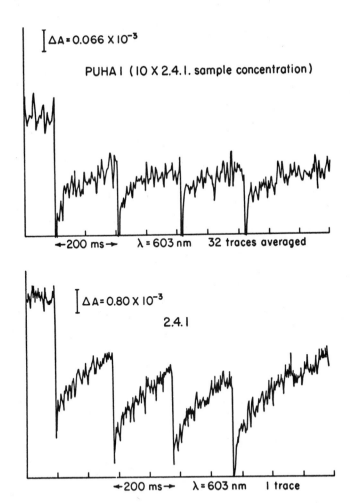

Fig. 1.   Reaction center signals in the puhA deletion
          mutation, PUHA1 (top panel) and in the wild type
          2.4.1 (bottom panel). Note that the mutant signal
          has been scaled upward over 100-fold relative to
          the wild type signal. These results were kindly
          provided by Dr. Colin Wraight.

to the other polypeptides routinely found in the three spectral complexes
in that it is the only polypeptide present in aerobically grown cells
(Chory, et al., 1984)[12]. Further, in photoheterotrophically growing
cells, the whole cell level of the H polypeptide relative to the level of
RC-L and RC-M is approximately 1.3-1.5:1 (Kiley and Kaplan, 1988)[8].
These, as well as the results described above, led us to suggest that the
H-polypeptide of the RC plays a critical role in the assembly of the RC
by serving a "docking" function in which the other polypeptides of the RC
are directed to the membrane at the site of H, as well as a "scaffolding"
function by which the L and M polypeptides are then aligned and
stabilized by H to form a functional RC (Kiley and Kaplan, 1988)[8]. We
imagine that other, as yet undiscovered factors, are involved in the
"preparation" of the specific small molecule ligands of the RC.

However, of additional significance here is the fact that as shown
in Figure 2, the B875 complex is also missing from the H mutant strain.
When the mutant is complemented in trans with a 1.45 kb BamHI fragment

107

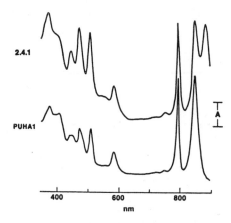

Rb. sphaeroides
Glucose/DMSO

2.4.1

PUHA1

400   600   800
nm

Fig. 2. Low temperature spectral profiles of
equivalent membrane concentrations (protein)
of wild type 2.4.1 and the puhA deletion
mutation, PUHA1. These spectra were kindly
provided by Dr. Roger Prince.

containing puhA and very little upstream or downstream DNA sequences the
H polypeptide can be restored, as well as RC function and photosynthetic
growth. However, B875 complexes as well as its apopolypeptides are still
not present despite the fact that transcription from the puf operon is
normal (Sockett, et al., 1989)[11].

Thus, we concluded that DNA sequences linked to puhA are involved in
B875 apopolypeptide synthesis and/or assembly. By complementing the puhA
deletion in trans with a cosmid containing both upstream and downstream
DNA sequences relative to puhA we were able to restore both RC and B875
complexes (Fig. 3). Subsequent studies have revealed that the region
upstream of puhA is critical to the post-transcriptional synthesis and/or
assembly of the B875 complexes.

However, additional results reveal that the synthesis and/or
assembly of the B875 complex involves a multicomponent system since, as
shown previously (Sockett, et al., 1989)[11], DNA sequences extending
approximately 1 kb upstream of pufBA, when present in trans
(approximately 4-6 copies), are able to restore B875 spectral complexes
to the puhA deletion strain. This restoration in B875 spectral complexes
is not accompanied by the restoration of RC's or photosynthetic growth
since the H polypeptide is still absent. The nature of the gene
product(s) encoded by the DNA sequence neighboring puhA and their
interaction with the gene products encoded near puf which serve to effect
B875 complex formation remain to be determined. These conclusions are
further confirmed by the fact that cosmids able to restore both RC-H and
B875 complexes to the puhA deletion are also able to restore B875
complexes to the B875 mutant RS103 (Fig. 3). Therefore, the defect in
B875 complex formation in this mutant strain is not the result of any
defect in the apopolypeptides but resides, instead, in the assembly
process apparently unique to the B875 complex.

The generality of the requirement for complex-specific assembly
factors for the expression of photosynthetic complex formation in Rb.

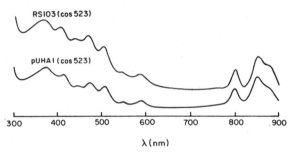

λ(nm)

Fig. 3.   Complementation of mutant RS103 and PUHA1 with cosmid
          523 containing a 10.2-kb insert of Rb. sphaeroides
          DNA encompassing puhA. Equivalent protein loads were
          employed to generate the absorption spectra. Reprinted
          from Socket, et al., 1989[11] with the kind permission
          of the Jounral of Bacteriology.

sphaeroides is further supported by recent results obtained from studies
of the puc operon of Rb. sphaeroides. We initially deleted the pucBA
structural genes and in our efforts to restore B800-850 complex formation
we were unsuccessful upon providing the structural genes as well as
suitable upstream and downstream DNA sequences in trans (Lee and Kaplan,
1989)[13]. We reasoned that the DNA was sufficient to restore complex
formation since as observed in Figure 4, we were able to restore the 0.5-
kb pucBA specific transcript encoding the B800-850 apopolypeptides.
However, when we complemented the deletion mutant with a 10.2 kb DNA
fragment encompassing pucBA, B800-850 complex formation was restored
(Fig. 5).

     One plausible explanation for these observations was that there are
additional DNA sequences required for B800-850 complex formation and the
deletion of pucBA together with the insertion of a Kn[R] cartridge created
a polar effect on the expression of DNA sequences downstream of pucBA.
These downstream DNA sequences would then be crucial to B800-850
formation. In Figure 4, lane 5, we can see the presence of a 2.3-kb
pucBA specific transcript which is less than 0.5% of the level of the 0.5
kb pucBA structural gene transcript.

     In order to investigate the role of the region downstream of pucBA
we inserted a Kn[R] cartridge approximately 0.2 kb 3' of pucBA. In this
mutant strain we found that no B800-850 complexes were observable (Fig.
5). Yet, the 0.5 kb pucBA-specific transcript was produced (Fig. 4, lane
1). These data unequivocally demonstrate that DNA sequences encompassing
approximately 1.8-kb downstream of pucBA are required for the assembly of
the B800-850 complex. By complementation in trans with DNA fragments
encompassing this entire region, we are able to restore B800-850 complex
formation to mutations in either the pucBA structural genes or in the
downstream DNA sequences (Fig. 5). We also know that transcription from
this downstream region is very complex and that promoters upstream as
well as downstream of pucBA must exist (Lee et al., 1989)[13].

     Very recently we have discovered that the absence of B800-850
complex formation resulting from the lack of expression of the DNA
sequences downstream of pucBA is not due to lack of translation of the
0.5 kb pucBA-specific mRNA but must be due to the process of assembly of
the B800-850 complex. This is a very important point because in these

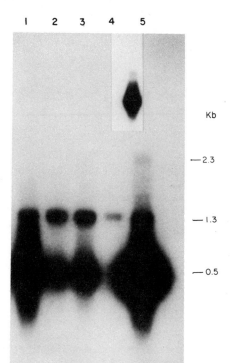

Fig. 4. Northern blot analysis of B800-850 mutant
strains and wild type 2.4.1. Lane 1 is
mutant PUC-Pv containing a Kn^R cartridge
inserted approximately 0.2-kb downstream
of pucBA. Lanes 2 and 3 are mutant PUC705-BA
complemented with an approximately 2.5-kb
PstI restriction endonuclease fragment,
extending approximately 0.8-kb upstream and
1.3-kb downstream of pucBA in both orientations
relative to the tet^R promoter of the vector.
Lane 4 is mutant PUC705-BA which contains a
deletion of pucBA coupled with the insertion of
a Kn^R cartridge into the site of the deletion.
Lane 5 is wild type 2.4.1 and the insert is a
shorter exposure of lane 5. Equivalent RNA
loads were used. Reprinted from Lee, et al.,
1989[12] with the kind permission of the Journal
of Bacteriology.

mutant cells growing photosynthetically at high light intensities, growth
rate is normal. Likewise, Bchl and Car are present but the
apopolypeptides are absent indicating rapid turnover of these
polypeptides although this has yet to be demonstrated. Thus, turnover of
these polypeptides can occur despite the presence of Bchl.

CONCLUSIONS

Taken together, these results suggest the following conclusions.
Each spectral complex possesses its unique array of "assembly" factors.
The mere presence of Bchl is insufficient, by itself, to stabilize the

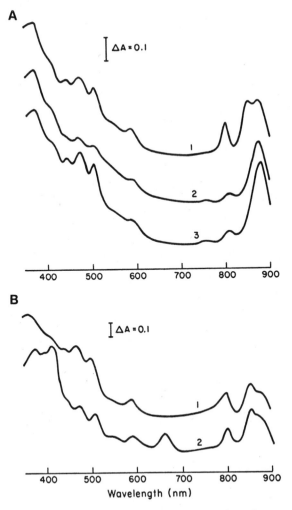

Fig. 5.   (A) Absorption spectra of strains 2.4.1 (spectrum 1),
PUC705-BA (spectrum 2), and PUC-Pv (spectrum 3) in the
visible and the near-infrared regions.   (B) Absorption
spectra of strains PUC705-BA(pRKE105) (spectrum 1) and
PUC-Pv(pRKE105) (spectrum 2) as above.   Cells were
grown photoheterotrophically at 100 W/m^2.   The
absorption spectrum for each strain was generated by
using identical amounts of protein.   See Figure 4 for
mutant designations.   Reprinted from Lee et al., 1989[12]
with the kind permission of the Journal of Bacteriology.

polypeptides in the ICM which comprise the spectral complexes.   Each
spectral complex can be assembled independent of the other spectral
complexes, assuming all of the appropriate information for assembly is
present.   Information for the assembly of the B800-850 complex is likely
to be expressed independently of the other two spectral complexes.
Conversely, information for the expression of the "assembly" factors for
the B875 complex is dependent upon the interaction of the puhA and puf
operon regions of the DNA.   This is reasonable since functioning of the
RC is dependent upon the B875 complex for light energy initially captured

by the B800-850 complex. Since the H polypeptide is essential for RC assembly, expression of the puhA region of the DNA is intimately involved in puf operon expression.

Thus, we can conclude that expression of spectral complex formation in Rb. sphaeroides involves the coordinated interplay of a large number of transcriptional and post-transcriptional activities. Many and perhaps all of these factors are encoded by genes closely linked to the structural genes encoding the polypeptides which comprise the spectral complexes.

ACKNOWLEDGEMENTS

This work was kindly supported by research grants from the USPHS and the USDA.

LITERATURE

1. K. Sauer, Photosynthetic light reaction-physical aspects. Photosythesis III: Photosynthetic membranes. in "Encyclopedia of plant physiology, new series, vol. 19." L.A. Staehelin and C.J. Arntzen, eds., Springer-Verlag, N.Y. (1986).
2. P.J. Kiley and S. Kaplan. Cloning, DNA sequence and expression of the Rhodobacter sphaeroides light-harvesting B800-850-α and B800-850-β genes. J. Bacteriol. 169:3268-3275 (1987).
3. S.W. Meinhardt, P.J. Kiley, S. Kaplan, A.R. Crofts, and S. Harayama. Characterization of light-harvesting mutants of Rhodopseudomonas sphaeroides 1. Measurement of the efficiency of energy transfer from light-harvesting complexes to the reaction center. Arch. Biochem. Biophys. 236:130-139.
4. P.J. Kiley, T.J. Donohue, W.A. Havelka, and S. Kaplan. DNA Sequence and in vitro expression of the B875 light-harvesting polypeptides of Rhodobacter sphaeroides. J. Bacteriol. 169:742-750 (1987).
5. B.S. DeHoff, J.K. Lee, T.J. Donohue, R.I. Gumport, and S. Kaplan. In Vivo Analysis of puf Operon Expression in Rhodobacter sphaeroides after Deletion of a Putative Intercistronic Transcription Terminator. J. Bacteriol. 170:4681-4692 (1988).
6. G. Feher and M.Y. Okamura. Chemical composition and properties of reaction centers. in: "The Photosynthetic Bacteria." R.K. Clayton and W.R. Sistrom, eds., Plenum Publishing Corp., N.Y. (1975).
7. H.-C. Yen and B.L. Marrs. Growth of Rhodopseudomonas capsulata under anaerobic dark conditions with dimethylsulfoxide. Arch. Biochem. Biophys. 181:411-418 (1977).
8. P.J. Kiley and S. Kaplan. Molecular Genetics of Photosynthetic Membrane Biosynthesis in Rhodobacter sphaeroides. Microbiol. Reviews 52:50-69 (1988).
9. G.S.L. Yen, B.D. Cain, and S. Kaplan. Cell-cycle specific biosynthesis of the photosynthetic membrane of Rhodopseudomonas sphaeroides; structural implications. Biochim. Biophys. Acta 777:41-45 (1984).
10. P.J. Kiley, A. Varga, and S. Kaplan. Physiological and Structural Analysis of Light-Harvesting Mutants of Rhodobacter sphaeroides. J. Bacteriol. 170:1103-1115 (1988).
11. R.E. Sockett, T.J. Donohue, A.R. Varga, and S. Kaplan. Control of Photosynthetic Membrane Assembly in Rhodobacter sphaeroides Mediated by puhA and Flanking Sequences. J. Bacteriol. 171:436-446 (1989).

12. J. Chory, T.J. Donohue, A.R. Varga, L.A. Staehelin, and S. Kaplan. Induction of the photosynthetic membranes of Rhodopseudomonas sphaeroides: Biochemical and Morphological Studies. J. Bacteriol. 159:540-554 (1984).

13. J.K. Lee, P.J. Kiley, and S. Kaplan. Posttranscriptional Control of puc Operon Expression of B800-850 Light-Harvesting Complex Formation in Rhodobacter sphaeroides. J. Bacteriol. 171:3391-3405 (1989).

ORGANIZATION OF GENES ENCODING HYDROGENASE ACTIVITY IN THE PHOTOSYNTHETIC BACTERIUM *RHODOBACTER CAPSULATUS*

Paulette M. Vignais, Pierre Richaud, Annette
Colbeau, Jean-Pierre Magnin, Béatrice Cauvin

Biochimie Microbienne (CNRS UA 1130), Département de
Recherche Fondamentale, Centre d'Etudes Nucléaires
de Grenoble, 85 X, 38041 Grenoble cedex, France

INTRODUCTION

Autotrophic bacteria, including the phototrophs, can use hydrogen gas as an energy source by means of H_2-uptake hydrogenase. Uptake hydrogenases (Hup) are intrinsic membrane proteins and transfer H_2 electrons either to the respiratory chain, under aerobic conditions, or, under anaerobic conditions, to redox carrier(s), still unidentified for the reduction of CO_2.

Membrane-bound uptake hydrogenases have been shown to be $\alpha\beta$ heterodimers and to contain Ni, as well as Fe, at their active site. Among the anoxygenic photosynthetic bacteria, membrane-bound hydrogenases were purified from *Chromatium vinosum*, *Rhodospirillum rubrum*, *Thiocapsa roseopersicina* and *Rhodobacter capsulatus* (cf. Vignais et al., 1985, Gogotov, 1986, for reviews). So far, only *R. capsulatus* hydrogenase has been studied by molecular biology techniques; these techniques were very recently extended to the study of *Rhodocyclus gelatinosus* hydrogenase (Uffen et al., 1989; Richaud et al., 1989).

In the present article, we summarize the results of our studies on *R. capsulatus* hydrogenase and describe the genetic organization of the genes involved in hydrogenase activity in *R. capsulatus*.

THE [NiFe] HYDROGENASE OF *R. CAPSULATUS*

Although hydrogenase catalyses the reversible oxidation of H_2 to protons and electrons physiologically, *R. capsulatus* hydrogenases functions for H_2 consumption. Consistent with this physiological role, the uptake hydrogenase of *R. capsulatus* is membrane associated. *R. capsulatus* chromatophores were shown by Paul et al. (1979) to oxidize H_2 with coupled ATP synthesis indicating that hydrogenase is part of the respiratory chain.

The hydrogenase enzyme was first isolated as a single subunit of 65 kDa (Colbeau et al., 1983). Subsequently, Seefeldt et al. (1987) provided immunological and molecular evidence for the presence of a second smaller subunit of 31 kDa. Since the enzyme had also been shown to contain Ni (Colbeau and Vignais, 1983), it became evident that *R. capsulatus* hydrogenase was a dimeric enzyme of the $\alpha\beta$ type, and belonged to the newly identified class of [NiFe] hydrogenases.

Such structural organization was confirmed after the structural genes of *R. capsulatus* hydrogenase had been cloned and sequenced (Leclerc et al., 1988). Structural genes of *R. capsulatus* hydrogenase were isolated from a 20 kb cosmid

Molecular Biology of Membrane-Bound Complexes in Phototrophic Bacteria
Edited by G. Drews and E. A. Dawes
Plenum Press, New York, 1990

115

library of *R. capsulatus* DNA by hybridization with the structural genes of H_2-uptake hydrogenase of *Bradyrhizobium japonicum* (Cantrell et al., 1983). The *R. capsulatus* genes were localized on a 3.5 kb *Hind*III-*Hind*III DNA fragment which could restore hydrogenase activity in the hydrogenase-deficient (Hup⁻) mutant JP91 (Leclerc et al., 1988). The nucleotide sequence of the 3.5 kb fragment revealed the presence of two open reading frames. The gene encoding the large subunit of hydrogenase (*hupL*) was identified from the size of its protein product (65.839 Da) and by alignment with the N-terminal protein sequence determined by Edman degradation. Upstream and separated by only three nucleotides was a gene capable of encoding a 34.256 Da polypeptide. The translation product of that gene showed nearly 80% identical amino acids with the small subunit of *B. japonicum* hydrogenase (Sayavedra-Soto et al., 1988) and could be aligned with the N-terminal amino acid sequence of the small subunit of *B. japonicum* and *Azotobacter vinelandii* hydrogenases (determined and communicated to us by L.C. Seefeldt and D.J. Arp). It was concluded that the second gene was the structural gene of the small subunit (*hupS*). The two genes were transcribed in the same direction; due to their close proximity, and the absence of recognisable promoter and termination signals, they probably constitute an operon (*hupSL*).

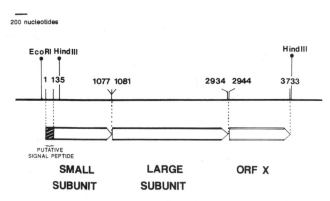

Fig. 1. Localization of ORFX in relation to the small and large subunit genes of *R. capsulatus* hydrogenase. Direction of transcription is indicated by arrow heads.

Downstream from *hupL*, starting 9 bases after the TAG stop codon of the large subunit, a third open reading frame (ORFX) was identified, capable of encoding a hydrophobic polypeptide of 262 amino acids (30.2 kDA) (Fig.1). ORFX not preceded either by any known consensus sequence, seemed to belong to the same *hupSL* operon (Richaud et al., 1989). The translation product of this gene has not yet been isolated and its function is still unknown. Comparison with other *hupSL* genes encoding membrane-bound [NiFe] hydrogenases indicated that only in *R. capsulatus* (Sayavedra-Soto et al., 1988; Ford et al., 1989; Uffen et al., 1989) was a third ORF present immediately downstream of *hupSL*. ORFX contains no Cys residues whereas 13 and 10 Cys residues are found in the *R. capsulatus* hydrogenase small and large subunits, respectively (Leclerc et al., 1988).

The small subunit of *R. capsulatus* hydrogenase was preceded by a putative leader sequence containing 45 amino acid residues. Identification of the leader peptide was based on sequence comparison. Fig.2 presents the alignement of the signal peptide sequences of four membrane-bound [NiFe] hydrogenases. The leader peptides' of *R. capsulatus* and *B. japonicum* are almost identical; relatively high amino acid identity is found in the signal peptides of the four hydrogenases (Fig.2)

```
              10         20         30         40 ↓
Rc   --LSDIETFYDVMRRQGITRRSFMKFCSLTAAALGLGPSFVPKIGEA M
Rg   -----METFYEVMRRQGISRRSFLKYCSLTATSLGLAPSFVPQIAHA M
Bj   -MGAATETFYSVIRRQGITRRSFHKFCCLTATSLGLGPLAASRIANA L
Ac   -----------MRRQGITRRSFLKYCSLTGRP-CLGPTFAPQIAHA M
            RRQGI RRSF K CSLT      L P      I A
     **** ************ * *****   ***** * * ** * *
```

Fig. 2. Putative leader sequence of *R. capsulatus* (Rc) hydrogenase proteins (Leclerc et al., 1988) aligned with those of *R. gelatinosus* (Rg) (Uffen et al., 1989), *B. japonicum* (Bj) (Sayavedra-Soto et al., 1988) and *A. chroococcum* (Ac) (Ford et al., 1989). Conserved residues in the four sequences are written in bold type and those found in only three of the four sequences are marked by asterisks on the last line. The arrow indicates the peptidase cleavage site.

The amino acid composition of the leader peptides shown on Fig.2 is consistent with the "residue distribution rules" deduced by von Heijne (1985) through analysis of 118 eukaryotic and 32 prokaryotic leader sequences. According to von Heijne (1985), even though the lengths and amino acid sequences may be very variable, all leaders comprise three regions: a positively charged amino-terminal region, n, of variable length; a central hydrophobic region, h, with minimal length of 8 residues in prokaryotes; and a polar carboxyl-terminal end, c, which includes the peptidase cleavage site (cf. von Heijne, 1983). The *R. capsulatus* signal peptide contains 25 amino acids in the N-terminal region, resulting in a net positive charge of about +2.0. The h-region comprises 13 residues, largely hydrophobic, and the c-region is 7 amino acids long including the cleavage site which presents the consensus site (Gly-X-Ala) of signal peptidase I (Perlman and Halvorson, 1983).

Periplasmic hydrogenases, such as the [NiFe] hydrogenase of *Desulfovibrio gigas* and the [NiFe] hydrogenase of *Desulfovibrio baculatus*, present a comparable organisation of their structural genes with the gene encoding the small subunit preceding the gene encoding the large one (the two of them forming an operon) and being preceded itself by a N-terminal signal sequence. Alignment of the signal sequences of *D. gigas* and *D. baculatus* hydrogenases (Voordouw et al., 1989a) with the four signal sequences of Fig.2 results in the following reduced consensus sequence: "$RRX_4RRXFXKXCX_{19}A$" (A at position -1, i.e. at the cleavage site). The [Fe] hydrogenases of *Desulfovibrio vulgaris*, which have no sequence homology with the [NiFe] and [NiFeSe] hydrogenases from *Desulfovibrio* nevertheless present a short consensus amino acid sequence "RRXFXK" in their signal peptide in common with the other leaders shown on Fig.2. That RRXFXK consensus was suggested by Voordouw et al. (1989b) to be specific for hydrogenase export; however the RRXFXK is also part of the larger consensus sequence found in the signal peptide of the four intrinsic membrane proteins of Fig.2. Therefore further work is required to understand the mechanisms involved in targeting a protein to a membrane or in directing export of the same type of protein to the periplasm.

If the same recognition sequence is used for both functions, what are the subsequent steps which determine whether a protein is retained in a membrane or extruded from it? There are certainly different possible answers to that question. The hydrophobicity of the N-segment of the signal peptides seems to be implicated in final localization of the protein, an increased hydrophobicity of the N-terminal segment being associated to an increased distance from the cytoplasm for the final location of the protein (Sjöström et al., 1987). Since four signal peptides of Fig.2 are practically identical, it may be anticipated that the corresponding hydrogenases have the same localization. Another possibility would be the co-synthesis of a hydrophobic protein which, by strong hydrophobic interactions with the phospholipid bilayer and the hydrophobic domains of hydrogenase subunits, would anchor hydrogenase to the membrane. The translation product of ORFX could be such a candidate; however, since it does not seem to be present in any of the other highly conserved respiratory membrane-hydrogenases (*B. japonicum, A. chroococcum, R. gelatinosus*), it is unlikely that this can really be its function. (Otherwise ORFX might be present but located elsewhere in the genome of the other bacteria).

REGULATION OF *hupSL* GENE EXPRESSION IN *R. CAPSULATUS*

The number of genes essential for hydrogenase biosynthesis has not yet been determined. Mutants of *R. capsulatus* unable to grow photoautotrophically with H_2 and CO_2 were isolated. Those lacking uptake hydrogenase activity, as measured by H_2-dependent methylene blue or benzyl viologen reduction, were used in complementation studies. Complementing DNA fragments were obtained from pLAFR1 cosmid libraries and analyzed genetically (Colbeau et al., 1986; Xu et al., 1989). Mutants RCC8 and RCC12 isolated by Colbeau et al. (1986) could be fully complemented by the 20 kb *Eco*RI insert of pAC57 or by the 9.6 kb *Hin*dIII fragment subcloned from pAC57 into pAC63, but not by the *hupSL* structural genes (Colbeau et al., 1989). Northern blot analyses using the 3.5 kb *Hin*dIII fragment carrying the structural genes as the probe showed no hybridizing band with mRNA isolated from RCC8 and RCC12 while the same RNA preparation could hybridize with the *nif* structural genes used in control experiments (M. Leclerc, 1988). The inability of the two strains, RCC8 and RCC12, to synthesize transcripts of the structural genes indicated that the two strains were regulatory mutants.

To obtain a more complete picture of the overall organization of *hup* gene in *R. capsulatus*, a second genomic library with 40 kb *Bam*HI fragments was constructed (Colbeau et al., 1989). Sixteen clones hybridizing with the structural genes were isolated; among these 16 clones, 3 hybridized also with the 9.6 kb *Hin*dIII fragment containing the regulatory genes. Two clones BC1 and BC2 which appeared to contain the structural and regulatory genes entirely were studied further. The DNA insertion of pBC1 overlapped over about 25 kb that of pBC2. The restriction map of pBC2 is shown on Fig.3 with the hydrogenase gene location. Hybridization analyses were performed to check that hydrogenase genes of *R. capsulatus* were localized in the same genomic environment on 40 kb DNA from the gene bank as on cell chromosome Restricted fragments from genomic DNA and from cosmids pBC1 and pBC2 were shown to contain fragments of the same size

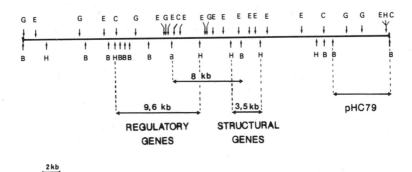

Fig. 3. Restriction map of *R. capsulatus* DNA insert in cosmid pBC2 with the location of identified *hup* vector pHC79 DNA. The 9.6 kb, 8 kb and 3.5 kb restricted fragments complement different sets of Hup⁻ mutants.
B: *Bam*HI; C: *Cla*I; E: *Eco*RI; H: *Hin*dIII; G: *Bgl*II.

hybridizing with the probes (e.g. a 17 kb *Cla*I fragment covering about half of the insertion of pBC2) (Colbeau et al., 1989).

Between the hup structural genes and the regulatory genes complementing RCC8 and RCC12 another *hup* region was identified when the 8 kb *Bam*HI fragment subcloned into vector pRK291 was shown to fully complement the Hup⁻ mutant RS20 (Colbeau et al., 1989). At the present time, the whole region downstream of the hup structural genes is being sequenced in our laboratory. The total DNA stretch on which *hup* genes were identified spanned 15-17 kb. At least 3 transcription units were identified on that stretch which does not seem to include all the *hup* DNA fragments described by Xu et al. (1989). On the other hand, another cosmid, pAG202, also isolated in our laboratory (Colbeau et al., 1986) has been shown to complement the autotrophic mutant IR4 (Willison et al., 1984) carrying the *aut-4* mutation. The *aut-4* mutation results in a reduced hydrogenase activity to 5 to 10% of that of the wild type (Willison et al., 1984). The DNA insert of pAG202 does not hybridize with pBC2. Genetic mapping of *hup* mutations by Magnin (1987) in our laboratory showed that all our *hup* markers mapped at the same locus on the genetic map of the *R. capsulatus* chromosome (Willison et al., 1984; 1989) except for the *aut-4* locus which was found near the *nif* structural gene operon (Magnin, 1987; Colbeau et al., 1989).

Before the nature and real function of the regulatory genes is elucidated, a possible way to get an insight of the type of regulatory control on hydrogenase expression is to look for the presence of known consensus sequences in the promoter region of the hydrogenase structural genes. No sequence homologous to the generalized -35/-10 promoter structure for *Escherichia coli* was identified in the 5' non coding region upstream of *hupSL*.

Recently the mode of expression in *E. coli* of formate hydrogen lyase (which includes hydrogenase 3 whose structural gene is still unknown and formate dehydrogenase encoded by *fdhF*) (Birkman et al., 1987) was shown to resemble that of the *nif* genes (Buck et al., 1986). Two *cis*-acting DNA elements were characterized in the 5' flanking region of the *fdhF* structural gene: (i) a -24(GG)/-12(GC) promoter (Thöny and Hennecke, 1989) which requires the RpoN (NtrA) protein, σ^{54}, for transcription and (ii) an "upstream activating sequence" (UAS) suggested to bind a specific activator protein (Birkmann and Böck, 1989).

An *rpoN*-like gene was recently identified in *Alcaligenes eutrophus* H16 and *Pseudomonas facilis* J strains and shown to control hydrogen oxidation and restore hydrogenase activity in Hno⁻ pleiotropic mutants of those strains (Römermann et al., 1989). These are 2 supplementary cases showing that RNA polymerase using σ^{54} is involved in the transcription of hydrogenase genes.

However, it should be noted that in *E. coli* as well as in *A. eutrophus*, *rpoN*-dependence was mostly observed for soluble hydrogenases and not for membrane-bound hydrogenases. (Hydrogenase isoenzymes 1 and 2 of *E. coli* were shown to be *ntrA* independent (Birkmann et al., 1987) and the membrane hydrogenase of *A. eutrophus* H16 was only slightly restored by cloned *rpoN* gene in Hno⁻ mutant (Römermann et al., 1989). Additionally anaerobic expression of the membrane-bound hydrogenases 1 and 2 in *Salmonella typhimurium* has been shown to be under the control of *fnr* gene (Jamieson and Higgins, 1986). We have synthesized oligonucleotides derived from the amino acid sequence of Fnr (Shaw and Guest, 1982) according to the *R. capsulatus* codon usage; those oligonucleotides hybridized with restricted *R. capsulatus* genomic DNA fragments indicating that *fnr* gene may be present in *R. capsulatus* (A. Colbeau, unpublished results). However, a consensus sequence search did not allow us to identify an Fnr consensus sequence of the type TTGA----TATCAAT-A (Spiro and Guest, 1987) within the sequenced 980bp of the 5' flanking region of *hupSL* genes.

On the other hand, a computer search for conserved promoter sequence elements recognizable by σ^{54} led to the identification of the following sequences:

```
-280  -24         -12
    AGGGCCGC----CGGCC
-476  -24         -12
    CTGGCGGC----GAGCG
-382  -24         -12
    CCGGACAG----AGGCG
```

Although the GG and GC pairs have the right spacing (10 nucleotides) they are located about 300 bp to 500 bp upstream from a coding region. Determination of the size of transcripts of the start of transcription and of RpoN/NtrA binding and protection will be required to decide whether any of these consensus sequences plays a role in hydrogenase expression.

In *R. capsulatus* a regulatory gene *nifA*-like was identified dowstream from and adjacent to the *nifHDK* operon and isolated by Ahombo et al. (1986). The gene, shown to be identical to *nifR4* (Klipp et al., 1988), was sequenced by Jones and Haselkorn (1989) and, in our laboratory, by Alias et al. (1989). It showed significant homology with the *rpoN* (*ntrA*) gene of *Klebsiella pneumoniae*. Indeed *nifR4* mutants are regulatory mutants unable to synthesize any nitrogenase polypeptide (Willison et al., 1985) and to transcribe the *nifHDK* operon (Leclerc, 1988). However, they are not affected in hydrogenase activity (Magnin, 1987) and can grow photosynthetically although the -24/-12 consensus promoter for σ^{54} binding was found upstream of the photosynthetic *puf* operon (Bauer et al., 1988). Thus, although similar, *nifR4* is not quite identical to *ntrA*. Either another true *ntrA* (*rpoN*) is present in *R. capsulatus* or the NifR4 protein does not bind to DNA exactly as the *rpoN* product.

Finally, anaerobic expression of nitrogenase (Kranz and Haselkorn, 1986) and photosynthetic (Zhu and Hearst, 1988) genes was shown to be inhibited by DNA gyrase inhibitors such as coumermycin and novobiocin and therefore to depend on DNA supercoiling. Willison et al. (1989) could indeed observe that novobiocin affects nitrogenase synthesis in response to N starvation; however, in a parallel experiment the same concentration of novobiocin (100 μg/ml) had no effect on the induction of hydrogenase synthesis (A. Colbeau, unpublished results) in contrast to results reported for the induction of *B. japonicum* hydrogenase synthesis (Novak and Maier, 1987).

Acknowledgment: We thank Mrs J. Boyer for help in preparation of the manuscript. This investigation was supported by grants from the Centre National de la Recherche Scientifique (CNRS UA 1130).

REFERENCES

Ahombo, G., Willison, J.C., and Vignais, P.M., 1986, The *nifHDK* genes are contiguous with a *nifA*-like regulatory gene in *Rhodobacter capsulatus*, *Mol.Gen. Genet.*, 205:442.

Alias, A., Cejudo, F.J., Chabert, J., Willison, J.C., and Vignais, P.M., 1989, Nucleotide sequence of wild-type and mutant *nifR4* (*ntrA*) genes of *Rhodobacter capsulatus*: identification of an essential glycine residue, *Nucleic Acids Research*, 17:5377.

Bauer, C.E., Young, D.A., and Marrs, B.L., 1988, Analysis of the *Rhodobacter capsulatus puf* operon. Location of the oxygen-regulated promoter region and the identification of an additional *puf*-encoded gene, *J. Biol. Chem.*, 263:4820.

Birkmann, A., and Böck, A., 1989, Characterization of a cis regulatory DNA element necessary for formate induction of the formate dehydrogenase gene (*fdhF*) of *Escherichia coli*, *Mol. Microbiol.*, 3:187.

Birkmann, A., Sawers, R.G., and Böck, A., 1987, Involvement of the *ntrA* gene product in the anaerobic metabolism of *Escherichia coli*, *Mol. Gen. Genet.*, 210:535.

Buck, M., Miller, S., Drummond, M., Dixon, R., 1986, Upstream activator sequences are present in the promoters of nitrogen fixation genes, *Nature* (London), 320:374.

Cantrell, M.A., Haugland, R.A., Evans, H.J., 1983, Construction of a *Rhizobium japonicum* gene bank and use in isolation of a hydrogen uptake gene, *Proc. Natl. Acad. Sci. U.S.A*, 80:181.

Colbeau, A., and Vignais, P.M., 1983, The membrane-bound hydrogenase of *Rhodopseudomonas capsulata* is inducible and contains nickel, *Biochim. Biophys. Acta*, 748:128.

Colbeau, A., Chabert, J., and Vignais, P.M., 1983, Purification, molecular properties and localisation in the membrane of the hydrogenase of *Rhodopseudomonas capsulata, Biochim. Biophys. Acta,* 748:116.

Colbeau, A., Godfroy, A., and Vignais, P.M., 1986, Cloning of DNA fragments carrying hydrogenase genes of *Rhodopseudomonas capsulata, Biochimie,* 68:147.

Colbeau, A., Magnin, J.P., Cauvin, B., Champion, T., and Vignais, P.M., 1989, Genetic-physical mapping of a hydrogenase gene cluster from *Rhodobacter capsulatus, Mol. Gen. Genet.,* (submitted).

Ford, C.M., Garg, N., Garg, R.P., Tibelius, K.H., Yates, M.G., Arp, D.J., and Seefeldt, L.C., 1989, The identification, characterization and sequencing of the genes (*hupSL*) encoding the small and large subunits of the H_2-uptake hydrogenase of *Azotobacter chroococcum, Mol. Microbiol.,* (submitted).

Gogotov, I.N., 1986, Hydogenases of phototrophic microorganisms, *Biochimie,* 68:181.

Jamieson, D.J., and Higgins, C.F., 1986, Two genetically distinct pathways for transcriptional regulation of anaerobic gene expression in *Salmonella typhimurium, J. Bacteriol.,* 168:389.

Jones, R., and Haselkorn, R., 1989, The DNA sequence of the *Rhodobacter capsulatus ntrA, ntrB* and *ntrC* gene analogues required for nitrogen fixation, *Mol. Gen. Genet.,* 215:507.

Klipp, W., Masepohl, B., and Pühler, A., 1988, Identification and mapping of nitrogen fixation genes of *Rhodobacter capsulatus*: duplication of a *nifA-nifB* region, *J. Bacteriol.,* 170:693.

Kranz, R.G., and Haselkorn, R., 1985, Characterization of *nif* regulatory genes in *Rhodopseudomonas capsulata* using *lac* gene fusions, *Gene,* 40:203.

Leclerc, M., 1988, Génétique de l'hydrogénase chez *Rhodobacter capsulatus* : séquençage de gènes et identification des ARNm, Ph.D. Thesis, Université Joseph Fourier, Grenoble I, France.

Leclerc, M., Colbeau, A., Cauvin, B., and Vignais, P.M., 1988, Cloning and sequencing of the genes encoding the large and the small subunits of the H_2 uptake hydrogenase (*hup*) of *Rhodobacter capsulatus, Mol. Gen. Genet.,* 214:97 and 1989, 215:368.

Magnin, J.P., 1987, Isolement d'une souche Hfr de la bactérie photosynthétique *Rhodobacter capsulatus* et cartographie du chromosome, Ph.D. Thesis, Université Joseph Fourier, Grenoble I, France.

Novak, P.D., and Maier, R.J., 1987, Inhibition of hydrogenase synthesis by DNA gyrase inhibitors in *Bradyrhizobium japonicum, J. Bacteriol.,* 169:2708.

Paul, F., Colbeau, A., and Vignais, P.M., 1979, Phosphorylation coupled to H_2 oxidation by chromatophores from *Rhodopseudomonas capsulata, FEBS Lett.,* 106:29.

Perlman, D., and Halvorson, H.O., 1983, A putative signal peptidase recognition site and sequence in eukaryotic and prokaryotic signal peptides, *J. Mol. Biol.,* 167:391.

Richaud, P., Vignais, P.M., Colbeau, A., Uffen, R.L., and Cauvin, B., 1989, Molecular biology studies of the uptake hydrogenases of *Rhodobacter capsulatus* and *Rhodocyclus gelatinosus, FEMS Microbiology Reviews* (in press).

Römermann, D., Warrelmann, J., Bender, R.A., and Friedrich, B., 1989, An *rpoN*-like gene of *Alcaligenes eutrophus* and *Pseudomonas facilis* controls expression of diverse metabolic pathways, including hydrogen oxidation, *J. Bacteriol.,* 171:1093.

Sayavedra-Soto, L.A., Powell, G.K., Evans, H.J., and Morris, R.O., 1988, Nucleotide sequence of the genetic loci encoding subunits of *Bradyrhizobium japonicum* uptake hydrogenase, *Proc. Natl. Acad. Sci. U.S.A,* 85:8395.

Seefeldt, L.C., McCollum, L.C., Doyle, C.M., and Arp, D.J., 1987, Immunological and molecular evidence for a membrane-bound, dimeric hydrogenase in *Rhodopseudomonas capsulata, Biochim. Biophys. Acta,* 914:299.

Shaw, D.J., and Guest, J.R., 1982, Nucleotide sequence of the *fnr* gene and primary structure of the Fnr protein of *Escherichia coli, Nucleic Acids Research,* 10:6119.

Sjöström, M., Wold, S., Wieslander, A., and Rilfors, L., 1987, Signal peptide amino acid sequences in *Escherichia coli* contain informations related to final protein localization. A multivariate data analysis, *EMBO J.*, 6:823.

Spiro, S., and Guest, J.R., 1987, Regulation and over-expression of the *fnr* gene of *Escherichia coli*, *J. Gen. Microbiol.*, 133:3279.

Thöny, B., and Hennecke, H., 1989, The -24/-12 promoter comes of age, *FEMS Microbiology Reviews*, (in press).

Uffen, R.L., Colbeau, A., Richaud, P., and Vignais, P.M., 1989, Cloning and sequencing the genes encoding hydrogenase subunits of *Rhodocyclus gelatinosus*, *Mol. Gen. Genet.*, (submitted).

Vignais, P.M., Colbeau, A., Willison, J.C., and Jouanneau, Y., 1985, Hydrogenase, nitrogenase and hydrogen metabolism in the photosynthetic bacteria, *in*: "Advances in Microbial Physiology," A.H. Rose, D.W. Tempest, eds., Vol. 26 pp.155-24, Academic Press Inc., London.

von Heijne, G., 1983, Patterns of amino acids near signal sequence cleavage sites, *Eur. J. Biochem.*, 133:17.

von Heijne, G., 1985, Signal sequences: the limits of variation, *J. Mol. Biol.*, 184:99.

Voordouw, G., Menon, N.K., LeGall, J., Choi, E.S., Peck, H.D.,Jr., and Przybyla, A.E., 1989a, Analysis and comparison of nucleotide sequences encoding the genes for [NiFe] and [NiFeSe] hydrogenases from *Desulfovibrio gigas* and *Desulfovibrio baculatus*, *J. Bacteriol.*, 171:2894.

Voordouw, G., Strang, J.D., and Wilson, F.R., 1989b, Organization of the genes encoding [Fe] hydrogenase in *Desulfovibrio vulgaris* subsp. *oxamicus* Monticello, *J. Bacteriol.*, 171:3881.

Willison, J.C., Madern, D., and Vignais, P.M., 1984, Increased photoproduction of hydrogen by non-autotrophic mutants of *Rhodopseudomonas capsulata*, *Biochem. J.*, 219:593.

Willison, J.C., Ahombo, G., Chabert, J., Magnin, J.P., and Vignais, P.M., 1985, Genetic mapping of the *Rhodopseudomonas capsulata* chromosome shows non clustering of genes involved in nitrogen fixation, *J. Gen. Microbiol.*, 131:3001.

Willison, J.C., Ahombo, G., and Vignais, P.M., 1989, Genetic control of nitrogen metabolism in the photosynthetic bacterium *Rhodobacter capsulatus*, *in*: "Inorganic nitrogen metabolism," W. Ullrich, C. Rigano, A. Fuggi, eds., Springer Verlag, Heidelberg (in press).

Xu, H.W., Love, J., Borghese, R., and Wall, J.D., 1989, Identification and isolation of genes essential for H_2 oxidation in *Rhodobacter capsulatus*, *J. Bacteriol.*, 171:714.

Zhu, Y.S., and Hearst, J.E., 1988, Transcription of oxygen-regulated photosynthetic genes requires DNA gyrase in *Rhodobacter capsulatus*, *Proc. Natl. Acad. Sci. U.S.A.*, 85:4209.

RATE-LIMITING ENDONUCLEOLYTIC CLEAVAGE OF THE 2.7 KB *PUF* mRNA OF

RHODOBACTER CAPSULATUS IS INFLUENCED BY OXYGEN

Gabriele Klug[1] and Stanley N. Cohen[2]

Zentrum fur Molekulare Biologie, Im Neuenheimer Feld 282,
D 6900 Heidelberg, FRG[1] and Department of Genetics, Stanford
University, School of Medicine, Stanford, CA 94305, U.S.A.[2]

INTRODUCTION

Segmental differences in the stability of a polycistronic mRNA were
first described for the *R. capsulatus puf* operon (Belasco et al., 1985).
The proximal and distal segments of a 2.7 kb mRNA species encoding reaction
center (RC) and light harvesting I (LHI) proteins are degraded at different
rates, causing a tenfold molar excess of mRNA encoding LHI versus mRNA
encoding RC proteins. The higher stability of the LHI-specific 0.5 kb mRNA
segment results from protection against 3' to 5' exonucleases by an intercistronic mRNA region of secondary structure localized between the LHI
genes (*pufB, pufA*) and the RC genes (*pufL, pufM*). Removal of this hairpin
loop structure results in a change in the molar ratio of the 2.7 kb mRNA
species and its 0.5 kb derivative and an altered stoichiometry of LHI and
RC complexes in the membrane (Klug et al., 1987).

The results presented here provide evidence that endonucleolytic
cleavage is the rate-limiting step in the degradation of the 2.7 kb *puf*
mRNA. Furthermore, we demonstrate that the rate of decay of the 2.7 kb
puf mRNA but not of the 0.5 kb *puf* mRNA segment depends on the oxygen tension in the culture.

RESULTS AND DISCUSSION

Deletion of a segment within the RC coding region of the *puf* transcript
increases *puf* mRNA stability

Earlier observations have suggested that rate-limiting degradation of
the 2.7 kb *puf* mRNA segment is initiated between the intercistronic hairpin
loop and the 3' terminator structures of the *puf* transcript (Chen et al.,
1988). If this view is correct, deletion of mRNA segments that contain the
putative sites of rate-limiting endonucleolytic cleavage should yield an
mRNA species that decays less rapidly. As shown in Figure 1, deletion of
DNA segments from the RC coding region results in prolongation of the half-
life of the resulting transcript from 8 minutes (for the wild-type 2.7 kb
transcript encoded by pTX35) to about 20 min.

The smallest deletion that caused increased *puf* mRNA half-life was
the 1.4 kb *Bst*EII-*Bst*EII deletion of plasmid p△RB6 and no further

Molecular Biology of Membrane-Bound Complexes in Phototrophic Bacteria
Edited by G. Drews and E. A. Dawes
Plenum Press, New York, 1990

123

half-lives of mRNA transcribed from

plasmid: deletions:

Q B A L . M   X

pTX35

Xba   S | P   Bst| Kpn
         Bst   T

plasmid:	deletions:	LHI genes:	LHI genes and downstream sequences:
pTX35		33.0 ± 2.1 min	8.0 ± 0.7 min
pΔRB6	BstEII - BstEII	33.0 ± 3.4 min	19.0 ± 1.1 min
pBst1	BstEII - Tth111I	32.0 ± 2.4 min	21.0 ± 2.0 min
pStu1	StuI - Tth111I	34.0 ± 1.9 min	20.5 ± 1.4 min

abbreviations: Xba : *Xba* I; S : *Stu* I; Bst : *Bst* EII; T : *Tth111* I; Kpn : *Kpn* I

Figure 1. DNA segments of the *puf* operon of *R. capsulatus* were inserted into the *XbaI* and *KpnI* sites of a derivative of plasmid pTJS133 (Narro and Cohen, submitted for publication). All plasmids were transferred into strain *R. capsulatus* ΔRC6 (Chen et al., 1988), which has the *puf* operon deleted from the chromosome. Total RNA was isolated from the resulting strains as described (von Gabain et al., 1983), separated on formaldehyde agarose gels, transferred to nylon membranes and probed against a 1.7 kb *XbaI-PstI puf* DNA fragment. The optical density of the *puf*-specific mRNA bands on the autoradiograph was measured in order to determine the half-lives listed in the table.

alterations of mRNA stability resulted from extension of the deletions (i.e., in plasmids pBst1 and pStu1). No deletion within the RC coding region smaller than that found in pΔRB6 caused a change in the *puf* mRNA half-life (Klug and Cohen, submitted for publication).

In none of the deletion strains shown in Figure 1 was the half-life of the LHI-specific 0.5 kb *puf* mRNA affected by the removal of RC-specific segments. These data support the view that rate-limiting endonucleolytic degradation of the 2.7 kb *puf* mRNA occurs within a 1.4 kb *BstEII-BstEII* segment of the RC coding region. However, endonucleolytic cleavage at this site is not rate-limiting for decay of the 0.5 kb segment of *puf* mRNA, which is protected by the intercistronic hairpin loop structure from exonucleolytic degradation extending in the 5' direction from the endonuclease-sensitive sites in the RC-coding region.

In vivo isolated *puf* mRNA is degraded differentially when incubated with *R. capsulatus* cell-free extracts *in vitro*

In order to establish a system we can use for testing for specific RNase activity, we isolated total RNA from *R. capsulatus*, incubated it *in vitro* with cell-free extracts from *R. capsulatus* for different periods of time, and followed the decay of *puf*-specific mRNAs by Northern blots. As seen in Fig. 2, different rates of decay were observed for the 2.7 kb mRNA *puf* species and its 0.5 kb derivative. No bands representing intermediates of degradation were detected by Northern blotting following *in vitro* incubation of total RNA with extracts or during analysis of RNA isolated from cells after addition of rifampicin. The *puf* mRNA showed logarithmic decay only for an initial phase of 20 min. For this phase the half-life of the 2.7 kb mRNA was determined to be 7 min, that of the 0.5 kb mRNA 16 min, when a cell-free extract containing 15 µg of total protein was added.

This difference in the stability of the *puf* mRNA segments was less than that found *in vivo* (Fig. 1), but was still significant. These data

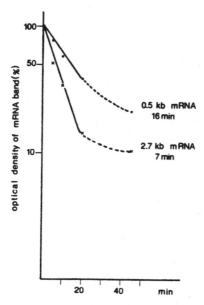

Figure 2. Total RNA was isolated (von Gabain et al., 1983) from
low oxygen cultures of strain ΔRC6(pTX35). The ethanol precipitated RNA was resuspended in 20mM Tris, pH 7.6, 10mM MgCl$_2$,
0.1mM EDTA, 1mM DTT, 5μg/ml tRNA. Equal aliquots of the total
RNA were incubated with cell-free extracts (containing 15μg of
total protein) from low oxygen cultures of *R. capsulatus* for
different time periods at 34°C. The samples were phenol-treated
and precipitated in ethanol. After treatment with DNAse and
ethanol-precipitation the samples were analysed by Northern
blots as described in Fig. 1, and the optical density of *puf*
specific mRNA bands was determined.

suggest that the stability observed for the 0.5 kb *puf* mRNA segment encoding the LHI genes is a consequence of its structural features.

The decay of the *puf* mRNA transcribed from RC genes is influenced by the oxygen tension in the culture

Oxygen partial pressure, which is the major factor regulating the
formation of the photosynthetic apparatus in *R. capsulatus* (for review see
Drews and Oelze, 1983), influences the amounts of *puf*-specific mRNA in the
cell (Clark et al., 1984; Klug et al., 1985) and specifically affects the
rate of transcription of the *puf* operon (Bauer et al., 1988; Adams et al.,
1989). Because not only the rate of transcription, but also the differential stability of the *puf* mRNA, is important in regulating the expression
of the *puf* genes (Klug et al., 1987), we determined whether the oxygen tension in the culture influences degradation of *puf*-specific mRNA.

R. capsulatus strain B10 (Marrs, 1974) was incubated at high (20% oxygen) or low (1-2% oxygen) oxygen tension. The amount of mRNA at various
time points after the addition of rifampicin was measured in order to determine the half-lives of the *puf* mRNA species (Fig. 3).

The half-life for the 2.7 kb mRNA was found to be 9-10 min in cultures
grown under low oxygen tension and 3-4 min in cultures grown under high oxygen tension. In contrast, the LHI-specific 0.5 kb *puf* mRNA segment was

degraded at a similar rate under high and low oxygen tension (half-life of 26-30 min under high oxygen and 28-33 min under low oxygen).

This result suggested that the endoribonucleolytic cleavage responsible for the rate-limiting degradation of the 2.7 kb *puf* mRNA, is affected by oxygen tension, whereas the mechanisms responsible for the rate-limiting step in degradation of the LHI-specific 0.5 kb *puf* mRNA are not. To test this hypothesis, the effect of oxygen on the degradation of *puf* mRNA transcribed from the deletion plasmids presented in Fig. 1 and from additional constructs are under investigation. Additional studies are underway to determine whether oxygen also affects the biological half-life of the 2.7 kb *puf* mRNA or only the chemical half-life as measured in the experiments described briefly above.

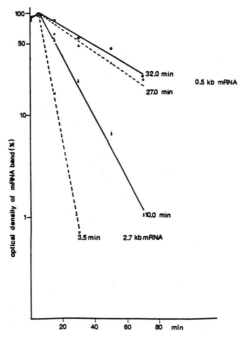

Figure 3. Total RNA from strain B10 was isolated at various time points after addition of rifampicin to the culture. The *puf*-specific RNA was analysed by Northern blots as described in legend to Fig. 1. The optical density of the mRNA bands was measured in order to determine the relative amounts of *puf* mRNA.
——— RNA was isolated from cultures grown at 1-2% oxygen.
– – – RNA was isolated from culture grown at 20% oxygen.

References

Adams, C.W., Forrest, M.E., Cohen, S.N., Beatty, T. (1989) J. Bacteriol. 171, 473-482.
Bauer, C.E., Young, D.A., Marrs, B.L. (1988) J. Biol. Chem. 263, 4820-4827.
Belasco, J.G., Beatty, J.T., Adams, C.W., von Gabain, A., Cohen, S.N. (1985) Cell 40, 171-181.
Chen, C.-Y.A., Beatty, J.T., Cohen, S.N., Belasco, J.G. (1988) Cell 52, 609-619.
Clark, W.G., Davidson, E., Marrs, B.L. (1984) J. Bacteriol. 157, 945-948.
Drews, G., Oelze, J. (1981) Adv. Microbiol. Physiol. 22, 1-97.

Klug, G., Kaufmann, N., Drews, G. (1985) Proc. Natl. Acad. Sci. USA 82, 6485-6489.

Klug, G., Adams, C.W., Belasco, J.G., Dorge, B., Cohen, S.N. (1987) EMBO J. 6, 3515-3520.

Marrs, B.L. (1974) Proc. Natl. Acad. Sci. USA 71, 971-973.

Schmidthauser, T.J., Helinski, D.R. (1985) J. Bacteriol. 164, 446-455.

von Gabain, A., Belasco, J.G., Schottel, J., Chang, A.C.Y., Cohen, S.N. (1983) Proc. Natl. Acad. Sci. USA 80, 653-657.

PHYCOBILISOMES OF THE CYANOBACTERIUM SYNECHOCOCCUS SP. PCC 7002:

STRUCTURE, FUNCTION, ASSEMBLY, AND EXPRESSION

Donald A. Bryant, Jianhui Zhou, Gail E. Gasparich, Robert
de Lorimier, Gerard Guglielmi, and Veronica L. Stirewalt

Department of Molecular and Cell Biology
The Pennsylvania State University
University Park, PA   16802   USA

INTRODUCTION

The cyanobacterial photosynthetic apparatus is remarkably similar in structure and function to that found in the chloroplasts of eucaryotic algae and higher plants (Bryant, 1987). Four major multiprotein complexes of the thylakoids--the Photosystem II complex = the water-plastoquinone photo-oxidoreductase; the cytochrome b6/f complex = the plastoquinol-plastocyanin (cytochrome c553) oxidoreductase; the Photosystem I complex = plastocyanin (cytochrome c553)-ferredoxin (flavodoxin) photo-oxidoreductase; and the ATP synthase--have been shown to be rather similar in all oxygenic procaryotes and eucaryotes studied. The predominant differences among the photosynthetic apparatuses in the various algae and higher plants derives from the considerable diversity that exists in the light-harvesting antennae systems among these organisms. In eucaryotic algae and higher plants, the light-harvesting complexes for Photosystem I and Photosystem II are a diverse collection of carteno-chlorophyll protein complexes that in general are integral membrane components (Owens, 1988; Thornber et al., 1988). Such antenna systems are also found in certain procaryotes such as Prochloron sp. and Prochlorothrix hollandica (Bullerjahn et al., 1987). However, in the cyanobacteria, in the chloroplasts of the eucaryotic red algae, and in the cyanelles of certain phylogenetically ambiguous eucaryotes such as Cyanophora paradoxa, the light-harvesting antenna complexes for Photosystem II are large, multiprotein complexes composed of water-soluble proteins, the phycobilisomes, which are attached to the thylakoid surface in close proximity to the Photosystem II reaction centers (Bryant, 1987). Phycobilisomes are largely composed of the brilliantly colored phycobiliproteins, a family of proteins which carry covalently attached, linear tetrapyrrole (phycobilin) chromophores (Glazer, 1985, 1989). Phycobilisomes additionally contain smaller amounts of non-chromophore-bearing polypeptides, termed linker polypeptides, which are required for the assembly and attachment of the structure to the thylakoid (Bryant, 1987, 1988; Glazer, 1985, 1989; Gantt, 1988). Phycobiliproteins are extremely abundant constituents of the cyanobacterial cell, and can account for up to 50% of the total suluble protein of such cells. Because of the relative ease of purification of both intact phycobilisomes and the phycobiliproteins, these photosynthetic antennae have been the objects of intensive study for over 150 years! The purpose of this article is to describe recent progress towards a complete molecular description of the phycobilisomes of the unicellular, marine cyanobacterium Synechococcus sp.

Molecular Biology of Membrane-Bound Complexes in Phototrophic Bacteria
Edited by G. Drews and E. A. Dawes
Plenum Press, New York, 1990

129

PCC 7002. A discussion of the current model for the structure, function, and assembly of hemidiscoidal phycobilisomes, as deduced from the analysis of the genes encoding the protein constituents of these structures, will be presented. Finally, aspects of the regulation of expression of the genes encoding phycobilisome components will be described.

SYNECHOCOCCUS SP. PCC 7002: GENERAL PROPERTIES

Synechococcus sp. PCC 7002 (formerly Agmenellum quadruplicatum strain PR-6) provides many advantages for a molecular analysis of photosynthetic functions. This unicellular, marine cyanobacterium is naturally competent for the uptake of DNA (i.e., it is naturally transformable; Porter, 1986) and has a rather active homologous recombination system (Murphy et al., 1987). The organism grows well at relatively high light intensities (100 to 250 μE m^{-2} s^{-1}) and temperatures (30 to 39° C). Under optimal conditions, doubling times of slightly less than four hours can be achieved. Synechococcus sp. PCC 7002 is a photoheterotroph and is capable of relatively rapid growth (doubling time of approximately 10 hours) under conditions where Photosystem II is not functional (e.g., in the presence of 10 μM DCMU; Lambert and Stevens, 1986) if 10 mM glycerol is added to the growth medium.

The transformation phenomenon is highly efficient for both linear and closed-circular plasmid DNAs, and this process has been well characterized (Porter, 1986). The genetic complexity (4000-5000 kbp) of this organism, as well as its moles G + C content (49%) are similar to the values for E. coli (Herdman et al., 1979a, 1979b). A wide variety of biphasic shuttle-vector plasmids have been developed for genetic analyses in this cyanobacterium (Buzby et al., 1983, 1985; Gasparich et al., 1987; Gasparich, 1989). These vectors carry replicons to allow their maintenance in either E. coli or Synechococcus sp. PCC 7002; carry a variety of selectable drug-resistance markers (e.g., Ap, Km, Em, Cm); and in some cases allow for the expression of cloned DNA from a cyanobacterial promoter with insertional selection on appropriate media by lacZ (β-galactosidase) inactivation.

Although the recA gene of Synechococcus sp. PCC 7002 has been cloned (Murphy et al., 1989) and characterized in some detail, it appears that the recA product provides a function required for viability in Synechococcus sp. PCC 7002. Thus, although this function can be provided by the E. coli recA gene product, it may not be possible to obtain recombination-deficient mutants of this cyanobacterium. Promoters for some Synechococcus sp. PCC 7002 genes are functional in E. coli (Murphy et al., 1987; Bryant et al., 1985; Porter et al., 1986), while some other promoters (e.g., psaA) are not (Gasparich, 1989). Correspondingly, some E. coli promoters appear to be functional in this cyanobacterium (Buzby et al., 1985). In summary, Synechococcus sp. PCC 7002 has many physiological and genetic properties which could greatly facilitate analyses of its photosynthetic apparatus. It was with these properties in mind that we began to analyze the phycobilisomes of this cyanobacterium in 1983.

PHYCOBILISOMES IN SYNECHOCOCCUS SP. PCC 7002

Hemidiscoidal phycobilisomes are composed of eight or nine cylindrical substructures approximately 11 nm in diameter (Bryant et al., 1979; Cohen-Bazire and Bryant, 1982; Glazer, 1982, 1984, 1985). Six of these are largely composed of phycocyanin, and phycoerythrin or phycoerythrocyanin (when present), and are referred to as "peripheral rods;" peripheral rods vary in length from about 12 to 30 nm. The other two or more commonly three cylinders are largely composed of the phycobiliprotein allophycocyanin and are stacked along their long axes to form a box- or pyramidal-shaped substructure referred to as the "core" (Bryant et al., 1979; Cohen-Bazire and Bryant, 1982; Glazer, 1982, 1984, 1985); the cylinders which comprise the core are usually 12 to 14 nm in length. The six peripheral rods are attached to two of the three sides of the core substructure; the third side of the core is closely appressed to the thylakoid membrane surface and to the Photosystem II reaction centers in vivo (Bryant, 1987).

The phycobilisomes of Synechococcus sp. PCC 7002 are typical hemidiscoidal phycobilisomes with tricylindrical core substructures and contain only phycocyanin and allophycocyanin as major phycobiliproteins (Bryant, 1988; Bryant et al., 1989). When examined in the electron microscope, these phycobilisomes were observed to have 5.72 ± 0.57 peripheral rods per phycobilisome; each of these peripheral rods was composed, on average of 2.04 ± 0.34 11 X 6 nm disc-shaped phycocyanin hexamers; and the average number of discs of phycocyanin per phycobilisome was observed to be 11.66 ± 1.33 (Bryant et al., 1989).

Biochemical analyses of the phycobilisomes of Synechococcus sp. PCC PCC 7002 indicate that these phycobilisomes are comprised of only eleven polypeptides. A listing of these polypeptides, of the gene loci which encode them, of their relative copy number in the phycobilisome, and of other information is shown in Table 1. Although a number of investigators have suggested that polypeptides in the 45-55 kDa mass range might be phycobilisome components, this is unlikely to be true for Synechococcus sp. PCC 7002. In some preparations of phycobilisomes such a component has been observed; however, it is not generally observed and even when present is typically in substoichiometric amount (less than one copy per phycobilisome). The phycobilisomes of Synechococcus sp. PCC 7002 were shown to contain allophycocyanin and phycocyanin in a molar ratio of approximately 1 : 2; the molar ratio of the linker polypeptides of 99 kDa : 33 kDa : 29 kDa: 9 kDa : 8.5 kDa is about 0.32 : 1 : 1 : 1 : 1; and the molar ratio of the αAP-B and β18 minor subunits of the phycobilisome core to either of the major allophycocyanin subunits was estimated to be 1 : 10-12 (Bryant et al., 1989). Hence, the molecular composition and structure of these phycobilisomes is similar to that of other extensively studied species (e.g., Synechococcus sp. PCC 6301, Synechocystis sp. PCC 6701, Mastigocladus laminosus, and Calothrix sp. PCC 7601; see Glazer, 1982, 1984, 1985; Zuber, 1987; Grossman et al., 1988; Tandeau de Marsac et al., 1988).

ORGANIZATION AND TRANSCRIPTION OF THE GENES ENCODING PHYCOBILISOME COMPO-

NENTS IN SYNECHOCOCCUS SP. PCC 7002

The genes encoding the peripheral rods of the phycobilisomes of Synechococcus sp. PCC 7002 are arranged into two transcriptional units. The first transcriptional unit, the cpcBACDEF operon, encodes all components of the peripheral rods except for the 29 kDa rod-core linker polypeptide (LRC29). The cpcA gene, encoding the α subunit of phycocyanin, was the first gene encoding a phycobilisome component to have been cloned from Synechococcus sp. PCC 7002 (de Lorimier et al., 1984; Pilot and Fox, 1984). Expression studies in E. coli subsequently demonstrated the presence of the cpcB gene, encoding the β subunit of phycocyanin, on the same DNA fragment (Bryant et al., 1985). The cpcC gene, which encodes the phycocyanin-associated, peripheral rod linker of 33 kDa, was located 3' to the cpcA gene by sequence analysis and mutagenesis (de Lorimier et al., 1989a; see below). The cpcD gene, which encodes the phycocyanin-associated, peripheral rod linker polypeptide of 9 kDa, was located 3' to cpcC by the same methods (de Lorimier et al., 1989b; see below). The cpcE and cpcF genes do not encode structural components of the phycobilisome but are apparently re-

TABLE 1. COMPOSITION OF THE PHYCOBILISOMES OF <u>SYNECHOCOCCUS</u> SP. PCC 7002

Subunit	Copies/PBS[a]	Gene Locus	Length[b]	Cloned	Sequenced	Mutants[c]
PERIPHERAL RODS:						
α^{PC}	72	<u>cpcA</u>	162	+	+	D,I
β^{PC}	72	<u>cpcB</u>	172	+	+	D,I
LR33	6	<u>cpcC</u>	290	+	+	I
LR9	6	<u>cpcD</u>	80	+	+	D,I
LRC29	6	<u>cpcG</u>	248	+	+	D
CORES:						
α^{AP}	32	<u>apcA</u>	161	+	+	D,I
β^{AP}	34	<u>apcB</u>	161	+	+	D,I
α^{AP-B}	2	<u>apcD</u>	161	+	+	I
$\beta18$	2	<u>apcF</u>	169	+	+	D
LC8.5	6	<u>apcC</u>	67	+	+	I
LCM99	2	<u>apcE</u>	886	+	+	I

[a] Copies per phycobilisome. Numbers reflect measured values for the ratio of allophycocyanin to phycocyanin, and the relative ratios of the linker polypeptides. The actual numbers for the αAP-B and β18 subunits could be slightly higher than listed. These numbers would reflect the composition of an "idealized phycobilisome" as predicted by the model of Glazer (1984, 1985) for a tricylindrical-core phycobilisome with a 1 : 2 ratio of allophycocyanin to phycocyanin.

[b] Length, in amino acids, of the deduced translation product of the gene.

[c] D, deletion; I, insertion.

quired for attachment of phycocyanobilin to the phycocyanin α subunit (J. Zhou and D. A. Bryant, unpublished results).

Primer extension and S1 nuclease protection mapping of the 5' endpoints of the <u>cpcBACDEF</u> transcripts have identified two mRNA endpoints which occur at -160 bases and -322 bases relative to the translational start codon for the <u>cpcB</u> gene (Gasparich, 1989). These endpoints occur in roughly equimolar amounts. The endpoint at -322 bases lies immediately 3' to sequence motifs (TTTAAA--17 bp--TAACAT) which closely resemble the <u>E. coli</u> consensus promoter recognized by the sigma-70 form of RNA polymerase. In fact, primer extension mapping experiments indicate that this promoter is recognized and utilized when <u>cpcB-lacZ</u> translational fusions are expressed in <u>E. coli</u>. This observation provides very strong evidence that <u>Synechococcus</u> sp. PCC 7002 has an RNA polymerase with a sigma factor with recognition specificity similar to the sigma-70 RNA polymerase of <u>E. coli</u>. The 5' sequence adjacent to the mRNA endpoint at -160 bases does not resemble any known procaryotic promoter and is not recognized as a promoter in <u>E. coli</u>. Whether this endpoint results from mRNA processing or from a secondary promoter cannot be definitively stated at this time.

However, sequence similarities between this putative promoter and the 5' flanking sequences for other transcript endpoints suggest that there may be a second, distinctive promoter class in Synechococcus sp. PCC 7002.

The cpcBACDEF operon produces a family of transcripts which correspond to the following species in order of decreasing abundance: cpcBA, cpcBAC, cpcBACD, cpcBACDEF in the approximate ratio 90:8:2:≤1 (Gasparich, 1989; R. de Lorimier, unpublished results). The relative abundance of these mRNA species could largely account for the ratio of their polypeptide products in the cell and phycobilisome (about 12 : 1, see Table 1). For the first three of these transcript species, there are actually two distinct transcripts of slightly different length, consistent with a 162 bases length difference, but approximately equal abundance which are detectable. These results suggest that the transcript pairs have different 5' endpoints but probably the same 3' endpoints. Computer analyses of the intergenic sequences between various coding sequences (cpcA-cpcC; cpcC-cpcD; cpcD-cpcE) demonstrate the potential for energetically favorable hairpin (stem-loop) structures with free-energy of formation values from -45 to -56 kCal mole^{-1}. These potential stem-loop structures could play a role in transcription termination, mRNA stabilization, or perhaps more likely both processes. Since the mRNAs corresponding to these various 3' endpoints do not accumulate to equal steady-state levels, factors other than the stability conferred by the 3' hairpins must be involved in transcript accumulation.

The LRC29 linker polypeptide is the product of the cpcG gene and is transcribed as a monocistronic mRNA (J. Zhou, V. L. Stirewalt, and D. A. Bryant, unpublished results). Primer extension mapping indicates the 5' end of the transcript occurs at -43 bases relative to the translational start codon, which interestingly is a TTG codon. The start codon TTG is used for only about 1% of genes in E. coli. The putative cpcG promoter region does not resemble the E. coli consensus promoter, but does have some similarity to the cpcBACDEF promoter at -160 bases (Gasparich, 1989).

The genes encoding the components of the cores of the phycobilisomes of Synechococcus sp. PCC 7002 are arranged in four transcriptional units, three of which are monocistronic (apcABC//apcD, apcE, apcF). The apcD (allophycocyanin-B α subunit) and apcE (linker phycobiliprotein of 99 kDa) promoters have weak homology to the consensus E. coli promoter and Northern-blot hybridization analyses indicate that the transcripts from these genes are relatively low-abundance (J. Zhou and D. A. Bryant, unpublished results). The promoter for the apcF gene (β18, the allophycocyanin-β-like subunit) does not exhibit any clear homology to known promoter sequences; two transcripts of slightly different length (about 50 bases) but approximately equal (and low) abundance are observed by Northern-blot hybridization.

The apcABC operon encodes the α and β subunits of allophycocyanin and the allophycocyanin-associated, core linker polypeptide (LC8.5) of 8.5 kDa. Three transcripts are observed from this operon by Northern-blot hybridization. The first of 1800 bases corresponds to apcABC; two transcripts of slightly different length (approximately 1400 and 1500 bases) encode only apcAB (Gasparich, 1989). The relative abundances of these three transcripts, 1800 : 1500 : 1400 = 20 : 40 : 40, could largely explain the relative levels of these gene products in the cell and phycobilisome (see Table 1). Primer extension mapping of the 5' endpoints of these transcripts suggest that multiple promoters exist as observed for the cpcBACDEF operon. Two minor endpoints, at -45 bases and at -112 bases relative to the ATG start codon of apcA, map 3' to sequences having homology to the E. coli consensus promoter for the sigma-70 subunit of RNA polymerase. The promoter adjacent to the -45 endpoint is the major one recognized by RNA polymerase in E. coli cells harboring an apcA-lacZ translational fusion. The major mRNA endpoint in Synechococcus sp. PCC 7002, however, occurs at -231 bases relative to the apcA start codon. The sequences 5' to this endpoint do not resemble any known procaryotic promoter sequences.

MUTATIONAL ANALYSES OF THE GENES ENCODING PHYCOBILISOME COMPONENTS IN SYNECHOCOCCUS SP. PCC 7002

In order to understand better structural, functional and assembly aspects of phycobilisomes, the genes encoding each phycobilisome component of Synechococcus sp. PCC 7002 have been inactivated by interposon mutagensis (Table 1). In this method mutations, either insertions or deletions, are created by cloning a DNA fragment carrying a gene encoding a protein capable of conferreing antibiotic resistance upon susceptable cells is cloned into or used to replace the coding sequence for a gene of interest in E. coli. Flanking sequences both 5' and 3' to the gene are left unaltered. This construction is ued to effect gene replacement in the target organism by transformation, homologous recombination, and antibiotic selection. After streak purification of transformants and confirmation by Southern-blot hybridization analysis that the putative mutation has segregated (is homozygous), the phenotype of the resultant mutant can be determined.

As mentioned above, the cpcBACDEF operon encodes all but one of the structural components of the peripheral rods of the phycobilisome. A region 7.488 kbp in length, surrounding and including the cpcBACDEF operon, has been sequenced. The cpcB and cpcA genes encode the phycocyanin apoproteins which are 172 and 162 amino acids in length, respectively. Amino-terminal amino acid sequence analyses of the mature phycocyanin subunits has shown that these polypeptides are not proteolytically altered after translation (Gardner et al., 1980). The structure of the hexameric $(\alpha PC\ \beta PC)_6$ form of phycocyanin has been determined at 2.5 Å resolution (Schirmer et al., 1986, 1987). The phycocyanin hexamer is a toroidal structure approximately 11 nm in diameter and 6 nm thick with a central hole approximately 3.5 nm in diameter. The hexamer is formed by the face-to-face stacking of two phycocyanin trimers $(\alpha PC\ \beta PC)_3$. It is presumed that the linker polypeptides interact with phycocyanin by occupying the central cavity. Finally, it is also presumed that the allophycocyanin trimer $(\alpha AP\ \beta AP)_3$, with dimensions of approximately 11 X 3.5 nm and considerable amino acid similarity to phycocyanin, has a similar three-dimensional structure (Bryant et al., 1976).

A mutant devoid of phycocyanin has been constructed by deleting the cpcBA coding sequences (Bryant, 1988; Bryant et al., 1989). This mutant does not produce detectable phycocyanin, and has a doubling time that is twice that of the wild-type at high light intensity (240 $\mu E\ m^{-2}\ s^{-1}$). At medium light intensity (110 $\mu E\ m^{-2}\ s^{-1}$) the deletion mutant grew seven times more slowly than the wild-type (doubling time about 56 hours). In direct competition experiments with the wild-type strain, the phycocyanin-less mutant cells were quickly overgrown by the wild-type strain. Hence, although phycocyanin is not required for photoautotrophic growth of Synechococcus sp. PCC 7002, this protein greatly enhances the light-harvesting potential of the cells. Although intact phycobilisome cores could not be isolated from this mutant, it is probable that intact cores are assembled in vivo (Bryant et al., 1989). The allophycocyanin content of this mutant was not significantly lower than that of the wild-type. This suggests that phycocyanin levels do not regulate allophycocyanin levels in this cyanobacterium.

The cpcC gene, encoding the LR33 linker polypeptide, occurs 118 bp 3' from the cpcA gene (de Lorimier et al., 1989a). The identity of this gene was proven by matching the amino-terminal sequence of the mature polypeptide to codons 2 through 21 deduced from the nucleotide sequence of the gene (de Lorimier et al., 1989a; Bryant et al., 1989). The deduced amino acid sequence indicates that the mature protein should contain 289 amino acids, since the amino-terminal methionine is removed post-translationally. The cpcC gene has been insertionally inactivated by the interposon method. The effect of this mutation was to prevent the assembly of 50% of the total phycocyanin into phycobilisomes, although phycocyanin was synthesized in the mutant cells in amounts comparable to those in the wild-type (de Lorimier et al., 1989a). As expected, the LR33 linker polypeptide was not present in the phycobilisomes of this mutant; additionally, the mutant phycobilisomes were totally devoid of the 9 kDa, phycocyanin-associated rod linker (see below). The phycocyanin which was not assembled was highly fluorescent and apparently free in the cytoplasm of the mutant cells. When examined by electron microscopy, the peripheral rods of the mutant phyco- bilisomes were never observed to contain more than one phycocyanin hexamer (one 11 X 6 nm disc) while those of the wild-type contain two hexamers per rod. Hence, the 33 kDa rod linker polypeptide is required for the attach- ment of the core-distal phycocyanin hexamer to the core-proximal hexamer. It is presumed that this linker interacts with two phycocyanin trimers to produce the tail-to-tail joining of the two trimers.

The cpcD gene, encoding the LR9 linker polypeptide, lies 178 bp 3' from the cpcC gene; the deduced amino acid sequence predicts a protein of 80 residues. The amino-terminal sequence of the protein confirms the identity of the gene and indicates that the amino-terminal methionine is not removed from the mature polypeptide (de Lorimier et al., 1989b; Bryant et al., 1989). Two mutations, a deletion and an insertion, have been contructed. In the deletion mutant, the LR9 linker polypeptide was missing; 75% of the phyco- cyanin was assembled onto phycobilisomes, while 25% of the phycocyanin was unassembled and free in the cytoplasm; the phycobilisomes were 35-40% de- ficient in the LR33 linker polypeptide. In the insertion mutant, the LR9 linker polypeptide was absent; phycocyanin levels were identical to those in the wild-type, and all phycocyanin was assembled onto phycobilisomes; the levels of the cpcC gene product were identical to those in the wild- type; however, the peripheral rods were much more variable in length than those of the wild-type (de Lorimier et al., 1989b). These results are interpretted as follows. In the deletion construction, cpcC-cpcD inter- genic sequences with the potential to form stem-loop structures were de- leted. This caused a destabilization of the cpcBAC transcripts resulting in a deficiency of the LR33 linker polypeptide. The function of the LR9 linker is to reduce rod-length variation by binding to the core-distal phycocyanin trimer and thereby preventing a rod extension reaction by the addition of a second LR33-phycocyanin complex. In principle, this should increase the efficiency of light-energy harvesting by minimizing the average number of energetically similar chromophore-chromophore transfer steps required for the energy of a photon entering a rod to reach the core.

The cpcE gene lies 136 basepairs downstream from the cpcD gene. The cpcE gene predicts a protein of 269 amino acids and does not encode a structural component of the phycobilisome (Bryant, 1988). The cpcF gene is found 53 basepairs 3' to the cpcE gene. This gene predicts a protein of 205 amino acids which similarly is not a structural component of the phycobilisome. Similar genes have been noted in Calothrix sp. PCC 7601 (Grossman et al., 1988; Tandeau de Marsac et al., 1988) and in Anabaena sp. PCC 7120 (Belknap and Haselkorn, 1987). Mutations in cpcE, cpcF, or double mutations, have similar phenotypes (J. Zhou and D. A. Bryant, unpublished results). These mutants grow more slowly than the wild-type and their coloration suggests that they do not accumulate normal levels of phycocyanin. However, more detailed analyses indicate that the mutants do contain a very small proportion of normal phycocyanin. However, most of the phycocyanin α subunit present in these cells does not carry any phyco-cyanobilin chromophore, and a considerable amount of unassembled phycocyanin β subunit is present in cell extracts (perhaps with α subunit apoprotein). All other phycobiliproteins appear to carry the appropriate chromophores. These results indicate that the cpcE and cpcF products are not required for phycocyanobilin synthesis and suggest that these gene products are required for the normal attachment of phycocyanobilin to the phycocyanin α subunits. However, it does appear that another chromophore attachment enzyme(s) can add phycocyanobilin ineffeciently to a small proportion of the α apoprotein. Efforts to overexpress the cpcE and cpcF gene products and to establish an in vitro chromophore attachment reaction using proteins prepared in E. coli are underway in collaboration with Dr. A. N. Glazer, University of California, Berkeley.

The cpcG gene, encoding the LRC29 linker polypeptide, predicts a protein of 248 amino acids (J. Zhou, V. L. Stirewalt, and D. A. Bryant, unpublished results). The deduced amino acid sequence of the gene matches the amino-terminal amino acid sequence of the purified protein, except for the absence of the initiator methionine (Bryant et al., 1989). A 120 amino-acid domain of the LRC29 linker protein is distantly homologous to the LR33 rod-linker polypeptide; however, outside this domain, the LRC29 linker has little if any homology to other linker polypeptide classes. A mutant in which the cpcG gene is insertionally inactivated does not produce detectable LRC29 polypeptide; produces normal phycocyanin levels; is highly fluorescent with emission wavelength characteristic of phycocyanin; is apparently incapable of assembling phycocyanin onto allophycocyanin cores; and produces abnormally large phycocyanin-LR33 linker protein polymers. Intact cores cannot be isolated from the cpcG mutant cells; however, core subassemblies can be isolated which are similar to those isolated from the cpcBA deletion mutant. This suggests that cores are assembled in vivo in this mutant. These results are all consistent with the role postulated for this linker polypeptide from in vitro assembly experiments--viz., the attachment of the peripheral rods to the phycobilisome cores (Glazer, 1982, 1984, 1985). Spectroscopic analyses of LRC29-phycocyanin complexes indicate that this linker protein causes a substantial red-shift in the absorbance properties of some phycocyanin chromophores; this is especially obvious in LRC29 (αPC βPC)$_3$ complexes which have an absorbance maximum at 639 nm (de Lorimier et al., 1989a).

The apcA and apcB genes encode the α and β subunits of allophycocyanin; both genes predict polypeptides of 161 amino acids, and it is not known whether any processing occurs (Bryant et al., 1989). A mutant in which the apcA and apcB genes are deleted has been constructed (V. L. Stirewalt and D. A. Bryant, unpublished results). The mutant is totally devoid of allophycocyanin, but synthesizes normal levels of phycocyanin. The cells are highly fluorescent at a wavelength characteristic for phycocyanin; wavelengths absorbed by phycocyanin are not effective at sensitizing chlorophyll fluorescence. The phycocyanin is not assembled into peripheral rod substructures, although LR33-phycocyanin complexes occur. Similar phenotype has recently been observed for mutants in which either the apcA or apcB genes were insertionally inactivated (J. Zhou and D. A. Bryant, unpublished results). These mutants, along with the mutants deficient in peripheral rod linkers, indicate that unassembled phycocyanin does not affect the expression of the cpcBA genes encoding its apoproteins. The doubling time of the apcAB deletion mutant is about 23 hours at high light (5.8 X that of wild-type) and 84 hours at medium light intensity (10.8 X that of wild-type). The growth rate of this mutant is markedly slower than that of the cpcBA deletion mutant (Bryant, 1988). This indicates that allophycocyanin does play a significant role in light-energy harvesting and that it is not simply present to facilitate energy-transfer to chlorophyll.

A spontaneous secondary mutant devoid of all phycobiliproteins was isolated from the apcAB deletion mutant (Bruce et al., 1989). In such a mutant, the relatively small number of chlorophylls (40–50) associated with the reaction center core is the only antenna for Photosystem II. Despite the absence of all phycobiliproteins, the mutant was capable of very slow photoautotrophic growth. The mutant does not grow photoautotrophically at medium to low light intensity and requires the addition of 10 mM glycerol to the growth medium (Bryant, 1988). These results demonstrate the importance of phycobiliproteins to light-energy harvesting; however, the phycobiliproteins are not absolutely required for photoautotrophic growth.

The apcC gene lies 226 bp 3' from the apcB gene and predicts a protein of 68 amino acids (Bryant, 1988). The identity of the gene has been confirmed by the amino-terminal amino acid sequence of the LC8.5 polypeptide (Bryant et al., 1989). An insertion mutant in the apcC gene was constructed by the interposon technique (Maxson et al., 1989). This mutant synthesizes normal levels of allophycocyanin and phycocyanin, but grows about 30% more slowly than the wild-type (Bryant, 1988). This mutant assembles functional phycobilisomes that are devoid of the LC8.5 linker; however, these mutant phycobilisomes are less stable than those of the wild-type (Maxson et al., 1989). Additionally, the mutant phycobilisomes exhibit greater fluorescence emission from allophycocyanin than those of the wild-type. Hence, the LC8.5 linker is not absolutely required for phycobilisome assembly, but its presence improves the stability and energy-transfer properties of the phycobilisomes.

The apcD gene, encoding the allophycocyanin-B α subunit, predicts a polypeptide of 161 amino acids which shares considerable sequence homology with the allophycocyanin α subunit (Bryant, 1988). The apcF gene, encoding the allophycocyanin β-subunit-like polypeptide denoted β18, predicts a polypeptide of 169 amino acids which shares considerable sequence homology with the allophycocyanin β subunit (Bryant, 1988) and which is not processed at its amino terminus (Bryant et al., 1989). Both genes encode minor components of the phycobilisome core (Table 1) which are believed to play important roles in energy transfer and in the structural asymmetry required to assemble the core on the thylakoid membrane surface (for a detailed discussion of the model for the phycobilisome core, see Bryant, 1988). Surprisingly, mutations in these genes do not greatly affect phycobilisome assembly, stability, nor energy transfer properties (Maxson et al., 1989; J. Zhou and D. A. Bryant, unpublished results) and do not produce detectable growth rate changes at high or low light intensity (Bryant, 1988). Hence, the apcD and apcF gene products are not required for phycobilisome assembly and function, and their absence has not yet produced a detectable phenotype. It has been suggested that the apcD gene product is replaced by the structurally similar apcA gene product in the mutant phycobilisomes (Maxson et al., 1989). A similar replacement of the β18 subunit by the structurally related β subunit of allophycocyanin could occur in the apcF mutant.

The apcE gene encodes the LCM99 linker phycobiliprotein, two copies of which occur in each phycobilisome (Table 1; Bryant et al., 1989). Studies by Gantt and coworkers (Gantt, 1988) have demonstrated that this polypeptide plays important roles in the attachment of phycobilisomes to the thylakoids and in the transfer of energy to chlorophyll a. The nucleotide sequence of the apcE gene predicts a protein of 886 amino acids (V. L. Stirewalt and D. A. Bryant, unpublished results). The identity of the gene has been confirmed by comparison with the amino-terminal sequence of the protein, from which the initiator methionine is post-translationally removed (Bryant et al., 1989). The deduced amino acid sequence for the protein reveals four-five structural domains. The amino-terminal 220 amino acids constitute a "phycobiliprotein domain." Although a mini-domain of 50–65 amino acids is inserted in this region relative to typical phycobiliproteins, this domain is homologous to all known phycobiliprotein subunits and contains the universal chromophore-binding subdomain. The position of the chromophore-binding cysteine is altered relative to its position in other phycobiliproteins (see Bryant, 1988), but the binding pocket is nonetheless quite similar overall. The remaining three domains of the LCM99 polypeptide are similar in sequence to one another. In each case a domain of approximately 120 amino acids is present which is conserved; these are connected by "connector" sequences of variable length which are not conserved. The conserved domains are also homologous to the similar domains of the cpcC and cpcG gene products, the peripheral-rod, phycocyanin-associated linkers. This homology suggests that each of the three "linker" domains of the apcE gene product provides the structural element for a tail-to-tail joining of two allophycocyanin trimers $(\alpha AP \ \beta AP)_3$. Since each phycobilisome has two copies of the LCM99 polypeptide, and since each phycobilisome core has the equivalent of twelve allophycocyanin trimers, the two copies of the LCM99 protein could provide the scaffolding necessary for the assembly of the entire core substructure (see Bryant, 1988).

The apcE gene of Synechococcus sp. PCC 7002 has been insertionally inactivated (J. Zhou and D. A. Bryant, unpublished results). The resulting mutant grows much more slowly than the wild-type and is highly fluorescent (Bryant, 1988). Intact phycobilisomes cannot be isolated from this mutant. This result confirms the critical role of the LCM99 linker phycobiliprotein in phycobilisome assembly. In future experiments we hope to modify by site-directed mutagenesis the cysteine to which the chromophore is normally attached to assess the role of the LCM99 protein in the energy-transfer path to chlorophyll a.

EFFECTS OF NUTRIENT AVAILABILITY AND LIGHT INTENSITY ON PHYCOBILIPROTEIN

GENE EXPRESSION

In cyanobacteria the phycobiliproteins represent not only antenna pigments but they also serve as storage materials for reduced carbon and nitrogen (Bryant, 1987). Under conditions of carbon or nitrogen starvation, 90% of the phycobiliproteins can be degraded in as little as 24 hours. A cpcB-lacZ translational fusion has been used to monitor cpcBACDEF expression during nutrient starvation (Gasparich et al., 1987; Gasparich, 1989). In Synechococcus sp. PCC 7002 β-galactosidase is unstable (half-life, 90-120 min) and is rapidly degraded when replacement synthesis is stopped. Removal of combined nitrogen does not cause β-galactosidase levels to decline for about three hours, after which time enzyme levels decline steadily and rapidly. This lag presumably represents the time necessary to deplete internal pools of nitrogen storage compounds such as cyanophycin and amino acids. Carbon starvation causes similar changes, although the time-course is slightly delayed. Sulfur and iron starvation did not produce similar responses, although phosphorous starvation produced an intermediate and slow decline (Gasparich, 1989). The results obtained from the translational fusion were confirmed by Northern-blot hybridization analyses with total RNAs isolated from cells undergoing starvation. Transcripts from the cpcBACDEF and apcABC operons declined and became undetectable in cells 4-5 hours after removal of combined nitrogen from the cells (Gasparich, 1989). Carbon starvation produced similar results but required slightly longer time.

Cyanobacteria, like other photosynthetic organisms, increase their contents of antennae proteins and Photosystem I reaction centers in response to decreasing light intensity (Bryant, 1987). In Synechococcus sp. PCC 7002, decreasing light intensity causes the phycocyanin:allophycocyanin ratio to increase; this increase is paralleled by an increase in the mRNA ratios for the two proteins (R. de Lorimier, personal communication). The relative abundance of the cpcBACDEF transcripts, as well as those for psaAB (encoding the apoproteins of the Photosystem I reaction center), increase as the light intensity decreases (from 282 to 119 $\mu E\ m^{-2}\ s^{-1}$; Gasparich, 1989). Similar results were obtained when cpcB-lacZ and psaA-lacZ translational fusions were used to monitor light-intensity-dependent gene expression. The mechanism by which the abundance of these transcripts is regulated is not known.

ACKNOWLEDGEMENTS

The research described from the laboratory of D. A. B. was supported by grants GM31625 from the National Institutes of Health (U. S. Public Health Service), DMB-8504294 from the National Science Foundation, and 83-CRCR-1-1336 from the U. S. Department of Agriculture.

REFERENCES

Bruce, D., Brimble, S., and Bryant, D. A., 1989, State transitions in a phycobilisome-less mutant of the cyanobacterium Synechococcus sp. PCC 7002, Biochim. Biophys. Acta, 974:66–73.

Bryant, D. A., 1987, The cyanobacterial photosynthetic apparatus: comparison to those of higher plants and photosynthetic bacteria, in "Photosynthetic Picoplankton," T. Platt and W. K. W. Li. eds., Canadian Bulletin of Fisheries and Aquatic Sciences, Vol. 214, Department of Fisheries and Oceans, Ottawa, Canada, pp. 423–500.

Bryant, D. A., 1988, Genetic analysis of phycobilisome biosynthesis, assembly, structure, and function in the cyanobacterium Synechococcus sp. PCC 7002, in: "Light-Energy Transduction in Photosynthesis: Higher Plant and Bacterial Models," S. E. Stevens, Jr. and D. A. Bryant, eds., American Society of Plant Physiologists, Rockville, pp. 62–90.

Bryant, D. A., de Lorimier, R., Guglielmi, G., and Stevens, S. E. Jr., 1989, Structural and compositional analyses of the phycobilisomes of Synechococcus sp. PCC 7002. Analyses of the wild-type strain and a phycocyanin-less mutant constructed by interposon mutagenesis, Arch. Microbiol., submitted for publication.

Bryant, D. A., Dubbs, J. M., Fields, P. I., Porter, R. D., and de Lorimier, R., 1985, Expression of phycobiliprotein genes in Escherichia coli, FEMS Microbiol. Lett., 29:343–349.

Bryant, D. A., Glazer, A. N., and Eiserling, F. A., 1976, Characterization and structural properties of the major biliproteins of Anabaena sp., Arch. Microbiol., 110:61–75.

Bryant, D. A., Guglielmi, G., Tandeau de Marsac, N., Castet, A.-M., Cohen-Bazire, G., 1979, The structure of cyanobacterial phycobilisomes: a model, Arch. Microbiol., 123:113–127.

Bullerjahn, G. S., Matthijs, H. C. P., Mur, L. R., and Sherman, L. A., 1987, Chlorophyll-protein composition of the thylakoid membrane from Prochlorothrix hollandica, a prokaryote containing chlorophyll b, Eur. J. Biochem., 168:295–300.

Buzby, J.S., Porter, R. D., and Stevens, S. E. Jr., 1983, Plasmid transformation in Agmenellum quadruplicatum PR-6: construction of biphasic plasmids and characterization of their transformation properties, J. Bacteriol., 154:1446–1450.

Buzby, J. S., Porter, R. D., and Stevens, S. E. Jr., 1985, Expression of the Escherichia coli lacZ gene on a plasmid vector in a cyanobacterium, Science, 230:805–807.

Cohen-Bazire, G. and Bryant, D. A., 1982, Phycobilisomes: composition and structure, in: "The Biology of the Cyanobacteria," N. G. Carr and B. A. Whitton, eds., Blackwell Scientific, Oxford, pp. 143–190.

de Lorimier, R., Bryant, D. A., Porter, R. D., Liu, W.-Y., Jay, E., and Stevens, S. E. Jr., 1984, Genes for the α and β subunits of phycocyanin. Proc. Natl. Acad. Sci. USA, 81:7946–7950.

de Lorimier, R., Guglielmi, G., Bryant, D. A., and Stevens, S. E. Jr., 1989a, Structure and mutation of a gene encoding a 33 kDa phycocyanin-associated linker polypeptide, Arch. Microbiol., submitted for publication.

de Lorimier, R., Guglielmi, G., Bryant, D. A., and Stevens, S. E. Jr., 1989b, Genetic analysis of a 9 kDa phycocyanin-associated linker polypeptide, Biochim. Biophys. Acta, submitted for publication.

Gantt, E., 1988, Phycobilisomes: assessment of the core structure and thylakoid interaction, in: "Light-Energy Transduction in Photosynthesis: Higher Plant and Bacterial Models," S. E. Stevens, Jr. and D. A. Bryant, eds., American Society of Plant Physiologists, Rockville, pp. 91-101.

Gardner, E. E., Stevens, S. E. Jr., and Fox, J. L., 1980, Purification and characterization of the C-phycocyanin from Agmenellum quadruplicatum, Biochim. Biophys. Acta, 624:187-195.

Gasparich, G. E., 1989, The effects of various environmental stress conditions on gene expression in the cyanobacterium Synechococcus sp. PCC 7002, The Pennsylvania State University, Ph. D. dissertation.

Gasparich, G. E., Buzby, J., Bryant, D. A., Porter, R. D., and Stevens, S. E. Jr., 1987, The effects of light intensity and nitrogen starvation on the phycocyanin promoter in the cyanobacterium Synechococcus sp. PCC 7002, in: "Progress in Photosynthesis Research, Vol. IV," J. Biggins, ed., Martinus-Nijhoff, Dordrecht, pp. 761-764.

Glazer, A. N., 1982, Phycobilisomes: structure and dynamics, Ann. Rev. Microbiol., 36:173-198.

Glazer, A. N., 1984, Phycobilisome. A macromolecular complex optimized for light energy transfer, Biochim. Biophys. Acta, 768:29-51.

Glazer, A. N., 1985, Light harvesting by phycobilisomes, Ann. Rev. Biophys. Biophys. Chem., 14:47-77.

Glazer, A. N., 1989, Light Guides. Directional energy transfer in a photosynthetic antenna, J. Biol. Chem., 264:1-4.

Grossman, A. R., Lemaux, P. G., Conley, P. B., Bruns, B. U., and Anderson, L. K., 1988, Characterization of phycobiliprotein and linker polypeptide genes in Fremyella diplosiphon and their regulated expression during complementary chromatic adaptation, Photosyn. Res., 17:23-56.

Herdman, M., Janvier, M., Waterbury, J. B., Rippka, R., Stanier, R. Y., and Mandel, M., 1979, Deoxyribonucleic acid base composition of cyanobacteria, J. Gen. Microbiol., 111:63-71.

Herdman, M., Janvier, M., Rippka, R., and Stanier, R. Y., 1979b, Genome size of cyanobacteria, J. Gen. Microbiol., 111:73-85.

Lambert, D. H. and Stevens, S. E. Jr., 1986, Photoheterotrophic growth of Agmenellum quadruplicatum PR-6, J. Bacteriol., 165:654-656.

Maxson, P., Sauer, K., Zhou, J., Bryant, D. A., and Glazer, A. N., 1989, Spectroscopic studies of cyanobacterial phycobilisomes lacking core polypeptides, Biochim. Biophys. Acta, in press.

Murphy, R. C., Bryant, D. A., Porter, R. D., and Tandeau de Marsac, N., 1987, Molecular cloning and characterization of the recA gene from the cyanobacterium Synechococcus sp. strain PCC 7002, J. Bacteriol., 169: 2739-2747.

Murphy, R. C., Bryant, D. A., and Porter, R. D., 1989, Nucleotide sequence and further characterization of the Synechococcus sp. PCC 7002 recA gene: complementation of a cyanobacterial recA mutation by the E. coli recA gene, J. Bacteriol., submitted for publication.

Owens, T. G., 1988, Light-harvesting antenna systems in the chlorophyll a/c-containing algae, in: "Light-energy Transduction in Photosynthesis: Higher Plant and Bacterial Models," S. E. Stevens and D. A. Bryant, eds., American Society of Plant Physiologists, Rockville, pp. 122-136.

Pilot, T. J. and Fox, J. L., 1984, Cloning and sequencing of the genes encoding the α and β subunits of C-phycocyanin from the cyanobacterium Agmenellum quadruplicatum, Proc. Natl. Acad. Sci. USA, 81:6983-6987.

Porter, R. D., 1986, Transformation in cyanobacteria, CRC Crit. Rev. Microbiol., 13:111-132.

Porter, R. D., Buzby, J. S., Pilon, A., Fields, P. I., Dubbs, J. M., and Stevens, S. E. Jr., 1986, Genes for the cyanobacterium *Agmenellum quadruplicatum* isolated by complementation: characterization and production of perodiploids, *Gene*, 41:249-260.

Schirmer, T., Huber, R., Schneider, M., Bode, W., Miller, M, and Hackert, M. L., 1986, Crystal structure analysis and refinement at 2.5 Å of hexameric C-phycocyanin from the cyanobacterium *Agmenellum quadruplicatum*. The molecular model and its implications for light-harvesting, *J. Mol. Biol.*, 188:651-676.

Schirmer, T., Bode, W., Huber, R., 1987, Refined three-dimensional structures of two cyanobacterial C-phycocyanins at 2.1 and 2.5 Å resolution. A common principle of phycobilin-protein interaction, *J. Mol. Biol.*, 196:677-695.

Tandeau de Marsac, N., Mazel, D., Damerval, T., Guglielmi, G., Capuano, V., and Houmard, J., 1988, Photoregulation of gene expression in the filamentous cyanobacterium *Calothrix* sp. PCC 7601: light-harvesting complexes and cell differentiation, *Photosyn. Res.*, 18:99-132.

Thornber, J. P., Peter, G. F., Chitnis, P. R., Nechushtai, R., and Vainstein, A., 1988, The light-harvesting complex of photosystem II of higher plants, *in*: "Light-Energy Transduction in Photosynthesis: Higher Plant and Bacterial Models," S. E. Stevens, Jr. and D. A. Bryant, eds., American Society of Plant Physiologists, Rockville, pp. 137-154.

Zuber, H., 1987, The structure of light-harvesting pigment-protein complexes, *in*: "The Light Reactions," J. Barber, ed., Elsevier Biomedical, Amsterdam, pp. 197-259.

GENETIC ANALYSIS OF THE CYANOBACTERIAL

LIGHT-HARVESTING ANTENNA COMPLEX

Nicole Tandeau de Marsac, Didier Mazel, Véronique Capuano, Thierry Damerval and Jean Houmard

Unité de Physiologie Microbienne (C.N.R.S, U.R.A. D1129) Département de Biochimie et Génétique Moléculaire, Institut Pasteur 28 rue du Docteur Roux, 75724 Paris cedex 15, France.

To harvest light energy the cyanobacteria have developed supramolecular structures called phycobilisomes. These fan-like structures, regularly arrayed and perpendicularly attached to the protoplasmic surface of the photosynthetic membranes (thylakoids), funnel photons primarily into the photosystem II reaction centers which contain the chlorophyll a-protein complexes. Phycobilisomes are water soluble complexes made up of at least twelve different structural proteins accounting for up to 50% of the total cell protein. In these complexes, the major proteins are phycobiliproteins whose correct assembly into functional phycobilisomes depends upon linker polypeptides.[1]

Numerous excellent reviews have been published on the structure, composition and biophysical properties of phycobilisomes, which the reader is invited to refer to for more detailed information.[2-10] Here, we will focus on recent advances in our knowledge of the organization of the genes involved in phycobilisome formation and of the regulation of their expression in response to various environmental factors, as well as during cell differentiation. Most of our work has been carried out with the filamentous cyanobacterium *Calothrix* sp. PCC 7601 (= *Fremyella diplosiphon* UTEX 481) which presents the most advanced pattern of adaptation to light (intensity and wavelength) and to nutritional factors; adaptation is sometimes accompanied by the induction of differentiation into heterocysts or hormogonia.[1,11,12] We have also started a genetic analysis of phycobilisome components in the unicellular cyanobacterium *Synechococcus* sp. PCC 6301 which has a two-cylinder phycobilisome core instead of the three-cylinder core which is found in most other cyanobacteria including *Calothrix* sp. PCC 7601.[6,13] Thus, we have compared core components from both strains in an attempt to understand the differences in architecture.

1. FUNCTIONAL ORGANIZATION OF THE GENES INVOLVED IN THE FORMATION OF PHYCOBILISOMES IN *CALOTHRIX* SP. PCC 7601

Calothrix sp. PCC 7601 is a cyanobacterium able to undergo complementary chromatic adaptation, a phenomenon which results in the modulation of the synthesis of the phycobiliproteins and of their associated linker polypeptides in response to the spectral quality of the light available during growth.[1,14] Biochemical and structural studies have provided detailed information about the structure and composition of the phycobilisomes extracted from cells grown under red or green light. Table 1 summarizes the protein composition of the phycobilisomes purified from cells grown under these two light regimes. The abbreviations and nomenclature used for the various polyeptides are also detailed in Table 1.

Molecular Biology of Membrane-Bound Complexes in Phototrophic Bacteria
Edited by G. Drews and E. A. Dawes
Plenum Press, New York, 1990

Table 1. Protein composition of the phycobilisomes[a]

A. *Calothrix* sp. PCC 7601

Core component		copy number/phycobilisome [b]	
		Red light	Green light
α allophycocyanin 1	(α^{AP1})	32	32
β allophycocyanin 1	(β^{AP1})	34	34
α allophycocyanin B	(α^{APB})	2	2
β 18.3	($\beta^{18.3}$)	2	2
core linker	($L_C^{7.8}$)	6	6
core-membrane linker	(L_{CM}^{92})	2	2

Rod component			
α phycocyanin 1	(α^{PC1})	36	36
β phycocyanin 1	(β^{PC1})	36	36
α phycocyanin 2	(α^{PC2})	72	–
β phycocyanin 2	(β^{PC2})	72	–
α phycoerythrin	(α^{PE})	–	108
β phycoerythrin	(β^{PE})	–	108
rod-core linker	(L_{RC}^{30})	6	6
PC2 rod linker	($L_R^{9.7}$)	6	–
PC2 rod linker	(L_R^{38})	6	–
PC2 rod linker	(L_R^{39})	6	–
PE rod linker	(L_R^{35})	–	6
PE rod linker	(L_R^{36})	–	6

B. *Synechococcus* sp. PCC 6301

Core component		copy number/phycobilisome [c]
α allophycocyanin	(α^{AP})	20
β allophycocyanin	(β^{AP})	22
α allophycocyanin B	(α^{APB})	2
β 18.3	($\beta^{18.3}$)	2
core linker	($L_C^{7.8}$)	4
core-membrane linker	(L_{CM}^{75})	2

Rod component		
α phycocyanin	(α^{PC})	108
β phycocyanin	(β^{PC})	108
rod-core linker	(L_{RC}^{27})	6
PC rod linker	(L_R^{12})	6
PC rod linker	(L_R^{30})	6
PC rod linker	(L_R^{33})	6

a) Abbreviations are as proposed in ref. 6; b) By analogy with published models;[6,26] c) Data taken from ref. 6 and 13.

Recently, we have characterized most of the genes encoding phycobilisome components (Table 1), as well as ten additional open reading frames which are also involved in the formation of functional phycobilisomes (Table 2). Their expression has been determined by hybridization of total RNA with specific DNA probes and most of the 5' extremities, as well as some of the 3' ends, of the transcripts have been mapped.[12,15-17]

The physical organization of these genes and their transcription pattern (number and size of the different mRNA species, as well as their relative abundance) are presented in figure 1. The genes encoding the α and β subunits of the major phycobiliproteins from the rods, PC1 (*cpcA1* and *cpcB1*), PC2 (*cpcA2* and *cpcB2*) or PE (*cpeA* and *cpeB*), are clustered on the genome, and cotranscribed as dicistronic units.[18-22] In addition, the genes encoding the linker polypeptides associated with PC2 (*cpcH2*, *cpcI2* and *cpcD2*) are located downstream from cpcA2 and are cotranscribed with the structural subunit genes as a long mRNA.[23] In contrast, neither the genes encoding the core-rod nor the PE-associated linker polypeptides were found in the vicinity of the genes encoding the α and β subunits of PC1 and PE, respectively. With the exception of the *apcD* gene (α^{APB}) which is transcribed independently,[24] the genes coding for the core components, *apcA1* (α^{AP1}), *apcB1* (β^{AP1}), *apcC* (L_C) and *apcE* (L_{CM}), are clustered and give rise to six different mRNA species (see Fig. 1 and Table 2).[12,16]

Based on the very high homology of their predicted amino acid sequences with those deduced from *cpcB1A1* and *cpcB2A2H2I2D2*, five clustered open reading frames (*cpcB3A3H3I3D3*) were identified as corresponding to a complete additional set of PC and linker polypeptide genes. These genes are organized similarly to the *cpcB2A2H2I2D2* operon, but, in addition to the *cpcB3A3* and *cpcB3A3H3I3D3* transcripts, a third mRNA species corresponding to the transcription of *cpcB3A3H3* has been detected (Fig. 1 and Table 2). The remarkable feature of the amino acid sequences predicted from this third *cpc* operon is a systematic eradication of the sulfur-containing residues with the exception of the three cysteinyl residues necessary for the binding of the bilin chromophores to the PC subunits.[12,17,22] As discussed later, these genes play an important role in cells grown under sulfur limiting conditions.

Among the five additional open-reading frames *apcA2*, *cpcE*, *cpcF*, *orfY* and *orfZ*, the *apcA2* gene most probably encodes a structural component of the phycobilisome.[12,16] This gene is transcribed as a monocistronic unit (Fig. 1 and Table 2). Its predicted amino acid sequence shares 59% and 43% identity with those of α^{AP1} and α^{APB}, respectively.[16] Although its function is not clear yet, this polypeptide might be accommodated within the phycobilisome core substructure and may account for the heterogeneity in the AP subunit complexes isolated from different cyanobacteria.[25-30] The open reading frames *cpcE* and *cpcF* are located downstream from *cpcA1*, while *orfY* and *orfZ* are downstream from *cpeA*. The *cpcE* gene is cotranscribed with the *cpcB1* and *cpcA1* genes (Fig. 1).[22] The three other open reading frames *cpcF*, *orfY* and *orfZ* are transcribed as monocistronic units (Table 2 and D. Mazel, unpublished results).[12,22]

Three green-yellow mutants, produced either by a deletion of the *cpcF* gene (mutant GY4) or the insertion of an endogenous mobile DNA element into the same gene (IS*701* and IS*703* in the mutants GY3 and GY1, respectively) were analysed and revealed a similar and very complex phenotype (D. Mazel, unpublished results).[12] Biochemical and genetic studies of these mutants indicate that the inactivation of the *cpcF* gene results in phycobilisomes almost devoid of rods as a consequence of a dramatically reduced level of PCs (<10%), although the transcription of the corresponding genes is unaffected. Most interestingly, the few α^{PC} subunits synthesized do not carry any detectable phycocyanobilin chromophore and the chromophore content of the β^{PC} subunits might also be reduced. This suggests that the *cpcF* gene is involved in posttranscriptional events. Since the AP subunits, while carrying the same phycocyanobilin chromophores as the PC subunits, appear normally synthesized and unmodified in the GY-type mutants, the *cpcF* gene is probably not implicated in the chromophore biosynthesis *per se*, but rather in its attachment to the PC polypeptide chains or in its isomerization. Thus, one can tentatively propose that, in the GY-type mutants, PC subunits are unable to form stable hexameric complexes and that the unassembled subunits are rapidly turned over. In contrast to the *apc* and *cpc* operons, the transcription of the *cpe* operon is almost completely stopped in these mutants. Our working hypothesis is that PE hexamers cannot

Table 2. Genes encoding phycobilisome components, expression and products

Gene synthesized nomenclature	Gene size (bp)	Operon	Transcript size (kb)	Gene product	Mr (kDa)	RL	GL
A. *Calothrix* sp. PCC 7601							
apcE	3243			L_{CM}^{92}	120.0	+	+
apcA1	486	apc1	5.4, 5.1,	α^{AP1}	17.2	+	+
apcB1	489		3.7, 1.7,	β^{AP1}	17.1	+	+
apcC	207		1.4, 0.3	$L_C^{7.8}$	7.8	+	+
apcD	486	-	0.55	α^{APB}	17.8	+	+
apcA2	486		0.65	α^{AP2}	17.8	+	+
cpcB1	519			β^{PC1}	17.8	+	+
cpcA1	489	cpc1	2.4, 1.6	α^{PC1}	17.2	+	+
cpcE	885			?	32.2	+	+
cpcF	786	-	1.1	?	28.5	+	+
cpcB2	522			β^{PC2}	18.1	+	–
cpcA2	489			α^{PC2}	17.4		
cpcH2 [a]	810	cpc2	3.8, 1.6	L_R^{38}	30.5	+	–
cpcI2 [a]	870			L_R^{39}	32.4	+	–
cpcD2 [a]	258			$L_R^{9.7}$	9.7	+	–
cpeB	555	cpe	1.5	β^{PE}	19.1	–	+
cpeA	495			α^{PE}	17.6	–	+
orfY	1290	-	1.5	?	48.1	+	+
orfZ	618	-	?	?	21.9	?	?

						S-deprivation	
cpcB3	519			β^{PC3}	18.1	+	
cpcA3	489			α^{PC3}	17.4	+	
cpcH3	816	cpc3	5.0, 3.3,	L_R	30.8	+	
cpcI3	858		2.0	L_R	31.6	+	
cpcD3	213			L_R	8.1	+	

B. *Synechococcus* sp. PCC 6301

Gene	Gene size	Operon	Transcript	Gene product	Mr		
apcE	2052			L_{CM}^{75}	72.4		
apcA	486	apc	2.3, 1.6,	α^{AP}	17.4		
apcB	486		1.3	β^{AP}	17.4		
apcC	204			$L_C^{7.8}$	7.8		

a) Data taken from ref. 23

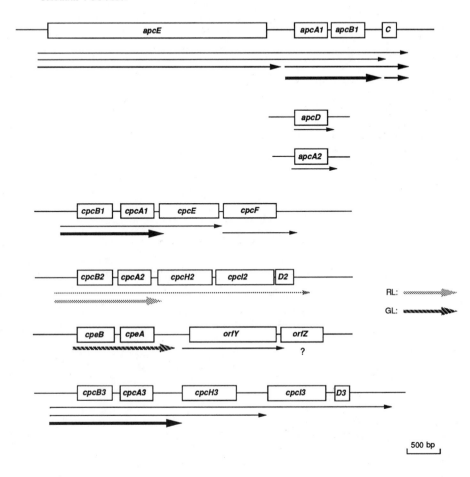

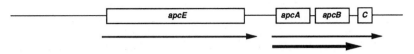

Fig. 1. Organization and Transcription of Genes Involved in Phycobilisome Formation. The thickness of the arrows is representative of the relative abundance of the different mRNAs.

attach directly to the phycobilisome core in the absence of PC1 hexamers and that free PE could prevent the transcription of the *cpe* operon. Although the GY-type mutants have a very low content of PC2 and synthesize only traces of PE, we observed that the synthesis of these phycobiliproteins is still under chromatic control (D. Mazel, unpublished results).[12] The occurrence of a *cpcF* homologue in *Synechococccus* sp. PCC 7002,[31] a strain unable to undergo complementary chromatic adaptation, strenghtens our conclusion that the product of this gene is not directly involved in this adaptation mode. Genes homologous to *cpcE* were also found in both *Synechococccus* sp. PCC 7002 [31] and *Anabaena* sp. PCC 7120.[32] According to D.A Bryant, CpcE⁻, CpcF⁻ and CpcEF⁻ mutants from *Synechococcus* sp. PCC 7002 exhibit a phenotype similar to that of the GY-type mutants with regards to the chromophore content of the PC subunits.[31] In addition, the predicted amino acid sequences of the four open reading frames *cpcE*, *cpcF*, *orfY* and *orfZ* show significant homologies (30 %) between themselves suggesting that these four gene products share some similar functional properties and are all needed for the formation of functional phycobiliproteins (D. Mazel, unpublished results). Obviously, inactivation of the *cpcE*, *orfY* and *orfZ* genes in *Calothrix* sp. PCC 7601, as well as more detailed biochemical and genetic analyses of such mutants, have to be performed to confirm this hypothesis.

As already observed in both *Synechococcus* sp. PCC 7002[31] and PCC 6301[33] and in *Anabaena* sp. PCC 7120,[32] most of the phycobiliprotein genes from *Calothrix* sp. PCC 7601 belong to operons which allows a coordinated synthesis of the α and β subunits in a 1:1 molar ratio. Segmented transcripts were often observed and the ratio of the different mRNA species generally reflects the relative abundance of the gene products within the phycobilisome.[12,34] In addition, we observed that the gene encoding αAPB, a minor component of the phycobilisome is very poorly transcribed, as is the *apcA2* gene (Fig. 1).[12,16,24] This suggests that the promoters of the different genes or operons have different efficiencies and/or require specific effectors to modulate transcription. At present, the comparison of the promoter sequences of the different genes characterized did not allow us to define a consensus sequence nor to relate promoter sequences with the known conditions of gene expression. Interestingly, however, sequence conservation is greater for promoter regions of a given operon between cyanobacterial species than between the various *Calothrix* sp. PCC 7601 promoters. In particular, the 65 nucleotides preceding the start site of transcription of the cpe operons from the two chromatically adapting strains *Calothrix* sp. PCC 7601 and *Pseudanabaena* sp. PCC 7409 have 81 % identity (D. Mazel, unpublished results and D.A. Bryant, personnal communication).[12] Important homologies were also observed between the 5' flanking regions of the *cpc* operons from *Calothrix* sp. PCC 7601,[12] *Anabaena* sp. PCC 7120 [32] *Synechococcus* sp. PCC 6301 [38] and PCC 7002.[31] These conserved regions might represent at least some of the target sequences involved in the binding of the effectors which control the regulation of gene expression.

While the effect of light intensity on the pigment content of *Calothrix* sp. PCC 7601 cells has been poorly studied so far, the adaptation to the spectral quality of the light, known as complementary chromatic adaptation, has been examined by different research groups over the last decades.[1,12,14,15,34] The recent development of genetic tools for cyanobacteria offers new possibilities for unraveling the molecular basis of this complex and fascinating problem of the regulation of cell metabolism by light.[35]

As first proposed by Bogorad and coworkers,[36,37] regulation of gene expression during complementary chromatic adaptation operates at the level of transcription. This regulation mode is, at present, directly demonstrated for the *cpc2* and *cpe* operons which are transcribed specifically in cells grown in red or green light, respectively[19,21,23]. It does not affect the genes encoding either the PC1 subunits or the core components.[12,16,22,24]

The results on the regulation of the expression of the genes coding for phycobiliprotein subunits upon light intensity changes were unexpected. In all photosynthetic organisms, the size of the light-harvesting antennae is inversely correlated with the photon density; the less energy being available, the more pigment being synthesized per cell unit. At the mRNA level, we found more transcripts for the phycobiliproteins when cells were grown under high light energy than when they were grown under low light energy (D. Mazel, unpublished results). This result indicates that most probably posttranscriptional controls occur under these conditions. Under high light energy, translation has to be deeply reduced or the excess of PC rapidly degraded, since more mRNAs but less gene products are synthesized. Similar results are obtained for the expression of the PC subunits in other cyanobacteria [32,38] and of the light-harvesting complex polypeptides from the purple bacterium *Rhodobacter capsulatus*.[39]

To our knowledge, *Calothrix* sp. PCC 7601 is the only cyanobacterium which has been shown to possess three complete sets of PC genes. One of them, named the *cpc3* operon, is expressed at a very low level under standard conditions, whatever the spectral quality of the light. What could be the function of this third gene cluster? Sequencing of the *cpc3* cluster predicted that the PC3 products would have no Met or Cys residues apart from the three Cys directly involved in chromophore binding, and the N-terminal Met.[17,22] Thus a role in adaptation to sulfur deprivation was immediately suggested.

When cells are grown in a medium containing a limiting amount of sulfate, they rapidly turn blue-green. Under these conditions, their generation time is roughly unaffected. Cells can be indefinitely subcultured under sulfate limiting conditions and can recover their normal pigmentation upon sulfate addition. As revealed by northern hybridization experiments performed by using DNA probes specific for the different phycobiliprotein genes, the reversible adaptation controlled by sulfur availability only affects the transcription of the operons encoding rod components. Under sulfate deprivation, the *cpc3* operon is rapidly turned on, the transcription of the *apc1* operon persists, but the *cpc1*, *cpc2* and *cpe* operons are completely switched off.[17] The molecular mechanism governing the transcription switch between *cpc3* and the other phycobiliprotein operons is unkown. No run of Met or Cys codons suggestive of an attenuation mechanism is present in the 5' region flanking the *cpc3* gene cluster and there is no homology between this region and the promoter regions of the other genes encoding phycobilisome constituents.[12] Emission fluorescence spectroscopy performed on whole cells grown under sufate limitation showed that light is efficiently transferred to the reaction centers of photosystem II and biochemical analyses of the purified phycobilisomes revealed that the new phycocyanin, PC3, is the only phycobiliprotein species present in their rods (D. Mazel and J.C. Thomas, unpublished results). Apart from the *cpc3* gene products, the other phycobilisome components which are also expressed under sulfur limiting conditions contain only a few sulfur-containing amino acids in their primary sequences, and, as evidenced by the determination of the N-terminal sequences of both the α and β subunits of PC3 and AP1, the initiator Met residues are posttranslationally removed.[17] Eradication of sulfur-containing residues in amino acid sequences is probably not restricted to phycobilisome components, since the N-terminus Met content of total cellular proteins is about 21% in cells grown under standard conditions versus 10% in sulfate-limited cells. Moreover, the content of Met and Cys residues of all the available PC and AP amino acid sequences is correlated with the sulfur abundance in the natural habitats from which the corresponding cyanobacterial species were isolated. These results clearly indicate that the natural environmental conditions, to which these various organisms were able to adapt, can be directly imprinted in the sequences of their most abundant proteins.[17]

Two types of cell differentiation can occur in *Calothrix* sp. PCC 7601.[11,40] In the absence of combined nitrogen in the culture medium, some cells within the vegetative filaments differentiate into heterocysts, which are cells specialized in the fixation of molecular nitrogen. On the other hand, in response to stress conditions, such as a drastic change of the light wavelength (transfer from green to red light), vegetative filaments differentiate into motile and buoyant hormogonia in less than 24 h. The regeneration of vegetative filaments occurs after this time. This transient differentiation process affects each cell within a filament and is characterized by a massive synthesis of gas vesicles which provide cells with buoyancy.[12]

In order to study this differentiation process at a molecular level, we have characterized the genes involved in gas vesicle formation (*gvp*) and determined their kinetics of transcription during the first 24 h. In parallel, we have also examined the transcription of the genes encoding AP1, PC1, PC2 and PE.

Transcription of the *gvp* genes starts 1.5 h after the transfer of the cells to red light, the maximum being reached between 6 to 9 h. Simultaneously, the genes encoding

AP1, PC1 and PC2 are immediately turned off, but the PE transcripts can be detected up to 9h later (T. Damerval, unpublished results).[12,41-43] This is consistent with the results of Oelmüller et al. [44] who recently reported that 70 to 80% of the PE transcripts are still present 8 h following the transfer of green-adapted cells to red light. Therefore, in contrast to the transcription of the PC genes, transcription of PE genes seems to persist for several hours after switching from inducing to noninducing light conditions. When the degradation of the *gvp* transcripts occurs after 9 to 12 h, the *apc1*, *cpc1* and *cpc2* operons are turned on, but the *cpe* operon remains off as expected for cells continuously maintained under red light (T. Damerval, unpublished results). The fact that some genes are switched on, while others are off during hormogonium differentiation suggests that different sigma factors are involved in this regulatory process. At present, no sigma factors have been characterized in cyanobacteria, but a look for such transcriptional factors is undoubtedly necessary to gain more insight into hormogonium differentiation in *Calothrix* sp. PCC 7601. Another important question which remains to be solved is the mechanism by which light wavelengths can act on both types of cell differentiation occurring in *Calothrix* sp. PCC 7601. Red light promotes hormogonium and supresses heterocyst differentiation, while the converse situation is observed under green light. Moreover, hormogonia stop synthesizing the red light-induced PC2, while heterocysts are devoid of the green light-induced PE (T. Damerval, unpublished results). Thus, these multiple photoregulatory mechanisms probably involve a very intricate network of antagonistic and/or cooperative effectors acting in both complementary chromatic adaptation and cell differentiation processes in *Calothrix* sp. PCC 7601.

2. COMPARISON OF THE ORGANIZATION AND EXPRESSION OF THE GENES ENCODING CORE COMPONENTS OF THE PHYCOBILISOMES IN *CALOTHRIX* SP. PCC 7601 AND *SYNECHOCOCCUS* SP. PCC 6301

One approach for studying the architecture of the phycobilisome core could be the comparative analysis of mutants or of chimaeric cores constructed by genetic means. Thus, it is of particular interest to compare the organization and expression of the *apc* operon which encodes most of the components of the three-cylinder phycobilisome core of *Calothrix* sp. PCC 7601 and of the two-cylinder core of *Synechococcus* sp. PCC 6301. In both organisms, the physical organization of the *apcE*, *apcA*, *apcB* and *apcC* genes is identical. However, their pattern of transcription differs, since six mRNA species are found in *Calothrix* sp. PCC 7601 but only three in *Synechococcus* sp. PCC 6301 (see Fig. 1; V. Capuano, unpublished results).[16,33]

The most interesting feature that emerged from the comparative studies of the *apc* genes from these two organisms concerns the *apcE* gene which encodes the L_{CM} polypeptide. This phycobiliprotein of high molecular weight (72 kDa for *Synechococcus* and 120 kDa for *Calothrix*, as calculated from the deduced amino acid sequences) plays a multifunctional role as the terminal energy acceptor and as a linker polypeptide which contributes to the core substructure and anchors the phycobilisome to the thylakoid membrane.[7] From existing models of the core substructure, the L_{CM} is part of one of the two central trimers which constitute the two antiparallel basal cylinders.[13,26,28]

As deduced from the nucleotide sequences, the *Synechococcus* sp. PCC 6301 L_{CM} is 74% homologous to that of Calothrix sp. PCC 7601. Both L_{CM} polypeptides were found to have a similar N-terminal domain, about 240 amino acid long, which carries the unique phycocyanobilin chromophore; but interestingly the C-terminal regions are distinguished by two (*Synechococcus*) or four (*Calothrix*) internal repeats. The N-terminal domains share 45 to 53% homology with the amino acid sequences of the α and β subunits of the various phycobiliproteins and the internal repeats are 50 to 60% homologous to the N-terminal region of the rod linker polypeptides (J. Houmard, unpublished results). Based on their homology, it is likely that each repeat interacts with similar structures. However, in agreement with their spatial location in the core, the region that spaces out these different domains varies from 48 to 141 amino acids in *Calothrix* sp. PCC 7601. Since the whole sequence of the L_{CM} from *Synechococcus* sp. PCC 6301 matches all of the N-terminal sequence of *Calothrix* sp. PCC 7601, it is likely that no part of this sequence interacts with the additional top cylinder of the *Calothrix* phycobilisome core. While the spacing of the different domains of the two L_{CM} is conserved, very little homology could be detected within the spacing arms.

With the aim of examining if the length and/or the number of internal repeats of the L_{CM} polypeptide do determine the number of cylinders in the phycobilisome core, the *apcE* gene from *Calothrix* sp. PCC 7601 has been transferred into Synechococcus sp. PCC 7942, a strain identical to *Synechococcus* sp. PCC 6301,[45] but which has retained its capability of being transformed by DNA. The system developed by J. van der Plas et al. (personal communication) which allows integration of any gene into the *Synechococcus* sp. PCC 7942 *metl* gene was used. Two types of transformants were obtained, one in which only the *apcE* gene has been integrated and another in which the cloning vector (pBR322) together with *apcE* have been inserted (V. Capuano, unpublished results). Both *apcE* genes are expressed and further biochemical and genetic analyses of the transformants are in progress.

3. CONCLUSIONS AND FURTHER PROSPECTS

Several regulatory mechanisms control the expression of the genes for light-harvesting antenna in *Calothrix* sp. PCC 7601. These mechanisms govern phycobilisome biosynthesis, structure and assembly, allowing cells to adapt to extreme environmental conditions. The information so far obtained provide an important basis for a better understanding of the sequence of events which leads to the building of functional phycobilisomes from their constituents, phycobiliproteins and linker polypeptides, and emphasises the need for the development of research on the biosynthesis, isomerization and attachment of the chromophore(s) to the apoproteins, aspects which are still poorly studied. Indeed, chromophores might play an important role in the stabilization of their partner proteins and, indirectly, in the regulation of other phycobilisome components.

All the regulatory mechanisms controlled by light, including pigment adaptation and differentiation, or by sulfur deprivation, probably result from a cascade of interconnected molecular events in which the synthesis of the phycobilisome components is involved. These complex regulations are exerted at all possible levels of gene expression (transcriptional, posttranscriptional, translational and/or posttranslational). None of the molecular effectors involved are known yet, but *in vitro* transcription analyses and *in vivo* genetic experiments are now being undertaken using both the wild type strain and mutants in order to try to identify them. Such approaches should permit us to define more precisely the promoter regions, to characterize and isolate the sigma factors involved in cell differentiation, as well as the cis- and/or trans-acting elements (activators or repressors) which govern gene expression in response to light and to sulfur deprivation. Hopefully, they should also provide a means for the discovery of the components of the signal transduction chain(s) for the perception of light (intensity and wavelength) and possibly lead to the identification of the proposed photoreversible pigment involved in the adaptation to chromatic illumination.[1] It will thus be possible to establish the homologies and differences, if any, between this photoperception system and the phytochrome in higher plants.[46]

ACKNOWLEDGEMENTS

We are grateful to Pr. G. Cohen-Bazire for constant support, to Dr. G. Guglielmi for helpful discussions and to A.-M. Castets and T. Coursin for their skilled technical assistance. We would like also to thank Pr. D.A. Bryant for sharing data with us prior to publication and Dr. R. Ford for critical reading of the manuscript.This work was supported by the Institut Pasteur, by the C.N.R.S. (U.R.A. D 1129) and by an A.I. C.N.R.S. grant 990019.

REFERENCES

1. N. Tandeau de Marsac, Phycobilisomes and complementary chromatic adaptation in cyanobacteria, Bull. Inst. Pasteur 81:201-254 (1983).
2. E. Gantt, Phycobilisomes, Annu. Rev. Plant Physiol. 32:327-347 (1981).
3. E. Gantt, Phycobilisomes, in: "Encyclopedia of Plant Physiology, New Series," A. Staehelin, and C. Arntzen, eds., pp 1-18, Springer-Verlag (1984).
4. A. N. Glazer, Phycobilisomes: structure and dynamics, Annu. Rev. Microbiol. 36:173-198 (1982).

5. A. N. Glazer, Phycobilisome. A macromolecular complex optimized for light energy transfer, Biochim. Biophys. Acta 768:29-51 (1984).
6. A. N. Glazer, Light harvesting by phycobilisomes, Annu. Rev. Biophys. Biophys. Chem.14:47-77 (1985).
7. A. N. Glazer, Phycobilisomes: assembly and attachment, in: "The Cyanobacteria," P. Fay, and C. van Baalen, eds., pp 69-94, Elsevier Science Publishers (1987)
8. A. N. Glazer, and A. Melis, Photochemical reaction centers: structure, organization, and function, Annu. Rev. Plant Physiol. 38:11-45 (1987).
9. A. N. Glazer, Light guides: Directional energy transfer in a photosynthetic antenna, J. Biol. Chem. 264:1-4 (1989).
10. B. A. Zilinskas, and L.S. Greenwald, Phycobilisome structure and function, Photosynth. Res. 10:7-35 (1986).
11. R. Rippka, and M. Herdman, Division patterns and cellular differentiation in cyanobacteria, Ann. Microbiol. (Inst. Pasteur) 136A:33-39 (1985).
12. N. Tandeau de Marsac, D. Mazel, T. Damerval, G. Guglielmi, V. Capuano, and J. Houmard, Photoregulation of gene expression in the filamentous cyanobacterium Calothrix sp. PCC 7601: light-harvesting complexes and cell differentiation, Photosynth. Res. 18:99-132 (1988).
13. A. N. Glazer, D.J. Lundell, G. Yamanaka, and R.C. Williams, The structure of a "simple" phycobilisome. Ann. Microbiol. (Inst. Pasteur) 134 B:159-180 (1983).
14. L. Bogorad, Phycobiliproteins and complementary chromatic adaptation, Annu. Rev. Plant. Physiol. 26:369-401 (1975).
15. A. R. Grossman, P.G. Lemaux, and P.B. Conley, Regulated synthesis of phycobilisome components, Photochem. Photobiol. 44:827-837 (1986).
16. J. Houmard, V. Capuano, T. Coursin, and N. Tandeau deMarsac, Genes encoding core components of the phycobilisome in the cyanobacterium Calothrix sp. strain PCC 7601: occurrence of a multigene family, J. Bacteriol. 170: 5512-5521 (1988).
17. D. Mazel, and P. Marlière, Adaptative eradication of methionine and cysteine from cyanobacterial light-harvesting proteins, Nature, in press.
18. V. Capuano, D. Mazel, N. Tandeau de Marsac, and J. Houmard, Complete nucleotide sequence of the red-light specific set of phycocyanin genes from the cyanobacterium Calothrix PCC 7601, Nucl. Acids Res. 16:1626 (1988).
19. P. B. Conley, P.G. Lemaux, and A.R. Grossman, Molecular characterization of evolution of sequences encoding light-harvesting components in the chromatically adapting cyanobacterium Fremyella diplosiphon, J. Mol. Biol. 199:447-465 (1988).
20. P. B.,Conley, P.G. Lemaux, T.L. Lomax, and A.R. Grossman, Genes encoding major light-harvesting polypeptides are clustered on the genome of the cyanobacterium Fremyella diplosiphon, Proc. Natl. Acad. Sci. USA 83:3924-3928 (1986).
21. D. Mazel, G. Guglielmi, J. Houmard, W. Sidler, D.A. Bryant, and N. Tandeau de Marsac, Green light induces transcription of the phycoerythrin operon in the cyanobacterium Calothrix 7601, Nucl. Acids Res. 14:8279-8290 (1986).
22. D. Mazel, J. Houmard, and N. Tandeau de marsac, A multigene family in Calothrix sp. PCC 7601 encodes phycocyanin, the major component of the cyanobacterial light harvesting antenna, Mol. Gen. Genet. 211:296-304 (1988).
23. T. L. Lomax, P.B. Conley, J. Schilling, and A.R. Grossman, Isolation and characterization of light-regulated phycobilisome linker polypeptide genes and their transcription as a polycistronic mRNA, J. Bacteriol. 169: 2675-2684 (1987).
24. J. Houmard, V. Capuano, T. Coursin, and N. Tandeau Marsac, Isolation and molecular characterization of the gene encoding allophycocyanin B, a terminal acceptor in cyanobacterial phycobilisomes, Mol. Microbiol. 2 : 101-107 (1988).
25. W. Reuter, and W. Wehrmeyer, Core substructure in Mastigocladus laminosus phycobilisomes : I. Microheterogeneity in two of three allophycocyanin core complexes, Arch. Microbiol. 150:534-540 (1988).
26. L.K. Anderson, and F.A. Eiserling, Asymmetrical core structure in phycobilisomes of the cyanobacterium Synechocystis 6701, J. Mol. Biol. 191:441-451 (1986).
27. O.D. Canaani, and E. Gantt, Circular dichroism and polarized fluorescence characteristics of blue-green algal allophycocyanins, Biochemistry 19:2950-2956 1980).
28. J. C. Gingrich, D.J. Lundell, and A.N. Glazer, Core substructure in cyanobacterial phycobilisomes, J. Cell Biochem. 22:1-14 (1983).

29. G. Guglielmi, and G. Cohen-Bazire, Etude taxonomique d'un genre de cyanobactéries Oscillatoriacae: le genre *Pseudanabaena* Lauterborn. II. Analyse de la composition moléculaire et de la structure des phycobilisomes, Protistologica XX:393-413 (1984).

30. B. A. Zilinskas, B.K. Zimmerman, and E. Gantt, Allophycocyanin forms isolated from *Nostoc* sp. Phycobilisomes, Photochem. Photobiol. 27:587-595 (1978).

31. D.A. Bryant, Genetic analysis of phycobilisome biosynthesis, assembly, structure, and function in the cyanobacterium *Synechococcus* sp. PCC 7002, in: "Light-energy transduction in photosynthesis: Higher plants and bacterial models" S.E. Stevens, Jr., and D.A. Bryant, eds., pp 62-90, The American Society of Plant Physiology (1988).

32. W. R. Belknap, and R. Haselkorn, Cloning and light regulation of expression of the phycocyanin operon of the cyanobacterium *Anabaena*, EMBO J. 6:871-884 (1987).

33. J. Houmard, D. Mazel, C. Moguet, D.A. Bryant, and N. Tandeau de Marsac, Organization and nucleotide sequence of genes encoding core components of the phycobilisomes from *Synechococcus* 6301, Mol. Gen. Genet. 205:404-410 (1986).

34. A. R. Grossman, P.G. Lemaux, P.B. Conley, B.U. Bruns, and L.K. Anderson, Characterization of phycobiliprotein and linker polypeptide genes in *Fremyella diplosiphon* andtheir regulated expression during complementary chromatic adaptation, Photosynth. Res. 17:23-56 (1988).

35. N. Tandeau de Marsac, and J. Houmard, Advances in cyanobacterial molecular genetics, in: "The cyanobacteria," P. Fay, and C. van Baalen, eds., pp 251-302, Elsevier, Amsterdam (1987).

36. L. Bogorad, S.M. Gendel, J.H. Haury, and K.-P. Koller, Photomorphogenesis and complementary chromatic adaptation in *Fremyella diplosiphon*, in: "Photosynthetic prokaryotes: cell differentiation and function, G.C. Papageorgiou, and L. Packer, eds., pp119-126, Elsevier Biomedical, New York (1983).

37. S. Gendel, I. Ohad, and L. Bogorad, Control of phycoerythrin synthesis during chromatic adaptation, Plant Physiol. 64:786-790 (1979).

38. S. R. Kalla, L.K. Lind, J. Lidholm, and P. Gustafsson, Transcriptional organization of the phycocyanin subunit gene clusters of the cyanobacterium *Anacystis nidulans* UTEX 625, J. Bacteriol. 170:2961-2970 (1988).

39. A. P. Zucconi, and J.T., Beatty, Posttranscriptional regulation by light of the steady-state levels of mature B800-850 light-harvesting, J. Bacteriol. 170:877-882 (1988).

40. M. Herdman, and R. Rippka, Cellular differentiation: hormogonia and baeocytes, Meth. in Enzymol. 167:232-242 (1988).

41. K. Csiszár, J. Houmard, T. Damerval, and N. Tandeau de Marsac, Transcriptional analysis of the cyanobacterial *gvpABC* operon in differentiated cells: occurrence of an antisense RNA complementary to three overlapping transcripts, Gene 60: 29-37 (1987).

42. T. Damerval, J. Houmard, G. Guglielmi, K. Csiszàr, and N. Tandeau de Marsac, A developmentally regulated *gvpABC* operon is involved in the formation of gas vesicles in the cyanobacterium *Calothrix* 7601. Gene 54:83-92 (1987).

43. N. Tandeau de Marsac, D. Mazel, D. A. Bryant, and J. Houmard, Molecular cloning and nucleotide sequence of a developmentally regulated gene from the cyanobacterium *Calothrix* PCC 7601: a gas vesicle protein gene, Nucl. Acids Res. 13:7223-7236 (1985).

44. R. Oelmüller, P.B., Conley, N., Federspiel, W.R., Briggs, and A.R., Grossman, Changes in accumulation and synthesis of transcripts encoding phycobilisome components during acclimation of *Fremyella diplosiphon* to different light qualities, Plant Physiol. 88:1077-1083 (1988).

45. S. S. Golden, M.S. Nalty, and D.-S.C. Cho, Genetic relationship of two highly studied *Synechococcus* strains designated *Anacystis nidulans*, J. Bacteriol. 171:24-29 (1989).

46. F. Nagy, S.A. Kay, and N-H. Chua, Gene regulation by phytochrome, Trends Genet. 4 : 37-42 (1988).

PIGMENT COMPLEXES IN PHOTOSYNTHETIC PROKARYOTES; STRUCTURE AND FUNCTION

J.Amesz

Department of Biophysics
University of Leiden
Leiden, The Netherlands

INTRODUCTION

The photosynthetic apparatus of prokaryotes, in contrast to that of higher plants, shows an almost bewildering variety of pigmentation and organization. Nevertheless, as in all photosynthetic organisms, two major functions of the pigments can be discerned. The first one, the so-called antenna function, consists of the absorption of light and the transfer of the energy of the excited pigment molecules to the reaction center. The second, equally important function concerns electron transfer in the reaction center; the primary photochemical reaction consists of the transfer of an electron from an excited chlorophyll or bacteriochlorophyll (BChl) molecule to a neighboring acceptor molecule. Subsequent electron transfer reactions serve to stabilize the energy of the radical pair thus formed and make it available for secondary electron transfer and biosynthetic reactions. More than 90 % of the pigments present in photosynthetic organisms have an antenna function only. Originally it was thought that these would be dispersed in the lipid bilayer of the photosynthetic membrane, but nowadays it is firmly established that they are either bound to intrinsic proteins or, as in cyanobacteria and green bacteria, contained in extramembraneous structures, the phycobilisomes and chlorosomes.

Five different divisions of eubacterial phototrophs may be distinguished (Stackebrandt *et al.*, 1988): the cyanobacteria and Prochlorophytes, the purple bacteria, the green filamentous bacteria, the green sulfur bacteria and the heliobacteria. The last four groups are commonly called photosynthetic bacteria. In the following sections a brief overview will be given of the similarities and differences in pigment organization of these five groups.

PURPLE BACTERIA

The primary photosynthetic reactions and pigment systems of purple bacteria have been studied in more detail than those of any other group of photosynthetic bacteria. This is especially true of the so-called α-subdivision (Stackebrandt *et al.*, 1988) which includes well-known species such as *Rhodobacter sphaeroides* and *Rhodopseudomonas viridis*. With the exception of *Rps. viridis* and a few other species they all contain BChl *a*.

The pigment-protein complexes are contained in invaginations of the cytoplasmic membrane. Compared to other groups the organization of the pigment system is relatively simple and based on arrays of antenna and reaction center complexes. These complexes can be solubilized by means of detergents and obtained in purified form. The study of primary electron transport has been strongly stimulated by the use of isolated reaction center complexes, which have been available for more than 20 years now. A major development in this field was the crystallization and subsequent X-ray diffraction analysis of the reaction center of *Rps. viridis* by Michel, Deisenhofer and coworkers (Deisenhofer *et al.*, 1984),

Molecular Biology of Membrane-Bound Complexes in Phototrophic Bacteria
Edited by G. Drews and E. A. Dawes
Plenum Press, New York, 1990

together with the determination of the amino acid sequences of the constituent polypeptides (see review by Parson, 1987). Crystalline reaction centers have now also been obtained from various strains of *R. sphaeroides* and X-ray analyses have been performed on these crystals also (Allen *et al.*, 1987, Reiss-Husson *et al.*, these Proceedings). The reaction center contains three peptides (the so-called L, M and H subunits), four bacteriochlorophylls (two of which form a dimer which constitutes the primary electron donor), two bacteriopheophytins, two quinones and one carotenoid. The complex of *Rps. viridis* also contains a *c*-type cytochrome. Except for the location of the carotenoid, the structure shows an approximately two-fold symmetry. Electron transfer, however, is asymmetrical and proceeds only via the so-called A or L chain, which comprises the primary donor, together with one bacteriochlorophyll, one bacteriopheophytin and one quinone (Q_A). Due to the orientation of the reaction center with respect to the membrane the charge separation creates an electrical potential over the intracytoplasmic membrane. Measurement of the flash-induced electrical signal has been applied to study trapping of the antenna excitations (Trissl *et al.*, these Proceedings) and electron transfer and protonation in the reaction center (Semenov, these Proceedings). Site-directed mutagenesis has become a powerful tool to study structure-function relationships of reaction center processes (Bylina *et al.*; Feher *et al.*, these Proceedings). Perhaps the most striking example is the replacement of histidine M200, which binds one of the BChls of the primary electron donor, by leucine or phenylalanine. This results in a replacement of BChl by bacteriopheophytin and the formation of a heterodimer which still shows photochemical activity (Bylina *et al.*, these Proceedings). A relatively new optical method to study the dynamics and interactions of reaction center components is Fourier transform infrared difference spectroscopy (Mäntele *et al.*, these Proceedings).

The antenna complexes of purple bacteria have been extensively characterized by optical and biochemical methods (see review by Zuber, 1987). All BChl *a* containing purple bacteria studied so far contain a long-wave absorbing complex, B875* or B880, which is closely associated with the reaction center. Most species, however, possess one or more additional complexes absorbing at shorter wavelengths, like B800-850 in *e.g. R. sphaeroides, R. capsulatus* and *Rps. palustris*, B800-820 in *Rps. acidophila* and *Chromatium vinosum*, B830 in *C. purpuratum* (Cogdell *et al.*, these Proceedings) and B806 in *Erythrobacter* sp. (Mimuro *et al.*, these Proceedings). Mutant studies indicate that these 'peripheral' antenna complexes may play an important role in the morphogenesis of the intracytoplasmic membrane (Sturgis *et al.*, these Proceedings). Many of the antenna complexes have been isolated, and some of those have been obtained in crystalline form. The quality of X-ray diffraction patterns obtained recently gives hope that a determination of their structure may not be too far away. In the mean time, optical methods, including circular and linear dichroism and resonance Raman measurements (Robert and Lutz, these Proceedings) are the main sources of information about the orientation of the pigments and their interactions with each other and with the proteins.

A general feature of the complexes is the occurrence of two homologous subunits of about 6 kD, the α and β peptides, in a 1 : 1 ratio (Zuber, 1987). The smallest functional unit is possibly an α-β trimer, containing six (in B875) or nine (in B800-850) BChl molecules. Dissociation of the B875 trimer leads to a blue-shift of the near-infrared absorption maximum. Successful reconstitution of dissociated complexes (Ghosh *et al.*, these Proceedings) and of the isolated polypeptides with BChl *a* (Loach *et al.*, these Proceedings) has been reported. Both the α and the β polypeptides are neccessary to restore the optical properties of the complex. Several studies indicate that the complexes are arranged in the membrane in such a way as to facilitate down-hill energy transfer from B800-850 to B875 and hence to the reaction centers (van Grondelle and Sundström, 1988; Mimuro, these Proceedings), but the amount of order in such an arrangement is still a matter of conjecture (van Mourik *et al.*; Zuber, these Proceedings).

GREEN BACTERIA

The absorption spectra of green sulfur as well as of green filamentous bacteria are dominated by strong absorption bands near 460 and 720-760 nm. These bands belong to BChl *c*, *d* or *e* of the major antenna in these bacteria, the chlorosomes, oblong bodies of several

---

*The numbers refer to the approximate location of the infra-red absorption bands.

hundred Å diameter that are bound to the cytoplasmic membrane. Chlorosomes may account for 10-20 % of the total volume of the cell. Electron microscopic and spectral studies (Amesz, 1987; Olson 1981) indicate a similar structure for chlorosomes of green sulfur and of green filamentous bacteria, and the primary structure of at least one of the chlorosome polypeptides (Redlinger et al., these Proceedings) shows a distinct homology for species of both divisions (Wechsler et al., 1985; Wagner-Huber et al., 1988). In addition to BChl c, chlorosomes contain small amounts of BChl a, which presumably functions as intermediate in energy transfer to the membrane (van Dorssen et al., 1988). Rapid energy transfer has been observed between BChl c molecules and from BChl c to BChl a (Fetisova et al., 1988; Gillbro et al., 1988; Holzwarth et al., these Proceedings).

The in vivo absorption maxima of BChl c, d and e are considerably red-shifted with respect to those in most organic solvents. However, in solvent mixtures such as hexane-methylchloride, considerable red shifts have been observed, which are thought to be due to the formation of oligomers (Smith et al., 1983; Brune et al., 1987; Olson, these Proceedings). These observations, as well as resonance Raman studies (Lutz and van Brakel, 1988) have led to the concept that the organization of BChl c in the chlorosomes is determined by interaction between pigment molecules, rather than by binding to proteins, as in other antenna complexes. Evidence that removal of protein does not basically alter the chlorosome structure supports this idea (Holzwarth et al., these Proceedings).

The cytoplasmic membrane of the green filamentous bacterium *Chloroflexus aurantiacus* shows a clear similarity to that purple bacteria. It contains an antenna BChl a-protein complex B808-866 (Feick and Fuller, 1984) which is structurally related to antenna complexes of purple bacteria. The same applies to the reaction center (Amesz, 1987). However, the structure of the membrane of green sulfur bacteria appears to be fundamentally different. Most of the BChl a is contained in a water soluble BChl a protein which is attached to, rather than imbedded in the membrane (Olson, 1981). About one-fourth of the BChl a, however, forms part of a reaction center-core complex, which contains approximately 20 BCls a and about 15 molecules of BChl c or a closely related pigment. Studies of electron transport (see Amesz, 1987) and of peptide composition (Hurt and Hauska, 1984) suggest a functional and structural relationship to the core of photosystem I of plants.

HELIOBACTERIA

The heliobacteria have been discovered only recently (Gest and Favinger, 1983). They contain the 'new' pigment BChl g as their major pigment. Several, apparently closely related (Stackebrandt et al., 1988) species are known now, most of which were isolated from rice paddies (J. Ormerod, personal communication).

The presently available evidence indicates that the heliobacteria contain a single antenna-reaction center complex. Although there are similarities in the electron transport pathways (see Amesz, 1989), the complex appears to be structurally different from the core complexes of photosystem I and of green sulfur bacteria, as judged from its peptide composition (van Kan et al., these Proceedings; van de Meent et al., 1990; Trost and Blankenship, 1990).

CYANOBACTERIA

The cyanobactera resemble higher plant chloroplasts in their ability to evolve oxygen. However, instead of the chlorophyll a/b light harvesting complexes of the chloroplast thylakoid membrane, they contain phycobilisomes as their major antenna complexes. Phycobilisomes are highly ordered organelles containing phycobiliproteins. They are bound to the thylakoid membrane, which contains the chlorophyll a-protein complexes and the reaction centers of photosystems I and II. The phycobilisomes account for up to 50 % of the total protein content of the cell. They are water soluble and are easily detached from the membrane. Recent reviews are given by Zuber (1987) and Glazer (1985).

The phycobiliproteins are composed of α and β subunits of about 20 kD. The blue pigments allophycocyanin and phycocyanin are always present in the phycobilisome. The core of the phycobilisome is formed by allophycocyanin and consists of an array of α-β trimers mainly. Rods containing phycocyanin extend from this core. Depending on the species, the distal ends of these rods may contain the red pigments phycoerythrocyanin and phycoeryhtrin.

The rods consist of α-β hexamers made up of two trimers each; the various trimeric and hexameric units of core and rods are joined by specific linker proteins. Specific mutations affecting the synthesis of these linker proteins may result in the formation of shortened rods (Bryant *et al.*, these Proceedings). The primary structures of the various biliprotein subunits have been determined for some species, as well as of some of the linker polypeptides (see Zuber, 1987). Three-dimensional structures have been determined by X-ray diffraction analysis for the α-β trimer and hexamer of phycocyanin (Schirmer *et al.*, 1986, 1987).

REFERENCES

Allen, J.P., Feher, G., Yeates, T.G., Komiya, H., and Rees, D.C., 1987, Structure of the reaction center from *Rhodobacter sphaeroides* R-26: The protein subunits, Proc. Natl. Acad. Sci. USA, 84:6162.

Amesz, J., 1987, Primary electron transport and related processes in green photosynthetic bacteria, Photosynthetica, 21:225.

Amesz, J., 1989, Energy transfer and electron transport in *Heliobacterium chlorum*, Photosynthetica, 23, in press.

Brune, D.C., Nozawa, T., and Blankenship, R.E., 1987, Antenna organization in the green photosynthetic bacterium *Chloroflexus aurantiacus*. 1. Oligomeric bacteriochlorophyll *c* as a model for the 740 nm absorbing bacteriochlorophyll *c* in chlorosomes, Biochemistry, 26:8644.

Deisenhofer, J., Epp, O., Miki, K., Huber, R., and Michel, H., 1984, Structure analysis of a membrane protein complex. Electron density map at 3 Å resolution and a model of the chromophores of the photosynthetic reaction center from *Rhodopseudomonas viridis*, J. Mol. Biol., 180:385.

Feick, R.G., and Fuller, R.C., 1984, Topography of the photosynthetic apparatus of *Chloroflexus aurantiacus*, Biochemistry, 23:3693.

Fetisova, Z.G., Freiberg, A.M., and Timpmann, K.E., 1988, Long-range molecular order as an efficient strategy for light-harvesting in photosynthesis, Nature, 334:633.

Gest, H., and Favinger, J.L., 1983, *Heliobacterium chlorum*, an anoxygenic brownish-green photosynthetic bacterium containing a 'new' form of bacteriochlorophyll, Arch. Microbiol., 136:11.

Gillbro, T., Sandström, A., Sundström, V., and Olson, J.M., 1988, Picosecond energy transfer kinetics in chlorosomes and bacteriochlorophyll *a*-proteins of *Chlorobium limicola*, in: "Green Photosynthetic Bacteria", J.M. Olson, J.G. Ormerod, J. Amesz, E. Stackebrandt, and H.G. Trüper, eds., Plenum Press, New York, p. 91.

Glazer, A.N., 1985, Light harvesting by phycobilisomes, Annu. Rev. Biophys. Biophys. Chem., 14:47.

Hurt, E.C., and Hauska, G., 1984, Purification of membrane-bound cytochromes and a photoactive P840 protein complex of the green sulfur bacterium *Chlorobium limicola* f. *thiosulfatophilum*, FEBS Lett., 168:149.

Lutz, M., and Van Brakel, G., 1988, Ground-state molecular interactions of the bacteriochlorophyll *c* in chlorosomes of green bacteria and in model systems: A resonance Raman study, in: "Green Photosynthetic Bacteria", J.M. Olson, J.G. Ormerod, J. Amesz, E. Stackebrandt, and H.G. Trüper, eds., Plenum Press, New York, p. 23.

Olson, J.M., 1989, Chlorophyll organization in green photosynthetic bacteria, Biochim. Biophys. Acta, 549:33.

Parson, W.W., 1987, The bacterial reaction center, in: "Photosynthesis", J. Amesz, ed., Elsevier, Amsterdam, p. 43.

Schirmer, T., Huber, R., Schneider, M., Bode, W., Miller, M., and Hackert, M.L., 1986, Crystal structure analysis and refinement at 2.5 Å of hexameric *c*-phycocyanin from the cyanobacterium *Agmenellum quadruplicatum*. The molecular model and its implications for light-harvesting, J. Mol. Biol., 188:651.

Schirmer, T., Bode, W., and Huber, R., 1987, Refined three-dimensional structures of two cyanobacterial *c*-phycocyanins at 2.1 and 2.5 Å resolution, A common principle of phycobilin-protein interaction, J. Mol. Biol., 196:677.

Smith, K.M., Kehres, L.A., and Fajer, J.., 1983, Aggregation of the bacteriochlorophylls *c, d* and *e*. Models for the antenna chlorophyll of green and brown photosynthetic bacteria,

        J. Am. Chem. Soc., 105:1387.
Stackebrandt, E., Embley, M., and Weckesser, J., 1988, Phylogenetic, evolutionary and
        taxonomic aspects of phototrophic eubacteria, in: "Green Photosynthetic Bacteria", J.M.
        Olson, J.G. Ormerod, J. Amesz, E. Stackebrandt, and H.G. Trüper, eds., Plenum Press,
        New York, p. 201.
Trost, J.T., and Blankenship, R.E., 1990, Isolation of a reaction center particle and a small c-
        type cytochrome from *Heliobacillus mobilis*, in: "Proc. 8th Intern. Congr.
        Photosynthesis, Stockholm, 1989", Kluwer, Dordrecht, in press.
Van de Meent, E.J., Kleinherenbrink, F.A.M., and Amesz, J., 1990, Properties of a solubilized
        and purified antenna-reaction center complex from heliobacteria, in:" Proc. 8th Intern.
        Congr. Photosynthesis, Stockholm, 1989", Kluwer, Dordrecht, in press.
Van Dorssen, R.J. and Amesz, J., 1988, Pigment organization and energy transfer in the green
        photosynthetic bacterium *Chloroflexus aurantiacus*. III. Energy transfer in whole cells,
        Photosynth. Res., 15:177.
Van Grondelle, R., and Sundström, V., 1988, Excitation energy transfer in photosynthesis, in:
        "Photosynthetic light-harvesting systems. Organization and function", H. Scheer, and
        S. Schneider, eds., Walter de Gruyter, Berlin, p. 403.
Wagner-Huber, R., Brunisholz, R., Frank, G., and Zuber, H., 1988, The BChl *c/e*-binding
        polypeptides from chlorosomes of green photosynthetic bacteria, FEBS Lett., 239:8.
Wechsler, L., Suter, F., Fuller, R.C., and Zuber, H., 1985, The complete amino acid sequence
        of the bacteriochlorophyll *c* binding polypeptide from chlorosomes of the green
        photosynthetic bacterium *Chloroflexus aurantiacus*, FEBS Lett., 181:173.
Zuber, H., 1987, Structure and function of light-harvesting pigment-protein complexes, in:
        "Photosynthesis", J. Amesz, ed., Elsevier, Amsterdam, p. 233.

CONSIDERATIONS ON THE STRUCTURAL PRINCIPLES OF THE ANTENNA COMPLEXES OF

PHOTOTROPHIC BACTERIA

H. Zuber

Institut für Molekularbiologie und Biophysik
Eidg. Technische Hochschule, ETH-Hönggerberg
CH - 8093 Zürich, Switzerland

INTRODUCTION

The various photosynthetic organisms show a multiplicity of different antenna structures, antenna complexes, types of pigments and polypeptides. This is mainly due to the diversity and selectivity of environmental conditions, particularly light, and metabolic restrictions which dominated the evolution of antenna structures. Metabolic and structural constraints also determined and coordinated evolution of pigments and antenna polypeptides. Because the antennae are part of or are linked to highly differentiated membranes, the structure and diversity of the antenna polypeptides adjusted structurally to the various types of lipid membranes. At the present time we can distinguish on this basis between 1.) the antennae of anoxygenic and oxygenic organisms, living in the spectral range above and below 700 nm, respectively, and 2.) antenna systems within and outside the membrane (cytoplasmic-, intracytoplasmic-, thylakoid-membrane). Thus the light-harvesting antennae, which evolved parallel with the reaction center from possibly related precursor molecules, represent not only one but many types of antenna molecules. We may assume, however, that in the multiplicity of antenna structures certain principles exist, valid in all antenna systems for the physical processes of light energy absorption and the energy transfer to the reaction center. In this respect it is generally assumed that all photons absorbed by pigment molecules form mobile electronic excited singlet (S_1) states (excitons) [1-4]. The excitation energy migrates within the antenna system and to the reaction center between the pigment molecules by a very fast random walk within $\sim 10^{-12}$ S (mean intermolecular distances $\simeq$ 20 Å). Intermolecular excitation migration obeys Förster theory for jumptimes $\gtrsim 10^{-12}$ S. Heterogeneous energy transfer between pigment molecules with different absorption maxima ($\simeq$ 10 nm difference in absorption maxima) will improve and focus energy migration by 20 - 30%. More rapid excitation migration of non-Förster type, may also take place between pigments $\gtrsim$ 15 Å apart. In this case, excitation energy is spread over a pair (or oligomers) of pigments (exciton-coupled pigments). Furthermore, these exciton coupled pigment pairs (oligomers) can be considered as "new" molecules with longer transition dipoles or π-electronic systems [5]. The critical distance between these pigment pairs (oligomers) will possibly be considerably enlarged and the corresponding jumptimes greatly (3-6 fold) decreased. It may be postulated that the

Molecular Biology of Membrane-Bound Complexes in Phototrophic Bacteria
Edited by G. Drews and E. A. Dawes
Plenum Press, New York, 1990

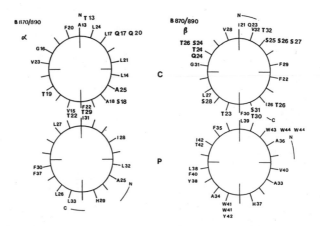

Fig. 1a Crossection of the α-helix of the α- and β-polypeptides of B870/890 at the cytoplasmic (C) and periplasmic (P) side. Distribution of polar residues (S, T, Q, N) at C and position of His (H) at P; Compilation of polar residues of the polypeptides from *Rs. rubrum*, *Rp. viridis*, *Rb. sphaeroides*, *Rb. capsulatus*, *Rc. gelatinosus* and *Rp. acidophila*.

hexamer

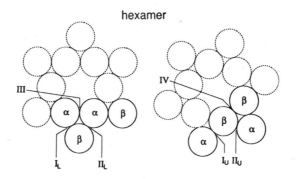

trimer

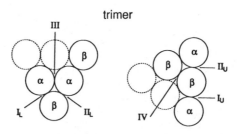

Fig. 1b Possible helix-helix interaction sites between the tilted helices of the α- and β-polypeptides in the cyclic hexamer or trimer I_L, II_L, III or I_U, II_U, IV: Interaction sites at the periplasmic or cytoplasmic side, respectively.

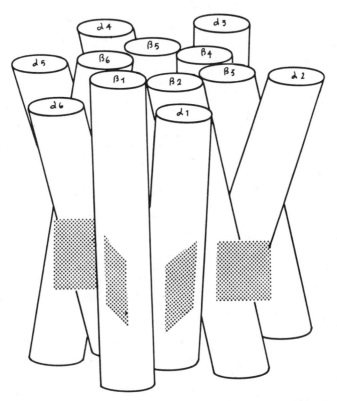

Fig. 1c   Model of the cyclic hexamer $\alpha_6\beta_6$ with tilted α-helices (e. g. B800-850 complex).

heterogeneous energy transfer and the non-Förster mechanisms are the basis for the efficient, directed and focused energy transfer to the reaction center.

HIERARCHICAL ORGANIZATION OF ANTENNA POLYPEPTIDES IN THE ANTENNA SYSTEMS IS THE BASIS FOR THE FORMATION OF SPECIFIC PIGMENT CLUSTER AND THE HETEROGENEOUS, DIRECTED ENERGY TRANSFER SYSTEM

These physical, i.e. functional principles of directed energy transfer and coupled excitons should correspond to certain structural principles. A structural prerequisite for heterogeneous energy transfer (energy traps) and the formation of exciton coupled pigments is the strict organization of the pigment molecules [6-8]. This specific organization may comprise: 1.) the formation of pigment clusters (domains) spatially separated to optimize random walk, 2.) well ordered pigment clusters having different absorption maxima (domains), and 3.) formation of small pigment clusters (micro-domains, energy traps) within the larger clusters (domains). Structural data available today indicate the existence of functionally important pigment clusters [6-12]. In the heterogeneous energy transfer system of the various photosynthetic organisms we can differentiate the following types: 1.) The intramembrane antenna system (pigment cluster close to the reaction center) is extended towards the blue spectral range by extramembrane antennae (pigment cluster: phycobilisome, chlorosome). 2. In the intramembrane antenna system core- and peripheral antenna complexes (pigment cluster) exist. 3.) Pigments (BChl, bilins) bound specifically to α- or ß-polypeptides (heterodimers) represent functionally different α- or ß-types (small pigment cluster, micro-domain: α-ß BChl heterodimer). Fundamental to the formation of specific, i.e. functionally active, pigment cluster are the polypeptides. The structural and functional role of the polypeptides with respect to the hierarchical organization of the whole antenna can be described as follows:

1.) All pigment molecules are bound specifically to polypeptides and form defined antenna complexes.

2.) Polypeptides determine the position, distance, orientation and environment of the pigment molecules.

3.) Polypeptides have specific association characteristics in forming oligomeric and multimeric units. This is the basis of the three-dimensional structure of the antenna and of the specific arrangement of the pigments.

4.) The antenna polypeptides associate in a hierarchic order:

   a.) Formation of micro-domains (small pigment clusters) of 2-4 polypeptides (pigments).

   b.) Formation of antenna complexes with functional clusters of pigments (domains) through association of the micro-domains.

   c.) Formation of larger, intermediate macro-domains and/or of the entire antenna through association of antenna complexes.

This hierarchical organization of polypeptides and pigment clusters which is the basis for the heterogeneous energy transfer, should be present in all antenna systems. The size and number of the antenna polypeptides, however, and therefore the complexity of the antenna systems and of the hierarchical organization of the domain structures increases from photosynthetic bacteria (prokaryotes) to higher plants (eukaryotes).

POLYPEPTIDE STRUCTURES AND FUNCTIONAL BCHL CLUSTER (DOMAINS) OF THE ANTENNA SYSTEM OF RS. RUBRUM, RB. SPHAEROIDES, RB. CAPSULATUS AND RP. VIRIDIS (CLASSICAL RHODOSPRILLACEAE)

General Features

Relatively simple antenna systems with the most clearly recognizable structural principles determining heterogeneous energy transfer are found in some purple bacteria (Rhodospirillaceae, Rs. rubrum, Rb. sphaeroides, Rb. capsulatus, Rp. viridis. Their heterogeneous energy transfer system is based on 1.) defined core (B870/890, B 1015) and peripheral (B800-850, B800-820) antenna complexes (domains), and 2.) individual complex-specific α- and ß- antenna polypeptides in the ratio 1:1. These complex-specific α- and ß-polypeptides have been characterized structurally by primary structure analysis. They show typical structural elements which are of structural and functional importance. The following features have been found [6-8]:

1.) A conserved, central His residue within a cluster of hydrophobic amino acid residues is probably the binding site for BChl, both in the α- and ß-polypeptides. In the ß-polypeptides an additional His residue (BChl binding site) exists.

2.) The α- and ß-polypeptides of the individual core or peripheral antenna complexes and the α- and ß-polypeptides between the core and peripheral antenna complexes differ by a typical conserved amino acid cluster. These α- and ß-polypeptide- or complex-specific amino acid clusters are most probably related to the specific role of the α- or ß-polypeptides a.) in the aggregation of the antenna polypeptides to form the various domain structures or cluster of BChl in the entire antenna, and b.) with respect to the functional (spectral) properties of the various polypeptides.

3.) In the environment of the BChl molecule (His residue) specific amino acid residues are present which interact structurally (additional BChl interaction sites) or functionally, e.g. aromatic or charged amino acid residues (red shift in absorption, variability of CD signal) with the BChl molecule [13,14].

These specific structural-functional features of the antenna polypeptides are ultimately the basis of the three-dimensional structure of the domains, antenna complexes and of the entire antenna. The most promising method to determine the three-dimensional structures of, for instance, defined (isolated) antenna complexes is X-ray structure analysis [15-18]. As experience of the last few years has demonstrated, it is extremely difficult to isolate defined antenna complexes from the "lake" of antenna complexes surrounding the reaction center. Different types of complexes are set free from the entire antenna depending on the conditions of solubilization. It is therefore also difficult to make inferences as to the types of antenna complexes or domains existing in the entire antenna. Thus it seems to be reasonable to estimate possible and probable structures and arrangements of the antenna polypeptides in the domains, antenna complexes and in the entire antenna on the basis of the primary structures of these "classical" antenna polypeptides with their most simple α : ß = 1 : 1 stoichiometry [7,8].

Fig. 2a 'Lake' of cyclic hexamers (six hexamers surrounding a central hexamer) optimally packed by specific helix-helix interactions between the hexamers. Circles: α-helices of the α- and β-polypeptide, Bars: BChl molecules.

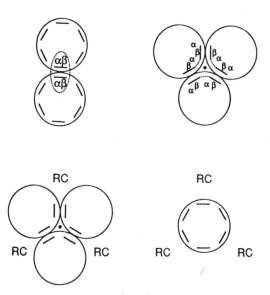

Fig. 2b Upper row: two or three hexamers connected by the basic structural and functional unit $\alpha_2\beta_2$. Lower row: three reaction centres are coupled to the three-hexamer unit or to a hexamer

Possible Arrangement of the Antenna Polypeptides in the Membrane to Form Basic Structural and Functional Domains Important for Heterogeneous and Directed Energy Transfer

Starting point for these considerations is the specific three-domain structure of the antenna polypeptides with a central hydrophobic domain and polar N- and C-terminal domains. It represents the minimum folding unit with the central hydrophobic domain forming a transmembrane α-helix and the N- and C-terminal domains folded in the polar head region or at the cytoplasmic or periplasmic surface, respectively, of the membrane [7,8]. The binding sites for the BChl at the conserved His residues are in the central hydrophobic domain of the α- and ß-polypeptide or are close to the N-terminus of the ß-polypeptide. The first step in the association processes leading finally to the formation of the entire antenna is most probably the formation of the α-ß-heterodimer (helix pair) representing the minimum structural-functional unit (micro-domain) [7,8]. The specific interaction of the α-helices, which is the most important contribution to the formation of the antenna, arises by means of the side chains of amino acid residues distributed in four rows in a helical distribution. In a cross-section of the helix it becomes apparent that at the cytoplasmic side polar residues are most frequent (Fig. 1a). They are concentrated at 3-4 sites forming possible regions for specific interactions of the α-helices. On the periplasmic side the BChl binding sites of the α- and ß-polypeptide chains are located in approximately the same position in the hydrophobic domain, i.e. at the same level within the membrane, which should be favorable for most efficient two-dimensional energy transfer to the special pair of the reaction center, which is located at the same periplasmic level. The specific formation of the α-ß-heterodimer-helix pair (micro-domain) results by means of interaction in region I (Fig. 1b). With this, the other three specific interaction regions II, III, and IV for the specific association of the α-ß heterodimer into larger arrays (domains, antenna complexes, entire antenna: "lake" of antenna complexes) are given (Fig. 1b). The tilt of the helices (30-35°) results in an optimum packing of the helices via the side chains, whereby the arrangement (packing) of the helices on the cytoplasmic side is reciprocal to that on the periplasmic side (Fig. 1b, Fig. 1c). This specific arrangement of the α-helices leads primarily to the formation of cyclical polypeptide aggregates, e.g. cyclical hexamers, and thus to a cyclical energy transfer system in these domains (Fig. 1c). Further, by means of the same type of interactions, including also the BChl molecules, optimally-packed polypeptide aggregates form between the cyclical aggregates, which are the basis for the formation of the "lake" of antenna complexes of the entire antenna and for energy transfer between the cyclical aggregates (domains) (Fig. 2a).

Fundamental to the size, structure and arrangement of the cyclical aggregates (domains) is the structure and organization of the heterogeneous and directed energy transfer system to the reaction center. In this respect the evolution of the antenna system probably occurred from inside (from a precursor of the reaction center) to the diverse antenna complexes outside having absorption maxima in the order from long wave-length absorbing (inside) to short wave-length absorbing (outside) complexes, respectively. During these evolutionary processes functional and therefore structural basic units developed which are the basis for heterogeneous and directed energy transfer.

The smallest cyclical aggregates of α-ß-heterodimers are cyclical trimers $(\alpha_3\beta_3)$ and cyclical hexamers $(\alpha_6\beta_6)$ (Fig. 1b) of which the hexamer, in contrast to the trimer, does fulfil all criteria as an optimally packed basic unit (domain) focusing energy migration in the lake

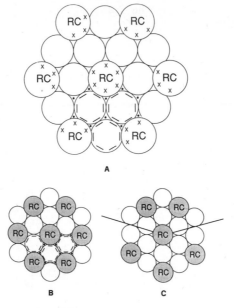

A

B   C

Fig. 3a Possible structure of the B870/B890 core complex. Cyclical arrangement of three three-hexamer units surrounding a central reaction centre. The three-hexamer units are also connected with 6/2 reaction centres (photoreceptor complex I). Bars: α-β-heterodimer, (X): possible intereraction site at the special pair ('entrance') with the α-β-heterodimer of the hexamer.

Fig. 3b Cyclical arrangement of six hexamers connected with a central and 6/2 peripheral reaction centres (photoreceptor complex II).

Fig. 3c Cyclical arrangement of three-hexamer units and hexamer (mixed type) connected with the reaction centres (photoreceptor III).

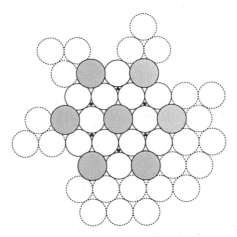

Fig. 3d Possible arrangement of the peripheral antenna complexes surrounding the core-complex and the reaction centre. Circles: hexamers of the core complex. Circles shadowed: reaction centres.

of the entire antenna to the reaction center (Fig. 2a). It may correspond to or may be part of the domains of the peripheral B800-850 (B800-820) or of the B880 core complex, respectively. In the hexamer, as determined by the tilt in the orientation to the membrane plane, the helices of the α-polypeptide lie on the cytoplasmic side and those of the ß-polypeptide on the periplasmic side, arranged together in a cycle (Fig. 1c, Fig. 2a). Six pairs of α-ß-BChl of a central hexamer and of six peripheral hexamers, surrounding the central hexamer, are arranged at the periplasmic side (Fig. 2a). It is via these possibly exciton-coupled α-ß-BChl pairs that energy transfer takes place within the hexamers. This is obviously the special energy transfer system between strongly exciton-coupled BChl pairs [5]. The contact regions between the cyclical hexamers are optimally packed structural $\alpha_2\beta_2$ units with four helices from each of two hexamers (Fig. 2a, Fig. 2b). Through these repeating functional basic units with two exciton coupled α-ß-BChl pairs (micro-domains), cyclical energy transfer is converted to energy transfer between the hexamers, i.e. via a coupled ß-ß-BChl pair. The cyclical hexamer $\alpha_6\beta_6$ (domain) and the dimer $\alpha_2\beta_2$ (micro-domain) are the basis for the formation of larger antenna units (macro-domains) or of the entire antenna. The next larger macro-domain could be two (two-hexamer unit) or three hexamers (three-hexamer unit) connected by the $\alpha_2\beta_2$ functional units (Fig. 2b). In these macro-domains the cyclic energy transfer between the exciton coupled α-ß-BChl pairs in the hexamers is transformed via one or three micro-domains $\alpha_2\beta_2$, respectively. These micro-domains $\alpha_2\beta_2$ with two exciton coupled α-ß-BChl pairs are the basis for the formation of the BChl cluster of directed energy transfer to the reaction center. In the entire antenna three reaction centers may be functionally connected with one hexamer or with the three-hexamer units (Fig. 2b). In these cases an exciton coupled α-ß-BChl pair is possibly coupled functionally with the special pair. It may be assumed that this structural and functional principle of associated hexamers or of three-hexamer units and the BChl cluster exists both in the core complexes and the peripheral antenna complexes of these classical Rhodospirillaceae, particularly since the primary structures of their antenna polypeptides are related.

Possible Structures of the Core Complexes with BChla or BChlb

For the various core complexes of Rhodospirillaceae variable ratios of BChla : reaction center between 21 : 1 and 41 : 1 have been found [19]. There are three possible reasons for this variablility:

1.) Approximately 36 BChl, which are bound to the central His residue of the hydrophobic domain of the α- and ß-polypeptides (36 polypeptides), surround the reaction center. The variability of the number of BChl per reaction center results from the variability of losses of BChl during isolation of the core complexes.

2.) 24 polypeptide (12 α- and 12 ß-polypeptides) with 24 BChl bound to the central His, and hypothetically 12 BChl bound to the His residue of the ß-polypeptide, surround the reaction center. The variability of the number of BChl results from the variable number of BChl bound to the ß-polypeptides or losses of these BChl during isolation of the core complex.

3.) The variability of the BChl molecules per reaction center is caused by the variability of the ratio of hexamers to the reaction center in the various core complexes.

These three possible reasons for the variable ratios of BChla : reaction center in the various core complexes can be explained on the basis of the two- or three-hexamer units.

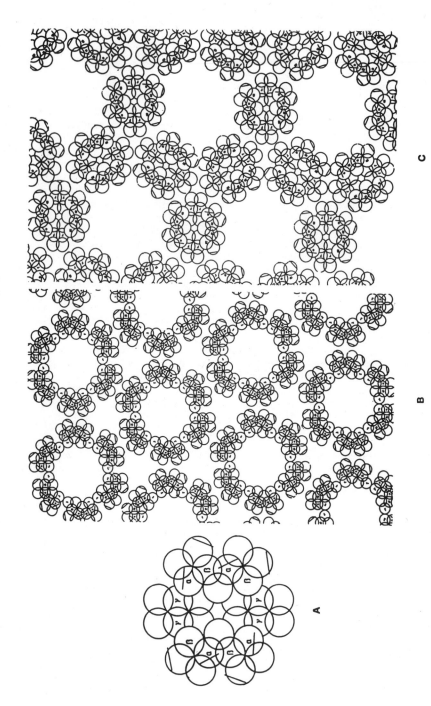

Fig. 4   Possible structure of the B1015 core complex, (A): Substitutions of two heterodimers in the hexamer by 2 x 2 γ-polypeptides. Six $\alpha_6\beta_6\gamma_6$ units of the core complex B1015 surrounding the reaction centre associate primarily via (B) the β-BChl pairs or (C) the γ-polypeptides (formation of γ_6 complexes).

1.) When we apply the basic principle of the three-hexamer unit to the core complex B880 with 36 BChl surrounding the reaction center, the result is the cyclical arrangement of six three-hexamer units of which each hexamer is part of two other three-hexamer units and interacts with two reaction centers (Fig. 3a). Six hexamers with 36 BChl (BChl cluster: 6 x 12/2 BChl) per reaction center surround the reaction center. The six three-hexamer units with their functional contact to three reaction centers each, i.e. totally 12 hexamers, form a directed energy transfer system in which excitation energy is trapped by four (1 + 6/2) reaction centers (photoreceptor complex I).

2.) Applying the principle of the two-hexamer units to the core complex B880, six hexamers again surround the reaction center (Fig. 3b). Each hexamer interacts with three hexamers and three reaction centers. This corresponds to two hexamers or 24 polypeptides or 24 central BChl (6 x 12/3) and hypothetically up to 12 additional BChl of the ß-polypeptides per reaction center (photoreceptor complex II).

3.) A mixed type of core complex with the variable ratio of hexamers to the reaction center corresponding to 24 to 36 BChl is caused by the variable number of two- and three-hexamer units (photoreceptor complex III) (Fig. 3c). The BChl molecules are bound only to the central His residues in the hydrophobic domain.

The efficiency of the directed energy transfer is related to the cooperative action of the core antenna and the reaction center. Their functional relationship determines the size of the photoreceptor complex. Efficient directed energy transfer and favorable size of the photoreceptor complex therefore is also a question of the state of the reaction centers (open and closed reaction centers). With respect to the relative rate of energy transfer within the antennae and to the reaction center, and of energy transduction in the reaction center (saturation of reaction centers with excitation energy), there should be an optimum number of BChl of the antenna relative to the special pair. On the other hand, there are structural (steric) conditions for optimum energy transfer to the reaction center. This is determined firstly by the two-fold symmetry of the special pair of the L- and M-subunits of the reaction center, as compared to the six-fold symmetry of the surrounding core complex.

The structure model of photoreceptor complex I (Fig. 3a) fits best to these structural conditions and also to the biochemical [19-26] spectroscopic [27-34] data, investigations with the electron microscope [35-38], X-ray structure analysis [39], and theoretical considerations on the core-reaction center complex. There are six possible orientations of the special pair within the core complex and relative to the six α-ß-BChl pairs of the hexamers in the environment (Fig. 3a). Each special pair has two "entrances" for the excitation energy transferred from two α-ß-BChl pairs. The central reaction center has two entrances per one orientation, and the six peripheral reaction centers have only one entrance per orientation and per 1/2 reaction center (the second half of the reaction center is part of the next photoreceptor complex. With the statistical orientation of the special pair, the photoreceptor complex has, with a maximum of 7 different orientations (1 + 6), a total of 8 entrances per 4 reaction centers (6 + 2, 2 entrances per reaction center). These 8 entrances correspond to 8 α-ß-pairs (16 BChl) which may represent the long-wave absorbing BChl 895 close to the special pair (final energy traps). 1/3 of the peripheral hexamers (6 x 2 x αß-BChl = 24 BChl) are part of the next photoreceptor complex. The hexamers in the photoreceptor complex contain a total of 144 BChl (12 x 12), of which 104 BChl (144 - 16 - 24) form the directed energy transfer system to the reaction center.

The structural and functional organization of the BChlb containing B1015 core complexes (Rp. viridis) with 12 α-, 12 β and 12 γ-polypeptides (24 BChl) per reaction center, shows some differences to the B880 core complex. The structural basis of the domain structure and the general principle of directed energy transfer, however, seem to be the same [7,8]. The cyclic structure of the B1015 core complex, visible in the electron microscope, seems to relate to the presence of γ-polypeptides. On the basis of the core complex with three-hexamer units, it may be postulated that through the substitution of two α-β-heterodimers with 2 x 2 γ-polypeptides per hexamer ($\alpha_4\beta_4\gamma_4$) the formation of hexamers ($\alpha_6\beta_6$) is hindered (Fig. 4a). The functionally important α-β-heterodimers are, however, conserved. The specific arrangement of the six $\alpha_2\beta_2$ micro-domains around the reaction center can be accomplished by new, specific interaction sites between the $\alpha_2\beta_2$-units and the γ-polypeptides. This would, however, require that in the formation of the polypeptide aggregate, primarily a.) α-β-heterodimers associate via the $\alpha_2\beta_2$ micro-domains and then bind 2 γ-polypeptides each (Fig. 4b), or b.) alternatively, six γ-polypeptides associate (γ_6), followed by rearrangement of the α-β-heterodimers (Fig. 4c). On the basis of the core complex with two-hexamer units, it is difficult to arrange additional six γ-polypeptides per hexamer without disturbing the interaction sites between the BChl of the antenna and the special pair.

Possible Structures of the Peripheral Antenna Complexes B800-850 and B800-820 of the Entire Antenna

The peripheral antenna system with the B800-850 (B800-820) complex should also be similarly organized, relative to the reaction center. It should contain the same structural and functional basic hexamers (antenna complexes) and three-hexamer or two-hexamer units as the core complex. Two different arrangements of the peripheral antenna complexes relative to the core complex are possible: a.) The peripheral and the core antenna are separated, and the peripheral antenna complexes surround the core complexes. b.) A mixed type of core and peripheral antenna complexes surrounds the reaction center. In the instance of the separated peripheral and core complexes a variable number of hexamers or three-hexamer units surrounds the core complex depending on the light conditions (e.g. 12 -36 hexamers, ratio of hexamers of the core to peripheral hexamers: 1:1 - 1:3) (Fig. 3d). In this arrangement also mixed types of three-hexamer units containing hexamers of the B800-850 and B800-820 or B800-850 and B880 may be formed. They may contain mixed types of micro-domains $\alpha_2\beta_2$ of BChl 850 or BChl 870, respectively, for heterogeneous energy transfer within or between the peripheral and core antenna system. The entire antenna is made up of a particular number of photoreceptor units of the core complexes (probably up to 8 photoreceptor units with > 1000 BChl and approximately 30 reaction centers, depending on the state of development of the photosynthetic membrane and light intensity). From a structural and functional point of view it can be expected that the individual photoreceptor complexes of the core complexes of variable size (minimum complex: 4 reaction centers, 12 hexamers) are connected by the "lake" of the peripheral antenna complexes.

The bulk of experimental data (biochemical studies, spectroscopic data, investigation with the electron microscope) is congruous with these structure models of the core and peripheral antennae. Interestingly, a similar principle of antenna structure (α-β-heterodimers, micro-domains $\alpha_2\beta_2$ and hexamers $\alpha_6\beta_6$) appears to exist in the phycobilisomes of cyanobacteria (Mastigocladus laminosus), demonstrating the general character of these structural and functional principles of antenna systems [7,8].

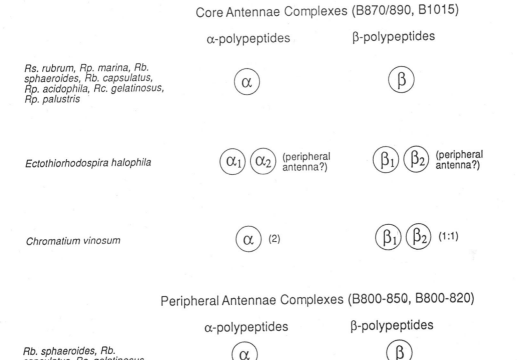

Core Antennae Complexes (B870/890, B1015)

α-polypeptides   β-polypeptides

Rs. rubrum, Rp. marina, Rb. sphaeroides, Rb. capsulatus, Rp. acidophila, Rc. gelatinosus, Rp. palustris

α   β

Ectothiorhodospira halophila

α_1 α_2 (peripheral antenna?) β_1 β_2 (peripheral antenna?)

Chromatium vinosum

α (2) β_1 β_2 (1:1)

Peripheral Antennae Complexes (B800-850, B800-820)

α-polypeptides   β-polypeptides

Rb. sphaeroides, Rb. capsulatus, Rc. gelatinosus

α   β

Rp. acidophila Ac10050

α   β

Rp. acidophila Ac7050

α_{850} α_{820} β_{850} β_{820}

Rp. acidophila Ac7750

α_{850} α_{820} β_{850} $\beta_{1\,820}$ $\beta_{2\,820}$

Rp. palustris 2.6.1

$\alpha_{1\,850}$ $\alpha_{2\,850}$ $\alpha_{3\,850}$ $\alpha_{4\,850}$ $\beta_{1\,850}$ $\beta_{2\,850}$ $\beta_{3\,850}$ $\beta_{4\,850}$

Chromatium vinosum

$\alpha_{1\,850}$ $\alpha_{2\,850}$ α_{820} β_{850} $\beta_{1\,820}$ $\beta_{2\,820}$ $\beta_{3\,820}$

Fig. 6   Multiplicity and variability of the α- and β-antenna polypeptides of the core- and peripheral antenna complexes of purple bacteria. Circle: α- or β-polypeptide, Number: absorption maximum of the corresponding antenna complex.

```
1   MWRIWQLFDPRQALVGLATFLFVLALLIHFILLSTERFNWLEGASTKPVQTS
2   MWKVWLLFDPRRTLVALFTLFVLALLIHFILLSTDRFNWMQGAPTAPAQTS
3   ATEYRTASWKLWLILDPRRVLTALFVYLTVIALLIHFGLLSTDRLNWWEFQRGLPKAA
4   MSKFYKIWMIFDPRRVFVAQGVFLFLLAVMIHLILLSTPSYNWLEISAAKYNRVAVAE
5   MSKFFKIWLVFDPRRVFVAQGVFLFLLAVLIHLILLSTPAFNWLTVATAHGYVAAAQ
6   MYKLIWLLFDPRRALVALSAFLFVLALLIIHFIALSTDRFNWLEGKPAVKAA
7   MYKIWLLFDPRRTLVALSAFLFVLGLLIIHFISLSTDRFNWLEGKPAVRA
8   MYKIWLLFDPRRTLVALSAFLFVLGLLIIHFISLSTDRFNWLEGKPAVRA
9   MWRIWKLYDPRRVLIGIFSWLAVLAVIHFILLSTDRFNWVGGAAN....
10  MWRMWKILDYRRTVVLAHVGMAVLALLIHFILLSTGSFNWLEGNPYG....
11  MHKIWQIEFDPRRTLVALFGELEVLGLLIHFILLSSPAFNWLSG...
12  MQPRSPVRTNIVIFTILGFVVALLIHFIVLSSPEIYNWLSNAEGG
13  MTNGKIWLVVKPTVGVPLFLSAAFIASVVIHAAVLTTTTLPAYYQGSAAVAAE
14  MNNAKIWTVVKPSTGIPLILGAVAVAALIVHAGLLTNTTWFANYWNGNPMATVVVAVAPAAQ
15  MNQGKIWTVVNPSVGLPLLLGSVTIAILVEHAAVLSHTTWFPAYWQGGLKKAA
16  MNQGKIWTVVPPAFGLPLMLGAVAITALLVHAAVLTHTTWYXAAFLQGGVKKAA
17  MNQGKIWTVVNPAVGLPLLLGSVTIAILVHLAILSHTTWFPAYWQGGVKKAA
18  MNQGKIWTVVNPAVGLPLLLGSVAITALLVHLAVLTHTTWFPAFTQGGLKKAA
19  MNQARIWTVVNPAIGIPALLGSVTIAILVHGAILSHTTWFPAYWQGGVKKAA
20  MNQARIWLVVKPSVGLPLLLGVVLLIALLVHGAILTNTSWYPTYFEGNW
21  MNQARIWTVVKPTVGLPLLLGSVTIAILVHFAVLSHTTWFSKYWNGPA
22  MNQGRIWTVVKPTVGLPLLLGSVAIMVFLVHFAVLTHTTWVAKFMNGKA
23  MNQGRIWTVVNPGVGLPLLLGSVTIVAILVHYAVLSNTTWFPKYWNGATVAAPAAA....
24  MNQGRIWTVVKPTVGLPLLLGSVTIAILVHFAVLSNTTWFPKYWNGKA

25  AIEFMGYKPLENDYFPWLVVNPATWLIPTLIAVALTAILHVVAFDLEGQGWHAPAAEAVEAAPAAQ
26  MNIEFMGYKPLEQDHRFWMVVNPATWLMPILIAVALVLVHFYAFSLPGQGFSAAPAEAAPAAAAPAQ...
27  SNVAKPKNPEDDWKIWLVVNPATWLMPIFYALVVPAIAVHAVVFLV...
```

1   Rhodospirillum rubrum B870-α
2   Rhodopseudomonas marina B880-α
3   Rhodopseudomonas viridis 1015-α
4   Rhodobacter sphaeroides B870-α
5   Rhodobacter capsulatus B870-α
6   Rp. acidophila Ac7050 B880-α
7   Rp. acidophila Ac7750 B880-α
8   Rp. acidophila Ac10050 B880-α
9   Ectothiorhodospira halophila B890$_1$-α
10 Ectothiorhodospira halophila B890$_2$-α
11 Chromatium vinosum B890-α
12 Chloroflexus aurantiacus J-10fl B806-866-α
13 Rhodobacter sphaeroides B800-850-α
14 Rhodobacter capsulatus B800-850-α
15 Rp. acidophila Ac7050 B800-850-α
16 Rp. acidophila Ac7050 B800-820-α
17 Rp. acidophila Ac7750 B800-850-α
18 Rp. acidophila Ac7750 B800-820-α
19 Rp. acidophila Ac10050 B800-850-α
20 Ectothiorhodospira halophila B800-850-α
21 Rp. palustris 2.6.1 B800-850-α$_1$
22 Rp. palustris 2.6.1 B800-850-α$_2$
23 Rp. palustris 2.6.1 B800-850-α$_3$
24 Rp. palustris 2.6.1 B800-850-α$_4$
25 Chromatium vinosum B800-850-α$_1$
26 Chromatium vinosum B800-820-α
27 Chromatium vinosum B800-850-α$_2$

Fig. 5a Primary structures of the α-polypeptides of the core- (B870/B890, B1015) and peripheral (B800-850, B800-820) antenna complexes of purple bacteria. Boxes: conserved complex specific structure elements (cluster of amino acid residues).

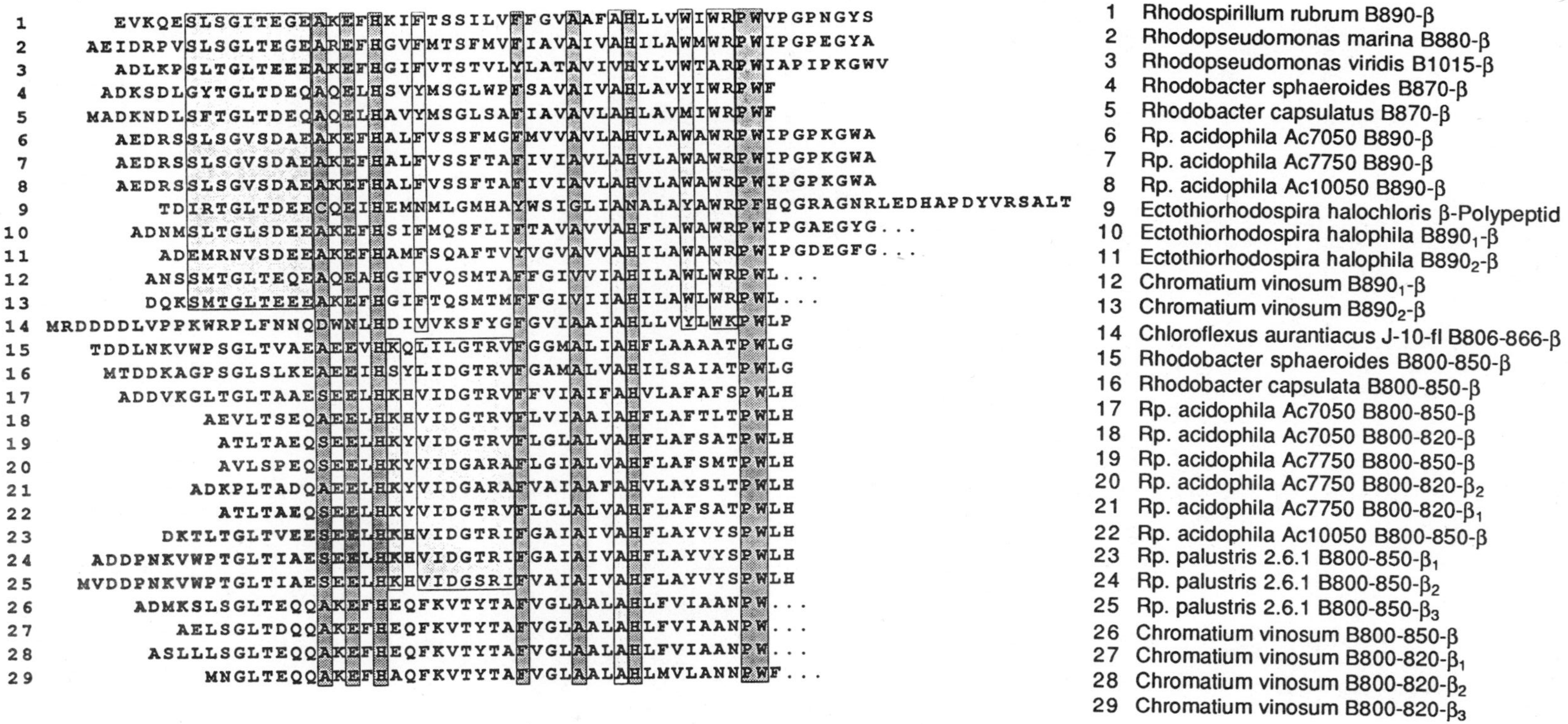

Fig. 5b Primary structures of the β-polypeptides of the core- (B870/B890, B1015) and peripheral (B800-850, B800-820) antenna complexes of purple bacteria. Boxes: conserved complex specific structure elements (cluster of amino acid residues).

1 Rhodospirillum rubrum B890-β
2 Rhodopseudomonas marina B880-β
3 Rhodopseudomonas viridis B1015-β
4 Rhodobacter sphaeroides B870-β
5 Rhodobacter capsulatus B870-β
6 Rp. acidophila Ac7050 B890-β
7 Rp. acidophila Ac7750 B890-β
8 Rp. acidophila Ac10050 B890-β
9 Ectothiorhodospira halochloris β-Polypeptid
10 Ectothiorhodospira halophila B890$_1$-β
11 Ectothiorhodospira halophila B890$_2$-β
12 Chromatium vinosum B890$_1$-β
13 Chromatium vinosum B890$_2$-β
14 Chloroflexus aurantiacus J-10-fl B806-866-β
15 Rhodobacter sphaeroides B800-850-β
16 Rhodobacter capsulata B800-850-β
17 Rp. acidophila Ac7050 B800-850-β
18 Rp. acidophila Ac7050 B800-820-β
19 Rp. acidophila Ac7750 B800-850-β
20 Rp. acidophila Ac7750 B800-820-β$_2$
21 Rp. acidophila Ac7750 B800-820-β$_1$
22 Rp. acidophila Ac10050 B800-850-β
23 Rp. palustris 2.6.1 B800-850-β$_1$
24 Rp. palustris 2.6.1 B800-850-β$_2$
25 Rp. palustris 2.6.1 B800-850-β$_3$
26 Chromatium vinosum B800-850-β
27 Chromatium vinosum B800-820-β$_1$
28 Chromatium vinosum B800-820-β$_2$
29 Chromatium vinosum B800-820-β$_3$

175

STRUCTURAL VARIABILITY AND MULTIPLICITY OF THE ANTENNA POLYPEPTIDES OF PURPLE BACTERIA WITH CONSERVED BASIC STRUCTURAL-FUNCTIONAL ELEMENTS (α-ß-HETERODIMERS, $\alpha_2\beta_2$-MICRO-DOMAINS AND HEXAMER-DOMAINS)

In recent years we have extended the primary structure analysis of the antenna polypeptides of photosynthetic bacteria to include the antenna polypeptides of Rp. marina [40], Rc. gelatinosus [41], Rp. acidophila [42,43], Rp. palustris [44], E. halophila [46], E. halochloris [45,46] and C. vinosum [46] (Fig. 5a, Fig. 5b). These analyses revealed more details on a.) structurally and functionally important amino acid residues, and b.) in particular the variability and multipilicity of the primary structures of antenna polypeptides present in the same antenna system. This structural multiplicity should be of functional relevance. For instance, specific structures, e.g. aromatic amino acid residues of α- and ß-polypeptides, are related to specific spectral properties (shift of absorption maxima) of the B800-850 and B800-820 complexes of Rp. acidophila [14] and C. vinosum [47]. In the instance of Rp. acidophila, in addition to the α- and ß-polypeptides of the B800-850 complex, specific α- and ß-polypeptides of the B800-820 complex were found (Fig. 6). The structural differences, i.e. substitution of a Trp residue for Leu, Phe, Thr and Pro may be related to the spectral shift from 850 nm to 820 nm. A multiplicity of different types of α- or ß-polypeptides was found, e.g. in Rp. palustris [44] and C. vinosum [46]. It is reasonable to assume that all of the structural variants of the α- or ß-polypeptides (Fig. 6) are related to functional (energy transfer) or structural (arrangement of the polypeptides) differences between the antenna complexes (domains). In spite of the multiplicity of the various antenna polypeptides, these polypeptides represent in all cases structurally either α- or ß-polypeptide types. This indicates that they are still involved in the basic α-ß-heterodimer structure with all consequences for the formation of specific higher aggregates (domains), similar to the antenna system of the classical Rhodospirillaceae. It is, however, a question whether the individual α- or ß-polypeptide types form specific antenna complexes (influencing heterogeneous energy transfer between specific domain structures), or form hybrid antenna complexes containing several types of α- or ß-polypeptides (α-ß-heterodimers, influencing heterogeneous energy transfer between micro-domains). The multiplicity (heterogeneity) of the antenna polypeptide pattern of C. vinosum or Rp. palustris, which is based on the heterogeneity of genes (gene families) [48], corresponds, for similar functional reasons, to the multiplicity of antenna polypeptides or genes found in higher plants, e.g. in the peripheral LHC II antenna complex.

STRUCTURAL AND FUNCTIONAL RELATIONSHIP BETWEEN THE ANTENNA POLYPEPTIDES (COMPLEXES) OF PURPLE BACTERIA AND HIGHER PLANTS (?)

The photosynthetic apparatus in the thylakoid membrane of higher plants and algae with PS I and II is, for regulatory reasons, highly differentiated and significantly more complex than that of bacteria. This is based on the one hand on the lateral heterogeneity of organization and function (appressed grana region with PS II, and non-appressed stroma regions with PS I, variable in size depending upon light conditions). On the other hand, a larger number of antenna complexes and antenna polypeptides, also larger in size, exists. Despite this complexity, the PS I and PS II antenna system is made up of functionally important domain structures (core and peripheral antenna complexes) for heterogeneous, directed energy transfer to the reaction center. The core and peripheral antenna complexes form 3-11 membrane spanning α-helices which bind the Chl molecules specifically, and are the basis for the specific association processes needed to build up the entire antenna and the heterogeneous energy transfer system. We may ask here, if the specific structure of

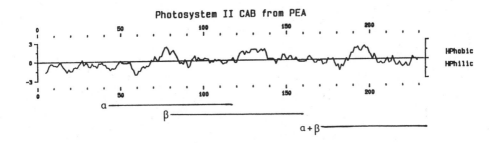

Photosystem II CAB from PEA

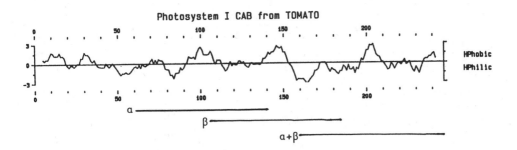

Photosystem I CAB from TOMATO

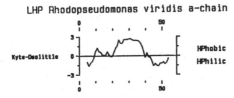

LHP Rhodopseudomonas viridis a-chain

Fig. 7   Regions of sequence homology between the LHC II (PSII pea) and LHC I (PSI tomato) antenna polypeptide (Cab protein) and the α- and β-antenna polypeptides from purple bacteria. Upper part: hydropathy plot of the Cab protein. Lower part (bars): regions of sequence homology with mainly α, or β or α+β polypeptides.

these large antenna polypeptides is structurally, functionally and perhaps phylogenetically related to the much smaller antenna polypeptides of photosynthetic bacteria. Do the same functional, and therefore structural, general principles exist both in bacterial and plant antenna? In that case the primary structures of the antenna polypeptides of both antenna systems should contain similar structural and functional information, i.e. the antenna polypeptides should be sequence homologous in the functional-structural important regions. We compared the amino acid sequences of the small bacterial antenna polypeptides with those of the large antenna polypeptides of plants, searching with a computer program for maximum homology between a large number of small bacterial α- and β-polypeptides and one large antenna polypeptide of plants. With this method three sequence homologous regions (17-30% homology) were found (Fig. 7), for instance with the peripheral antenna complex LHC II [49-51] (or LHC I [52,53]). They correspond to the N-, middle and C-terminal regions of LHC II and involve the transmembrane helices and the N- and C-terminal parts of each region (three-domain structure) [51,54]. These regions also cover possible BChl binding sites of His, Gln, Asn residues inside and outside the membrane. Sequence homology between His, Asn or Gln residues of LHC II and His residues of the bacterial antenna polypeptides in these regions reveals also possible Chla (b) binding sites in LHC II, confirming these hypothetical sites. This data points either to a structural-functional relationship or, in addition, also to a phylogenetic relationship. In both instances the formation of a heterogeneous, directed energy transfer system (Chla/b) to the core complex and the reaction center is of decisive importance. Interestingly, we find similar sequence homologous regions between the antenna polypeptides from bacteria and those of the core complexes (CC I and CC II) and with the reaction center polypeptides of Rp. viridis or Rb. sphaeroides [55]. The sequence homology to the reaction center polypeptides is particularly high in the regions of the functionally important D and E helices of the L and M subunits. Surprisingly, in these helices the His residues of the central BChl binding site of the bacterial antenna polypeptides and the His residues of the Fe binding site of the L- and M-subunits, and not the His residue of the special pair, are sequence homologous. Both His residues are located at the C-terminal region of the transmembrane helix, but the orientation of both helices in the membrane (cytoplasmic or periplasmic side) is inverted. This resembles the inverse orientation of the specific and related cluster of aromatic amino acid residues in the antenna polypeptides around the central His residue and of the special pair [14]. Both findings point possibly to the same evolutionary process in which the phylogenetically related antenna and reaction center polypeptides may have been inversely inserted in the cytoplasmic membrane and structurally and functionally modified.

ACKNOWLEDGEMENTS

I wish to thank Ms. E. Zollinger and Dr. I. Bissig for their help in the preparation of the manuscript. Our work on the structure and function of antenna polypeptides and antenna complexes of photosynthetic bacteria was supported by the Swiss National Foundation, (projects 3.286-0.82, 3.207-0.85) and by the Eigenössische Technische Hochschule, Zürich.

REFERENCES

[1] Knox, R.S, in: Topics in Photosynthesis, Vol. II, Barber, J., ed., Elsevier, Amsterdam, 55, (1977).
[2] Pearlstein, R.M., in: Phtosoynthesis: Energy conversion by plants and bacteria, Govijndjee, ed., Vol. 1, Academic Press, New York, 293 (1982).

[3]   Pearlstein, R.M., in: Antennas and Reaction Centers of Photosynthetic Bacteria, Chemical, Physics Series, 42, Michel-Beyerle, M.E., ed., Springer, Berlin, Heidelberg, 53 (1985).

[4]   Pearlstein, R.M., in: Antennas and Reaction Centers of Photosynthetic Bacteria, Chemical Physics Series, 42, Michel-Beyerle, M.E., ed., Springer, Berlin, Heidelberg, 53 (1985).

[5]   Borisov, A.Y. and Zuber, H., paper in preparation.

[6]   Zuber, H., Photochem. Photobiol., 42:821 (1985).

[7]   Zuber, H., in: The Light Reactions, Barber, J., ed., Elsevier Science Publ., 197 (1987).

[8]   Zuber, H., Brunisholz, R., and Sidler, W., in: Photosynthesis, Amesz, J., ed., Elsevier Science Publ., 233 (1987).

[9]   Cogdell, R.J. and Valentine, J., Yearly Review, Bacterial Photosynthesis, Photchem, Photobiol., 769 (1983).

[10]   Drews, G., in: Membranes of Phototrophic Bacteria, Microbiol. Rev., 59, (1985).

[11]   Glazer, A.N., Annu. Rev. Biochem., 52:152 (1983).

[12]   Glazer, A.N., Annu. Rev. Biophys. Biophys., Chem., 14:47 (1985).

[13]   Brunisholz, R.A, Workshop on structure, function and formation of membrane-bound complexes in Phototrophic Bacteria, Freiburg, FRG, I-P-3, 17 (1987).

[14]   Brunisholz, R.A. and Zuber, H., in: Photosynthetic Light-Harvesting Systems, Scheer, H. and Schneider, S., eds., Walter de Gruyter & Co., Berlin, New York, 103 (1988).

[15]   Welte, W., Wacker, T., Leis, M., Kreutz, W., Shiozawa, J., Gad'on, N., and Drews, G., FEBS Letters, 182:260 (1985).

[16]   Wacker, T., Gad'on, N., Becker, A., Maentele, W., Kreutz, W., Drews, G., and Welte, W., FEBS Letters 197:267 (1986).

[17]   Cogdell, R.J., Woolley, K., MacKenzie, R.C., Lindsay, J.G., Michel, H., Dobler, J., and Zinth, W., in: Antennas and Reaction Centers of Photosynthetic Bacteria, Michel-Beyerle, M.E., ed., Springer, Berlin, Heidelberg, 85 (1985).

[18]   Wacker, T., Gad'on, N., Steck, K., Welte, W., and Drews, G., Biochim. Biophys. Acta, 933:299 (1988).

[19]   Dawkins, D.J., Ferguson, L.A., and Cogdell, R., in: Photosynthetic Light-Harvesting Systems, Scheer, H. and Schneider, S., eds., Walter de Gruyer, Berlin, New York, 115 (1988).

[20]   Hunter, C.N., Pennoyer, J.D., Sturgis, J.N., Farrelly, D., and Niedermann, A., Biochemistry 27:3459 (1988).

[21]   Miller, J.F., Hinchingeri, P.S., Parkes-Loach, P.M., Callahan, J.R., Sprinkle, J.R., Roccobono, J.R., and Loach, P.A., Biochemistry, 26:5055, (1987).

[22]   Ghosh, R., Hauser, H., and Bachofen, R., Biochemistry, 27:1004 (1988).

[23]   Peters, J., Welte, W., and Drews, G., FEBS Letters, 171:267 (1984).

[24]   Peters, J., Takemoto, J., and Drews, G., Biochemistry, 22:5660 (1983).

[25]   Peters, J. and Drews, G., Eur. J. Cell Biol., 29:115 (1983).

[26]   Ludwig, F.R. and Jay, F.A., Eur. J. Biochem, 151:83, (1985).

[27]   Breton, J. and Nabedryk, E., in: Light Reactions, Barber, J., ed., Elsevier Science Publ., 159 (1987).

[28]   Kramer, H.J.M., van Grondelle, R., Hunter, C.N., Westerhuis, W.H.J., and Amesz, J., Biochim. Biophys. Acta, 765:156 (1984).

[29]   Van Grondelle, R., Hunter, C.N., Bakker, J.G.C., and Kramer, H.J.M., Biochim. Biophys. Acta, 723:30 (1983).

[30]   Hunter, C.N., Kramer, H.J.M., and van Grondelle, R., Biochim. Biophys. Acta, 807:44 (1985).

[31]   Van Grondelle, R., Biochim. Biophys. Acta, 811:147 (1985).

[32]   Vos, M., van Grondelle, R., van der Kooij, F.W., van de Poll, D., Amesz, J., and Duysens, L.N.M., Biochim. Biophys. Acta, 850:501 (1986).

[33] Hunter, C.N. and van Grondelle, R., in: Photosynthetic Light-Harvesting Systems, Scheer, H. and Schneider, S., eds., Walter de Gruyter, Berlin, New York, 247 (1988).

[34] Van Dorssen, R.J., Hunter, C.N., van Grondelle, R., Kovenhof, A.H., and Amesz, J., Biochim. Biophys. Acta, 932:179 (1988).

[35] Engelhardt, H., Baumeister, W. and Saxton, W.O., Arch. Microbiol., 135:169 (1983).

[36] Stark, W., Kühlbrandt, K., Wildhaber, I., Wehrli, E., and Mühlethalter, K., The EMBO J., 3/4:777 (1984).

[37] Jay, F., Lambillotte, M., Stark, E., and Mühlethaler, K., The EMBO J., 3/4:773 (1984).

[38] Stark, W., Jay, F., and Mühlethaler, K., Arch. Microbiol., 146:130, (1986).

[39] Cogdell, R.J., Papiz, M.Z., Woolley, K.J., Ferguson, L.A. Wightman, P., and Lindsay, J.G., Abstracts VI Symposium on Photosynthetic Prokaryotes, Noordwijkerhout, The Netherlands, 59 (1988).

[40] Brunisholz, R.A., Bissig, I., Wagner-Huber, R., Frank, G., Suter, F., Niederer, E., and Zuber, H., Z. Naturforschung, 44c:132 (1989).

[41] Brunisholz, R.A., Suter, F., and Zuber, H., Eur. J. Biochem, submitted (1989).

[42] Bissig, I., Brunisholz, R.A., Suter, F., Cogdell, R.J., and Zuber, H., Z. Naturforschung 43c:77 (1988).

[43] Brunisholz, R.A., Bissig, I., Niederer, E., Suter, F., and Zuber, H., Photosynthesis Research, submitted (1989)

[44] Brunisholz, R.A., Evans, M.B., Cogedell, R.J., Frank, G., and Zuber, H., FEBS Letters, submitted (1989).

[45] Wagner-Huber, R., Brunisholz, R.A., Bissig, I., Frank, G., and Zuber, H., FEBS Letters 233:7 (1988).

[46] Bissig, I., Wagner-Huber, R., Brunisholz, R.A., Frank, G., and Zuber, H., Symposium on Molecular Biology of Membrane-Bound Complexes in Phototrophic Bacteria, Freiburg, FRG (1989)

[47] Bissig, I, Thesis # 8945, Eidgenössische Technische Hochschule, Zürich (1989).

[48] Tadros, M.H. and Waterkamp, K., The EMBO J., 8/5:1303 (1989).

[49] Coruzzi, G., Broglie, R., Cashmore, A., and Chua, N.H., J. Biol. Chem., 258:1399 (1983).

[50] Dunsmuir, P., Smith, S.M., and Bedbrook, J., J. Mol. Appl. Genet, 2:285 (1983).

[51] Karlin-Neumann, G.A., Kohorn, B.D., Thornber, J.P., and Tobin, E.M., J. Mol. Appl. Genet, 3:45 (1985).

[52] Stayton, M.M., Brosio, P., and Dunsmuir, P., Plant Mol. Biol., 10:127 (1987).

[53] Hoffman, N.E., Pichersky, E., Malik, V.S., Castresana, C., Ko, K., Darr, S.C., and Cashmore, A.R., Proc. Natl. Acad. Sci. USA, 84:8844 (1987).

[54] Bürgi, R., Suter, F., and Zuber, H., Biochim. Biophys. Acta, 890:346 (1987).

[55] Zuber, H., Suter, F., Sidler, W, and Brunisholz R., paper in preparation.

PIGMENT-PROTEINS OF ANTENNA COMPLEXES FROM PURPLE NON-SULFUR

BACTERIA: LOCALIZATION IN THE MEMBRANE, ALIGNMENTS OF PRIMARY

STRUCTURE AND STRUCTURAL PREDICTIONS

Monier Habib Tadros+*   Gerhart Drews+

+Institute of Biology II   *European Molecular
 Microbiology   Biology Laboratory
 Albert-Ludwigs-University   6900 Heidelberg
 D-7800 Freiburg   Fed. Rep. of Germany
 Fed. Rep. of Germany

Reaction center (RC) and light-harvesting (LH) or antenna
complexes are the major pigment-proteins of the photosynthetic apparatus of non sulfur purple bacteria. They are
localized on the intracytoplasmic membranes. The LH-complexes
serve to gather light-energy and funnel it to the photochemical RC where the excitation energy is transduced into a
charge separation state and a redox potential difference.
Reaction centers are surrounded by a constant number of core
antenna complexes (B870 or B1020). Most species have a second
and variable light-harvesting (LH) complex (B800-850) which
interconnects the core complexes (Drews 1985). The LH
complexes are oligomers of basic subunits which consist of
two different small pigment-binding polypeptides α and β,
having M_r of about 5000 to 7000. These polypeptides are
amphiphilic proteins and span the membrane only once by a
central hydrophobic domain (Fig. 1). The N- and C-terminal
domains consist of polar, charged and hydrophobic amino acid
residues and are exposed on the membrane surface. Two or
three bacteriochlorophyll (Bchl) and one or two carotenoid
molecules are bound non-covalently to the α and β polypeptides.

Recently a multigene family has been found coding for α
and β polypeptides of the B800-850 complex of *Rhodopseudomonas palustris* (Tadros and Waterkamp 1989).
The significance of these results and their importance for
other species is under investigation.

In this article recent work on the localization and orientation of the α and β LH-polypeptides will be summarized and
discussed. The alignment of amino acid sequences of α and β
LH polypeptides will be a basis for discussion of the function of conserved amino acyl residues.

Molecular Biology of Membrane-Bound Complexes in Phototrophic Bacteria   181
Edited by G. Drews and E. A. Dawes
Plenum Press, New York, 1990

Localization of pigment proteins

The axes of the pigment molecules bound to the LH polypeptides have a strict orientation relative to the plane of the membrane (Breton and Vermeglio 1982, Breton and Nabedryk 1987, Wacker et al. 1986, 1988). Since the orientation of pigment molecules depends upon binding to specific amino acids of the α and β polypeptides the localization and siteness of polypeptides in the membrane must be regular. The localization of polypeptides has been studied by proteolytic treatment of inside-out (chromato-phores) and right side-out membrane vesicles (spheroplast), respectively, from phototrophically grown cells and subsequently amino acid sequence analysis of the isolated polypeptides (Tadros et al. 1987). In control experiments it was previously demonstrated that isolated complexes were completely digested by protease K treatment whilst the parts embedded in the lipid layer are protected. The data of recent work on that aim are summarized in Table 1. They show that the N-terminal domains of L and M subunits of RC and of α and β polypeptides of both LH complexes are exposed on or point to the cytoplasmic surface of the membrane and that the C-termini are exposed (α) or point to the periplasmic surface of the membrane (β). The length of peptides, which are "split off" from the N- or C-termini by proteolytic treatment, do not often correspond to the total N- or C-terminal regions predicted to be outside of the hydrophobic central domain on the membrane surface (see Table 1 and Fig. 1). Thus 6 and 9 amino acid residues, respectively, were removed from B870 α in the mutant strains *R. rubrum* G-9 and *Rb. capsulatus* A_{1a}^+, although 10 and 15 residues were outside the hydrophobic central core. Furthermore, in the wild type strains of the same species, no amino acids were proteolytically removed from the N-termini of B870 α (Table 1). It is obvious that these N-termini are protected from proteolytic attack possibly due to formation of secondary structures and protein-protein or protein-lipid interactions. The C-termini of all investigated β polypeptides were not disintegrated, presumably because they are hydrophobic and buried in or attached to the membrane. However, parts of the N-termini of all β polypeptides were "split off". Not only the short LH polypeptides, which span the membrane once, but also the M and L subunits of RC, which span the membrane five time, have the N-termini exposed on the cytoplasmic membrane surface (Table 1). Amino acids were "split off" only when inside-out vesicles were treated with protease K. The same orientation, i.e. N-termini on the cytoplasmic membrane surface, of all pigment-binding polypeptides, supports the idea that the mechanism of insertion of the polypeptides into the membrane and the translocation across the lipid double layer is the same for all of these pigment-binding polypeptides and that the N-terminal regions have a function in membrane targeting and insertion.

Alignments of primary structure and sequence correlations

When prediction algorithms were used (Argos et al. 1982) to detect transmembrane spanning helices and averaged over the aligned polypeptides, a single central hydrophobic core region was indicated for α and β LHI and LHII polypeptides. The predictions are in agreement with the results on

```
RR B870α       M---WRIWQLFDPRQALVGLATFLFVLALLIHFILLSTERFNWLEGASTKPVQTS
RV B870α       ATEYRTASWKLWLILDPRRVLTALFVYLTVIALLIHFGLLSTDRLNWWEFQRGLPKAA
RC B870α       MSKFYKIWLVFDPRRVFVAQGVFLFLLAVLIHLILLSTPAFNWLTVATAKHGYVAAAQ
RS B870α       MSKFYKIWMIFDPRRVFVAQGVFLFLLAVMIHLILLSTPSYNWLEISAAKYNRVAVAE
RM B870α       M---WKVWLLFDPRRTLVALFTFLFVLALLIHFILLSTDR---FNWMQGAPTAPAQTS
RC B800-850α   MNNA-KIWTVVKPSTGIPLILGAVAVAALIVHAGLLTNTT--WFANYWNGNPMATVVAVAPAQ
RS B800-850α   MTN-GKIWLVVKPTVGVPLFLSAAVIASVVIHAAVLTTTT--WL-PAYYQGSAAVAAE
RA B800-850α   MNQ-GKIWTVVNPAIGIPALLGSVTVIAILVHLAILSHTT--WFPAYWQGGVKKAA
RP B800-850αa  MNQ-ARIWTVVKPTVGLPLLLGSVTVIAILVHFAVLSHTT--WFSKYWNGKAAAIESSVNVG
RP B800-850αb  MNQ-GRIWTVVNPGVGLPLLLGSVTVIAILVHYAVLSNTT--WFPKYWNGATVAAPAAAPAPAAPAAKK
RP B800-850αc  MNQ-GRIWTVVSPTVGLPLLLGSVAAIAFAVHFAVLENTS--WVAAFMNGKSVAAAPAPAAPAAPA-KK
RP B800-850αd  MNQ-GRIWTVVKPTVGLPLLLGSVAIMVFLVHFAVLTHTT--WVAKFMNGKAAAIESSIKAV
RP B800-850αe  MNQ-GRIWTVVKPTVGLPLLLGSVTVIAILVHFAVLSNTT--WFSKYWNGK--AAAIR
                         W      P                 H  L         W
                            ααααααααααααααααααααααα
```

```
RR B870β       EVKQE-SLSGITEGEAKEFHKIFTSSILVFFGVAAFAHLLVWIWRPWVPGPNGYS
RV B1020β      ADL-KPSLTGLTEEEAKEFHGIFVTSTVLYLATAVIVHYLVWTARPWIA
RC B870β       ADKNDLSFTGLTDEQAQELHAVYMSGLSAFIAVAVLAHLAVMIWRPWF
RS B870β       ADKSDLGYTGLTDEQAQELHSVYMSGLWPFSAVAIVAHLAVYIWRPWF
RM B880β       AEIDRPVSLSGLTEGEAREFHGVFMTSFMVFIAVAIVAHILAWMWRPWIPGPEGYARV
RC B800-850β   MTDDK--AGPSGLSLKEAEEIHSYLIDGTRVFGAMALVAHILSAIATPWLG
RS B800-850β   TDDLNKVWPSGLTVAEAEEVHKQLILGTRVFGGMALIAHFLAAAATPWLG
RA B800-850β   ATLTAEQSEELHKYVIDGTRVFLGLALVAHFLAFSATPWLH
RA*B800-850β   AEVLTSEQAEELHKHVIDGTRVFLVIAAIAHFLAFTLTPWLH
RA*B800-820β   ADDVK--GLTGLTAAESEELHKHVIDGTRVFFVIAIFAHVLAFAFSPWLH
RP B800-850βa  MADKTL----TGLTVEESEELHKHVIDGTRIFGAIAIVAHFLAYVYSPWLH
RP B800-850βb  MADDPNKVWPTGLTIAESEELHKHVIDGTRIFGAIAIVAHFLAYVYSPWLH
RP B800-850βc  MVDDSKKVWPTGLTIAESEEIHKHVIDGARIFVAIAIVAHFLAYVYSPWLH
RP B800-850βd  MVDDPNKVWPTGLTIAESEELHKHVIDGSRIFVAIAIVAHFLAYVYSPWLH
RP B800-850βe  MADDPNKVWPTGLTIAESEELHKHVIDGTRIFGAIAIVAHFLAYVYSPWLH
                       E H           A   H      PW
                          ααααααααααααααααααααααα
```

Fig.1 Alignments of the amino-acid-sequences of the α (a)
and the β (b) LH polypeptides in the one-letter code,
determined by automatic amino acid sequence analysis or
deduced from the DNA sequence (Rps. palustris). The
central domain of hydrophobic amino acyl residues
predicted to be in a α-helical structure is marked by
ααααααααα. Conserved residues are marked by the respective amino acid in the one-letter code.

Abbreviations: RV, *Rhodopseudomonas viridis*;
RR, *Rhodospirillum rubrum*; RC, *Rhodobacter capsulatus*;
RS, *Rhodobacter sphaeroides*; RM, *Rhodopseudomonas
marina*; RA, *Rhodopseudomonas acidophila*;
RP, *Rhodopseudomonas palustris*. The sequences are taken
from the literature.

Table 1 Localization of pigment-binding polypeptides in membranes of purple non sulfur bacteria.

Species	strain and phenotype	Number of proteolytically removed amino acids from											
		the N-terminus on the cytoplasmic surface						the C-terminus on the periplasmic surface					
		RC		B870 (B1020)		B800-850		RC		B870 (B1020)		B800-850	
		M	L	α	β	α	β	M	L	α	β	α	β
R. rubrum[1]	G-9 (crt.-)	-	16	6	16								
R. rubrum[2]	S1 (crt.+) wild type	-	-	-	5								
Rb. sphaeroides[3,4]	NCIB 8253 wild type	46	15	-	4	-	7			13	-	9	-
Rb. capsulatus[5]	A1a+ (LHII-) (crt.)	-	27	9	22			-	-	-	-	-	-
Rb. capsulatus[6]	37b4 wild type	48	26	-	4	-	9	-	-	16	-	16	-
Rps. viridis[7]	wild type	48	28	-	6			-	-	-	-	-	-

(-) no amino acids were split off; in the cases where no value is given either the respective complex is not present in the strain under study or the sequence has not been determined.
R., Rhodospirillum; Rb., Rhodobacter; Rps., Rhodopseudomonas.

1) Brunisholz et al (1984)
2) Brunisholz et al. (1986)
3) Takemoto et al. (1987)
4) Tadros et al. (1988)
5) Tadros et al. (1986)
6) Tadros et al. (1987)
7) Tadros et al. (1988)

localization of the N- and C-termini of α and β polypeptides
on opposite sides of the membrane as only one helix traverses
the lipid layer. The α-helical structure of the hydrophobic
central domain was not only predicted by the helical

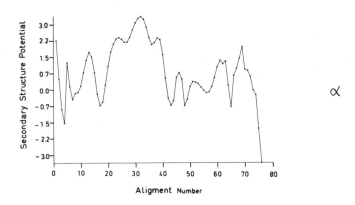

α

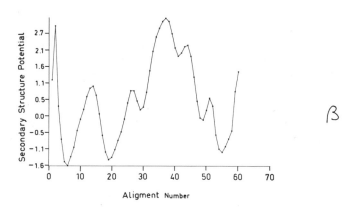

β

Fig. 2 Helical predictions.
      The membrane-buried helical prediction algorithm
      (Argos et al. 1982) were applied to each of the
      aligned sequences and the respective curves averaged.

predictions (Fig. 2) but also confirmed by UV-CD measurements
(Cogdell and Scheer 1985). The transmembrane helices were
found to be tilted to the plane of the membrane (Breton and
Nabedryk 1984).

There are conserved amino acid residues in the hydrophobic central domain. The most prominent conserved amino acid is the histidine residue in the lower part of the hydrophobic region (relative to the N-terminus). There is now good evidence, from resonance Raman spectroscopy (Robert and Lutz 1985), that this histidine binds the central magnesium atom of Bchl.

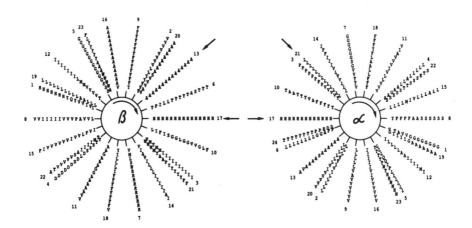

Fig. 3 Helical wheels of the hydrophobic cores of the α and β polypeptides of the LH pigment-protein complex of the species given in legend of Fig. 1.
"Wheels" are idealized projections of helices down their axes with successive Cα-atoms, associated with the side chains, consequentively displaced by 100° along the wheel (Schiffer and Edmundson, 1967, Argos et al. 1982). The side groups would be roughly positioned along spokes emanating from the wheel center.

X-ray diffraction studies of crystallized reaction centers (Yeates et al. 1988) and observations on directed mutations affecting the bacteriochlorophyll binding site (Bylina et al. 1988) have also shown clearly that histidine is the ligand for Mg-Bchl. Ala-X-X-X-His in α and β and Ala-X-X-X-His-X-X-X-Leu in α are also conserved. Alanine can be replaced by residues smaller than valine, the LHI complex of *Rb. capsulatus* is, however, lost from the membrane if alanine is replaced by residues larger than valine (Bylina et al. 1988).

The 2-acetyl and the 9 keto carbonyl groups of Bchl have been shown by Resonance Raman spectroscopy and by X-ray diffraction studies to interact with other amino acid residues (Robert and Lutz, this volume and 1985, Yeates et al. 1988).

A construction of helical wheels of the hydrophobic central domains indicates that the histidine residues of all polypeptides are at the same position of the α helix and therefore oppositely located on the α and β helices (Fig.3).

The positions of the conserved histidine in α and β polypeptide suggest that the two Bchl molecules ligated to histidine are in close distance and in the same orientation to each other. This idea is supported by the observation that the Q_y transitions of both the 850 and of the 870 nm absorption bands are aligned parallel to the membrane plane, while the Q_x transitions are tilted at a 55° away from the membrane plane (Breton and Nabedryk 1987). The strong CD signal of the 850 band indicates an exciton-type interaction of the two Bchl's in the LHII complex. The 870 nm CD was found to be weaker (Bolt et al. 1981).

The β polypeptides have a second conserved histidine in the central domain close to the N-terminal region (Fig. 1 and Tadros et al. 1985), which might be the ligand for the third Bchl molecule in the LHII complexes. In the model of Kramer et al. (1984) the third Bchl was placed at the end of the α helix with the Q_y Q_x transitions laying in the plane of the membrane.

It has been suggested (Schatz 1987) from a literature search that the signal sequence of N-terminal domains, which directs a protein into mitochondria is not a defined sequence but consists of a positively charged amphiphilic motif. A sequence of alternative hydrophobic, hydrophilic and charged amino acids with a positive net charge is present in the N-terminal domains of B870 α polypeptides. The N-terminal region of the β-polypeptide, however, has a negative net charge and contains more hydrophilic amino acids than the polypeptides (Table 2). The number of five negatively charged amino acid residues in the N-terminal domain of the β polypeptides are highly conserved (Table 2). The two oppositely charged amino acid residues (LHI α-Arg, LHI β-Glu), which are localized on the N-termini in opposite positions, close to the beginning of the central α helix, are of importance for the assembly process. This idea has been supported experimentally by replacing the charged amino acids by oppositely charged amino acids. The replacement of pairs of oppositely charged amino acyl residues at the end of the N-termini was less effective (B. Dörge, G. Klug, N. Gad'on, S.N. Cohen, G. Drews unpublished). Two other highly conserved amino acids are a Trp and a Pro residue in the α polypeptide (Fig. 1) and a Leu-Thr (Ile-Thr or Leu-Ser) motif in the β polypeptides (Fig. 1). It is proposed that these conservative motifs of the N-termini are important for targeting and for the assembling of α-β pairs in the functional pigment-protein complexes.

Table 2 Number of charged amino acid residues in the
N-terminal domains of α and β LH-polypeptides.

Species	complex	α (+)	α (-)	β (+)	β (-)	net charge
Rps. viridis	B1020	4	2	2	5	1-
R. rubrum	B870	2	1	2	5	2-
Rb. capsulatus	B870	4	1	1	5	1-
Rb. sphaeroides	B870	4	1	1	5	1-
Rb. capuslatus	B800-850	2	-	2	5	1-
Rb. sphaeroides	B800-850	2	-	1	5	2-
Rps. palustris	LH2 a	2	-	1	5	2-
" "	" b	1	-	1	5	3-
" "	" c	1	-	2	5	2-
" "	" d	2	-	1	4	1-
" "	" e	2	-	1	5	2-

Studies on the dynamics of energy transfer using
picosecond absorption recovery (van Grondelle et al. 1987
Hunter et al. 1989), separation of oligomeric states of
light-harvesting pigment-protein complexes by Li Dodecylsulfate polyacrylamide gel electrophoresis (Hunter et al.
1988) and reconstitution experiments (Loach et al., this
volume) indicate that the α and β polypeptides aggregate in a
defined and hierarchical sequence to form characteristic and
well organized oligomeric and multimeric units. This
assumption predicts that specific protein-protein interactions occur which result in the formation of functional
complexes and multimeric structures such as core complexes
and RC-LHI-LHII arrangements. No amphipathic sideness could
be detected from the wheels as the amino acid residues are
nearly all hydrophobic (Fig.3). Interaction between α and β
is proposed to be besides hydrophobic interaction of the α
helices through intensive binding between charged group and
H-bonds within the N-terminal region. The type of interaction
between RC and the core LH870 complexes has to be determined.
Possibly the H-subunit of RC is involved. Binding of Bchl to
these transmembrane polypeptides is a prerequisite for the
assembly process (Klug et al. 1986, Bylina et al. 1988).
Possibly molecular chaperones, these are proteins, which are
not functional constituents, assist the assembly of oligomeric protein structures (Hemmingsen et al. 1988). Reaction
center-core and LH complexes have been crystallized (Wacker
et al. 1986, 1988). These preparations would be good candidates to study the organization of oligomeric structures.
Unfortunately, the crystals were not good enough for high
resolution X-ray diffraction studies.

Although α and β polypeptides of LHI and LHII complexes
contain highly conserved amino acids they are different from
each other. The similarity between LHII α and β of *Rb.
capsulatus*, and the corresponding α and β polypeptides of
LHII complexes of other species is between 44 and 69 %
(Tab.3). The similarity between α, β of LHII from *Rb.
capsulatus* and α, β LHI of various strains is only 23-42 %
(Table 3).

Table 3 Similarity (in percent) between the amino acid
sequences of α- and β-polypeptides of the LHII-
complexes of *Rb. capsulatus* (Tadros et al. 1983,
1985) and the α- and β-polypeptides of the LHI- and
LHII- complexes of other non sulfur-purple bacteria
and *Chloroflexus*.
LHI = B870, B880 and B1020, respectively;
LHII = B800-850, B806-866, B800-820.
n.d. not determined.

	β	α	References
Rhodospirillum LHI *rubrum*	34,8	21,6	Brunisholz et al., 1984
Rps. viridis LHI	34,0	22,6	Brunisholz et al., 1985
Rb. capsulatus LHI	36,2	26,3	Tadros et al., 1985 Tadros et al., 1984
Rb. sphaeroides LHI	42,6	24,6	Theiler et al., 1984
Rps. marina LHI	34,8	45,1	Brunisholz et al., 1989
Rb. sphaeroides LHII	68,8	45,3	Theiler et al., 1984
Rps. acidophila LHII Stamm 7050	51,1	n.d.	Wagner-Huber et al. 1988
Rps. acidophila LHII Stamm 7050	47,6	n.d.	Wagner-Huber et al. 1988
Rps. acidophila LHII Stamm 7750	56,1	48,1	Bissig et al. 1988
Rps. palustris LHII-A	52,2	44,4	Tadros & Waterkamp 1989
Rps. palustris LHI-B	53,1	46,3	Tadros & Waterkamp 1989
Rps. palustris LHII-C	51,0	44,4	Tadros & Waterkamp 1989
Rps. palustris LHII-D	49,0	46,3	Tadros & Waterkamp 1989
Rps. palustris LHII-E	53,1	48,1	this volume
Chloroflexus *aurantiacus* LHII	26,5	n.d.	Wechsler et al., 1987

Multiple LHII α and β polypeptides have been found in *Rps. palustris* (Tadros and Waterkamp 1989). The N-terminal regions of all identified α-polypeptides are very similar in its amino acid sequence, whereas the C-terminal regions of α polypeptides are more variable (Fig. 1). The net charge of all α polypeptides remains positive. The C-termini of the five *Rps. palustris* LHII β polypeptides and the central domain are highly conserved (Fig. 1). The net charges of the β polypeptides are relatively constant. The *Rps. palustris* LH II amino acid sequences were deduced from DNA sequences (Tadros and Waterkamp 1989). Preliminary protein sequence studies have shown that the LHII α b and c polypeptides are presumably eight amino acids shorter than deduced from the coding DNA (M. Tadros, unpublished).

Literature

Argos, P., Rao, J.K.M. and Hargrave, P.A. (1982) Eur. J. Biochem. 128:565-575

Bissig, I., Brunisholz, R.A., Suter, F., Cogdell, R.J., Zuber, H. (1988) Z. Naturforschg. 43c:77-83.

Bolt, J.D., Sauer, K., Shiozawa, J. and Drews, G. (1981) Biochim. Biophysw. Acta 635:535-541.

Breton, J. and Nabedryk, E. (1984) FEBS Lett. 176:355-359.

Breton, J. and Nabedryk, E. (1987) in: The Light Reactions (Barber, J., ed.) Elsevier Publ. Amsterdam, pp. 159-195.

Breton, J. and Vermeglio, A. (1982) in: Photosynthesis, vol. 1, (Govindjee, ed.), Academic Press, New York, pp. 153-194.

Brunisholz, R.A., Frances, J., Suter, F. and Zuber, H. (1985) Biol. Chem. Hoppe-Seyler 366:87-98.

Brunisholz, R.A., Wiemken, V., Suter, F., Bachofen, R. and Zuber H. (1984) Hoppe Seyler's Z. Physiol. Chem. 365:689-701.

Brunisholz, R.A., Zuber, H., Valentine, J., Lindsay, J.G., Woolley, K.J. and Cogedell, R.J. (1986) Biochim. Biophys. Acta 849:295-303.

Brunisholz, R.A., Bissig, R., Wagner-Huber, R., Frank, G., Suter, F., Niederer, E. and Zuber H. (1989) Z. Naturforschg. 44c:407-414.

Bylina, E.J., Robbes, S.J. and Youvan, D.C. (1988) Israel I. Chem. 28:73-78.

Cogdell, R.J. and Scheer, H. (1985) Photochem. Photobiol. 42:669-678.

Drews, G. (1985) Microbiol. Rev. 49:59-70

Hemmingsen, S.M., Woolford, C., van der Vies, S.M., Tilly, K., Dennis, D.T., Georgopoulos, C.P., Hendrix, R.W. and Ellis, R.J. (1988) Nature (Lond.) 333:330-334.

Hunter, C.N., Pennoyer, J.D., Sturgis, J.N., Farrelly, D. and Niederman, R.A. (1988) Biochemistry 27:3459-3467.

Hunter, C.N., van Grondelle, R., Olsen, J.D. (1989) TIBS 14:72-76.

Klug, G., Liebetanz, R. and Drews, G. (1986) Arch. Microbiol. 146:284-291.

Kramer, H.J.M., van Grondelle, R., Hunter, C.N. Westerhuis, W.H.J. and Amesz, J. (1984) Biochim. Biophys. Acta 765:156-165.

Robert, B. and Lutz, M. (1985) Biochim. Biophys. Acta 807:10-23.

Schatz, G. (1987) Eur. J. Biochem. 165:1-6.

Tadros, M.H., Frank, G. Zuber, H. and Drews G. (1985) FEBS Lett. 190:41-44.

Tadros, M.H., Frank, R. and Drews, G. (1985) FEBS Lett. 183:91-94.

Tadros, M.H., Suter, F., Seydewitz, M.M., Witt, I., Zuber, H. and Drews, G. (1984) Eur. J. Biochem. 138:209-212.

Tadros, M.H., Frank, R. and Drews, G. (1986) FEBS Lett. 196:233-236.

Tadros, M.H., Frank, R., Dörge, B., Gad'on, N., Takemoto, J.Y. and Drews, G. (1987) Biochemistry 26:7680-7687.

Tadros, M.H., Spormann, D. and Drews G. (1988) FEMS Microbiol. Lett. 55:243-248.

Tadros, M.H., Suter, F., Drews, G. and Zuber, H. (1983) Europ. J. Biochem. 129:533-536.

Tadros, M.H. and Waterkamp, K. (1989) EMBO J. 8:1303-1308.

Takemoto, J.Y., Peterson, R.L., Tadros, M.H. and Drews, G. (1987) J. Bacteriol. 169:4731-4736.

Theiler, R., Suter, F., Wiemken, V. and Zuber, H. (1984) Hoppe Seyler's Ztschr. Physiol. Chem. 365:703-719

Van Grondelle, R., Bergström, H., Sundström, V. and Gillbro, T. (1987) Biochim. Biophys. Acta 894:313-326.

Wacker, T., Gad'on, N., Steck, K., Welte, W. and Drews, G. (1988) Biochim. Biophys. Acta 933:299-305.

Wacker, T., Gad'on, N., Becker, A., Mäntele, W., Kreutz, W., Drews, G. and Welte, W. (1986) FEBS Lett. 197:267-273.

Wagner-Huber, R., Brunisholz, R.A., Bissig, I., Frank, G. and Zuber, H. (1988) FEBS Lett. 233:7-11.

Wechsler, T., Brunisholz, R.A., Frank, G., Suter, F. and Zuber H. (1987) FEBS Lett. 210:189-194.

Yeates, T.O., Komiya, H., Chirino, A., Rees, D.C., Allen, J.P. and Feher, G. (1988) Proc. Natl. Acad. Sci. USA 85:7993-7997.

STRUCTURE OF THE ANTENNA COMPLEXES FROM PURPLE BACTERIA AS SEEN FROM RESONANCE RAMAN SPECTROSCOPY

Bruno ROBERT and Marc LUTZ

Service de Biophysique, Département de Biologie

CEN Saclay, 91191 Gif/Yvette Cédex, FRANCE

1. INTRODUCTION

Since 1984, X-ray diffraction studies of crystallized bacterial reaction centers (RC) allowed the structure of these protein-pigment complexes to be solved with an atomic resolution [1-2]. By contrast, none of the several attempts to crystallyse light-harvesting complexes of purple bacteria (*Rhodospirillales*) had resulted in highly diffracting crystals until only very recently. Only this year has there been a report of 3.5 angströms diffracting crystals [3], but no structural model has yet been obtained from the diffraction patterns. On the other hand, a large body of information has been obtained by different biochemical and biophysical methods (see below). Resonance Raman (RR) spectroscopy is still, currently, the only method capable of providing direct information about the interactions between the bacteriochlorophyll (BChl) molecules present in these complexes and their local, proteic environments [4], thus giving detailed information on the structures of the BChl host sites within the protein. In this work, we have tried to fit the RR results and the conclusions drawn from other methods, in order to build models of the membrane-embedded parts of the light-harvesting complexes.

2. STRUCTURE OF THE LIGHT HARVESTING COMPLEXES FROM RHODOSPIRILLALES : A SHORT OVERVIEW

Two types of light-harvesting complexes occur in *Rhodospirillales* : the core antenna (B 890/875, B 1015) closely connected to the RC, and the peripheral antenna (B 800-850). These complexes have been isolated from many bacteria : the basic structure appears to be that of a pair of α/β hydrophobic polypeptides carrying two and three BChl in the core and peripheral antenna, respectively [5]. The transmembrane arrangement of these polypeptides, that was predicted from hydropathy plots, has been supported by both proteolysis and labelling experiments [6,7]. In the hydrophobic regions of these polypeptides, which are ca. 20-

Molecular Biology of Membrane-Bound Complexes in Phototrophic Bacteria
Edited by G. Drews and E. A. Dawes
Plenum Press, New York, 1990

193

25 aminoacids long, a highly conserved His residue appears in the primary sequence, which is the presumed ligand for the central Mg of the BChl molecules [4,5]. Both the core and the peripheral antenna possess a high α-helical content (see e.g. [8]), which agrees with the prediction of two transmembane α-helices per α/β polypeptide, assuming ca. 25 residues per α-helix [5]. From IR dichroism studies, it was further concluded that the tilt of these α-helices with respect to the membrane normal was approximately 30° [9].

The orientations of the different BChls in both the core and the peripheral antenna complexes have been determined by linear dichroism studies (see e.g. [10]). It was shown that the Q_Y transitions of both the 850 nm- and the 880 nm-absorbing molecules are aligned parallel to the membrane plane, while their Q_X transitions are tilted at ca 55° away from the membrane plane. By contrast, both the Q_X and Q_Y transitions of the 800 nm-absorbing molecule are parallel to the membrane plane. The orientations of the α- and β-helical polypeptides as well as of the BChls relative to the membrane plane within light-harvesting complexes are thus known. However, no direct information concerning the orientations of these polypeptides and BChls relative to each other has yet been obtained. This lack of information forbids the construction of any atomic model of reasonable accuracy of the light-harvesting complexes.

3. RESONANCE RAMAN OF THE LIGHT-HARVESTING COMPLEXES FROM *RHODOSPIRILLALES*

As stated in the introduction, RR spectroscopy permits selective observation of the 'active sites' of the antenna, i.e. of those regions in the protein which bind the BChl molecules. Excitation of both BChl a and BChl b near the maxima of their Soret transitions provides RR spectra containing information about the interactions assumed by those carbonyl groups which are conjugated with the dihydrophorbin macrocycle, namely the 2-acetyl and 9-keto carbonyls. These groups are known to be predominantly involved in the intermolecular interactions of chlorophylls, both *in vivo* and *in vitro* [11]. The frequencies of these stretching modes are sensitive to the occurence and strength of intermolecular bonds in which these groups may be engaged : typically, the 2-acetyl groups of BChl a and b vibrate at 1665 and 1670 cm^{-1} respectively, when free from intermolecular bonding. These frequencies may shift down to ca. 1625 cm^{-1} upon formation of hydrogen bonds [11,12]. Similarly, the 9-keto groups of both BChl a and b vibrate near 1700 cm^{-1} when free from intermolecular interactions, and their stretching frequencies may shift down to 1660 cm^{-1} when these groups assume hydrogen bonding. [11,12]. In the following sections, we will discuss the interaction states of the 2-acetyl carbonyl groups only.

3.1. Core antenna

Among the various species of core antenna complexes studied using resonance Raman, eight (*Rhb sphaeroides, capsulatus, Rps palustris, acidophila, Rsp rubrum, Rcs gelatinosus, Chr vinosum, Thiocapsa roseopersicina*) exhibit close similarities in their spectra. It may thus be concluded that, in all these complexes, the proteic host sites are extremely similar, most probably providing the same binding sites to the conjugated carbonyls of the BChl molecules. In these spectra, a strong 1645 cm^{-1} band has been attributed to the stretching modes of both the acetyl groups of the BChls present in the complex. This frequency corresponds to intermolecularly interacting 2 acetyl C=O. In one species only, *Rps viridis*, the bonding of the 2 acetyl of the BChl b molecules appears to be different : in RR spectra of the B 1015 complex, two bands contribute in the 2-acetyl stretching region at 1639 and 1670 cm^{-1} respectively, most likely arising from a strongly H-bonded and from a non-interacting 2 acetyl C=O, respectively [13].

3.2 850 nm-absorbing BChls in peripheral antenna

RR spectroscopy has recently revealed that the host site of the 850 nm-absorbing BChl was strongly species-dependent [4]. However, the 2-acetyl carbonyl groups of these BChls appear most often intermolecularly bound. Indeed, a single stretching frequency is observed at 1633 cm^{-1} in *Rps palustris* whereas two bands contribute at 1633 and 1641 cm^{-1} in *Rhb sphaeroides*. The stretching frequencies of the 2-acetyl groups of most of the 850 nm-absorbing BChl of non sulfur *Rhodospirillales* (*Rhodospirillaceae*) are observed to be the same as those of *Rhb sphaeroides* or of *Rps palustris*.

4. MODELLING THE TRANSMEMBRANE SEGMENT OF THE LIGHT-HARVESTING COMPLEXES FROM RR DATA

We have attempted to construct a molecular model for the transmembrane segment of both the core and the peripheral antenna in which the following topological constraints would be satisfied :

1. the transmembrane part of both the α and β polypeptide is a regular α-helical structure, the axis of which is approximately linear. The length of this α-helix is 23 aminoacids, according to the model of Zuber [5]. We have built these α-helices by using the C$_{\alpha}$ backbone of the D helix of the L and M subunits of the bacterial RC [14], and by substituting the sidechain according to the published primary sequences of antenna complexes.

2. the central Mg of the 880 nm- and/or 850 nm-absorbing BChls are ligated by the conserved His residue within these segments. For clarity, this residue will be numbered as 0 in the following, the neighbour aminoacids being numbered as a function of their position relative to this residue.

3. the orientations of the polypeptides and of the BChls are as those described in section 2.

4. the 2-acetyl C=O of the 880 nm- and of the 850 nm-absorbing molecules must be H-bonded by a suitable, proteic side-chain. This proteic side-chain must be conserved in the complexes exhibiting the same RR characteristics, and interspecific changes observed in RR spectra must correspond to changes in the primary sequences of the α and/or β polypeptides.

In fact, no molecular model can be built in which all these constraints are satisfied. Because of the orientation of the BChl molecules whithin the membrane, and because of the hydropathic profiles of the α and β polypeptides, the vicinity of the 2-acetyl C=O appears not to provide any side-chain that could interact with these groups. Indeed, the α-helical structure of the polypeptide results in the fact that only a few aminoacid side-chains can be brought in proximity to the 2-acetyl carbonyls, whatever the geometry chosen for the BChls, within the range imposed by the dichroism measurements. Moreover, these few aminoacid side-chains are hydrophobic and thus are not capable of H-bonding to the carbonyls.

From this negative result, a first hypothesis can be drawn : The partner molecule interacting with the 2-acetyl C=O is not an aminoacid side-chain, but is an NH group of the peptide backbone. This hypothesis is not very likely for the following two reasons. First, it does not appear much easier to bring peptidic NH groups close to the acetyl groups of the BChl molecules, for the same reasons as for the side-chains. Second, the interspecific variations of the stretching frequencies of the acetyl carbonyls observed in RR spectra would be difficult to explain, if they were supposed all to be bound to these chemically identical groups. The hypothesis of the 2-acetyl groups being bound to peptidic NH groups however cannot be formally excluded until the several possible geometries offered by arranging two or four α/β polypeptides have actually been investigated.

In *Rps viridis* RCs, the central Mg atoms of the two BChl molecules constituting the primary electron donor each bind a His residue located in a transmembrane α-helical segment (L 173 and M 200 His, respectively), and the 2-acetyl groups of these molecules each interact with a side-chain located near these histidines (L 168 His and M 195 Tyr, respectively). However, in the case of the primary donor, the histidine residues that bind the central Mg of the BChls are located near the end of the transmembrane D α-helices. Indeed, these helices terminate at prolines L 171 and M 197, respectively. The breakage of the D α-helices allows the C_α backbone to fold so that the L 168 and M 195 side-chains are in contact with the 2-acetyl groups of the BChls constituting the primary electron donor. According to the secondary structure predictions of the α and β polypeptides of the light-harvesting complexes, the transmembrane, α-helical segment terminates four (respectively six) aminoacids away from the conserved histidine residue which supposedly binds the central Mg of the 850 nm- and/or 880 nm-absorbing molecules. However, these secondary sequence predictions are mainly based on hydropathy considerations. In *Rps viridis* RCs, the D helix of the M subunit terminates at the M 223 Ala residue, in the middle of a LAVA sequence. Such a LAVA sequence would be predicted as being part of the transmembrane α-helical segment by most of the commonly used methods for sequence predictions in membranar proteins. Moreover, in the β polypeptides, the conserved His residue is located in a HLAV sequence in the core antenna of both *Rhb sphaeroides* and *capsulatus*. The hypothesis according to which the transmembranar segment of the α and β polypeptides would stop only two or three residues after the conserved

His may be considered as possible. If this segment begins at the -18 position (relative to the conserved His) and terminates at the +2 or + 3 position, and does not run, as commonly admitted, from the -16 to the +4 position (+6 for the β polypeptides), the aminoacids located in the +4 - +7 range may be considered as possible H-bond donors for the 2-acetyl carbonyl of the 880 nm- and 850 nm-absorbing BChls. In the primary sequence of the peripheral antenna complexes, the threonine residues +5 in the α polypeptide and +7 in the β polypeptide of *Rhb sphaeroides* or *capsulatus* are likely candidates for interacting with the 2-acetyl carbonyls of the 850 nm-absorbing BChls. It is worth noting that these aminoacids have no equivalent in the sequence of the β polypeptide of *Rps palustris*, but that the presence of a +5 Tyr residue could well explain the interspecific variations observed in the RR spectra [15]. On the other hand, in the sequence of the β polypeptides of the core antenna, a +6 Trp appears to be conserved in all the bacterial species but one : *Rps viridis*, in which it is replaced by an Ala residue incapable of forming an H-bond with a C=O group. This Trp could thus be a likely candidate for H-bonding the 2 acetyl of one of the 880 nm-absorbing BChl in all the bacterial species but *Rps viridis*, as observed by RR spectroscopy.

It thus appears, that shifting the position of the transmembrane, α-helical segment of the α- and β-polypeptides by only a few positions may permit the RR spectroscopic data to be interpreted by only considering the interspecific aminoacid mutations. This hypothesis implies that the folding of the C_α backbone of the α and β polypeptides would be somewhat related to that of the D helices of the L and M subunits of the bacterial RCs, at least in the vicinity of the conserved histidine residues [16]. This hypothesis also implies that the BChls of the 850 nm- and 880 nm-absorbing complexes are at approximately the same depth in the membrane with respect to the primary electron donor.

REFERENCES

[1] Michel, H., Epp, O. and Deisenhofer, J. (1986) EMBO J. 5-10 2445-2451
[2] Feher, G., Allen, J.P., Okamura, M.Y. and Rees, D.C. (1989) Nature 339, 111-116
[3] Hawthornwaite, A.M., Papiz, M.Z., Cogdell, R.J., Ferguson, L.A., Wightman, P. and Lindsay, J.G. (1989) these Proceedings
[4] Robert, B. and Lutz, M (1985) Biochim. Biophys. Acta 807, 10-23
[5] Zuber, H. in: Photosynthetic Membranes and Light Harvesting Systems, Encyclopedia of Plant Physiology (Staehlin, L.A. and Arntzen, C.J. eds) Springer Verlag, Berlin, Vol 19, pp 238-251
[6] Wiemken, V., Brunisholz, R., Zuber, H. and Bachofen, R (1983) FEMS Microbiol. Lett. 16, 197-301
[7] Jay, F., Lambillotte, M. and Wyss, F. (1985) Eur. J. Cell Biol. 37, 14-20
[8] Cogdell, R.J. and Scheer, H.(1985) Photochem. Photobiol. 42, 669-678
[9] Breton, J and Nabedryk, E. (1984) FEBS Lett.176, 355-359

[10] Breton, J. and Nabedryk, E. (1987) in: The Light Reactions (Barber, J. ed.) Elsevier, Amsterdam, pp 159-195

[11] Lutz, M. (1984) in: Advances in Infrared and Raman Spectroscopy (Clark, R.J. and Hester, R.E. eds) Wiley Heyden, London, Vol. 11, pp 211-280

[12] Robert, B., Nabedryk, E. and Lutz, M (1989) in: Time-resolved Spectroscopy (Clark, R.J. and Hester, R.E. eds) Wiley Heyden, London, Vol. 18, pp 301-334

[13] Robert, B., Verméglio, A., Steiner, R., Scheer, H. and Lutz, M. (1988) in : Photosynthetic Light-Harvesting Systems (Scheer, H and Schneider, S eds), de Gruyter, New York, pp 355-363

[14] Deisenhofer, J., Epp, O., Miki, K., Huber, R. and Michel, H. (1985) Nature, 318, 618-624

[15] Tadros, M.H. and Waterkamp, K (1989) EMBO J., 1303-1308

[16] Brunisholz, R.A. and Zuber, H. (1988) in: Photosynthetic Light-Harvesting Systems (Scheer, H and Schneider, S eds), de Gruyter, New York, pp 103-114

MULTIPLE ANTENNA COMPLEXES IN VARIOUS
PURPLE PHOTOSYNTHETIC BACTERIA

Iwan Bissig, Regula Verena Wagner-Huber, René A.
Brunisholz, Herbert Zuber

Institut für Molekularbiologie und Biophysik
ETH-Hönggerberg
CH-8093 Zürich, Switzerland

INTRODUCTION

Under anaerobic conditions purple photosynthetic bacteria are capable of
phototrophic growth. They can convert light energy to chemical energy (to a
proton gradient) with the help of membrane bound photochemical reaction
centres and an electron recycling system (e. g. cytochrom bc_1 complex). In bright
day light the absorption capacity of bacteriochlorophyll pigments of reaction
centres is limited to 1 to 10 photons per second, although some orders more
could be processed[1]. Therefore, it is inevitable that organisms increase the
number of absorbing pigments in order to prevent light saturation of reaction
centres even at optimal light conditions. Purple bacteria have solved this
problem by the synthesis of light-harvesting pigment-protein-complexes which
are specialized in absorbing light energy and delivering it to the photochemical
reaction centre where charge separation takes place.

These antenna complexes are composed of small hydrophobic
polypeptides (40 to 70 amino acid residues)[2-16], carotenoids and
bacteriochlorophyll (BChl) molecules which are always of the same type as
those of the special pair in the reaction centre. However the optical
characteristics (absorption, CD, fluorescence etc.) of the antenna BChl are
significantly different. The environment and the associate function of antenna
and reaction centre BChl molecules are not the same.

There are two groups of antenna complexes[17-25] in purple bacteria:

- core complexes (B890, B880, B875 or B870) are situated adjacent to
  the reaction centre delivering excitation energy directly to it. The ratio
  of core complex polypeptides to reaction centre polypeptides is $\approx$ 12
  α/β polypeptides each to 1 L-, M- and H-polypeptides (and Cyt c_{558})
  each in all examined organisms[26]. In the case of the photoreceptor
  unit of *Rhodopseudomonas viridis* and *Ectothiorhodospira*

Molecular Biology of Membrane-Bound Complexes in Phototrophic Bacteria
Edited by G. Drews and E. A. Dawes
Plenum Press, New York, 1990

199

halochloris electromicroscopic investigations and computer assisted image reconstruction[27-30] led to the following model: the antenna complexes are hexagonally arrayed around the reaction centre complex, each of these six regions consists of two smallest units (α/β) of the core complex[20].

- peripheral complexes (B800-850, B800-820 etc.) are supposed to be located around the photoreceptor unit. They probably form a lake wherein the reaction centre/core complexes are swimming. The amount of peripheral complexes being expressed is often dependant on environmental conditions like temperature and light intensity[31].

With the exception of the antenna complexes of some BChl b containing bacteria[6, 32] the smallest unit of all examined antenna complexes consists of an α/β-polypeptide heterodimer. The primary structure analysis of different light-harvesting polypeptides[2-16] revealed some general features: all of them exhibit a typical three domain structure with hydrophilic N- and C-termini and a highly hydrophobic central part which is most probably spanning the membrane. The N-terminal part of the polypeptides was shown to be cytoplasmically located while the C-Terminus extends to the periplasma[4, 33]. Highly conserved His residues in the hydrophobic stretch of the polypeptides are the postulated fifth ligands[2, 3, 8, 11, 17-20] to the Mg^{2+} cation of the bacteriochlorophyll molecule. While α-polypeptides contain one conserved His residue only, two histidines[19] are found at specific positions in β-polypeptides. Characteristic properties of either α- or β-polypeptides[20-21] have been elucidated by comparative primary structure studies. Core and peripheral antenna polypeptides can be distinguished by typical elements on the basis of primary structure characteristics[21].

Up to now most of the bacteria investigated (*Rhodospirillum rubrum*, *Rhodopseudomonas viridis*, *Rhodobacter capsulatus*, *Chloroflexus aurantiacus*, *Rhodopseudomonas marina* and *Rhodobacter sphaeroides*[2-7, 10-12, 14-16]) were characterized by a simple antenna composition (at most two antenna complexes). Each of these antenna complexes is composed of an α/β heterodimer (exception: the antenna complex of *Rhodopseudomonas viridis* consists of a $\alpha/\beta/\gamma$ heterotrimer) with a fixed 1 to 1 stoichiometry. Recently it was shown for the first time that antenna complexes can have a different polypeptide pattern. *Rhodopseudomonas acidophila* strain Ac7750 which synthesizes two peripheral antenna complexes (B800-850 and B800-820) shows in the latter complex two β-polypeptides and one α-polypeptide in a 1/1/2 ratio[13].

These experimental data are limited to members of the family of purple non-sulphur bacteria (Rhodospirillaceae) and to a green non-sulphur bacterium. Hardly any information is available on primary structures of antenna polypeptides from purple sulphur bacteria (Chromatiaceae and Ectothiorhodospiraceae). Therefore we extended our investigations on antenna polypeptides to that from *Chromatium vinosum* and from two halophilic species of the family of Ectothiorhodospiraceae. This article will briefly discuss new data obtained from our recent research.

MATERIAL AND METHODS

Cells were broken at 4°C by ultrasonication or disrupted with a French pressure cell in the presence of DNAse and $MgCl_2$ to prevent interference by

DNA. A low spin centrifugation was carried out to remove cell debris. The membranes in the supernatant were spun down in the ultracentrifuge. For the antenna complex preparations, membranes were solubilized with different detergents (LDAO, Triton X-100, DOC) and applied to sucrose gradient centrifugation (0.3, 0.6, 1.2 M sucrose in buffer) or ion exchange chromatography (Whatman DE-52). The purified antenna complexes were dialyzed and lyophilized[13].

Either lyophylized antenna complexes, membranes or cells were extracted with an organic solvent mixture (methanol/dichloromethane/ammonium acetate = 1/1/0.1 M) and separated on a Sephadex LH-60 size exclusion column [Fig. 1].

Further purification of the respective light-harvesting polypeptide fraction was achieved by reversed phase chromatography on HPLC or FPLC systems [Fig. 2]. The primary structures of the purified polypeptides were determined by N-terminal Edman degradation on a Applied Biosystems A470 protein sequencer[37] and with the help of amino acid analysis, hydrazinolysis and carboxypeptidase digestion[13].

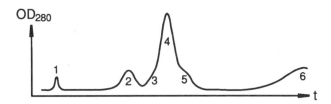

Fig. 1 Elution profile of an organic solvent extract of *Chromatium vinosum* cells grown in low light separated on a Sephadex LH-60 gel permeation column.
1   void volume with RC-polypeptides,
2   polypeptides with N-terminal sequences MNIE and AIEF,
3   green coloured oxidation products of BChl a with N-terminal sequence SNVA,
4   β-polypeptides and two α-polypeptides with N-terminal sequences MHKI and SNVA,
5   β-polypeptides,
6   pigments, lipids, salt.

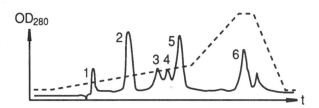

Fig. 2 Purification of pool 5 of Fig. 1 by reversed phase HPLC chromatography on C_8-resign with a gradient from 30% to 50% acetonitrile in water with 1‰ trifluoroacetic acid[30].
1   salt peak,
2   polypeptide with the N-terminal sequence MNGL,
3+4   aggregates,
5   polypeptides with N-terminal sequences AELS and ASLL,
6   aggregates.

RESULTS AND DISCUSSION

Chromatium vinosum

The absorption spectra of cultures of *Chromatium vinosum* are dependent on light conditions [Fig. 3] during growth[31, 35]. Cells grown under high light condition typically show, beside the absorption of the core antenna (B890), peripheral antenna complexes with absorption maxima at 800 and 850 nm whereas low light cells are characterized by dominant absorption bands at 800 and 820 nm.

Three different antenna complexes (B890, B800-850 and B800-820[35]) were isolated by the methods described above. In the case of the B890 antenna core complex, protein fractionation and purification resulted in the isolation of three light-harvesting polypeptides. According to specific primary structure elements

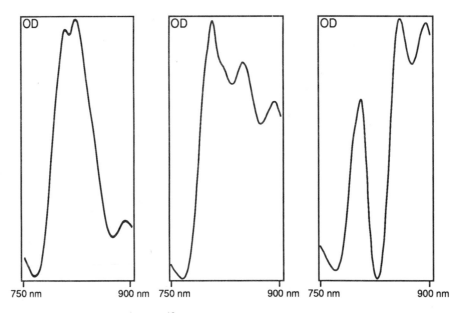

Fig. 3   Absorption spectra[13] between 750 nm and 900 nm of cell suspensions of *Chromatium vinosum* grown at low light (200 lux, left), high light (2000 lux, right) or at 500 lux (middle).

two of these were assigned as β- and one as an α-polypeptide. The peripheral antenna complexes (B800-850 and B800-820) were cross-contaminated with polypeptides of each other to a very high degree. For sequence analysis we omitted the complex isolation procedure and polypeptides were obtained directly from either high light or low light cells. The polypeptides were classified as B800-850- or B800-820 antenna complex polypeptides according to their distribution in high or low light cells. The B800-850 complex was shown to contain at least two α- and one β-polypeptide. One α- and three β-polypeptides minimally were found in the B800-820 antenna complex [Fig. 4, 5].

```
                    MHKIWQIFDPRRTLVALFGFLFVLGLLIHFILLSSPAFNWLSG...
                     |   |   |           |||    |             |
       AIEFMGYKPLENDYFPWLVVNPATWLIPTLIAVALTAILIHVVAFDLEGQGWHAPAAEAVEAAPAAQ
       | | | | | | | |   |   | | | | | | | |   | | | | | |   |   | | |   | |   | |
       MNIEFMGYKPLEQDHRFWMVVNPATWLMPILIAVALVAVLVHFYAFSLPGQGFSAAPAEAAPAAAAPAQ...
         |   | |   | | | | | | | | | | |   |     | | |     |
        SNVAKPKNPEDDWKIWLVVNPATWLMPIFYALLVWAIAVHAVVFLV...
                                                 ↑
```

Fig. 4 Amino acid sequences of the antenna α-polypeptides of *Chromatium*
        vinosum. The strictly conserved His residue is indicated by an arrow.
        B890-α: N-terminal sequence MHKI,
        B800-850-α$_1$: N-terminal sequence AIEF,
        B800-820-α: N-terminal sequence MNIE,
        B800-850-α$_2$: N-terminal sequence SNVA.

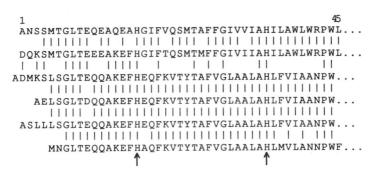

```
       1                                          45
       ANSSMTGLTEQEAQEAHGIFVQSMTAFFGIVVIAHILAWLWRPWL...
        | | | | | | |   | |   | | | | | | | | | | | | | |
       DQKSMTGLTEEEAKEFHGIFTQSMTMFFGIVIIAHILAWLWRPWL...
        | |   | |     | | | |     | | | | |         | |         | |
       ADMKSLSGLTEQQAKEFHEQFKVTYTAFVGLAALAHLFVIAANPW...
        | | | | |   | | | | | | | | | | | | | | | | | | | | | | | | | | | |
       AELSGLTDQQAKEFHEQFKVTYTAFVGLAALAHLFVIAANPW...
        | | | | |   | | | | | | | | | | | | | | | | | | | | | | | | | | | |
       ASLLLSGLTEQQAKEFHEQFKVTYTAFVGLAALAHLFVIAANPW...
        | | | | | | | | | | |   | | | | | | | | | | | | |   |   |   | | |
       MNGLTEQQAKEFHAQFKVTYTAFVGLAALAHLMVLANNPWF...
                         ↑                      ↑
```

Fig. 5 The β-polypeptides of *Chromatium vinosum*: ANSS and DQKS are core
        antenna (B890) polypeptides, ADMK is a B800-850 polypeptide and AELS,
        ASLL and MNGL were found in low light cells (absorption spectrum in Fig. 3,
        left) which produced great amounts of B800-820 antenna complex. The
        strictly conserved His residues are indicated by arrows.

The core antenna α-polypeptides of *Chromatium vinosum* show a high
homology (up to 72%) to the corresponding core antenna polypeptides of purple
non-sulphur bacteria whereas the homology between the α-polypeptides of the
peripheral antennas is significantly lower (≈ 25–35%) [Tab. 1].

Tab. 1 Homologies in percent between the α-polypeptides of the antenna of
        Chromatium vinosum and core and peripheral polypeptides of
        Rhodopseudomonas (Rp.) acidophila.

antenna polypeptide	1	2	3	4	5	6	7
1 *Rp. acidophila* Ac7750 B890-α	100	72.1	16.3	12.2	22.9	19.6	19.6
2 *Chromatium vinosum* MHKI	$\frac{31}{43}$	100	18.6	20.9	25.7	25.6	25.6
3 *Chromatium vinosum* AIEF	$\frac{8}{49}$	$\frac{8}{43}$	100	64.2	43.5	28.3	32.1
4 *Chromatium vinosum* MNIE	$\frac{6}{49}$	$\frac{9}{43}$	$\frac{43}{67}$	100	43.5	30.2	30.2
5 *Chromatium vinosum* SNVA	$\frac{8}{35}$	$\frac{9}{35}$	$\frac{20}{46}$	$\frac{20}{46}$	100	35.1	35.1
6 *Rp. acidophila* Ac7750 B800-850-α	$\frac{10}{51}$	$\frac{11}{43}$	$\frac{15}{53}$	$\frac{16}{53}$	$\frac{13}{37}$	100	77.3
7 *Rp. acidophila* Ac7750 B800-820-α	$\frac{10}{51}$	$\frac{11}{43}$	$\frac{17}{53}$	$\frac{16}{53}$	$\frac{13}{37}$	$\frac{41}{53}$	100

Tab. 2 Homologies in percent between the β-polypeptides of the antenna of *Chromatium vinosum* and core and peripheral polypeptides of *Rhodopseudomonas (Rp.) acidophila* strain Ac7750[8, 12-13].

antenna polypeptide	1	2	3	4	5	6	7
1 *Rp. acidophila* Ac7750 B890-β	100	53.3	43.2	43.9	24.4	31.7	32.6
2 *Chromatium vinosum* ANSS	$\frac{24}{45}$	100	40.9	46.3	32.5	35.0	35.7
3 *Chromatium vinosum* ASLL	$\frac{19}{44}$	$\frac{18}{44}$	100	85.0	35.9	30.8	36.6
4 *Chromatium vinosum* MNGL	$\frac{18}{41}$	$\frac{19}{41}$	$\frac{34}{40}$	100	32.5	30.0	36.6
5 *Rp. acidophila* Ac7750 B800-850-β	$\frac{10}{41}$	$\frac{13}{40}$	$\frac{14}{39}$	$\frac{13}{40}$	100	82.9	65.9
6 *Rp. acidophila* Ac7750 B800-820-β	$\frac{11}{41}$	$\frac{14}{40}$	$\frac{12}{39}$	$\frac{12}{40}$	$\frac{34}{41}$	100	68.3
7 *Rp. acidophila* Ac7750 B800-820-β	$\frac{14}{43}$	$\frac{15}{42}$	$\frac{15}{41}$	$\frac{15}{41}$	$\frac{27}{41}$	$\frac{28}{41}$	100

A slightly different picture is created by the β-polypeptides: the core antenna polypeptides show homologies in the range of ≈ 50% while the homologies between the peripheral antenna polypeptides is 35% and less [Tab. 2].

In *Chromatium vinosum* and purple non-sulphur bacteria, core antenna polypeptides are more homologous than peripheral antenna polypeptides[19]. One could assume that the ancestor of purple sulphur and non-sulphur bacteria contained only a core antenna complex. Peripheral antenna complexes which appeared later in evolution by duplication of the genes coding for core complex polypeptides have developed individually in different branches of the ancestral tree of photosynthetic bacteria.

Ectothiorhodospira halochloris and *Ectothiorhodospira halophila*

The extremely halophilic photosynthetic bacteria *Ectothiorhodospira halochloris* and *Ectothiorhodospira halophila* are members of the new family of Ectothiorhodospiraceae[36] originally belonging to the Chromatiaceae. The purple sulphur bacteria *Ectothiorhodospira* spec. deposit elemental sulphur extracellularly in contrast with the purple sulphur bacteria of the family of Chromatiaceae which deposit sulphur intracellularly. *Ectothiorhodospira halophila* is a BChl a containing organism whereas *Ectothiorhodospira halochloris* (like *Rhodopseudomonas viridis* and *Thiocapsa pfennigii*) contains BChl b which causes a far red shifted absorption maximum around 1020 nm in the NIR. For both halophilic bacteria spectral properties seem to be independant of growth conditions (temperature, light intensity) [Fig. 6].

The absorption maxima at 890 and 1020 nm respectively, presumably the core antenna complexes, dominate the spectra. The absorption bands of the peripheral antenna complexes at 800, 850 nm and 800, 830 nm respectively, are significantly weaker.

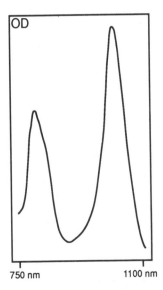 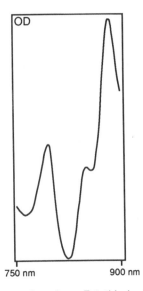

| 750 nm | 1100 nm | 750 nm | 900 nm |

Fig. 6 Absorption spectra of membrane preparations from *Ectothiorhodospira halochloris*[38] (left) from 750 nm to 1100 nm and from *Ectothiorhodospira halophila* (right) between 750 nm and 900 nm.

```
MWRLWKLYDPRRVLIGIFSWLAVLALVIHFILLSTDRFNWVGGAAN...
MWRMWKILDYRRTVVLAHVGMAVLALLIHFILLSTGSFNWLEGNPYG...
MNQARIWLVVKPSVGLPLLLGVVLLIALLVHGAILTNTSWYPTYFEGNW

ADNMSLTGLSDEEAKEFHSIFMQSFLIFTAVAVVAHFLAWAWRPWIPGAEGYG...
ADEMRNVSDEEAKEFHAMFSQAFTVYVGVAVVAHILAWAWRPWIPGDEGFG...
                  ↑                        ↑
```

Fig. 7 Amino acid sequences of five light-harvesting polypeptides from *Ectothiorhodospira halophila*. Upper three are α-polypeptides, lower two β-polypeptides.

Antenna complex isolation was performed only in the case of *Ectothiorhodospira halophila*. B890 complex (violet) and B800-850 complex (brown) were obtained by detergent solubilization (Triton X-100, LDAO) and subsequent sucrose gradient centrifugation[13]. However as the spectrally pure complexes showed an identical polypeptide composition, this indicated a specific destruction of the missing spectral component by the detergent. We therefore had to rely on primary structure comparison to assign the antenna polypeptides of both investigated bacteria to the different antenna complexes.

At least five light-harvesting polypeptides were isolated from *Ectothiorhodospira halophila*, three of them with α and two of them with β characteristics [Fig. 7].

Tab. 3 Comparison of the antenna α-polypeptides of *Ectothiorhodospira halophila* with the core antenna polypeptides of *Rhodospirillum rubrum*, *Rhodopseudomonas acidophila* and *Chloroflexus aurantiacus* and the peripheral antenna polypeptides of *Rhodopseudomonas acidophila* strain Ac7750.

antenna polypeptide	1	2	3	4	5	6	7	8
1 *Rhodospirillum rubrum* B890-α	100	61.2	63.0	55.3	38.6	23.4	21.6	19.6
2 *Rhodopseudomonas acidophila* B890-α	$\frac{30}{49}$	100	56.5	51.1	31.8	17.0	22.4	16.3
3 *Ectothiorhodospira halophila* B890$_1$-α	$\frac{29}{46}$	$\frac{26}{46}$	100	54.3	29.5	19.6	17.4	17.4
4 *Ectothiorhodospira halophila* B890$_2$-α	$\frac{26}{47}$	$\frac{24}{47}$	$\frac{25}{46}$	100	38.6	19.1	17.0	17.0
5 *Chloroflexus aurantiacus* B806-865-α	$\frac{17}{44}$	$\frac{14}{44}$	$\frac{13}{44}$	$\frac{17}{44}$	100	11.4	18.2	15.9
6 *Ectothiorhodospira halophila* B800-850-α	$\frac{11}{47}$	$\frac{8}{47}$	$\frac{9}{46}$	$\frac{9}{47}$	$\frac{5}{44}$	100	55.1	57.1
7 *Rhodopseudomonas acidophila* B800-850-α	$\frac{11}{51}$	$\frac{11}{49}$	$\frac{8}{46}$	$\frac{8}{47}$	$\frac{8}{44}$	$\frac{27}{49}$	100	77.3
8 *Rhodopseudomonas acidophila* B800-820-α	$\frac{10}{51}$	$\frac{8}{49}$	$\frac{8}{46}$	$\frac{8}{47}$	$\frac{7}{44}$	$\frac{28}{49}$	$\frac{41}{53}$	100

Tab. 4 Comparison of the antenna β-polypeptides of *Ectothiorhodospira halophila* with the corresponding core complex polypeptides of *Rhodospirillum rubrum*, *Rhodopseudomonas acidophila* and *Chloroflexus aurantiacus* and the peripheral antenna β-polypeptides of *Rhodopseudomonas acidophila* strain Ac7750.

antenna polypeptide	1	2	3	4	5	6	7	8
1 *Rhodospirillum rubrum* B890-β	100	50.0	52.8	43.1	31.3	31.7	26.8	32.6
2 *Rp. acidophila* Ac7750 B890-β	$\frac{27}{54}$	100	60.4	56.9	31.3	24.4	31.7	32.6
3 *Ectothiorhodospira halophila* B890$_1$-β	$\frac{28}{53}$	$\frac{32}{53}$	100	62.7	29.8	34.1	36.6	30.2
4 *Ectothiorhodospira halophila* B890$_2$-β	$\frac{22}{51}$	$\frac{29}{51}$	$\frac{32}{51}$	100	24.4	31.7	31.7	25.6
5 *Chloroflexus aurantiacus* B806-866-β	$\frac{15}{48}$	$\frac{15}{48}$	$\frac{14}{47}$	$\frac{11}{45}$	100	31.7	34.1	39.5
6 *Rp. acidophila* Ac7750 B800-850-β	$\frac{13}{41}$	$\frac{10}{41}$	$\frac{14}{41}$	$\frac{13}{41}$	$\frac{13}{41}$	100	82.9	65.9
7 *Rp. acidophila* Ac7750 B800-8$\frac{5}{2}$0-β	$\frac{11}{41}$	$\frac{13}{41}$	$\frac{15}{41}$	$\frac{13}{41}$	$\frac{14}{41}$	$\frac{34}{41}$	100	68.3
8 *Rp. acidophila* Ac7750 B800-820-β	$\frac{14}{43}$	$\frac{14}{43}$	$\frac{13}{43}$	$\frac{11}{43}$	$\frac{17}{43}$	$\frac{27}{41}$	$\frac{28}{41}$	100

Four polypeptides (two α and two β) are proposed to be the most probable components of core antenna complexes due to their high degree of homology ($\approx$ 40-60%) to core complex polypeptides of purple non-sulphur bacteria [Tab. 3 and 4]. Interestingly, the homology between the α- and the β-polypeptides is in the same range. The remaining third α-polypeptide found in *E. halophila* appears only in minor amounts, approximately ten times less than the other α-polypeptides. It is significantly homologous (≈ 55%) to the peripheral antenna polypeptides of purple non-sulphur bacteria, but not to any of the core complex polypeptides [Tab. 3].

In the case of *Ectothiorhodospira halochloris* six light-harvesting polypeptides were extracted from whole cells. Based on their primary structure analyses, four of them were assigned unequivocally as two α- and two β-polypeptides [Fig. 8]. The fifth polypeptide caused a lot of difficulties during preparation and was therefore obtained only once and not in a pure state. Nevertheless the first ten amino acids which have been sequenced point to the existence of a third β-polypeptide (not shown). Another polypeptide (29 amino acid residues) is homologous to the B1015-γ polypeptide of *Rhodopseudomonas viridis*[6, 32].

Both α-polypeptides of *Ectothiorhodospira halochloris* can be described as components of a core complex according to their high homology to the corresponding polypeptides of core complexes of purple bacteria[13]. However, one of them (N-terminal sequence MWRI) always exhibits a higher homology (42%–64%) than the second polypeptide (N-terminus: MWKL; 27%–47%). Interestingly, the homology to *Rhodopseudomonas viridis*, another BChl b containing species, is unexpectedly low (45% and 27%, respectively). The homology to α–polypeptides of peripheral antenna complexes is rather poor (9%–26%).

In the case of the β–polypeptides of *Ectothiorhodospira halochloris* the assignment can not be performed with certainty. Although both of them seem to be polypeptides of the core antenna complex[13]. The homologies to the core complex antenna polypeptides is significantly higher (ANDI: 24%–49%, TDIR: 22%–41%) than to the peripheral antenna polypeptides (18%–31% and 20%–33%, respectively) of purple bacteria. The third β–polypeptide can not be characterized further due to the little sequence data available up to now.

In Tab. 5 and 6 sequence homologies between antenna polypeptides of *Ectothiorhodospira halophila* and *halochloris* are presented. The highest homology (61%) is found between the MWRL α–polypeptide of *E. halophila* and MWRI of *E. halochloris*. This homology is even higher than the homology among the α–polypeptides of each organism [Tab. 5]. These polypeptides contain a sequence element (DPR) at the N–terminus which is strongly conserved in all core complex α–polypeptides of purple bacteria. Unfortunately the β–polypeptides cannot be classified in the same way [Tab. 6].

```
MWRIWKVFDPRRILIATAXWLIIIALTIHVILMXTERFNWLEGAPAAEYYS...
MWKLWKFVDFRMTAVGFHLFFALLAFAVHFACISSERFNWLEGAPAAEYYMDENPGIWKRTSYDG

ANDIRPLRDFEDEEAQEFHQAAVQAFFLYVAVAFVAHLPV...
 TDIRTGLTDEECQEIHEMNMLGMHAYWSIGLIANALAYAWRPFHQGRAGNRLEDHAPDYVRSALT
              ↑                           ↑
```

Fig. 8 Primary structures of four light-harvesting polypeptides isolated from *Ectothiorhodospira halochloris*. Upper two are α-polypeptides, lower two β-polypeptides.

Tab. 5 Comparison of the light-harvesting α-polypeptides of *Ectothiorhodospira halophila* and *Ectothiorhodospira halochloris*.

antenna polypeptide	1	2	3	4	5
1 *Ectothiorhodospira halophila* B890$_1$-α (MWRL)	100	54.3	60.9	43.5	19.6
2 *Ectothiorhodospira halophila* B890$_2$-α (MWRM)	$\frac{25}{46}$	100	48.9	46.8	19.1
3 *Ectothiorhodospira halochloris* B890$_1$-α (MWRI)	$\frac{28}{46}$	$\frac{25}{51}$	100	47.1	21.3
4 *Ectothiorhodospira halochloris* B890$_2$-α (MWKL)	$\frac{20}{46}$	$\frac{22}{47}$	$\frac{23}{47}$	100	14.9
5 *Ectothiorhodospira halophila* B800-850-α	$\frac{9}{46}$	$\frac{9}{47}$	$\frac{10}{47}$	$\frac{7}{47}$	100

Tab. 6 Comparison of the light-harvesting β-polypeptides of *Ectothiorhodospira halophila* and *Ectothiorhodospira halochloris*.

antenna polypeptide	1	2	3	4
1 *Ectothiorhodospira halophila* B890$_1$-β (ADNM)	100	62.7	43.6	35.3
2 *Ectothiorhodospira halophila* B890$_2$-β (ADEM)	$\frac{32}{51}$	100	48.6	33.3
3 *Ectothiorhodospira halochloris* B890-β (ANDI)	$\frac{17}{39}$	$\frac{18}{37}$	100	21.6
4 *Ectothiorhodospira halochloris* (TDIR)	$\frac{18}{51}$	$\frac{17}{51}$	$\frac{8}{37}$	100

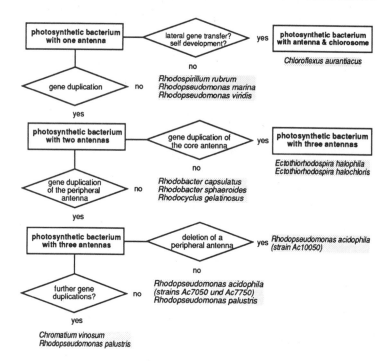

Fig. 9 Possible routes of evolutionary development of antenna complexes of different purple bacteria.

One could assume that there are two core-type antenna complexes in *Ectothiorhodospira* species absorbing at 890nm and 1020nm respectively. The first complex would be adjacent to the reaction center and delivering excitation energy directly to it, whereas the second antenna complex would surround the reaction center/core complex to which it transfers its energy.

How these kind of core-type antenna complexes could have been emerged during evolution is presented in [Fig. 9].

Summarizing the data we have obtained by investigation of previously unstudied organisms we would like to state the following conclusions:

- antennna complexes are not necessarily composed of one α- and one β-polypeptide building up an α/β-heterodimer in a 1:1 ratio as it has been found for the antenna systems investigated so far. However it appears that the total amount of α-polypeptides is equal to that of the β-polypeptides in any given antenna complex.
- core complex polypeptides of purple sulphur bacteria exhibit relatively high homologies to the corresponding polypeptides of purple non sulphur bacteria,
- whereas the peripheral antenna polypeptides can either be related rather closely (Ectothiorhodospiraceae and Rhodospirillaceae species) or less in the case of *Chromatium vinosum*.
- additional antenna complexes could have been evolved by gene duplication of peripheral antennas (*Rhodopseudomonas acidophila* strains Ac7050 and Ac7750, *Chromatium vinosum*) or core antennas (*Ectothiorhodospira* species).

REFERENCES

1   Glazer, A. N., J. Biol. Chem. **264** (1), 1-4 (1989).
2   Brunisholz, R. A., Cuendet, P. A., Theiler, R., and Zuber, H., FEBS Letters **129** (1), 150-154 (1981).
3   Brunisholz, R. A., Suter, F., and Zuber, H., Hoppe-Seyler's Z. Physiol. Chem. **365**, 675-688 (1984).
4   Brunisholz, R. A., Wiemken, V., Suter, F., Bachofen, R., and Zuber, H., Hoppe-Seyler's Z. Physiol. Chem. **365**, 689-701 (1984).
5   Wechsler, T., Brunisholz, R., Suter, F., Fuller, R. C., and Zuber, H., FEBS Letters **191** (1), 34-38 (1985).
6   Brunisholz, R. A., Jay, F., Suter, F., and Zuber, H., Biol. Chem. Hoppe-Seyler **366**, 87-98 (1985).
7   Wechsler, T. D., Brunisholz, R. A., Frank, G., Suter, F., and Zuber, H., FEBS Letters **210**, 189-194 (1987).
8   Bissig, I., Brunisholz, R. A., Suter, F., Cogdell, R. C., and Zuber, H., Z. Naturforsch. **43c**, 77-83 (1988).
9   Wagner-Huber, R., Brunisholz, R. A., Bissig, I., Frank, G., and Zuber, H., FEBS Letters **233** (1), 7-11 (1988).
10   Brunisholz, R. A., Bissig, I., Wagner-Huber, R. V., Frank, G., Suter, F., Niederer, E., and Zuber, H., Z. Naturforsch. **44c** (5/6), 407-414 (1989).
11   Theiler, R., Suter, F., Wiemken, V., and Zuber, H., Hoppe-Seyler's Z. Physiol. Chem. **365**, 703-719 (1984).
12   Brunisholz, R. A., Bissig, I., Niederer, E., Suter, F., and Zuber, H., Progress in Photosynth. Research (Ed. Biggins), Martinus Nijhoff Publ., II.1.13.

13 Bissig, I., 'Die Primärstrukturanalyse der Antennenpolypeptide von *Rhodopseudomonas acidophila, Chromatium vinosum* und *Ectothiorhodospira halophila*', Dissertation ETH Zürich Nr. 8945 (1989).

14 Tadros, M. H., Suter, F., Drews, G., and Zuber, H., Eur. J. Biochem. **129**, 533-536 (1983).

15 Tadros, M. H., Frank, G., Zuber, H., and Drews, G., FEBS Letters **190** (1), 41-44 (1985).

16 Tadros, M. H., Suter, F., Seydewitz, H. H., Witt, I., Zuber, H., and Drews, G., FEBS Letters **138**, 209-212 (1984).

17 Zuber, H., Brunisholz, R., Sidler, W., Photosynthesis (Ed. J. Amesz), Elsevier Science Publishers B. V., (Biomedical Division), 233-271 (1987).

18 Zuber, H., TIBS **11**, 414-419 (1986).

19 Zuber, H., The Light Reactions (Ed. J. Barber), Elsevier Science Publishers B. V. (Biomedical Division), 198-251 (1987).

20 Zuber, H., Photochem. Photobiol. **42** (6), 821-844 (1985).

21 Brunisholz, R. A., and Zuber, H., Photosynthetic Light-Harvesting Systems, Ed. Scheer and Schneider, de Gruyter, 103-114 (1987).

22 Thornber, J. P., Trosper, T. L., and Strouse, C. E., The Photosynthetic Bacteria (Clayton, R. K., and Sistrom, W. R, eds.), Plenum Press, New York, 133-160 (1978).

23 Thornber, J. P., Cogdell, R. J., Pierson, B. K., and Seftor, R. E. B., J. Cell. Biochem. **23**, 159-169 (1983).

24 Cogdell, R. J., and Thornber, J. P., FEBS Letters **122** (1), 1-8 (1980).

25 Cogdell, R. J., and Thornber, J. P., Chlorophyll Organization and Energy Transfer in Photosynthesis, CIBA Foundation Symposium **61**, 61-79 (1979).

26 Dawkins, D.J., Ferguson, L.A., and Cogdell, R., The structure of the "core" of the purple bacterial photosynthetic unit, in Photosynthetic Light-Harvesting Systems, Scheer, H. and Schneider, S., Eds., Walter de Gruyter, Berlin, New York, ll5 ff. (l988).

27 Engelhardt, H., Guckenberger, R., Hegerl, R., and Baumeister, W., Ultramicroscopy **16**, 395-410 (1985).

28 Engelhardt, H., Baumeister, W., and Saxton, W. O., Arch. Microbiol. **135**, 169-175 (1983).

29 Engelhardt, H., Engel, A., and Baumeister, W., Proc. Natl. Acad. Sci. **83**, 8972-8976 (1986).

30 Stark, W., Kühlbrandt, W., Wildhaber, I., Wehrli, E., and Mühletaler, K., EMBO J. **3**, 777-783 (1984).

31 Mechler, B., and Oelze, J., Arch. Microbiol. **118**, 91-114 (1978).

32 Brunisholz, R. A., Steiner, R., Scheer, H., and Zuber, H., FEBS Lett. (submitted).

33 Brunisholz, R. A., Zuber, H., Valentine, J., Lindsay, J. G., Woolley, K. J., and Cogdell, R. J., Biochem. Biophys. Acta **849**, 295-303 (1986).

34 Brunisholz, R. A., and Zuber, H., Experientia **43**, 672 (1987).

35 Thornber, J. P., Biochemistry **9** (13), 2688-2698 (1970).

36 Imhoff, J. F., Int. J. Syst. Bacteriol. **34** (3), 338-339 (1984).

37 Frank, G., in: Methods in Protein Sequence Analysis. Proceedings of the 7th Int. Conf., Ed. Brigitte Wittmann-Liebold, Springer-Verlag, Berlin, 116-121 (1989).

38 Steiner, R., 'In vivo und in vitro Untersuchungen an den Bakteriochlorophyll b haltigen Organismen *Ectothiorhodospira halochloris, Ectothiorhodospira abdelmalekii* und *Rhodopseudomonas viridis*', Inaugural-Dissertation, Ludwig-Maximilian-Universität München (1984).

THE STRUCTURE AND FUNCTION OF SOME UNUSUAL VARIABLE ANTENNA COMPLEXES

[1]RJ Cogdell [1]AM Hawthornthwaite [1]LA Ferguson [1]MB Evans [1]M Li
[1]A Gardiner [1]RC Mackenzie [2]JP Thornber [3]RA Brunisholz [3]H Zuber
[4]R van Grondelle [4]F van Mourik

[1]Botany Department Glasgow University Glasgow G12 8QQ UK
[2]Biology Department UCLA Los Angeles Cal 90024 USA
[3]Molecular Biology & Biophysics ETH Hönggerberg CH-8093
 Zürich Switzerland
[4]Biophysics Department Free University of Amsterdam
 The Netherlands

INTRODUCTION

In purple photosynthetic bacteria the extent of the development of the
photosynthetic apparatus is regulated by environmental conditions, such as
the light-intensity at which the cells are grown.[1,2] In the two most
commonly studied species, Rhodobacter sphaeroides and Rhodobacter capsulatus,
the effect of growing cells at different light-intensities has been well
documented.[1-3] The major response to lowering the light-intensity is the
increased synthesis of the variable B800-850-complexes (the LH2 complexes).
There are however a number of less well-studied species that not only regulate the size of their photosynthetic units in response to changes in the
ambient environment, but also have the ability to alter the type of LH2
complex which is synthesised.[4-6] Chromatium vinosum, Rhodopseudomonas
palustris and Rhodopseudomonas acidophila are examples of species which show
this type of response. The ability to synthesise these extra kinds of
antenna complexes appears to correlate with the capacity of these species to
grow at very low light-intensities. It is interesting that these different
types of LH2 complex have unusual absorption spectra.

In this paper we shall summarise our studies, both biochemical and
biophysical, which we have undertaken to investigate the structure and
function of these additional antenna complexes. We would like to understand
how the bacteria benefit by their presence and expect that a study of these
complexes will increase our understanding of exactly how the antenna
apoproteins control the spectral properties of the antenna complexes. In
addition a new type of antenna complex, B830, obtained from a marine
species, Chromatium purpuratum, will be described.

GROWTH OF CELLS

In each of the species of purple bacteria used in this study the photosynthetic apparatus is very plastic. It is important, therefore, to
provide some details on how to grow the cells so as to maximally enrich for

Molecular Biology of Membrane-Bound Complexes in Phototrophic Bacteria
Edited by G. Drews and E. A. Dawes
Plenum Press, New York, 1990

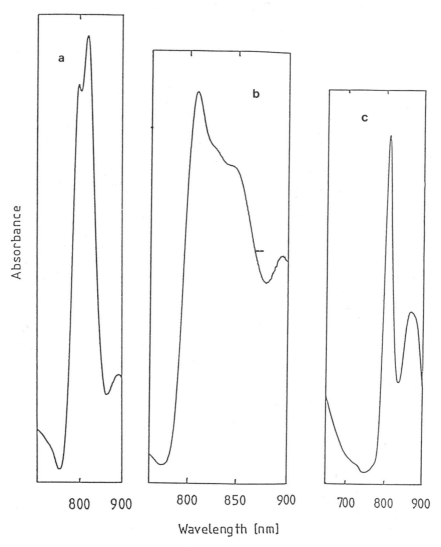

Absorbance

Wavelength [nm]

Fig. 1.   The absorption spectra of whole cells of the bacteria
used here grown under conditions so as to enrich for
the unusual LH2 complexes (a) Rps. acidophila strain
7750 (b) Chr. vinosum strain D and (c) Rps. palustris
strain 2·1·6.

the desired antenna complex. Fig. 1 shows absorption spectra of membranes from cells grown so as to induce the formation of the unusual LH2 antenna types, that is the B800-820-complexes from Rps. acidophila and Chr. vinosum and the 'high-800' B800-850-complex from Rps. palustris.

When Rps. acidophila strain 7750 is grown at temperatures between 22°C and 20°C the B800-820-complex predominates. At 30°C the normal sphaeroides-like B800-820-complex is synthesised. The 'high-800' B800-850-complex from Rps. palustris strain 2·1·6 is induced when the cells are grown at 30°C, but at light-intensities of about 0.2 watts m^{-2}. If the light-intensity is raised to 5-10 watts m^{-2} then again the more well-known sphaeroides-like B800-850-complex is synthesised. With Chr. vinosum strain D there is an interplay between the composition of the growth medium, the temperature at which the cells are grown and the incident light-intensity.[5] The B800-820-complex is preferentially synthesised when the cells are grown at 40°C with sodium thiosulphate as the electron donor.

Chr. purpuratum was grown anaerobically in the light at 30°C in the medium of Pfennig with the addition of 3% sodium chloride, 0.05% yeast extract and 0.15% sodium thiosulphate instead of sodium sulphide.[7]

THE COMPOSITION OF THE ANTENNA COMPLEXES

The absorption spectra of the four different types of antenna complexes used in the present study are shown in Fig. 2. The position and intensity of the bacteriochlorophyll's NIR absorption bands are quite variable.

Table 1.

Type of complex	Bchl:Car ratio	Major Car type	Efficiency of singlet-singlet Car ---> Bchl energy transfer
Rps. acidophila B800-820	2:1	rhodopinal-glucoside	70%
Rps. palustris B800-850	3:1	spirolloxanthin	36%
Chr. vinosum B800-820	2:1	rhodopin	30%
Chr. purpuratum B830	1:1	okenone	n.d.

Table 1 summarises our pigment analyses of the four antenna types. The two B800-820-complexes have Bchla:Carotenoid ratios of 2:1, the 'high-800' B800-850-complex has a ratio of 3:1, while the B830-complex appears to have a ratio of 1:1 (in the last case the exact ratio is a little uncertain due to the lack of a reliable extinction coefficient for the carotenoid okenone). The 3:1 ratio found with the B800-850-complex from Rps. palustris may indicate the presence of an additional molecule of bacteriochlorophyll as compare with other more usual LH2 complexes. Some spectroscopic evidence in favour of this will be presented in the paper by van Mourik et al.,[8] elsewhere in this book.

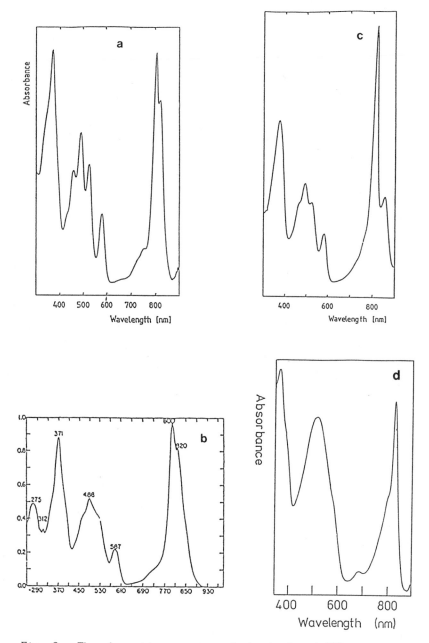

Fig. 2.   The absorption spectra of the isolated LH2 complexes
          (a) B800–820-complex from <u>Rps</u>. <u>acidophila</u>, (b) B800–
          820-complex from <u>Chr</u>. <u>vinosum,</u> (c) B800–850-complex
          from <u>Rps</u>. <u>palustris</u> and (d) B830-complex from <u>Chr</u>.
          <u>purpuratum</u>.

The polypeptide composition of these four antenna complexes was compared by SDS polyacrylamide electrophoresis. They all contain low molecular weight apoproteins in the 5-8 KD range. However, their polypeptide compositions are not simple. Most antenna complexes so far studied have been shown to be oligomers of two types of antenna apoprotein, the α- and β-apoproteins.[9,10] This may well also be true for the B830-complex from Chr. purpuratum, because even though it appears to show three bands on the SDS-gel more detailed investigation has shown that it only contains two organic solvent soluble apoproteins (data not shown). The third band is most likely an artefact on the gel due to aggregation.

However, in the case of the other complexes it is clear that they contain more than single α,β types. The antenna apoproteins from a purified sample of the B800-820-complex from Rps. acidophila have been extracted into organic solvent, separated by FPLC and sequenced.[11] At least one α-apoprotein and two β-apoproteins were found. We have repeated this for the B800-850-complex from Rps. palustris and have identified and sequenced four α- and four β-apoproteins, each present in approximately equimolar amounts. Similar studies on the B800-820-complex from Chr. vinosum are still in progress but again multiple antenna apoprotein types are present within the single complex.

It is quite clear that these unusual antenna types are constructed by the oligomerisation of more than single α- and β-apoprotein types. This is the first time that this more complicated composition has been demonstrated and it is tempting to speculate that this added complexity allows the greater variation of spectral forms to be produced. Recent genetic analysis of the genes for the antenna apoproteins in both Rps. palustris[12] and Rps. acidophila (Mackenzie et al., this meeting) have identified and sequenced multiple coding regions for the LH2 α- and β-apoproteins. Our present study suggests that all of these genes can be expressed under the correct environmental conditions. Comparison of the sequences of these apoproteins determined by amino acid sequencing with those deduced from the DNA sequence reveals that most of the α-apoproteins in both Rps. palustris and Rps. acidophila show C-terminal processing. An example of this is shown in Fig. 3. The significance of this C-terminal processing is as yet unclear.

Amino acid sequencing	Q G G L K K A A
DNA deduced sequence	Q G G L K K A A ↓ A I G H V V A L ↓

The sequence between the arrows is the C-terminal extension

Fig. 3. A comparison of the C-terminal region of the B800-820-α-apoprotein from Rps. acidophila strain 7750 as determined by amino acid sequencing with the sequence deduced from the DNA base-sequence.

CRYSTALLISATION OF THE ANTENNA COMPLEXES

It is still true that the only universal method for determining the complete high-resolution structure of a protein is X-ray crystallography. So far we have only tried to crystallise the B800-820-complex from Rps. acidophila and the 'high-800' B800-850-complex from Rps. palustris. We have obtained crystals with both, but so far only large ones with the B800–820–complex. This is illustrated in Fig. 4. The largest crystal in this picture is about 2.5 mm in length. We have however still to determine whether it will diffract X-rays to sufficiently high-resolution to allow the structure to be determined.

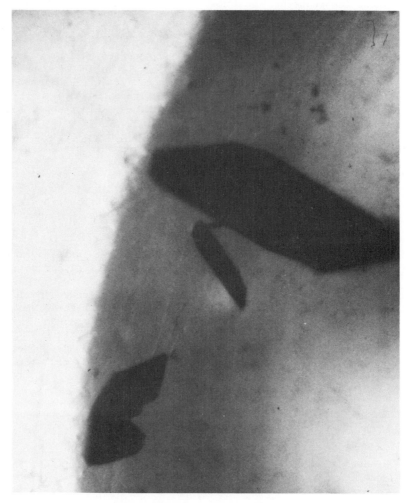

Fig. 4. Photograph of crystals of the B800–820–complex from Rps. acidophila strain 7750. The largest crystal is about 2 mm in length.

DETERMINATION OF THE EFFICIENCY OF THE CAROTENOID TO BACTERIOCHLOROPHYLL SINGLET-SINGLET ENERGY TRANSFER IN THESE UNUSUAL ANTENNA COMPLEXES

Instinctively one would suspect that antenna complexes whose biosynthesis is induced by low-light conditions would be more efficient at light-harvesting than those complexes produced at higher light-intensities. One aspect of their light-harvesting function can be readily investigated and this is how efficiently the carotenoids in these antenna complexes act as access ory light-harvesting pigments. We have therefore determined the efficiency of the carotenoid to bacteriochlorophyll singlet-singlet energy transfer to see if, in general, this process is more efficient in the 'low-light' complexes than in their 'high-light' counterparts. The results of these determinations are presented in Table 1.

The values of the efficiencies of energy transfer in these complexes are quite variable, and moreover there is no general increase in efficiency when these values are compared with the corresponding 'high-light' ones. If indeed these complexes are more efficient light-harvesters then this must be looked for in the behaviour of the bacteriochlorophylls.

ACKNOWLEDGMENTS

This work was supported by grants from the SERC, EEC, EMBO, the Chinese Government and the Schweizerische Nationalfonds.

REFERENCES

1. J. Aagaard and W.R. Sistrom, Control of the synthesis of reaction centre bacteriochlorophyll in photosynthetic bacteria, Photochem. Photobiol., 15:209 (1972).
2. J. Oelze and G. Drews, Membranes of phototropic bacteria, in: "Organisation of prokaryotic cell membranes," B.K. Ghosh, ed., CRC Press, Boca Raton, Florida (1981).
3. G. Drews, Structure and functional organisation of light-harvesting complexes and photochemical reaction centres in membranes of phototropic bacteria, Microbiol. Rev., 49:59 (1985).
4. H. Hayashi, T. Nozawa, M. Hatano and S. Morita, Circular dichroism of bacteriochlorophyll a in light-harvesting bacteriochlorophyll protein complexes from Chromatium vinosum, J. Biochem., 89:1853 (1981).
5. H. Hayashi, M. Nakono and S. Morita, Comparative studies of protein properties and bacteriochlorophyll contents of bacteriochloro-phyll-protein complexes from spectrally different types of Rhodopseudomonas palustris. J. Biochem., 92:1805 (1982).
6. R.J. Cogdell, I. Durant, J. Valentine, J.G. Lindsay and K. Schmidt, The isolation and partial characterisation of the light-harvesting pigment-protein complement of Rhodopseudomonas acidophila, Biochim. Biophys. Acta, 722:427 (1983).
7. J.F. Imhoff and H.G. Truper, Chromatium purpuratum, sp. nov., a new species of the Chromatiacae, Zbl. Baklt. I. Abt. Orig., C1:61 (1980).
8. F. van Mourik, A.M. Hawthornthwaite, C. Vonk, M.B. Evans, R.J. Cogdell and R. van Grondelle, Spectroscopic characterisation of the low light B800-850 light-harvesting complex of Rhodopseudomonas palustris, Biochim. Biophys. Acta, in press (1989).
9. H. Zuber, Primary structure and function of light-harvesting polypeptides, Ency. Pl. Physiol., 19:238 (1982).
10. R.J. Cogdell, Light-harvesting complexes in purple photosynthetic bacteria. Ency. Pl. Physiol., 19:252 (1982).
11. R.A. Brunisholz, I. Bissig, E. Niederer, F. Suter and H. Zuber, Structural studies on the light-harvesting polypeptides of Rhodopseudomonas acidophila, in: "Progress in Photosynthesis Research," J. Biggins ed., Martinus Nijhoff, The Hague (1987).
12. M.H. Tadros and K. Waterkamp, Multiple copies of the coding regions for the light-harvesting B800-850 α- and β- polypeptides are present in the Rhodopseudomonas palustris genome, EMBO J., 8: 1303 (1989).

ASSEMBLY OF INTRACYTOPLASMIC MEMBRANES IN *RHODOBACTER SPHAEROIDES* MUTANTS LACKING LIGHT-HARVESTING AND REACTION CENTER COMPLEXES

James N. Sturgis, C. Neil Hunter* and Robert A. Niederman

Department of Molecular Biology and Biochemistry, Rutgers University, Piscataway, NJ 08855-1059 (USA) and *Department of Molecular Biology and Biotechnology, Biochemistry Section, University of Sheffield, Sheffield S10 2TN (UK)

INTRODUCTION

The intracytoplasmic membrane (ICM) of photosynthetic bacteria contains a number of integral bacteriochlorophyll (BChl)-protein complexes. In *Rhodobacter sphaeroides,* these consist of photochemical reaction centers together with the B800-850 and B875 light-harvesting proteins[1] which function as peripheral and core antennae, respectively. The B875 and reaction center complexes form the fixed cores of photosynthetic units and are found in a constant molar ratio of ~25:1, respectively [2]. The B800-850 antenna comprises the variable portion of the photosynthetic unit and can reach levels more than three-fold greater than those of B875 under low illumination [2] or in the latter stages of induction of ICM formation at reduced oxygen tension [3]. The ICM is continuous with the cytoplasmic membrane [4, 5] and *in vivo* surface labeling has demonstrated that the interior of the ICM is accessible from the periplasmic space [6, 7].

The development of the ICM of *R. sphaeroides* is initiated at membrane invagination sites that can be isolated in an upper pigmented band by rate-zone sedimentation of cell-free extracts [8,9]. Pulse-chase studies have suggested that newly synthesized B875-reaction center core particles are assembled preferentially at these invaginations and become concentrated within the developing ICM by virtue of membrane growth. Spectral characterization of the upper pigmented fraction has revealed that components of the light-driven electron transport reactions are not fully assembled at these sites [10,11] and that the levels of B800-850 are reduced [10]. Examination of fluorescence yield properties indicated that while energy was transferred efficiently from B875 to the reaction center, the B800-850 antenna was only partially connected to these core structures [12]. Excitation annihilation measurements showed that domain sizes (i.e., the clusters of functionally interconnected antenna BChl molecules) were at least 3-fold smaller than in chromatophores [13]. Moreover, radiolabeling studies in synchronously dividing cells [14] have implied that the addition of B800-850 to further interlink the photosynthetic units is a dominant event in the expansion of the ICM.

Recently, a number of mutant strains which lack one or more of the pigment-protein complexes have been obtained by chemical and transposon mutagenesis techniques [15-17]. This approach has permitted both structural and functional studies of the antenna components

---

[1]The designations B800-850 and B875 refer to the positions of the maxima (in nm) of the near-IR absorption bands of these complexes. The B800-850 and B875 antennae are sometimes designated as LH2 and LH1, respectively, especially when other spectral components within these complexes are also considered. Since such components are not addressed here, the former designations will be used to remain consistent with the nomenclature recommendations proposed for the light-harvesting complexes by Cogdell et al. [1].

Molecular Biology of Membrane-Bound Complexes in Phototrophic Bacteria
Edited by G. Drews and E. A. Dawes
Plenum Press, New York, 1990

individually in their native membrane environments rather than in preparations isolated in the presence of detergents [16-19]. A more critical test of the role of B800-850 in ICM development has also been provided by rate-zone sedimentation and ultrastructural analyses of these mutant strains [20]. With strain NF57 (B875-, reaction center -), gradient profiles were similar to those of the wild type in which the majority of the photosynthetic pigments were found with the ICM-derived chromatophore fraction. In strain M21 (B800-850-), no chromatophore vesicles were observed and most of the BChl banded with the upper pigmented fraction. Electron micrographs revealed that vesicular ICM, similar to that of the wild type, was present in mutant NF57, whereas in M21, the internal membranes appeared mainly as enlarged tubular structures. These ultrastructural differences suggested that B800-850 is essential for the formation of fully vesicularized ICM and that the maturation process can be completed in the absence of B875-reaction center core particles [20].

Here, these studies have been extended to other mutant and revertant strains as well as to transconjugants in which the genetic defects have been corrected. The results, together with an investigation of the effects of various levels of the B800-850 complex on ICM morphogenesis in the wild-type parental strain, have indicated that the size of ICM vesicles is dependent upon the protein composition of the membrane. These findings are discussed in relation to the possible genetic deficiencies in the mutant strains and to a model for the role of the B800-850 antenna complex in ICM morphogenesis.

MATERIALS AND METHODS

The various *R. sphaeroides* strains used in this study are listed in Table 1. Mutant strains M21 and NF57 were isolated after mutagenesis of wild-type strain NCIB 8253 with N-methyl-N'-nitro-N-nitroso-guanidine [15]. Mutant M21 lacks the B800-850 complex and is capable of limited photosynthetic growth (see below), while NF57 is unable to grow photosynthetically and lacks both the reaction center and B875 complexes. The B800-850 complex has been restored to strain M21 by complementation with plasmid pMA81 containing the *puc*BA structural genes which encode the B800-850-α and -β polypeptides [15]. The missing complexes, as well as photosynthetic competence, were restored to strain NF57 by plasmid pSCN5H-1 which contains the structural gene encoding the reaction center-H subunit [16]. Strain M2192, which was derived from M21 after insertion of transposon Tn5 into the *puf*L gene [17] encoding the L subunit of the reaction center, lacks both the B800-850 and reaction center complexes. Reaction center activity was reestablished in strain M2192 by complementation with plasmid pSRC2 which contains an insert that includes *puf*LM [16]; *puf*M is the structural gene for the reaction center-M subunit.

The wild-type, mutant and transconjugant strains were grown semiaerobically as described previously [20], unless indicated otherwise. French pressure cell extracts of the various strains were subjected directly to rate-zone sedimentation on sucrose density gradients by previously described procedures [20]. Cell fixation, sectioning, poststaining and electron microscopy techniques are also presented in [20].

Table 1. *Rhodobacter sphaeroides* strains

Strain	Relevant characteristics	Reference
NCIB8253	Wild type	[3]
M21	NTG[a] mutant (B800-850-); contains lesion in *puc* operon	[15]
NF57	NTG mutant (B875-, reaction center-); contains lesion in *puh* operon	[15, 16]
M2192	Tn5 mutant of strain M21 (B800-850-, reaction center-); transposon inserted into *puf*L gene	[17]
M21pMA81	M21 complemented with plasmid pMA81 (B800-850 complex restored)	[15]
NF57pSCN5H-1	NF57 complemented with plasmid pSCN5H-1 (B875 and reaction center complexes restored)	[16]

[a]NTG, N-methyl-N'-nitro-N-nitroso-guanidine

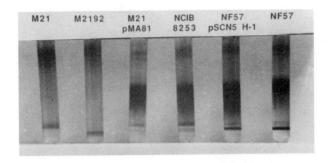

Fig. 1. Rate-zone sedimentation of mutant and complemented strains on sucrose density gradients. Cell-free extracts were layered onto 5-35% (w/w) sucrose gradients prepared over a 1.3-mL cushion of 60% sucrose and centrifuged at 4°C for 105 min in a Beckman SW40Ti rotor at 40,000 rpm. Pigmented membrane fractions (top to bottom): upper pigmented band; chromatophores; some pigment was also observed sedimenting with the cell envelope fraction near the 60% sucrose interface. Relevant characteristics of the various strains are listed in Table 1 and described in the text.

RESULTS AND DISCUSSION

The distribution of pigmented membranes after a rate-zone sedimentation analysis of French-pressure cell extracts of the mutant and complemented strains is shown in Fig. 1. In the transconjugant M21 strain M21pMA81 in which the B800-850 complex was reintroduced, a membrane banding pattern essentially identical to that of the wild-type was restored; photosynthetic pigments can be seen mainly in the ICM-derived chromatophore fraction with a lesser amount associated in the upper pigmented band. The absence of a chromatophore fraction was confirmed in strain M21, which together with strain M2192, exhibited most of their photopigments in an upper pigmented band. A wild-type sedimentation profile was observed in transconjugant NF57pSCN5H-1; this strain regained photosynthetic competence as well as the ability to form the B875 and reaction center complexes. In contrast, chromatophores sedimented slightly more slowly in the original NF57 mutant than in the parental and tansconjugant strains; results presented below demonstrate that this sedimentation behavior is apparently due to differences in vesicle sizes.

Electron micrographs of thin sections of the mutant strains were obtained in order to relate the results of ultracentrifugation studies to the morphology of intracellular membranes. Fig. 2 verifies previous observations of a vesicular type ICM in strain NF57 [20], but further analysis (see below) indicated that the mean diameters of the NF57 vesicles were significantly smaller than in the wild-type parental and the complemented NF57 strains [21]. The presence of abnormal internal membranes in M21 is also corroborated in the thin section of the M21G strain (Fig.2); these consisted mainly of large tubular structures that were usually oriented along the long axis of the cell, and were sometimes observed running between two cells that appeared to be arrested in division (see inset in Fig. 2 for an enlargement of these structures). In addition, the present ultrastructural analysis revealed large vesicular membranes in this M21 strain. The presence of these structures was not noted previously and the observation of fewer and shorter tubules together with a greater number of vesicles in more recent electron micrographs of strain M21 [21] may be related to a strong selection pressure as a result of the apparent inhibition of cell division by larger tubules. This would greatly favor survival of those cells that avoid formation of enlongated cylindrical membranes and the associated difficulties in partitioning them between daughter cells. Recent studies have been performed with semiaerobic cultures inoculated with cells propagated in liquid culture for more generations than in the earlier work, which may thus account for the lower frequency of tubules. It is also noteworthy that ultrastructural analysis of the transconjugant M21 strain has shown that the ability to form normal vesicular ICM was restored [21]. Because the transcon-

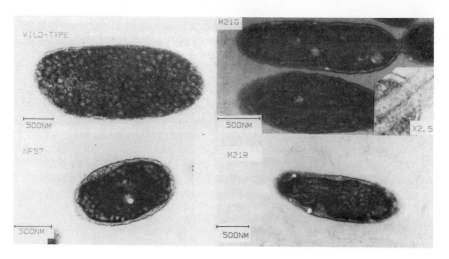

Fig. 2. Electron micrographs of thin sections of wild-type and mutant cells. M21G is a green derivative of strain M21 with a carotenoid composition similar to that of *R. sphaeroides* strain Ga in which neurosporene predominates and the absorption spectrum is typical of that exhibited by this carotenoid [20]. M21R represents a revertant cell that was present in a culture of strain M21 which reverted during photoheterotrophic growth.

jugant strains behaved in all respects like the parental strain, it appears that the introduction of DNA from the wild type was responsible for restoration of the normal ICM morphogenesis process.

Although the exact genetic lesions in strains M21 and NF57 have not been elucidated, DNA-RNA hybridization has demonstrated that the 0.5-kb transcript encoding the B875 polypeptides is present in NF57 [16], while low-temperature absorption spectra [18] and sodium dodecyl sulfate-polyacrylamide gel electrophoresis [16] suggested that both the completed complex and these polypeptide components are absent from the membrane. The lack of B875 in this strain is apparently a pleiotropic effect resulting from a point mutation of unknown nature that is corrected by DNA present on the complementing plasmid pSCN5H-1. Similar observations have been reported after mutagenesis of the *puh*A region in *R. capsulatus* [22], and in an *R. sphaeroides* strain constructed by an *in vitro* deletion in this region[23]. Together, these results are consistent with a role for the product of the *puh*A structural gene (the reaction center-H subunit) or a flanking region in the assembly of the B875-reaction center core complex.

The 0.55-kb transcript from the *puc* operon which encodes the B800-850 polypeptides could not be demonstrated in M21 and the mutation in this strain apparently lies somewhere within the 3.75 kb insert present in plasmid pMA81 which contains the *puc*BA structural genes [15]. Recently, apparent downstream coding sequences essential for expression of the *puc*BA-specific transcript have been described in studies with other *R. sphaeroides* mutants lacking the B800-850 complex [24]. Since the large insert which complemented the defect in M21 may also contain this newly described region as well as putative upstream regulatory sequences [15], the possibility that these may be related to the observed morphological changes requires further investigation. In this connection, it has been noted [24] that mutations downstream from *puc*BA are apparently correlated to grossly abnormal ICM produced in other *R. sphaeroides* strains [25].

When grown photoheterotrophically, strain M21 readily reverts to form low levels of the B800-850 complex; thin section of such revertant cells have exhibited both short tubular structures and apparently wild-type vesicular membranes within distinct intracellular regions (Fig.2). After rate-zone sedimentation of extracts from such partially revertant cultures, small amounts of chromatophores can be isolated in which B800-850 has been demonstrated in ab-

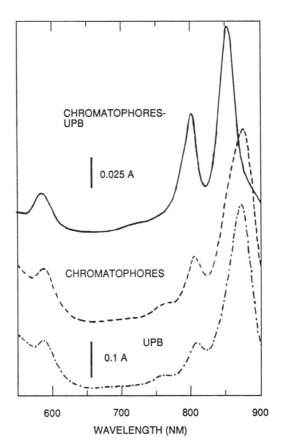

CHROMATOPHORES-UPB

0.025 A

CHROMATOPHORES

UPB

0.1 A

600   700   800   900

WAVELENGTH (NM)

Fig. 3. Near-IR absorption spectra of membrane fractions isolated from M21R. The M21R cells were obtained from a culture of strain M21 which reverted during photoheterotrophic growth. The absorption spectra were obtained at room temperature with a Perkin-Elmer Lambda 3B split-beam spectrophotometer equipped with a Hamamatsu R928 photomultiplier tube; levels of B875 in each membrane fraction were equalized. UPB, upper pigmented band. The absorption band with a maximum near 808 nm seen in the upper pigmented band represents that of monomeric reaction center BChl, while that observed at 800 nm in the chromatophores-minus-upper pigmented band absorption difference spectrum reflects the B800 band of the B800-850 complex.

sorption difference spectra obtained by subtraction of the near-IR absorption spectrum of this fraction from that of the upper pigmented band (Fig. 3). An essentially homogeneous B800-850 spectrum was obtained in this manner, while the spectrum of this complex appeared to be absent from the upper pigmented fraction. Although confirmation of this result requires immunoblotting with appropriate B800-850 antisera as well as pulse-chase measurements, it can be inferred that in these revertant cells, the fully assembled B800-850 antenna complex resides mainly in the vesicular ICM with isolation characteristics of chromatophores rather than in the tubular or peripheral membrane that gives rise to the upper pigmented band. These findings provide support for a proposal describing the development of the *R. sphaeroides* ICM in which B800-850 both drives the formation of vesicular structures and becomes localized within them (see below).

As noted above, the slowed sedimentation of NF57 chromatophores (Fig. 1) is apparently due to a decreased size in comparison to that of the parental and complemented strains. Measurements of the internal diameters of ICM vesicles observed in electron micrographs revealed that the average size of these structures in strain NF57 was about 10% smaller than their wild-type and NF57pSCN5H-1 counterparts [21]. This was confirmed by a gel-permeation chromatographic analysis of the chromatophores isolated from these strains using a Sepharose 2B column; the structures from NF57 were retarded significantly when compared to those of the parental and complemented strains which eluted in identical positions ahead of the NF57 vesicles. The internal diameters of the few vesicles observed in thin sections of strain M21 were ~45% larger than in the transconjugant M21pMA81 and NCIB 8253 strains.

The results of these sedimentation, ultrastructural and chromatographic analyses of the mutant and complemented strains imply that the size of ICM vesicles is dependent upon the protein composition of the membrane. Accordingly, the reduced vesicle size in strain NF57 appears to be due to the high levels of B800-850 which is the sole pigment-protein complex present in this strain. In the absence of this antenna complex, such as in strain M21, both tubular structures and vesicles of greatly increased size are formed, consistent with the possibility that B800-850 possesses properties essential for the ICM vesicularization process and that in the absence of this protein, morphogenesis is arrested at a tubular stage. Moreover, the presence of vesicular ICM in NF57 showed that the maturation process can be completed in the absence of B875-reaction center core particles. It is possible that the apparent role of the B800-850 antenna in ICM morphogenesis is enhanced by the aggregation state of this complex, since only higher B800-850 oligomers were observed when chromatophores of NF57 were subjected to lithium dodecyl sulfate-polyacrylamide gel electrophoresis [20]. Evidence that large supramolecular arrays of this complex are present in the membranes of strain NF57 was also obtained by excitation annihilation measurements [19]. These data not only confirmed the results of previous annihilation studies with solubulized B800-850 complexes [26], but also indicated further that units of ~30 B850 BChls are interconnected to form clusters of >350 molecules capable of intercomplex energy transfer.

Recently, the relation between vesicle size and the levels of the B800-850 complex within the ICM has been examined in the wild-type parental strain grown at different light intensities [27]. This permitted a comparison of the membrane maturation process in cells containing various levels of this antenna complex. Cells were grown at high, moderate and low illumination levels of ~1000, 300 and 100 WM^{-2}, respectively, which gave rise to chromatophores that differed in B800-850 content over a nearly two-fold range. During sucrose density gradient centrifugation, marked differences in the sedimentation rates of the chromatophore fractions were observed with those from extracts of the cells grown at the lowest light intensity and containing the highest B800-850 content sedimenting more slowly than the chromatophores from the other cells. Isoosmotic centrifugation revealed no significant differences in the equilibrium buoyant density of the chromatophore vesicles from each of the extracts while gel-permeation chromatography indicated that their distinct sedimentation properties were due to differences in size.

These results confirm the observations made with the NF57 strain and demonstrate further that the size of ICM vesicles is correlated to the levels of B800-850 within the membrane when B875-reaction center core particles are also present. Moreover, the rate-zone sedimentation analysis of the membranes from the wild-type cells revealed a size and compositional heterogeneity in which more of the B800-850 antenna was found in the smaller vesicles in the trailing edge of the chromatophore band than in the larger structures in the leading edge [27]. This implies that a compositional stability exists in the ICM such that once the individual vesicles are formed, the B800-850 and other integral proteins within these structures do not equilibrate over the entire ICM, but are instead restricted to the structures into which they were incorporated initially. Since the cells in these studies were subjected to self shading during the latter stages of photoheterotrophic growth, the ICM formed during this period would be expected to contain increased B800-850 levels and therefore give rise to the smaller vesicles observed in the upper portion of the chromatophore fraction that apparently arose from structures in the proximal portion of the ICM.

In order to explain the results presented here, it has been proposed that the B800-850 protein is capable of bending the membrane and altering the morphology of the bilayer [27]. In this proposal, it is assumed that these complexes are incorporated randomly into flat peripheral cytoplasmic membrane regions in which they exist as separate units. Support for this has been provided from the high fluorescence yield observed for the B850 BChls in the upper pigmented fraction [12] which has been shown to contain these sites [8, 9] and excitation annihilation measurements at 4 K which indicated that although individual B800-850 units containing ~40-55 B850 BChls are present [28], their interconnections are limited [13]. In the next stage of development, the B800-850 oligomers undergo local aggregation which results in the initiation of the membrane invagination process. The initiation sites then elongate into tubular structures [29] as more components are integrated; morphogenesis of internal membranes is apparently arrested at this stage in those mutants in which ultrastructural analysis has revealed the presence of tubular structures [20, 25]. When the local concentrations of the B800-850 aggregates in these structures reaches sufficiently high levels, the necks of the tubules become constricted and vesicles are formed which give rise to chromatophores upon cellular disruption. The composition of the individual ICM vesicles becomes fixed with a high protein to lipid ratio since the protein aggregates are unable to migrate through the intervesicular constrictions. The process of elongation and vesicularization then continues to generate chains of ICM vesicles containing high concentrations of the B800-850 protein.

Overall, the results presented here confirm that development of the ICM is impaired in mutant strains lacking the B800-850 complex and reveal further that ICM vesicles of reduced size are formed in strain NF57 and in wild-type cells which contain high levels of this antenna. In accordance with the above model, the latter findings can be explained by the increased curvature of the vesicles that would be expected to form when the concentration of B800-850 aggregates becomes highly elevated. In the related bacterium *Rhodospirillum rubrum*, vesicular ICM is formed even though this organism contains a single light-harvesting complex that is homologous to the B875 core antenna of *R. sphaeroides* and lacks an antenna equivalent to B800-850. Although this would appear not to be in accord with the role proposed here for the B800-850 complex in the ICM vesicularization process, excitaton annihilation measurements suggested that unlike the *R. sphaeroides* B875 complex, the B880 antenna of *Rs. rubrum* exists in large aggregates [30] that are present in connected arrays of >1000 BChl molecules over the surface of the membrane. This implies that under some circumstances, core antenna complexes may possess sufficient membrane bending activity to drive vesicle formation. The observation of tubular structures in mutant strains lacking B800-850 suggests that the *R. sphaeroides* B875 complex is also capable of distorting the membrane; however, the B875 clusters demonstrated in the membranes of strain M21 [19] are apparently neither large enough nor coalesce sufficiently to initiate the formation of ICM vesicles.

Acknowledgements: This work was supported by U. S. National Science Foundation grant DMB85-12587 to R. A. N. and grants from the U. K. Science and Engineering Research Council to C. N. H. J. N. S. was the recipient of a fellowship from the Charles and Johanna Busch Memorial Fund Award to the Rutgers Bureau of Biological Research.

REFERENCES

1. R. J. Cogdell, H. Zuber, J. P. Thornber, G. Drews, G. Gingras, R. A. Niederman, W. W. Parson, and G. Feher, Recommendations for the naming of photochemical reaction centres and light-harvesting pigment-protein complexes from purple photosynthetic bacteria, *Biochim. Biophys. Acta* 806:185 (1985).

2. J. Aagaard and W. R. Sistrom, Control of synthesis of reaction center bacteriochlorophyll in photosynthetic bacteria, *Photochem. Photobiol.* 15:209 (1972).

3. R. A. Niederman, D. E. Mallon, and J. J. Langan, Membranes of *Rhodopseudomonas sphaeroides*. IV. Assembly of chromatophores in low-aeration cell suspensions, *Biochim. Biophys. Acta* 440:429 (1976).

4. G. Drews and J. Oelze, Organization and differentiation of membranes of phototrophic bacteria, *Adv. Microbiol. Physiol.* 22:1 (1981).

5. S. Kaplan and C. J. Arntzen, Photosynthetic membrane structure and function, *in:* "Photosynthesis," Vol. 1, Govindjee, ed., p. 65, Academic Press, New York (1982).

6. G. A. Francis and W. R. Richards, Localization of photosynthetic membrane components in *Rhodopseudomonas sphaeroides* by a radioactive labeling procedure, *Biochemistry* 19:5104 (1980).

7. G. S. Inamine, J. van Houten, and R. A. Niederman, Intracellular localization of photosynthetic membrane growth initiation sites in *Rhodopseudomonas sphaeroides*, *J. Bacteriol.* 158:425 (1984).

8. R. A. Niederman, D. E. Mallon, and L. C. Parks, Membranes of *Rhodopseudomonas sphaeroides*. VI. Isolation of a fraction enriched in newly synthesized bacteriochlorophyll *a*-protein complexes, *Biochim. Biophys. Acta* 555:210 (1979).

9. R. A. Niederman, C. N. Hunter, G. S. Inamine, and D. E. Mallon, Development of the bacterial photosynthetic apparatus, *in:* "Photosynthesis,Vol. 5, Chloroplast Development," G. Akoyunoglou, ed., p. 663, Balaban, Philadelphia (1981).

10. C. N. Hunter, N. G. Holmes, O. T. G. Jones, and R. A. Niederman, Membranes of *Rhodopseudomonas sphaeroides.* VII. Photochemical properties of a fraction enriched in newly synthesized bacteriochlorophyll a-protein complexes, *Biochim. Biophys. Acta.* 548:253 (1979).

11. J. R. Bowyer, C. N. Hunter, T. Ohnishi, and R. A. Niederman, Photosynthetic membrane development in *Rhodopseudomonas sphaeroides*: Spectral and kinetic characterization of redox components of light-driven electron flow in apparent photosynthetic membrane growth initiation sites, *J. Biol. Chem.* 260:3295 (1985).

12. C. N. Hunter, R. van Grondelle, N. G. Holmes, O. T. G. Jones, and R. A. Niederman, Fluorescence yield properties of a fraction enriched in newly synthesized bacteriochlorophyll a-protein complexes from Rhodopseudomonas sphaeroides, Photochem. Photobiol. 30:313 (1979).

13. C. N. Hunter, H. J. M. Kramer, and R. van Grondelle, Linear dichroism and fluorescence emission of antenna complexes during photosynthetic unit assembly in Rhodopseudomonas sphaeroides, *Biochim. Biophys. Acta.* 807:44 (1985).

14. P. A. Reilly and R. A. Niederman, Role of apparent membrane growth initiation sites during photosynthetic membrane development in synchronously dividing *Rhodopseudomonas sphaeroides*, *J. Bacteriol.* 167:153 (1986).

15. M. K. Ashby, S. A. Coomber, and C. N. Hunter, Cloning, nucleotide sequence and transfer of genes for the B800-850 light-harvesting complex of *Rhodobacter sphaeroides*, *FEBS Lett.* 213:245 (1987).

16. C. N. Hunter and R. van Grondelle, The use of mutants to investigate the organisation of the photosynthetic apparatus of Rhodobacter sphaeroides, *in*: "Photosynthetic Light-Harvesting Systems," H. Scheer and P. Schneider, eds., p. 247, Walter de Gruyter, New York (1988).

17. C. N. Hunter, R. van Grondelle, and R. J. van Dorssen, The construction and properties of a mutant of *Rhodobacter sphaeroides* with the LH1 antenna as the sole pigment protein, *Biochim. Biophys. Acta* 973:383 (1989).

18. R. J. van Dorssen, C. N. Hunter, R. van Grondelle, A. H. Korenhof, and J. Amesz, Spectroscopic properties of antenna complexes of *Rhodobacter sphaeroides in vivo, Biochim. Biophys. Acta* 932:179 (1988).

19. M. Vos, R. J. van Dorssen, J. Amesz, R. van Grondelle, and C. N. Hunter, The organization of the photosynthetic apparatus of *Rhodobacter sphaeroides*: studies of antenna mutants using singlet-singlet quenching, *Biochim. Biophys. Acta* 933:132 (1988).

20. C. N. Hunter, J. D. Pennoyer, J. N. Sturgis, D. Farrelly, and R. A. Niederman, Oligomerization states and associations of light-harvesting pigment-protein complexes of *Rhodobacter sphaeroides* as analyzed by lithium dodecyl sulfate-polyacrylamide gel electrophoresis, *Biochemistry* 27:3459 (1988).

21. J. N. Sturgis, C. N. Hunter, and R. A. Niederman, In preparation.

22. D. C. Youvan, J. E. Hearst, and B. L. Marrs, Isolation and characterization of enhanced fluorescence mutants of *Rhodopseudomonas capsulata, J. Bacteriol.* 154:748 (1983).

23. R. E. Sockett, T. J. Donohue, A. R. Varga, and S. Kaplan, Control of photosynthetic membrane assembly in *Rhodobacter sphaeroides* mediated by puhA and flanking sequences, *J. Bacteriol.* 171:436 (1989).

24. J. K. Lee, P. J. Kiley, and S. Kaplan, Posttranscriptional control of puc operon expression of B800-850 light-harvesting complex formation in *Rhodobacter sphaeroides, J. Bacteriol.* 171:3391 (1989).

25. P. J. Kiley, A. Varga, and S. Kaplan, Physiological and structural analysis of light-harvesting mutants of *Rhodobacter sphaeroides, J. Bacteriol.* 170:1103 (1988).

26. R. van Grondelle, C. N. Hunter, J. G. C. Bakker, and H. J. M. Kramer, Size and structure of antenna complexes of photosynthetic bacteria as studied by singlet-singlet quenching of the bacteriochlorophyll fluorescence yield, *Biochim. Biophys. Acta* 723:30 (1983).

27. J. N. Sturgis and R. A. Niederman, Role of B800-850 light-harvesting pigment-protein complex in the morphogenesis of *Rhodobacter sphaeroides* membranes, *in*: "Proc. VIII Internat. Congr. Photosynth.", In press.

28. W. H. J. Westerhuis, M. Vos, R. J. van Dorssen, R. van Grondelle, J. Amesz and R. A. Niederman, Associations of pigment-protein complexes in phospholipid-enriched bacterial photosynthetic membranes, *in*: "Biological Role of Plant Lipids," P. A. Biacs, K. Gruiz, and T. Kremmer, eds., p. 227, Plenum, New York.

29. J. T. Chory, T. J. Donohue, A. R. Varga, L. A. Staehelin, and S. Kaplan, Induction of the photosynthetic membranes of *Rhodopseudomonas sphaeroides*: biochemical and morphological studies, *J. Bacteriol.* 159:540 (1984).

30. M. Vos, R. van Grondelle, F. W. van der Kooij, D. van de Poll, J. Amesz, and L. N. M. Duysens, Singlet-singlet annihilation at low temperatures in the antenna of purple bacteria, *Biochim. Biophys. Acta* 850:501 (1986).

ORGANIZATION OF CHLOROPHYLL AND PROTEIN IN CHLOROSOMES

J.M. Olson[a], D.C. Brune[a,b] and P.D. Gerola[c]

[a]Inst. of Biochemistry, Odense University
DK-5230 Odense M, Denmark
[b]Dept. of Chemistry and Center for the Study of Early
Events in Photosynthesis, Arizona State University
Tempe, AZ 85287-1604, USA
[c]Dept. of Ecology, University of Calabria
I-87030 Arcavacata di Rende (Cosenza), Italy

INTRODUCTION

Chlorosomes are membrane-attached antenna complexes containing BChl c (or d or e in some strains) and small amounts of BChl a that occur in both Chlorobiaceae and Chloroflexaceae (Blankenship et al., 1988a), even though these bacterial families are not closely related (Woese, 1987). Although the chlorosomes in the two families have identical functions, some structural differences have been noted in the size of the internal rod elements (Feick and Fuller,1984; Wechsler et al., 1985; Gerola et al.,1988; Wagner-Huber et al., 1988) and in their mode of attachment to the cytoplasmic membrane (Staehelin et al., 1980; Sprague et al., 1981).

There are two opposing views about the organization of chlorophyll and protein in these chlorosomes. According to one view each bacteriochlorophyll (BChl) c molecule is bound to a specific site on a specific protein (Wechsler et al., 1985). According to the other view the BChl c molecules first self-assemble to form an aggregate, and then the aggregate is stabilized by binding to the appropriate protein (Brune et al., 1987; Olson and Pedersen, 1988).

Wagner-Huber et al. (1988) have demonstrated that the 6.3-kDa glycine-rich proteins from chlorosomes of green sulfur bacteria are ca. 30% homologous to the 5.6-kDa protein from chlorosomes of Chloroflexus aurantiacus. These proteins are thought to be directly involved in binding BChl c, but the 7.5-kDa protein may also play a role in green sulfur bacteria (Gerola et al., 1988).

Self assembly of BChl c aggregates in solution has been demonstrated in at least 4 model systems completely lacking protein. In the first system natural mixtures of farnesyl BChl c homologues from green sulfur bacteria were dissolved in CH2Cl2 and then diluted (1:200) in hexane (Smith et al., 1983). The aggregate that formed had an absorption maximum at 748 nm and a CD spectrum similar to that of chlorosomes from Cf. aurantiacus (Olson et al., 1985). The second model system consisted of stearyl BChl c homologs from Cf. aurantiacus in hexane (Brune et al.,

Molecular Biology of Membrane-Bound Complexes in Phototrophic Bacteria
Edited by G. Drews and E. A. Dawes
Plenum Press, New York, 1990

1987; 1988). Again an aggregate formed, this time with absorption maximum at ca. 740 nm and a CD spectrum again similar to that of Chloroflexus chlorosomes (Blankenship et al., 1988b). The third model system consisted of an aggregate of 4-9 molecules of 4-isobutyl homologs of farnesyl BChl c in CCl₄ with absorption maximum at 747 nm and CD spectrum like that of Chloroflexus chlorosomes (Olson and Pedersen, 1988; submitted to Photosynth Res). In the fourth model system farnesyl BChl c was dissolved in 50% perdeuterated toluene in perdeuterated octane (v/v), and water added. Upon mild sonication, cylindrical micelles with absorption maximum at 750 nm formed (Worcester et al., 1986). The diameter of these cylindrical micelles (ca. 12 nm) closely matched the diameter of rod elements from Chlorobium (but not those from Chloroflexus).

Previously published results indicated that the circular dichroism (CD) spectra of Chlorobium and Chloroflexus chlorosomes were quite different (Betti et al., 1982; Olson et al., 1985; van Dorssen et al., 1986; Blankenship et al., 1988b). In particular CD spectra obtained by different researchers indicated that the derivative-shaped, near-infrared (NIR) spectral feature centered near the BChl c absorption maximum at 740-750 nm was of opposite chirality in chlorosomes from the two species. The longest wavelength band was found to be positive in Chlorobium while in Chloroflexus it was negative. Some of the experiments described here were undertaken to make a direct comparison between the CD spectra of Chlorobium and Chloroflexus chlorosomes prepared using the same procedure (Gerola and Olson, 1986). This procedure does not use detergents but rather uses the chaotropic agent NaSCN, which was shown to stabilize Chlorobium chlorosomes during their isolation.

Recently Griebenow and Holzwarth (1989) reported that Chloroflexus chlorosomes prepared with the detergent lithium dodecylsulfate do not contain any BChl a. Absorption spectra of these chlorosomes indicated that their BChl c component was unaffected by the detergent treatment. Here we report on the absorption, fluorescence and CD spectra of chlorosomes before and after treatment with SDS.

MATERIALS AND METHODS

Chloroflexus aurantiacus cultures were grown under continuous illumination at 55°C (Cox et al., 1988). Cells were harvested by centrifugation at 20,000 x g for 15 minutes. Cb. limicola f. thiosulfatophilum cultures were grown in stirred 20-liter carboys at room temperature under continuous illumination and cells were harvested as described by Olson et al. (1973). The harvested cells were stored frozen at -20°C. Chlorosomes from both organisms were prepared by breaking the cells in 10 mM, pH 7.5 phosphate buffer containing 10 mM ascorbate and 2M NaSCN. After centrifuging to remove large debris, the supernatant was ultracentrifuged on sucrose gradients containing 2M NaSCN (Gerola and Olson, 1986). The pigment band was removed and subjected to a second gradient centrigufation. Absorption spectra of the second gradient pigment band gave a typical chlorosome BChl c spectrum free of contaminating membrane pigments. Chlorosomes were stored frozen until use.

SDS-treated Chloroflexus chlorosomes were prepared by diluting chlorosomes isolated as described above in 10 mM, pH 8 Tris Cl buffer and adding 3% SDS in the same buffer to give a final SDS concentration of 0.3% and a chlorosome absorbance of 100 at 740 nm in a total volume of 2 ml. The sample was then loaded onto a Sepharose CL 6B column that was previously equilibrated with 10 mM, pH 8 Tris Cl containing 0.3% SDS and 20 mM NaCl (the eluting buffer). Elution was carried out with a flow rate of 0.85 ml/min and absorbance of the eluent monitored at 259 nm. Three well-resolved bands were collected. The first, a green fraction that probably ran in the void volume, contained BChl c from which BChl a had been removed. The second was orange and had the absorption spectrum of carotenoid contaminated with a trace of bacteriopheophytin c. The third band was colorless. The BChl c-containing material was collected by centrifuging at 200,000 x g for 2 hrs, discarding the colorless supernatant, and resuspending in a small volume of eluting buffer.

Absorption spectra were measured with a Perkin Elmer 330 recording spectrophotometer, and fluorescence spectra with a Spex Fluorolog 111A fluorometer in the front-face-emission mode. Circular dichroism spectra were measured with a Jobin Yvon dichrograph IV/V (Novo Research Institute, Bagsværd). Molecular masses were determined on a BioIon 10K ^{252}Cf plasma desorption mass spectrometer (Roepstorff et al., 1988).

Proteins were extracted from Chlorobium chlorosomes with 80% acetone and then purified by HPLC (Gerola et al., 1988). Elution of the proteins from the C18 reverse-phase column was monitored at 214 nm and 280 nm. Hydrolysis of proteins was carried out using 6M HCl containing 0.1% phenol for 18-24 hrs at 110°C under reduced pressure. Amino acid analyses were carried out on a Hewlett Packard HP 1090 Aminoquant amino acid analyzer.

RESULTS

Absorption and CD spectra of Chlorobium and Chloroflexus chlorosomes are shown in Fig. 1. Although the derivative-shaped NIR feature of the Chlorobium chlorosome spectrum is broader and shifted to longer wavelengths than that of the Chloroflexus chlorosomes, both spectra have the same chirality and about the same rotational strength. Thus this feature of the Chlorobium chlorosome spectrum appears to be sign-reversed from its counterpart in the chlorosome spectrum obtained previously by Olson et al. (1985). On the other hand, the rotational strength of this feature in the Chloroflexus chlorosome spectrum is considerably larger, both in absolute magnitude and in relation to the minor negative band near 810 nm (813 nm in Fig. 1) due to BChl a, than was reported by Blankenship et al. (1988b). It is also about twice as large as the NIR feature in the 77 K Chloroflexus chlorosome spectrum of van Dorssen et al. (1986). The negative band at 717 nm is less pronounced than the corresponding band in these previously reported Chloroflexus chlorosome CD spectra. In fact, the narrow and rather symmetrical chlorosome NIR CD spectrum in Fig. 1 is strikingly similar to that of the 740-nm BChl c oligomer reported by Blankenship et al. (1988b).

Reexamination of the absorption and CD spectra of earlier Chlorobium chlorosome preparations indicated that the apparent inversion of their NIR CD spectra was due to chlorosome instability. Gerola and Olson (1986) had noted previously that isolated Chlorobium chlorosomes had absorption maxima ranging from 730 to 750 nm, although intact cells absorbed maximally at 750 nm. The shift of the chlorosome absorption maximum to shorter wavelengths during isolation is accompanied by weakening or even loss of the negative CD band at 770 nm and strengthening of the negative band at ca. 720 nm, which causes the apparent inversion (Brune et al., submitted to Photosynth Res).

The absorption and fluorescence emission spectra of SDS-treated Chloroflexus chlorosomes showed that BChl a had been removed, in confirmation of the results of Griebenow and Holzwarth (1989). Surprisingly, removal of BChl a did not cause an increase in the steady-state yield of BChl c fluorescence at 750 nm, even though this procedure should have removed a major pathway for draining energy from excited states of BChl c.

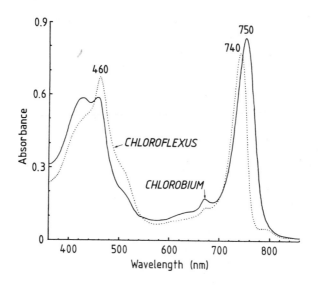

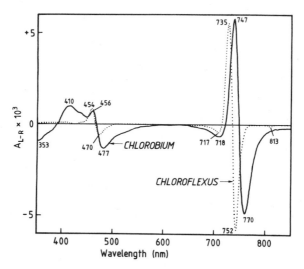

Fig. 1. Absorption (upper) and circular dichroism (lower) spectra of chlorosomes from Chlorobium (——) and Chloroflexus (.....). Chlorosomes were suspended in 10 mM Tris Cl buffer, pH 7.8. Samples for the CD measurements had absorbances of 1.0 at 750 nm (Chlorobium) and 0.8 at 740 nm (Chloroflexus).

The CD spectrum of SDS-treated chlorosomes was almost indistinguishable from that of the untreated sample, except for the expected loss of a weak negative band at 813 nm due to BChl a. Experiments on Chlorobium chlorosomes in which the suspending buffer contained 0.3% SDS also showed no effect on the CD spectrum, even though the chlorosomes lost absorbance at ca. 795 nm and fluorescence at 805 nm. However, these chlorosomes were exposed only briefly to SDS and were not subjected to sepharose chromatography, so that the experiments were not completely comparable. Nevertheless, these results imply that SDS has surprisingly little effect on BChl c organization in chlorosomes of either Chloroflexus or Chlorobium.

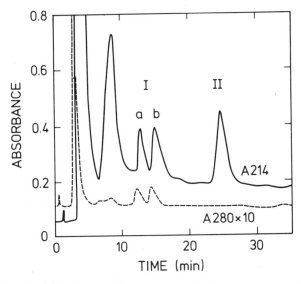

Fig. 2. Elution of <u>Chlorobium</u> chlorosome proteins from an RP18 column with an isopropanol gradient in 0.1% TFA and water.

The elution pattern of three proteins (Ia, Ib and II) extracted from <u>Chlorobium</u> chlorosomes are shown in Fig. 2. Their molecular masses are 6.3, 6.3 and 7.5 kDa respectively. A preliminary amino acid analysis clearly showed that peak II contained the 7.5-kDa protein of Gerola et al. (1988). Preliminary analyses of the proteins in peaks Ia and Ib suggested that they were related to the glycine-rich 6.3-kDa protein of Wagner-Huber et al. (1988). Since the protein in Ia contained histidine and the protein in Ib did not, it is likely that only peak Ia contained the Wagner-Huber protein. Figure 2 demonstrates that both the 6.3-kDa Wagner-Huber protein and the 7.5-kDa Gerola protein can be found in <u>Chlorobium</u> chlorosomes prepared in the presence of NaSCN by the method of Gerola and Olson (1986).

DISCUSSION

As noted above the previously reported difference in chiralty of the NIR CD spectra of <u>Chlorobium</u> and <u>Chloroflexus</u> chlorosomes is due to the instability of <u>Chlorobium</u> chlorosomes. Because BChl <u>c</u> in whole cells of the strain of <u>Chlorobium</u> used in our experiments has its absorption maximum at 750 nm, it is reasonable to assume that isolated chlorosomes with their absorption maximum at this wavelength are intact, while those with absorption maxima at shorter wavelengths are somehow damaged during isolation. None of the earlier <u>Chloroflexus</u> chlorosome CD spectra exhibited an apparent inversion in the NIR region, presumably because <u>Chloroflexus</u> chlorosomes are more stable than those of <u>Chlorobium</u>.

Previously published <u>Chloroflexus</u> chlorosome CD spectra (Betti et al., 1982; van Dorssen et al., 1986; Blankenship et al., 1988b) are generally similar to that in Fig. 1., but the low rotational strength of the spectrum measured by Blankenship et al (1988b) is anomalous. As the previous work was done on chlorosomes prepared with Miranol, the CD spectra imply that very similar chlorosomes can be obtained using either detergent or nondetergent methods. Although two of the earlier preparations had a larger negative CD band at ca. 720 nm (van Dorssen et al., 1986; Blankenship et al., 1988b) than did our preparation, in no cases was an inversion in the NIR CD spectrum observed.

Somewhat surprisingly, treatment of chlorosomes with the strong detergent SDS, which removes BChl a, apparently has little effect on the arrangement of BChl c molecules. Thus SDS-treated chlorosomes may be useful as a less complex system than untreated chlorosomes for investigating the arrangement and photophysical properties of BChl c in the green bacterial antenna. The BChl c-proteins in detergent-treated chlorosomes apparently remain associated with chlorosome-sized structures rather than existing as detergent-solubilized pigment-protein complexes. This is indicated by the finding of Griebenow and Holzwarth (1989) that they are too large to migrate into electrophoretic gels and by our observation that they can be precipitated quantitatively by a 2-hour ultracentrifugation.

Except for the minor contribution from BChl a, both Chlorobium and Chloroflexus chlorosome NIR CD spectra in Fig. 1 are similar to those previously published for BChl c aggregates in hexane (Olson et al., 1985; Blankenship et al., 1988b), supporting the hypothesis that these aggregates are good models for BChl c in chlorosomes. Previous reports disagree on whether the rotational strengths of the aggregates are larger (Blankenship et al., 1988b) or smaller (Olson et al., 1985) than those of chlorosomes, but recent measurements show that the rotational strengths of 4-isobutyl farnesyl BChl c aggregates in CCl4 are an order of magnitude lower than those of chlorosomes (Olson and Pedersen, submitted to Photosynth Res).

The proteins most directly involved in binding BChl c aggregates are probably the 5.6- and 6.3-kDa proteins of Chloroflexus and green sulfur bacteria respectively (Wechsler et al., 1985; Wagner-Huber et al., 1988), but the 7.5-kDa protein may also play a role in green sulfur bacteria (Gerola et al., 1988). In the 6.3-kDa proteins from chlorosomes of 4 green sulfur bacteria (Wagner-Huber et al., 1988) there is not a single BChl c-binding site conserved of the 7 proposed by Wechsler et al. (1985) for the 5.6-kDa protein from chlorosomes of Chloroflexus. This suggests that the 7 specific binding sites on these 5 homologous proteins do not exist.

There are two main regions of conserved residues in the 5.6-kDa Chloroflexus protein: GHW (24-26) and INRNAY (43-48). The first region (GHW) is part of a β-bend (QGHW) in the Chloroflexus protein (Gerola et al., 1988), and it seems reasonable to assume that the GHW region is conserved in all 5 proteins in order to conserve the β-bend. The second conserved region (INRNAY) appears in a "random coil" part of the Chloroflexus protein (Gerola et al., 1988). We suggest that this may be the chlorophyll binding site. Since there is only one such site per polypeptide, and since there are ca. 7 BChl c molecules per 5.6-kDa polypeptide in Chloroflexus, we propose that an aggregate of ca. 7 BChl c molecules is bound to each polypeptide.

The proposed aggregate in Chlorobium might be a cylinder as suggested by Worcester et al. (1986). In Chlorobium the rod-element diameter is ca. 10 nm, and the proteinaceous subunits can be represented as 3.3-nm spheres arranged in hexagons (Olson, 1980). In Chloroflexus a somewhat different structure would be required to fit the 5.2-nm diameter rod elements. We suggest an arrangement (Fig. 3) in which ca. 2.5-nm spheres represent the proteinaceous subunits of each rod element. Six subunits might form the core of each 6-nm segment seen in electron micrographs (Staehelin et al., 1978) with 36-42 BChl c molecules forming a cylinder around the outside.

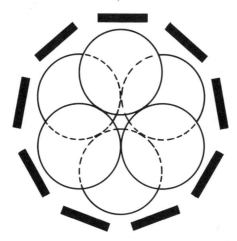

Fig. 3. Model of a cross-section of a chlorosome from <u>Chloroflexus</u> <u>aurantiacus</u>.

ACKNOWLEDGEMENTS

We thank Peter Højrup (OU) and Frede Hansen (Rigshospitalet) for performing the amino acid analyses, and Peter Roepstorff (OU) for carrying out the mass determinations. We also thank Steen G. Melberg and Yvonne B. Madsen (Novo Research Institute) for recording the CD spectra in Fig. 1. Jack Pedersen (OU) provided skillful technical assistance. One of us (DCB) was partially supported by the ASU Center for the Study of Early Events in Photosynthesis. The Center is funded by US DOE grant DE-FG02-88ER13969 as part of the USDA/DOE/NSF Plant Science Centers Program. This is publication no. 26 from the ASU Center for the Study of Early Events in Photosynthesis.

REFERENCES

Betti JA, Blankenship RE, Natarajan LV, Dickinson LC and Fuller RC (1982) Antenna organization and evidence for the function of a new pigment species in the green photosynthetic bacterium <u>Chloroflexus</u> <u>aurantiacus</u>. Biochim Biophys Acta 680:194-201.
Blankenship RE, Brune DC and Wittmershaus BP (1988a) Chlorosome antennas in green photosynthetic bacteria. In: Stevens, SE, Jr. and Bryant, DA (eds) Light-Energy Transduction in Photosynthesis: Higher Plant and Bacterial Models, pp 32-46, Rockville, MD, American Society of Plant Physiologists.
Blankenship RE, Brune DC, Freeman JM, King GH, McManus JD, Nozawa T, Trost JT and Wittmershaus BP (1988b) Energy trapping and electron transfer in <u>Chloroflexus</u> <u>aurantiacus</u>. In: Olson JM, Ormerod JG, Amesz J, Stackebrandt E and Trüper HG (eds) Green Photosynthetic bacteria, pp 57-68, New York: Plenum Press.
Brune DC, King GH and Blankenship RE (1988) Interactions between bacteriochlorophyll molecules in oligomers and chlorosomes of green bacteria. In: Scheer H and Schneider S (eds) Photosynthetic Light-Harvesting Systems, pp 141-151, Berlin: Walter de Gruyter.
Brune DC, Nozawa T and Blankenship RE (1987) Antenna organization in green photosynthetic bacteria. 1. Oligomeric bacteriochlorophyll <u>c</u> as a model for the 740 nm absorbing bacteriochlorophyll <u>c</u> in <u>Chloroflexus</u> <u>aurantiacus</u>. Biochemistry 26:8644-8652.
Cox RP, Jensen MT, Miller M and Pedersen JP (1988) Spin label studies on chlorosomes from green bacteria. In: Olson JM, Ormerod JG, Amesz J,

Stackebrandt E and Trüper HG (eds) Green Photosynthetic Bacteria, pp 15-21, New York: Plenum Press.

Feick RG and Fuller RC (1984) Topography of the photosynthetic apparatus of Chloroflexus aurantiacus. Biochemistry 23:3693-3700.

Gerola PD, Højrup P and Olson JM (1988) A comparison of the bacteriochlorophyll c-binding proteins of Chlorobium and Chloroflexus. In: Scheer H and Schneider S (eds) Photosynthetic Light-Harvesting Systems, pp 129-139, Berlin: Walter de Gruyter.

Gerola PD and Olson JM (1986) A new bacteriochlorophyll a-protein complex associated with chlorosomes of green sulfur bacteria. Biochim Biophys Acta 848:69-76.

Griebenow K and Holzwarth AR (1989) Pigment organization and energy transfer in green bacteria. 1. Isolation of native chlorosomes free of bacteriochlorophyll a from Chloroflexus aurantiacus by gel-electrophoretic filtration. Biochim Biophys Acta 973:235-240.

Olson JM, Gerola PD, van Brakel GH, Meiburg RF and Vasmel H (1985) Bacteriochlorophyll a- and c-protein complexes from chlorosomes of green sulfur bacteria compared with bacteriochlorophyll c aggregates in CH₂Cl₂-hexane. In: Michel-Beyerle ME (ed) Antennas and Reaction Centers of Photosynthetic Bacteria, pp 67-73, Berlin: Springer-Verlag.

Olson JM and Pedersen JP (1988) Bacteriochlorophyll c aggregates in carbon tetrachloride as models for chlorophyll organization in green photosynthetic bacteria. In: Scheer H and Schneider S (eds) Photosynthetic Light-Harvesting Systems, pp 365-373, Berlin: Walter de Gruyter.

Olson JM, Philipson KD and Sauer K (1973) Circular dichroism and absorption spectra of bacteriochlorophyll-protein and reaction center complexes from Chlorobium thiosulfatophilum. Biochim Biophys Acta 292: 206-217.

Roepstorff P, Nielsen PF, Klarskov K and Højrup P (1988) Applications of plasma desorption mass spectrometry in peptide and protein chemistry. Biomed. Environ. Mass Spectrom. 16:9-18.

Smith KM, Kehres LA and Fajer J (1983) Aggregation of the bacteriochlorophylls c, d, and e. Models for the antenna chlorophylls of green and brown photosynthetic bacteria. J Am Chem Soc 105:1387-1389.

Sprague SG, Staehelin LA, DiBartolomeis MJ and Fuller RC (1981) Isolation and development of chlorosomes in the green bacterium Chloroflexus aurantiacus. J Bacteriol 147:1021-1031.

Staehelin LA, Golecki JR and Drews G (1980) Supramolecular organization of chlorosomes (Chlorobium vesicles) and of their membrane attachment sites in Chlorobium limicola. Biochim Biophys Acta 589:30-45.

Staehelin LA, Golecki JR, Fuller RC and Drews G (1978) Visualization of the supramolecular architecture of chlorosomes (Chlorobium type vesicles) in freeze-fractured cells of Chloroflexus aurantiacus. Arch. Microbiol. 119:269-277.

van Dorssen RJ, Vasmel H and Amesz J (1986) Pigment organization and energy transfer in the green photosynthetic bacterium Chloroflexus aurantiacus. II. The chlorosome. Photosynth Res 9:33-45.

Wagner-Huber R, Brunisholz R, Frank G and Zuber H (1988) The bacteriochlorophyll c/e-binding polypeptides from chlorosomes of green photosynthetic bacteria. FEBS Lett 239:8-12.

Wechsler T, Suter F, Fuller RC and Zuber H (1985) The complete amino acid sequence of the bacteriochlorophyll c-binding polypeptide from chlorosomes of the green photosynthetic bacterium Chloroflexus aurantiacus. FEBS Lett 181:173-178.

Woese CR (1987) Bacterial evolution. Microbiol Revs 51:221-271.

Worcester DL, Michalski TJ and Katz JJ (1986) Small angle neutron scattering studies of chlorophyll micelles: models for bacterial antenna chlorophyll. Proc Natl Acad Sci USA 83:3791-3795.

COMPARISON OF STRUCTURAL SUBUNITS OF THE CORE LIGHT-HARVESTING

COMPLEXES OF PHOTOSYNTHETIC BACTERIA

Paul A. Loach[1], Pamela S. Parkes-Loach[1], Mary C. Chang[1],
Barbara A. Heller[1], Peggy L. Bustamante[1] and Tomasz Michalski[2]

[1]Department of Biochemistry, Molecular Biology and Cell
Biology, Northwestern University, Evanston, Il. 60208 and
[2]Argonne National Laboratory, Argonne, Il. 60439, USA.

INTRODUCTION

The photoreceptor complex (PRC) of photosynthetic bacteria consists of
a reaction center (RC) and one or more light-harvesting (LH) complexes.
All bacteria seem to have a core LH complex which is biosynthesized in
constant proportion to the RC and coded for by genes that are adjacent to
those for the L and M polypeptides of the RC (1-3). Among the more simple
bacteria, Rhodospirillum rubrum contains only the core LH complex which
has been named B881 for wildtype R. rubrum, or B873 for its car⁻ G-9 mutant
after the λ_{max} of the red-most absorbance band (Q_y band) in the organism.

The core LH complexes are usually composed of two polypeptides (α and β)
of about 6 kDa in size which are present in a 1:1 stoichiometry (4). They
have a pigment content of 2 BChl/$\alpha_1\beta_1$. Methodology has recently been
developed in our laboratory by which a structural subunit can be prepared
in near 100 % yield from the B881 or B873 of R. rubrum (5,6). This
subunit absorbs at 820 nm and has thus been named B820. It can be
reassociated to form the apparently native B873. B820 can also be further
reversibly dissociated to free BChl and the α- and β-polypeptides (7).
Importantly, it was demonstrated that both B820 and B873 could be
reconstituted from BChl $\underline{a}$ and the separately isolated α- and β-
polypeptides (7,8).

The methodology developed for preparation of a structural subunit of
R. rubrum has been applied to wildtype Rhodobacter sphaeroides,
Rhodobacter capsulatus and Rhodopseudomonas viridis, as well as to a
B800-850⁻ mutant of each of the first two organisms. In this paper, the
properties of the B820-type complexes prepared from each of these
organisms will be compared with that from R. rubrum. Also discussed are
the results of probing the BChl binding site in B820 and B873 by using
BChl analogs in a reconstitution assay. Preliminary results of
reconstitution of B820 and B873 using hybrid polypeptides are summarized.
Finally, our progress in reconstitution of the PRC from reconstituted B820
and the RC will be described.

Molecular Biology of Membrane-Bound Complexes in Phototrophic Bacteria
Edited by G. Drews and E. A. Dawes
Plenum Press, New York, 1990

METHODS AND MATERIALS

The method of preparation of B820 from R. rubrum has been described (6) as have the methods for the BChl analog and hybrid polypeptide reconstitution assays (7-9). The details for the method of preparation of the structural subunit from Rb. sphaeroides and Rb. capsulatus will be published elsewhere. A preliminary report of these latter preparations has been given (10).

The β-polypeptide and a combination of α- and β-polypeptides were titrated with BChl to determine the stoichiometry of pigment binding. Analysis of the data obtained in these experiments was carried out as follows. At those points in a titration when sufficient number of BChl species existed so that it was uncertain as to the absorbance due to B820 or B873, a deconvolution of the spectra was performed. An example of this treatment is shown in Figures 1 and 2. For this data, four Gaussian absorption bands were adjusted until the best fit was obtained. We are grateful to Drs. Randy Miller and Kenneth Spears of Northwestern University for use of their software program in performing this analysis. The relative contribution of each component used is indicated in Figure 1 and the values for the λ_{max} and half-width of each band are given in Table 1. Component B is free BChl (in detergent), component C represents B820 and component D is probably mostly due to a BChl aggregate (no protein) observed when the concentration of free BChl at this %OG exceeds about 1 X 10^{-5} M. Component A is probably an oxidized product of BChl. Figure 2 shows the original absorbance curve (solid) compared with the fitted curve (dashed). As may be seen, the fit of the data is sufficiently good to evaluate the absorbance of the B820 component, assuming the Q_y transition was a single Gaussian absorption band. However, because B820 has a shoulder at 780 nm (5), part of the absorbance attributed to free BChl (component B), is due to B820. This contribution could be estimated for each point in the BChl titration where overlap of the several species was a problem. As a result of these considerations, a procedure was adopted where the deconvoluted spectra were used to estimate B820 and B873 absorbance. For the case shown, 1/4 of the value for the measured absorbance at 777 nm was used to subtract from the measured absorbance at 816 nm. No correction was made for component D because part of this long wavelength contribution arises from a 9% contaminant of the β-polypeptide by the α-polypeptide which would give rise to a small amount of 865 absorbance from some B873 formation under these conditions.

A similar deconvolution of the absorbance spectrum and estimate of B820 and B873 components was conducted for the titration of the combined α- and β-polypeptides where significant overlap of these multiple species occurred. In this case, it was assumed that component D was entirely due to B873.

The α- and β-polypeptides were separately prepared and their UV absorption spectra recorded in hexafluoroacetone. The molar extinction coefficient of each in this solvent has been determined to be 10,200 cm^{-1} (7). BChl a with a phytyl esterifying alcohol was obtained from Sigma Chemical Co. and its purity checked by HPLC. It was dissolved in acetone and an aliquot added to 4.5 % OG where its molar extinction coeficient has been determined to be 55,000 cm^{-1} (11). For calculations of the concentration of B820 and B873, molar extinction coeficients of 87,000 cm^{-1} and 120,000 cm^{-1} were used (11).

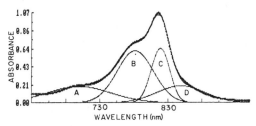

Figure 1. Gaussian components used to fit the absorption spectra of B820 formed with BChl a (phytyl) and the β-polypeptide (1.1 X 10^{-5} M) of R. rubrum. The total BChl concentration was 1.25 X 10^{-5} M.

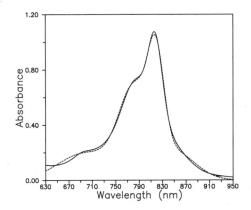

Figure 2. Absorption spectrum of reconstituted B820 compared with the fitted curve composed of 4 Gaussian bands (see Table 1).

TABLE 1
Parameters Used in Curve Fitting

Component	$\lambda_{max.}$ (nm)	Band Width (nm)
A	709	80
B	783	59
C	820	37
D	864	46

RESULTS AND DISCUSSION

Comparison of B820 Properties

    The absorption spectra of the structural subunit (B820) of the core
LH complexes of R. rubrum, Rb. capsulatus, and Rb. sphaeroides are
compared in Figure 3.   As may be noted, the spectra are very similar
although there is a little variation in the location of the Q_y band.   It
was found to be at 816, 820, and 825 nm in Rb. capsulatus, R. rubrum, and
Rb. sphaeroides, respectively.   These complexes have accordingly been
named B816, B820, and B825.

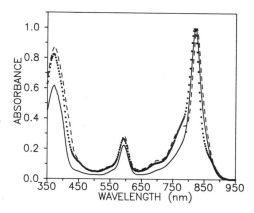

Figure 3.   Comparison of absorption spectra of B820 complexes prepared
from R. rubrum (——), Rb. capsulatus (···) and Rb. sphaeroides (---).   For
comparison purposes, the spectrum of each sample was normalized to an
absorbance of 1 at 820 nm.

    The circular dichroism (CD) spectrum of each complex was also very
similar, each with a minimum near the Q_y λ_{max} and a maximum at about 780
nm (10).   Thus, the CD data and the absorption spectra strongly support
the conclusion that the detailed structural features of each are nearly
identical.

    Each B820-type complex can be reversibly dissociated to free BChl a and
the α- and β-polypeptides, as well as reassociated to form a B875-like
complex.   The B820 from Rb. sphaeroides and R. rubrum are moderately
stable in the dark at room temperature, whereas that from Rb. capsulatus
is much less stable.   Since the core LH from both Rb. sphaeroides and Rb.
capsulatus have the same chromophore structure (BChl a with a phytyl
esterifying alcohol), the source of the relative   instability in Rb.
capsulatus B820 must lie in differences in the α- and β-polypeptides.

Reconstitution of B820 from Separately-Purified Components

    As demonstrated with the isolated α- and β-polypeptides from R. rubrum,
the B820- and B873-type complexes could be reconstituted with the
homologous polypeptides from Rb. capsulatus and Rb. sphaeroides under
appropriate conditions (9).   In each of these cases, BChl a failed to
show red-shifted species with only the α-polypeptide under B820- or B873-
forming conditions, but the corresponding B820-type complexes were easily
formed with the β-polypeptide only (see Figure 4).   This implies that   in

the B820 subunit complex, the β-polypeptide is primarily responsible for binding BChl. In Rb. capsulatus, as with R. rubrum, the α-polypeptide must also be present for B873 formation (9).

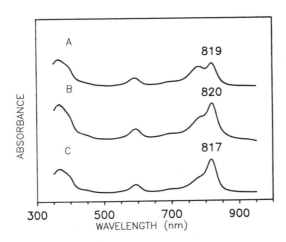

Figure 4. Comparison of reconstituted B 0 using the homologous β-polypeptide only from Rb. sphaeroides (A), Rb. capsulatus (B), or R. rubrum (C) and BChl a (phytyl).

Stoichiometry of BChl Binding

Because BChl is so tightly bound to the protein in B820 and B873 of R. rubrum, it is possible to determine the amount of BChl bound to the α- and β-polypeptides by titration with increasing amounts of BChl. It is, of course, important that a high percentage of the α- and β-polypeptides are native (that is, without denatured or unreactive protein present) if a reliable quantitative stoichiometry is to be obtained. A titration of a system containing both the α- and β-polypeptides (in a 3:2 ratio so that the β-polypeptide was limiting) using BChl a with a phytyl esterifying alcohol is shown in Figure 5. From the data, it is observed that as the BChl concentration is increased (at 0.73% OG), the population of B820 first increases, but then it shifts to B873 at higher concentrations. By visual inspection, it is apparent that the protein was saturated with BChl at about the last two additions. The last sample was chilled overnight (4° C) to shift all B820 to B873. A plot of the amount of B820+B873 formed versus the concentration of BChl added shows a linear increase until saturation of the complex (see Figure 6). The ratio of BChl to the β-polypeptide of the complex, when an excess of the α-polypeptide is present, approaches 2, as would be expected for completely active protein and full reconstitution. Thus, the result is gratifying in that a fully reconstituted native complex may be formed. Three other titrations (one using BChl a with a geranylgeranyl esterifying alcohol) gave similar results.

A similar titration of an identical sample of the β-polypeptide was conducted, but without the α-polypeptide present. These results are shown in Figure 7. Again, it is apparent that the first additions of BChl a (again the phytyl tail derivative) produce nearly pure B820-type spectra. As more BChl was added, the saturation of the complex was evident. The data were plotted (with the aid of deconvolution of the spectra as described in the methods section) as the amount of B820 formed versus the

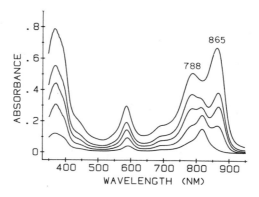

Figure 5. Titration of the α- and β-polypeptides (3:2 ratio) of R. rubrum with increasing amounts of BChl a (phytyl). The curves from the lowest to highest total BChl concentration are 6.2, 12.4, 18.6, 24.8, and 37.2 μM. The β-polypeptide was 1.1 X 10⁻⁵ M.

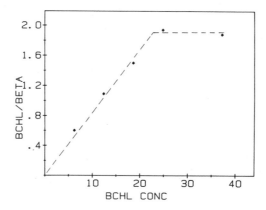

Figure 6. Combined concentration of BChl in B873 and B820 forms at increasing BChl concentrations. To obtain values for the Y-axis, the amount of BChl in the combined B820 and B873 forms was divided by the concentration of the β-polypeptide. The BChl concentrations given on the X-axis are in μM.

added BChl concentration (see Figure 8). Because no attempt was made to correct for aggregated BChl which occurs increasingly at higher free BChl concentrations at this % OG, the titration does not become completely flat at higher concentrations of BChl. Even so, it appears clear that 1 BChl is bound/β-polypeptide, that is, half the amount bound when the α-polypeptide is also present. These results are particularly interesting in view of the spectroscopic data (e.g., absorbance and CD) that indicate that 2 molecules of BChl interact to explain the properties of the B820 complex (11). Thus, B820 formed with β only may be a dimer (2 BChl·2β).

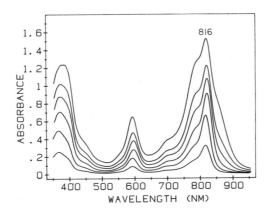

Figure 7. Titration of the β-polypeptide of R. rubrum with BChl a (phytyl). The curves from lowest to highest total BChl concentration are 3.1, 6.2, 9.3, 12.4, 15.5, and 21.7 μM. The β-polypeptide was 1.1 X 10^{-5} M.

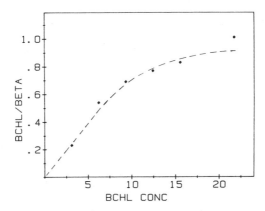

Figure 8. Concentration of BChl in B820 with increasing BChl concentration. To obtain the values for the Y-axis, the amount of BChl in B820 was divided by the concentration of the β-polypeptide present. The BChl concentration given on the X-axis are in μM.

Reconstitution with BChl Analogs

Using the separately-isolated α- and β-polypeptides of R. rubrum and a series of BChl a analogs, evidence was obtained for the importance of three functional groups and of the oxidation state of the macrocyclic ring for B820 and B873 formation, as well as for a role for the esterifying alcohol (tail) in stabilizing the B820 complex. The structure of BChl a is shown below with the changes indicated for the analogs tested. The functional groups that have so far been identified as being required for binding are enclosed within the solid ovals. It was expected that Mg^{++} would be a required component, as it provides a site for ligand coordination from the protein. The groups attached to the $C-13^2$ position seem to be quite important, as the two analogs with changes at that location (pyroBChl <u>a</u> and 13²-hydroxyBChl <u>a</u>) failed to form red-shifted species. A similar requirement for the C-3 acetyl group was found, as indicated by the lack of formation of red-shifted species with BChl g. Thus, the C-13² carbomethoxy group and the C-3 acetyl groups are probably involved in hydrogen bonding to the protein in B820 and B873.

R = farnesyl (f), geranylgeranyl (gg), or phytyl (p)

BChl a$_p$ = unmodified structure + p;

BChl a$_{gg}$ = unmodified structure + gg;

BChl b$_p$ = C + p;

BChl g$_f$ = A + C + f;

PyroBChl a = E + p;

13²-OH BChl a = D + p;

Et BChl a = unmodified structure + R = -OCH$_2$CH$_3$;

2-hydroxyethyl pyroBChl a = pyroBChl a + R = -O(CH$_2$)$_2$OH;

Chl a$_p$ = A + B + p;

3-desvinyl-3-acetyl Chl a$_p$ = B + p;

Chl b$_p$ = A + p + B* (where 7-CH$_3$ is replaced by CHO).

The oxidation state of the macrocyclic ring is also important in forming red-shifted species. Three analogs having chlorin oxidation states, 3-acetyl chlorophyll (Chl) <u>a</u>, Chl <u>a</u>, and Chl <u>b</u>, all failed to show spectral red shifts with the protein. The first of these structures is exactly like BChl <u>a</u>, except for the oxidation of ring II. These results suggest that the more flexible BChl ring system may be required for appropriate binding.

<u>Reconstitution Using Hybrid Mixtures of Polypeptides</u>

Reconstitution experiments using mixed polypeptides from different bacteria have been initiated (9). For example, with a β-polypeptide from <u>R</u>. <u>rubrum</u> and an α-polypeptide from <u>Rb</u>. <u>capsulatus</u>, the formation of a B820 complex is inhibited (Figure 9; top spectrum), but with the β-polypeptide from <u>Rb</u>. <u>capsulatus</u> and the α-polypeptide from <u>R</u>. <u>rubrum</u>, a B873 complex is formed (Figure 9; bottom spectrum). Similarly, with the β-polypeptide from <u>Rps</u>. <u>viridis</u> and the α-polypeptide from <u>R</u>. <u>rubrum</u>, a very nice B873-like complex was formed with a λ_{max} at 875 nm (9), but with the β-polypeptide from <u>R</u>. <u>rubrum</u> and the α-polypeptide from <u>Rb</u>. <u>sphaeroides</u> B800-850 complex no B820- or B873-type complexes formed (data not shown). Thus, a specificity of binding between the α- and β-polypeptides is indicated, and further hybrid reconstitution assays may indicate some of the important elements in this interaction.

<u>Formation of a Reassociated PRC Using Reconstituted B820 and RC</u>

The B820 complex of <u>R</u>. <u>rubrum</u> may be reassociated to form B873 in the presence of phospholipid and RC (prepared from the same bacteria) to form active PRC. The assay used for a functional PRC requires the measurement of the quantum yield for light energy utilization when it is absorbed by the reassociated LH complex (e.g., at 890 nm). Experimental details for the method of preparation and measurement will be published elsewhere. By varying the molar ratio of RC/LH, the specificity of interaction can be assessed (see Figure 10). Thus, a high efficiency is observed at low RC/LH ratios but a maximum quantum yield of .5 was found at high RC/LH ratios. The latter would be understandable if the RC always packs into a proteoliposome with one orientation, whereas the LH can be incorporated randomly. Similar reassociation experiments are being conducted in which the B820 complex is reconstituted, rather than that prepared directly from chromatophores. Thus, using reconstituted B820 complexes, it should be possible to begin to systematically examine those features of the LH polypeptides that are important for association of this complex with the RC to form the PRC.

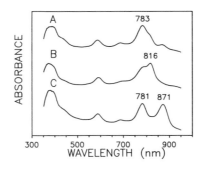

Figure 9. Attempted reconstitution of B820 and B873 using heterologous polypeptides. (A) β-polypeptide from R. rubrum with the α-polypeptide from Rb. capsulatus and BChl a (phytyl). (B) β-polypeptide from Rb. capsulatus and BChl a (phytyl). (C) β-polypeptide from Rb. capsulatus with the α-polypeptide from R. rubrum and BChl a (phytyl).

SUMMARY

Development of the biochemical methodology for preparation of structural subunits of the core LH complex and their reconstitution from individual components has provided a tool for probing structure-function relationships at three different levels: BChl binding, polypeptide interaction within the LH complex, and interaction between the LH complex and the RC. Further cultivation of this experimental approach should provide improved knowledge about how BChl is bound and how the LH complex performs its function.

ACKNOWLEDGEMENTS

This research was supported by grants to P.A.L. from the U.S. Public Health Service (GM 11741) and the National Science Foundation (DMB-8717997). Part of the work was supported by the U.S. Department of Energy, Office of Basic Energy Sciences, Division of Chemical Sciences under contract W-31-109-Eng-38 (T.J.M.).

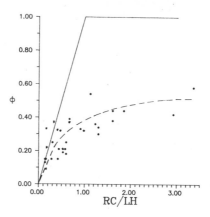

Figure 10. Quantum yield for light utilization in reassociated PRC. The solid line represents an ideal efficiency.

REFERENCES

1. Williams, J. C., Steiner, L. A., Ogden, R. C., Simon, M. I., and Feher, G. (1983) Proc. Natl. Acad. Sci. U.S.A. 80, 6505-6509.
2. Youvan, D. C., Bylina, E. J., Alberti, M., Begush, H., and Hearst, J. E. (1984) Cell 37, 949-957.
3. Michel, H., Weyer, K. A., Gruenberg, H., and Lottspeich, F. (1985) EMBO J. 4, 1667-1672.
4. Cogdell, R. J. (1986) Encyclopedia of Plant Physiol., Vol. 19, Eds. L. A. Staehlin and C. J. Arntzen, Ch 6.2, pp 252-259.
5. Loach, P. A., Parkes, P. S., Miller, J. F., Hinchigeri, S. B., & Callahan, P. M. (1985) in Cold Spring Harbor Symposium on Molecular Biology of the Photosynthetic Apparatus (Arntzen, C., Bogorad, L., Bonitz, S., & Steinback, K., Eds.) pp 197-209, Cold Spring Harbor Laboratory, Cold Spring Harbor, NY.
6. Miller, J. F., Hinchigeri, S. B., Parkes-Loach, P. S., Callahan, P.M., Sprinkle, J. R., & Loach P. A. (1987) Biochemistry 26, 5055-5062.
7. Parkes-Loach, P. S., Sprinkle, J. R., & Loach, P. A. (1988) Biochemistry 27, 2718-2727.
8. Loach, P. A., Michalski, T., and Parkes-Loach, P. S. (1989) VIII International Congress on Photosynthesis, Stockholm, Aug 6-11.
9. Parkes-Loach, P. S., Heller, B. A., Chang, M. C., Bass, W. J., Chanatry, J. A., and Loach, P. A. (1989) VIII International Congress on Photosynthesis, Stockholm, Aug 6-11.
10. Heller, B. A., Parkes-Loach, P. S., Chang, M. C., and Loach, P. A. (1989) VIII International Congress on Photosynthesis, Stockholm, Aug 6-11.
11. Chang, M. C., Callahan, P. C., Parkes-Loach, P. S., Cotton, T. C., and Loach, P. A. Biochemistry, submitted.

QUARTERNARY STRUCTURE OF THE B875 LIGHT-HARVESTING

COMPLEX FROM RHODOSPIRILLUM RUBRUM G9+

R. Ghosh[1], J. Kessi[2], H. Hauser[2], E. Wehrli[3] and R. Bachofen[1]

[1]Institute for Plant Biology, University of Zürich, Zollikerstr.107, CH-8008 Zürich,

[2]Laboratory for Biochemistry and [3]Department for Electron Microscopy, ETH-Zürich,

Universitätsstr. 16, CH-8092 Zürich, Switzerland

INTRODUCTION

The structure of the light-harvesting complexes in purple non-sulphur bacteria is a subject of intense research at present (see Zuber, 1985). A particularly well-characterized example is the B875 complex from *Rhodospirillum rubrum*. The single type of light-harvesting complex (B875) in this organism is formed from two non-identical polypeptides, α and β, respectively, and possess 2 mol bacteriochlorophyll (BChl) and 1 mol spirilloxanthin/mol αβ dimer, respectively (Picorel et al. 1983). The sequences and topology of the complex have been determined (Brunisholz et al. 1984, Meister et al. 1985, Brunisholz et al. 1986, Bachofen und Wiemken, 1986). However, the quarternary structure of the complex is a subject of some controversy (Miller et al. 1987, Ghosh et al. 1988, Hunter et al. 1988). Recently Loach and coworkers isolated a complex absorbing at 820nm (B820) which was suggested to be a subunit form of the B875 complex. However, the precise molecular weight of the complex and its *in vivo* significance remained unclear. In this paper we examine the molecular properties of the B820 complex in more detail and suggest a possible model for the role of the B820 complex *in vivo*.

MATERIALS AND METHODS

1. Isolation of the B875 complex

The B875 complex was isolated from chromatophores of *R. rubrum* G9+ essentially according to Picorel et al. (1983) with the following modifications: (a) the B875 was eluted from the DEAE-cellulose column using a mixture of detergents (0.05% LDAO and 0.1% β-octylglucoside (βOG); (b) the DEAE chromatography was performed twice. Finally the eluted B875 was dialysed extensively against 20mM TrisHCl pH 8.0 at 4°C and then collected by centrifugation. The pellet was resuspended in a small amount of buffer to give approx. 8mg protein/ml and aliquots frozen rapidly and stored at -70°C. As judged by their reassociation properties and spectra, samples stored at -70°C are stable for at least 2 yrs. Protein was determined by the modified Lowry method of Peterson et al. (1977) and the purity of the complexes was routinely checked by SDS-PAGE followed by silver staining or by FPLC chromatography.

2. Analytical ultracentrifugation

B820/B775 complexes were prepared by adding octyl-pentaoxyethylene (OPOE) to 0.48 mg B875 complexes to a final concentration of 0.4 or 0.8 % respectively. The concentration of

Molecular Biology of Membrane-Bound Complexes in Phototrophic Bacteria
Edited by G. Drews and E. A. Dawes
Plenum Press, New York, 1990

245

OPOE required to produce complete dissociation of B875 to B820 was determined by titration directly prior to ultracentrifugation. Analytical ultracentrifugation was performed according to Ghosh et al. (1988) with the exception that the solubilization buffer was 50mM NH_4HCO_3 pH 7.8 containing 10 mM Na ascorbate. Using this buffer photodegradation during the measurement is minimal.

3. Electron microscopy

B875 complexes and reconstituted membranes were examined with both negative contrast and freeze-fracture electron microscopy. Negative staining was performed using 1% uranyl acetate and for freeze-fracture analysis the membranes were prepared in buffer containing 20% glycerol.

4. Reconstitution of the B875 complexes into liposomes

Three methods of reconstitution were tried:

(a) sonication: B875 complexes (170µg) were sonicated at low power for various times (5-10 min) at room temperature in the presence of added phospholipids (DMPC, DOPC,5 µg -2 mg);

(b) dialysis: B875 complexes were dissolved in various detergents (βOG, 1%; LDAO, 0.05%; OPOE, 1%) and mixed with phospholipids solubilized in the same detergent, then dialysed at 4^oC against 5l buffer;

(c) dilution: B875 complexes (170µg) were suspended in 100µl 50mM NH_4HCO_3 pH 8.0 (50NC8) containing 1% βOG, then added to 5µg-2mg sonicated phospholipid suspended in 100µl 50NC8. The mixture was then diluted with 0.5ml 50NC8.

For all methods the reconstituted membranes were collected by centrifugation, the pellet suspended in 0.5ml 50NC8, loaded onto a sucrose gradient (0.4M - 2M) in the same buffer and centifuged for 22h at 26000rpm in a Beckman SW28 rotor. Membrane bands were collected using a Pasteur pipette, the sample centrifuged at 100000g for 1 h and the pellet resuspended in 50µl 50NC8 to use for electron microscopy and determination of the lipid-to-protein ratio.

5. 2-dimensional crystallisation of the B875 complex

B875 complexes (170 µg) were dissolved in 50NC8 containing 1% βOG in a final volume of 100µl and dialysed against various buffers in a microdialysis apparatus (McPherson, 1982) for 3-4 days at 4^oC. Microcrystals were then adsorbed to a carbon-coated copper grid and stained with 1% uranyl acetate for electron microscopy.

6. Isolation of the B868 photounit

Chromatophores were suspended in 10mM sodium phosphate buffer pH 7.5 containing 150 mM NaCl (PBS buffer) to a final concentration of approx. 5mg. protein/ml and diheptanoylphosphatidylcholine (DHPC) added dropwise to a final concentration of 40mM. The solubilized membranes were vortexed rapidly and centrifuged for 20min at 100000g. The supernatant was loaded onto a DEAE-Sephadex column (1cm x 8 cm) equilibrated with PBS buffer containing 1.5mM DHPC. The column was washed extensively with buffer until the absorption at 280 nm returned to the baseline value after the elution of a colourless protein fraction (Fraction I). The column was then eluted with the same buffer containing 1M NaCl (Fraction II) and the coloured fractions pooled and stored at -70^oC after freezing in liquid N_2.

RESULTS AND DISCUSSION

The B820 spectral unit originally isolated by Loach and coworkers (Miller et al. 1987) has been characterized in more detail and we show here its possible significance for the structure of the B875 light-harvesting complex.

Molecular weight of B820

Analytical ultracentrifugation with two concentrations of OPOE at 25^oC in 50NC8 enabled to determine the relative contributions from B820 and B775 to be varied (Fig. 1). Analysis of the sedimentation profiles showed that under both conditions only the molecular species of 26K ($\alpha_2\beta_2$) and 13K ($\alpha\beta$) are present and that the reduction of the B820 form leads to a corresponding loss of the 26K component with a compensating increase in the 13K component. We therefore

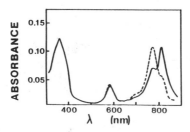

Fig. 1. Absorption spectra of B820/B875 complexes used for analytical ultracentrifugation. Complexes were dissolved in 0.4% (————) and 0.8% (-------) OPOE.

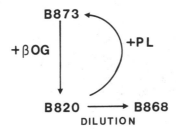

Fig.2. Schematic diagram summarizing the spectral transitions of the B875 complexes after dissolving with βOG to form the B820 complex and then diluting in the presence or absence of phospholipids (PL).

conclude that B820 is a tetramer, possibly $\alpha_2\beta_2(BChl)_4$. We also observed that when the temperature was reduced to 10°C during centrifugation, the complex rapidly aggregated parallel to an increase in absorption at 868nm.

Molecular transitions of the B820 oligomer
Dilution of B820 with 50NC8 to a final concentration below 0.2% leads to the appearance of a further spectral form, B868. Thus in βOG the dissociation of the B875 complex is *not* completely reversible. As noted by Parkes-Loach et al. (1988) prior addition of phospholipids before dilution leads to the reappearance of the B875 complex (Fig. 2).

Ultrastructure of B875 and B868 complexes

Elecron micrographs of negatively-stained preparations of dialysed B875 complexes show them to spontaneously form large vesicular structures, quasi-crystalline in appearance. Freeze-fracture electron micrographs of the B875 structures demonstrate that they contain hexagonally arrayed 10nm particles and are essentially identical in appearance to native chromatophores (Fig. 3a, b). Negatively-stained preparations of B868 however show a homogeneous population of particles with a size of approx. 20 - 50nm (Fig. 3c).
Rosenbach-Belkin et al. (1988) have suggested that the spectral shift and hyperchromic properties of B800-850 and B875 would be well-explained by a dodecamer ($\alpha_6\beta_6$) structure. We propose a rationallization of these ideas as follows: B820 tetramers aggregate spontaneously to form B868 dodecamers, which in the presence of phospholipids may aggregate further to form closed vesicular structures absorbing at 875nm (Fig. 4).

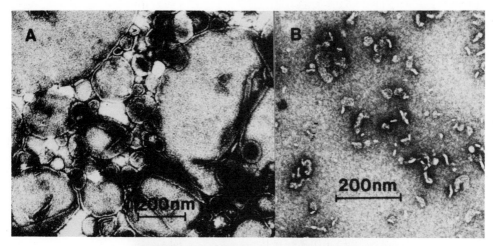

Fig. 3. Electron micrographs of (a) the isolated B875 complexes after negative staining; (b) the B868 complexes after negative staining; (c) freeze-fracture planes of the B875 complexes.

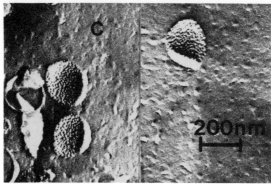

Fig. 4. A tentative model of the B875 structure *in vitro*. The tetramers represent the B820 oligomers which associate to form a B868 dodecamer (dotted line). The B868 structures aggregate to form a two-dimensional array which absorbs at 875nm.

Two-dimensional crystallisation of the B875 complexes

The two-dimensional array shown in Fig. 5 depicts three B868 complexes forming a central "hole" in which the reaction centre could be contained. Fig. 5 shows two-dimensional crystals of the B820 complex formed in the presence of 10mM $MgCl_2$ and 1% βOG. Clearly observable are the hexagonally arranged "holes" with a diameter of approx. 5nm. We are presently improving the quality of the crystals for a more detailled analysis.

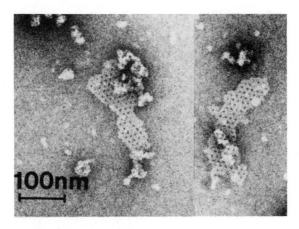

Fig. 5. A two-dimensional crystal of the B875 complex obtained with 10mM $MgCl_2$ and 1% βOG in 50mM NH_4HCO_3 pH8.

Reconstitution of the B875 complexes with phospholipids

We have also tested various methods for reconstitution of the B875 complexes with phospholipids. Sonication and dialysis both lead to reconstitution of the complexes but are generally unsuitable; the former method yields low and irreproducible amounts of reconstituted complexes, and the latter is too slow. In both cases large amounts of B775, corresponding to dissociated complexes, are observed in the final preparations. In contrast, dilution of the B820 complexes with added phospholipids is very efficient and reproducibly gives high yields of reconstituted membranes containing low quantities of B775. We have also found that reconstitution at high lipid-to-protein ratios requires unsaturated phospholipids and is rather insensitive to the nature of the head-group. Thus egg phosphatidylcholine or DOPC yield comparable results to *E. coli* phosphatidylethanolamine, leading to high amounts of B875 after reconstitution. In contrast DMPC yields larger amounts of B775 at high lipid-to-protein ratios. We have examined *both sides* of a B875 membrane reconstituted with DMPC at a lipid-to-protein ratio of 2:1 (mol/mol) using freeze-fracture electron microscopy (Fig. 6). This amount of lipid essentially only "fills the holes" of the preexisting B875 structure without diluting the protein within the lipid bilayer. Under these conditions it is apparent that the particle density is significantly less on the inner side of the vesicular structure and suggests that the particles are asymmetrically distributed and possibly conical in shape.

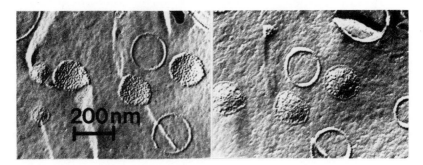

Fig. 6. Freeze-fracture electron micrographs showing *both sides* of the membrane fracture plane of B875 complexes reconstituted with DMPC at a molar lipid-to-protein ratio of 2/1.

Isolation of a B868 photounit

Using a new phospholipid detergent, diheptanoylphosphatidylcholine (DHPC) we have been able to isolate the photosynthetic unit in a homogenous and active form. Solubilization of the chromatophore membrane with DHPC followed by DEAE-Sephadex chromatography yields a coloured product (fraction II) absorbing at 868nm (Fig. 7). This component chromatographs as a single peak during Sepharose 6B gel filtration (data not shown). Negative-staining of the 868nm component shows a homogeneous distribution of particles essentially identical to the B868 units described above but with a length of approx. 30 nm (Fig. 8). However, the DEAE-Sephadex eluate also contains reaction centres, cytochrome c_1, and possibly cytochrome b. We believe that this new complex represents the basic photosynthetic unit of *R. rubrum* and it suggests a weak but definite physical link between the reaction centres, cytochrome bc_1 complex and B868 units. We are presently investigating the properties of this complex further.

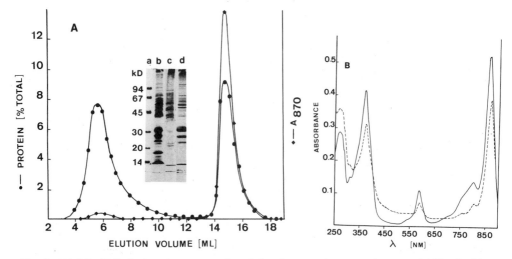

Fig. 7. (A) DEAE-Sephadex chromatography of the chromatophore membranes solubilized with DHPC as described in the text. The insert depicts the SDS-PAGE profiles of (a) the protein standard; (b) chromatophore membranes; (c) fraction I and (d) fraction II. (B) absorption spectrum of the B868 photounit (fraction II, solid line). The dotted line shows the absorption spectrum of isolated chromatophores (not normallized)

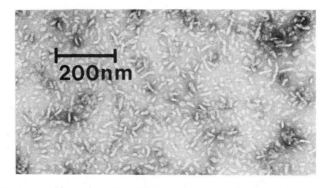

Fig. 8. Electron micrographs of the B868 photounit (fraction II) after negative staining.

SUMMARY

The data presented suggest that the B873 light-harvesting complexes contain most of the necessary information for the assembly of chromatophore-like structures. We present a model for chromatophore morphogenesis in which B868 dodecamers are formed from B820 tetramers, associate with the reaction centre and the cytochrome bc_1 complex to form large two-dimensional arrays, thus building the skeleton for the final chromatophore membrane.

Acknowledgments

The generous support of our research by the Swiss National Science Foundation is greatly acknowledged (Grant 3.243-0.85).

REFERENCES

Bachofen, R. and Wiemken, V. (1986) Encycl. Plant Physiol. 19, 620-631.

Brunisholz, R. A., Suter, F. and Zuber, H. (1984) Hoppe-Seyler's Z. Physiol. Chemie 365, 675-688.

Brunisholz, R. A., Zuber, H., Valentine, J., Lindsay, J. G., Woolley, K. J. and Cogdell, R. J. (1986) Biochim. Biophys. Acta 849, 295-303.

Ghosh, R., Hauser, H. and Bachofen, R. (1988) Biochemistry 27, 1004-1014.

Hunter, C. N., Pennoyer, J. D., Sturgis, J. N., Farrelly, D. and Niederman, R. A. (1988) Biochemistry 27, 3459-3467.

McPherson, A. (1982) Preparation and Analysis of Protein Crystals, Wiley-Interscience.

Meister, H.-P., Bachofen, R., Semenza, G. and Brunner, J. (1985) J. Biol. Chem. 260, 16326-16331.

Miller, J. F., Hinchigeri, S. B., Parkes-Loach, P. S., Callaghan, P. M., Sprinkle, J. R., Riccobono, J. R. and Loach, P. A. (1987) Biochemistry 26, 5055-5062.

Parkes-Loach, P. S., Sprinkle, J. R. and Loach, P. A. (1988) Biochemistry 27, 2718-2727.

Peterson, G.L. (1977) Analyt. Biochem. 83, 346-356.

Picorel, R., Belanger, G. and Gingras, G. (1983) Biochemistry 22, 2491-2497.

Rosenbach-Belkin, V., Braun, P. and Scherz, A. (1988) in: Photosynthetic Light-Harvesting Systems (Scheer H. and Schneider S., eds), Walter de Gruyter, Berlin, p. 323-325.

Zuber, H. (1985) Photochem. Photobiol. 42, 821-844.

STUDIES ON THE FUNCTION OF THE *pufQ* GENE

PRODUCT IN BACTERIOCHLOROPHYLL BIOSYNTHESIS

William R. Richards and Shafique Fidai

Department of Chemistry
Simon Fraser University
Burnaby, B.C., Canada V5A 1S6

INTRODUCTION

The *pufQ* gene is an open reading frame, located between an oxygen-regulated promoter region and the *pufB* gene (which codes for the B870β light-harvesting protein, in the *puf* operon of *Rhodobacter capsulatus*[1,2] and *Rhodobacter sphaeroides*.[3] Evidence has accumulated in a number of laboratories that expression of the *pufQ* gene is required for normal levels of bacteriochlorophyll (Bchl) biosynthesis. The *pufQ* gene was shown to employ frequently used codons and to be induced by the same environmental factors which led to the induction of the other known *puf* genes.[1] Although strains from which the entire *puf* operon had been deleted were still able to form some Bchl (appearing in the form of the B800-850 [LHII] complex), restoration of the *pufQ* gene in *trans* greatly increased Bchl synthesis.[4,5] In addition, Bauer and Marrs[4] have found that the amount of Bchl produced was directly proportional to the amount of the *pufQ* gene product expressed. These experiments were performed in strains of *R. capsulatus* genetically manipulated to first remove the *puf* operon, and then insert the *pufQ* gene on a plasmid in which its transcription could be regulated by the nature of the nitrogen source. It is not known, however, where the *pufQ* gene product (or Q-protein) exerts its effect. Bauer and Marrs[4] have shown that the expression of *pufQ* does not appear to regulate the transcription (or translation) of *bch* genes. Other possibilities, however, include: activation of the *bch* gene products (the biosynthetic enzymes), activation of their substrates (magnesium tetrapyrrole intermediates), transport of intermediates into (or across) biomembranes, or regulation of the assembly of light-harvesting (LH) and/or reaction center (RC) Bchl-protein complexes.

The Q-protein contains 74 amino acids (M_r 8556 Da)[4] similar in size to the Bchl-binding LH proteins.[6] Translation of the nucleotide sequence of the *pufQ* gene reveals that the gene product has little homology with any of the 4 LH proteins, however, and has instead some limited homology with portions of both the L- and M-subunits of the photosynthetic RC (both of which bind both Bchl and ubiquinone molecules).[4] A hydropathy plot reveals a hydrophobic region of 21 amino acids, indicating that it may be an intrinsic protein which either spans the membrane one time, or is at least firmly anchored in it.[2] It also has 2 His and 3 Gln residues in its N-terminal hydrophilic domain, the side chains of which could provide sites for chelation of magnesium tetrapyrroles. The *pufQ* gene

Molecular Biology of Membrane-Bound Complexes in Phototrophic Bacteria
Edited by G. Drews and E. A. Dawes
Plenum Press, New York, 1990

253

product might, therefore, act as a membrane-bound carrier protein for Bchl intermediates, with which membrane-bound (or even water-soluble) biosynthetic enzymes could complex during Bchl synthesis.

The idea of a carrier protein for Bchl synthesis was first proposed by Lascelles in 1966.[7] She based her proposal on the fact that intermediates excreted by mutants of *R. sphaeroides* unable to form Bchl appeared not to be free, but bound in the form of pigment-protein complexes. The presence of a detergent such as Tween 80 in the growth medium increased the yield of the pigment-protein complex.[7] The pigment-protein complexes of several of Lascelles' mutants were subsequently isolated and partially purified in our laboratory in 1975.[8] The pigments were shown to be firmly but not irreversibly bound to the protein, and the protein (which appeared to have a very small molecular weight of ca. 9000 Da) aggregated readily into a series of oligomers.[8] The protein was not characterized further at that time.

In order to test the hypothesis that the Q-protein may be a carrier protein for Bchl synthesis, we have studied Bchl synthesis in strain U43 of *R. capsulatus* from which the entire *puf* operon (from the 54th codon of the *pufQ* gene to just beyond the end of the *pufX* gene) has been deleted.[9] This bacterium, therefore, lacks both the B875 (LHI) and RC complexes, and it also has mutation(s) in the *puc* operon so that the B800-850 (LHII) complex is either missing or defective.[9] Strain U43 was compared to a derivative of U43 to which a plasmid containing a fragment of the *puf* operon (the *pufQ* gene, preceded by its oxygen-regulated promoter, and followed by the *Escherichia coli lacZ* gene fused to a portion of the *pufB* gene[2]) had been inserted (J.T. Beatty, unpublished). The possible accumulation of the Q-protein in the resulting strain, termed U43(pΔ4), was investigated. It and a *bchA*⁻ mutant (mutant 17; J.T. Beatty, unpublished) were also grown in the presence of precursors and inhibitors of tetrapyrrole synthesis. The synthesis of Bchl and/or tetrapyrrole intermediates and the presence of pigment-protein complexes was compared in both strains.

METHODS AND MATERIALS

R. capsulatus strain U43 (originally constructed by Youvan et al[9]), strain U43(pΔ4) (constructed by J. T. Beatty, unpublished, using a DNA segment [pΔ4] described by Adams et al.[2] containing the *pufQ* gene, and strain 17 (*bchA*⁻) (obtained by Dr. J. T. Beatty, unpublished) were grown as described by Bauer and Marrs[4] in either medium RCV or medium RCV⁺ (which contained 0.5% sodium pyruvate, 0.6% glucose, 50 mM dimethyl sulfoxide [DMSO], and, in the case of strain U43(pΔ4), 0.5 mg L⁻¹ tetracycline) as indicated in the Results and Discussion. After growth for 16 h, the bacteria were harvested by centrifugation at 8000 rpm and the bacteria from 200 mL of fully grown cultures were resuspended in 100 mL of fresh medium RCV⁺ with the following additions (as indicated in Table I): 1 mM δ-aminolevulinic acid (together with 0.1 mM L-methionine), 1.5 μM N-methylprotoporphyrin (together with 10 mM glycine, 10 mM sodium succinate, and 0.1 mM L-methionine), and 12 mM nicotinamide, and incubated for a further 24 h. The bacteria were again centrifuged and the cell mass was extracted with 15 mL of dimethylformamide in a hand-held homogenizer. The bacteria were not lysed by this extraction procedure; they were then recentrifuged and the spectrum of the dimethylformamide was recorded. The absorbances due to bacteriopheophytin (at 753 nm) or P631 (at 628 nm) were normalized by dividing by the absorbance at the carotenoid maximum at 495 nm, to correct for differences in yields during the extraction (on the assumption that the carotenoid content remained constant during the incubations). The spectra of the centrifuged media were also recorded in order to detect excreted porphyrins. In large scale (1-2 L) incubations, the bacteria were harvested, resuspended in 20 mL of 50 mM potassium phosphate buffer, pH 7.5, and disrupted by 2 passages through a French pressure cell at 16000 psi. The lysed cells were centrifuged for 20 min at 12000 rpm to remove unbroken cells and larger cellular debris, and the supernatant was then centrifuged for 90 min at 65000 rpm and 4 °C in a Beckman Model L5-75 ultracentrifuge with a Ti75 rotor. The resulting membrane fraction was resuspended in 50% (v/v) glycerol in 0.2 M Tris-Cl, pH 7.8, and stored at -20 °C. A pigment-protein complex containing P662 (2-devinyl-2-hydroxyethylchlorophyllide) was isolated from lysed cells of mutant 17 in a similar manner, except that the membrane fraction was first removed by centrifugation at 42000 rpm, followed by recentrifugation of the resulting supernatant at 65000 rpm. The pigment-protein complex which sedimented under these conditions was lyophilized and stored at -20 °C. Visible spectra were recorded on a Philips Model PU8720 and fluorescence spectra on a Perkin Elmer Model MPF-44B spectrometers. Sodium dodecyl sulfate (SDS)-polyacrylamide gel electrophoresis was carried out as previously described[10] in a Bio-Rad Mini-Protean II gel apparatus.

RESULTS AND DISCUSSION

Strains of *R. capsulatus* from which the entire *puf* operon has been deleted (but which still contain a functional *puc* operon), form small amounts of Bchl which accumulates in the membrane as the B800-850 (LHII) complex.[3-5] When a plasmid containing only the *pufQ* gene was inserted into such a strain (ΔRC6) by Klug and Cohen,[5] approximately 3.0 times as much Bchl was formed. By comparrison, when a plasmid containing the entire *puf* operon was inserted, 6.7 times as much Bchl was formed.[5] The U43 strain of *R. capsulatus*, on the other hand, forms no Bchl-binding LH or RC proteins, and indeed, both Youvan et al.[9] and Klug and Cohen[5] were unable to detect any Bchl synthesis at all when strain U43 was grown semiaerobically in medium RCV. However, when the strain was grown anaerobically in medium RCV+ (which contains DMSO as the terminal electron acceptor), we found that significant amounts of a Bchl derivative were formed (Table I). The *in vivo* spectrum of whole cells grown under such conditions showed only a maximum at 755 nm, indicating that the Bchl formed had been converted to bacteriopheophytin (Bpheo). When membranes (consisting mainly of fragments of the cytoplasmic membrane) were prepared from the bacteria, no absorbances due to either Bchl or Bpheo were detected in the membrane fraction. Hence, the Bchl and/or Bpheo must have either not been located in the membrane at all, or else must have been only loosely associated with the membrane such that it was lost during the membrane preparation.

Table I. Incubation of *R. capsulatus* strains in DMSO-containing media.

strain	additions[a]	absorbance ratios 753/495	628/495	relative absorbance Bpheo	P631[b]
U43	none	0.077	---	0.28	---
U43	NMP	1.182	---	4.36	---
U43	ALA	0.067	---	0.24	---
U43 (pΔ4)	none	0.271	---	1.00	---
U43 (pΔ4)	nicotinamide	0.234	0.091	0.86	1.0
U43 (pΔ4)	NMP	0.312	---	1.15	---
U43 (pΔ4)	NMP + nicotinamide	0.216	0.164	0.80	1.8
U43 (pΔ4)	ALA	0.285	---	1.05	---
U43 (pΔ4)	ALA + nicotinamide	0.243	0.132	0.90	1.4

[a]NMP = N-methylprotoporphyrin; ALA = δ-aminolevulinic acid
[b]A mixture of MgDVP and MgMVP

Strain U43 was also incubated in the presence of N-methylprotoporphyrin (NMP) and δ-aminolevulinic acid (ALA), in order to induce the
bacteria to produce more porphyrin intermediates, and, perhaps, more
Bchl. The presence of NMP, an inhibitor of ferrochelatase,[11] should
reduce the concentration of heme, thereby reducing the feedback
inhibition of ALA synthetase, while the presence of ALA should bypass
this regulation site. The results (Table I) demonstrated that Bchl
synthesis was greatly stimulated by the presence of NMP, but not by the
presence of ALA. Pigments were extracted from the cells by
dimethylformamide; however, the Bchl had again lost its magnesium and was
isolated in the form of Bpheo. The inclusion of ALA in the incubation
medium caused the bacteria to excrete a large amount of a porphyrin with
a spectrum identical to coproporphyrin. Little or no free porphyrin was
present in the centrifuged medium of the incubation with NMP.

When strain U43(pΔ4) was grown in medium RCV[+] (containing DMSO),
approximately 3.6 times more Bchl was formed than in strain U43 (Table
I). The *in vivo* spectrum of the Bchl formed by strain U43(pΔ4) had a
maximum at *ca.* 855 nm, indicating that the Bchl was in an aggregated or
complexed form of some sort. This absorbance was predominant in freshly
grown cells, but changed in older cultures to include maxima at *ca.* 800
and 755 nm. The latter was likely due to uncomplexed Bpheo; however, the
absorbance at 800 nm may have been due to an aggregated or complexed form
of Bpheo (or Bchl). The same absorbances were also found in membranes
prepared from strain U43(pΔ4) which had been grown semiaerobically in
medium RCV: fresh preparations showed only an absorbance at 855 nm,
while older preparations exhibited bands at 800 and 755 nm as well.
Membranes were prepared from both strains U43 and U43(pΔ4) after having
been grown in medium RCV[+]. Analysis of both membrane fractions by SDS-
polyacrylamide gel electrophoresis indicated that there were no obvious
differences in the polypeptides visible in the two samples, and in
particular, no bands with apparent M_r's of ca. 10000 Da appeared in
U43(pΔ4) which were not also present in U43.

Incubations of U43(pΔ4) were also carried out in fresh medium
containing either NMP or ALA, both in the presence and absence of
nicotinamide. Nicotinamide has been shown to cause the parent strains of
Rhodocyclus gelatinosa[12] and *R. sphaeroides*[13] to excrete magnesium 2,4-
divinylpheoporphyrin a_5 (MgDVP or P631), plus traces of magnesium 2-
vinylpheoporphyrin a_5 (MgMVP or chlorophyllide), into the growth medium.
This was interpreted by Shioi et al.[13] to be due to the inhibition of the
enzyme responsible for the reduction of the 4-vinyl group of MgDVP to the
4-ethyl group found in MgMVP (ie. the 4-vinylreductase). The results are
also shown in Table I. In this case, the formation of Bchl was not
greatly affected by either NMP or ALA; however, nicotinamide caused a
small depression in Bchl synthesis. The inclusion of nicotinamide also
led to the accumulation of a pigment (P631) with absorbance maxima at 628
and 440 nm (most likely a mixture of MgDVP and MgMVP). In this case, NMP
and ALA both had a stimulatory effect on the production of P631.

A large scale incubation of strain U43(pΔ4) was carried out for 16 h
in medium RCV[+] containing NMP and nicotinamide. Membranes were prepared
from some of the bacteria following this incubation. The remainder were
centrifuged at 4°C, washed with cold (4°C) buffer, resuspended in fresh
medium containing no nicotinamide, and incubated for an additional 2.5 h.
The cells were recentrifuged, and a second membrane fraction prepared
from them. Spectra of the membrane fractions indicated that P631 (with
prominent absorbances at 635 and 446 nm due to MgDVP and/or MgMVP) had
accumulated in the membrane after incubation in the presence of NMP and

nicotinamide. Also, maxima at 416 and 690 nm were present, together with the same bands (at 755, 800, and 850 nm) visible in membranes isolated from U43(pΔ4) grown in the absence of nicotinamide. The fluorescence emission spectra of the membrane exhibited the following fluorescence maxima (wth probable assignments in parentheses): (a) excitation at 443 nm: 640 nm (MgDVP) and 698 nm (unknown); (b) excitation at 400 nm: the above two maxima plus 765 nm (Bpheo) and a large emission band at 800 nm (Bchl); and (c) excitation at 357 nm: 714 (2-desacetyl-2-hydroxyethyl-bacteriochlorophyllide or P720). No emission at 850-860 nm was observed by excitation at either 400 or 357 nm. The fluorescence spectra of the membrane fraction isolated from cells reincubated in nicotinamide-free media revealed that the 640 nm peak had shifted to 635 nm and that, while the ratio of the intensities of the 635 nm peak to the 765 nm peak had not changed, the intensity of the 698 nm peak had diminished with respect to both of the other bands. The cause of these changes and the exact nature of the pigments responsible for the observed fluorescence maxima are under current investigation. However, it is possible that the nicotinamide not only inhibited the reduction of the 4-vinyl group of MgDVP by 4-vinylreductase (as suggested by Shioi et al.[13]), but also more severely inhibited the reduction of MgMVP (protochlorophyllide) to chlorophyllide by protochlorophyllide reductase (PCR). Incomplete removal of nicotinamide from the cells during the reincubation might have allowed at least partial conversion of MgDVP to MgMVP (responsible for the shift in the fluorescence maximum from 640 to 635 nm), while completely inhibiting the reduction of MgDVP and/or MgMVP by PCR to divinylchlorophyllide and/or chlorophyllide, respectively. Fluorescence maxima at 637-638 nm have been previously observed in mutants of R. capsulatus which excrete P631 (predominantly MgDVP) while maxima at 628 nm were detected in mutants which excreted P720 or P730 (G. Ross and W. R. Richards, unpublished). It is not known what pigment is responsible for the fluorescence maximum at 698 nm. Such an emission was previously observed in a P631 mutant, and it appeared that it could have been due to MgDVP in the form of a tetrapyrrole-carotenoid-protein complex, since it had excitation maxima at 442, 468, and 484 nm (G. Ross and W. R. Richards, unpublished). While it could possibly have been due to a biosynthetically active form of MgDVP, no further information is available on it at this time.

    Mutant 17 of R. capsulatus is deficient in the bchA gene product (the chlorin reductase enzyme) and accumulates 2-devinyl-2-hydroxyethyl-chlorophyllide (P662) in the growth medium. It was also grown in medium RCV+ and incubated in fresh medium containing NMP in the presence of nicotinamide. In this case, both P631 and P662 were produced. Much of the pigment was excreted from the cells; however, a good portion remained in the cells. After preparation of membranes from such cells, it was found that while some of the pigments had sedimented with (and were presumably contained in) the membrane, a portion was still present in the supernatant. The supernatant was recentrifuged at a higher speed (cf. Methods) and the remaining pigment sedimented in the form of a pigment-protein complex. The membranes and pigment-protein complex from mutant 17 and the membranes (which had accumulated P631 and other intermediates) from strain U43(pΔ4) after it was incubated in NMP plus nicotinamide, were run on SDS-polyacrylamide gel electrophoresis and compared with membranes prepared from strain U43. In no case were any significant differences observed in the polypeptides contained in any of the membrane fractions, and no polypeptide was present in strain U43(pΔ4) or mutant 17 which was not also present in strain U43 which could be attributed to the presence of detectable amounts of the Q-protein. In the case of the pigment-protein complex from mutant 17, however, two major bands were observed at 31000 and 66000 Da. However, bands of both these apparent M_r-values were also both present in the membranes of strain U43, albeit at much reduced levels.

The above results have confirmed the observation[4,5] that, although the presence of the *pufQ* gene product does facilitate increased Bchl synthesis in *R. capsulatus,* it is not absolutely required for it. The results have also demonstrated that no Bchl-binding proteins need be present in the membrane to accept the newly synthesized Bchl. Although strain U43 can be induced to form fairly large amounts of Bchl (always, however, isolated as its magnesium-free derivative, Bpheo), the Bchl was never found in the membrane fraction, due, no doubt, to the complete lack of Bchl-binding proteins. Strain U43(pΔ4), on the other hand, accumulated Bchl in the membrane. Since the *in vivo* spectrum of this Bchl had a λ_{max} at 855 nm, it apparently existed in an aggregated or complexed form, although this Bchl form was quite unstable and decayed to Bpheo (the latter appearing in either a complexed or free form). Since the only difference between the two strains was that the *pufQ* (and *lacZ*) genes had been added to the latter via a plasmid, it is tempting to assume that it was the Q-protein to which the Bchl was bound, although this has not been directly demonstrated since the Q-protein has never been isolated or identified within the membrane, and no accumulation of a low molecular weight putative Q-protein was observed. Adams *et al.*[2] have demonstrated a relatively low abundane of *pufQ*-encoding mRNA segments, and have speculated that the *pufQ* gene product may have a catalytic, rather than structural function, and be present in very low amounts.

Fluorescence spectroscopy has been used in the past to detect intermediates of Bchl synthesis in both *R. sphaeroides*[14] and *R. capsulatus*[15]. Also, Beck *et al.*[16] have concluded that many of these intermediates were bound in the cytoplasmic membrane within pigment-protein-lipid complexes. Oelze[17] has recently shown that magnesium protoporphyrin monomethyl ester (MgPME or P590) accumulated in a subcellular membrane fraction of *R. sphaeroides* which contained a high content of B875 complexes and may have been a photosynthetic intracytoplasmic membrane precursor fraction derived from the cell membrane.[18] Oelze[17] postulated that this fraction might be the location of Bchl synthesis. While our results have demonstrated that Bchl intermediates can accumulate in the membrane and are presumed to be in some form of a complex, there was no indication that it was the Q-protein to which they were bound, as membranes which had accumulated elevated amounts of intermediates still did not exhibit elevated amounts of a putative Q-protein. A catalytic, rather than a stoichiometric function for the Q-protein is still possible, however, and it may simply facilitate the transport of intermediates into or across the cytoplasmic membrane. Such a transport might occur only very slowly completely unaided (eg. in strain U43), but might be facilitated somewhat by the presence of Bchl-binding (eg. LHII) proteins in the membrane.

Nothing is known about the actual location of the biosynthetic enzymes, other than the S-adenosylmethionine:magnesium protoporphyrin methyltransferase which has been shown to be bound to the cell membrane in mutants unable to form Bchl (W. R. Richards, unpublished). Wellington and Beatty[19] have recently completed sequencing the *bchC* gene, which codes for a late enzyme in the Bchl biosynthetic pathway responsible for oxidation of the 2-hydroxyethyl group of 2-desacetyl-2-hydroxyethyl-bacteriochlorophyllide (P720) to the 2-acetyl group of bacteriochlorophyllide (P770). They have shown that the deduced gene product is less hydrophobic than known integral proteins of *R. capsulatus,* but that there are two more pronounced hydrophobic regions that could interact with a membrane.[19]

The question still arises as to how it was possible to induce fairly substantial levels of Bchl synthesis in strain U43 which completely lacks the *pufQ* gene. The level of Bpheo detected was *ca.* 4 times that produced in U43(pΔ4) which *does* contain the gene (*cf.* Table I). It is possible that the DMSO in medium RCV⁺ in which the U43 was incubated might be mimicing the effect of the *pufQ* gene product by facilitating increased synthesis of Bchl by enzymes already present and active. If the DMSO is indeed mimicing the effect of the Q-protein, it may have been acting to ferry tetrapyrrole intermediates into (or across) the cytoplasmic membrane. DMSO is itself quite permeable to biological membranes and has been used medically in drug delivery systems. In contrast, however, Davis *et al.*[3] found a *decrease* in Bchl synthesis in DMSO media (in comparison to growth in low oxygen) in a *pufQ⁻* mutant of *R. sphaeroides* (PUFB1). This mutant did, however, still contain the genes for LHII and accumulated the Bchl as the B800-850 complex.

Finally, it has been demonstrated that a pigment-protein complex which accumulates inside the cell does not appear to contain a putative Q-protein. This pigment-protein complex may, therefore, be quite different from the pigment-protein complexes excreted into the growth medium by mutants of *R. sphaeroides*.[8]

ACKNOWLEDGEMENTS

Supported by the President's Research Fund of Simon Fraser University and grant A5060 from the NSERC of Canada. We are grateful to Dr. J. T. Beatty for a gift of the bacterial strains and also for helpful discussions. We are also grateful to Drs. T. J. Borgford and Gabriel Kalmar for suggestions, aid, and helpful discussions concerning molecular biology, and to Rajinder Singh for isolation of the pigment-protein complex from mutant 17.

REFERENCES

1. C. E. Bauer, D. A. Young, and B. L. Marrs, Analysis of the *Rhodobacter capsulatus puf* Operon. Location of the Oxygen-Regulated Promoter Region and the Identification of an Additional *puf*-Encoded Gene, J. Biol. Chem. 263:4820 (1988).
2. C. W. Adams, M. E. Forrest, S. N. Cohen, and J. T. Beatty, Transcriptional Control of the *Rhodobacter capsulatus puf* Operon: A Structural and Functional Analysis, J. Bacteriol. 171:473 (1989).
3. J. Davis, T. J. Donohue, and S. Kaplan, Construction, Characterization, and Complementation of a Puf⁻ Mutant of *Rhodobacter sphaeroides*, J. Bacteriol. 170:320 (1988).
4. C. E. Bauer, and B. L. Marrs, The *Rhodobacter capsulatus puf* Operon Encodes a Regulatory Protein (PufQ) for Bacteriochlorophyll Biosynthesis, Proc. Natl. Acad. Sci. U.S.A. 85:7074 (1988).
5. G. Klug and S. N. Cohen, Pleitropic Effects of Localized *Rhodobacter capsulatus puf* Operon Deletions on Production of Light-Absorbing Pigment-Protein Complexes, J. Bacteriol. 170:5814 (1988).
6. G. Drews, Structure and Functional Organization of Light-Harvesting Complexes and Photochemical Reaction Centers in Membranes of Phototrophic Bacteria, Microbiol. Rev. 49:59 (1985).
7. J. Lascelles, The Accumulation of Bacteriochlorophyll Precursors by Mutant and Wild-Type Strains of *Rhodopseudomonas sphaeroides*, Biochem. J. 100:175 (1966).

8. W. R. Richards, R. B. Wallace, M. S. Tsao, and E. Ho, The Nature of a Pigment-Protein Complex Excreted from Mutants of *Rhodopseudomonas sphaeroides*, Biochemistry 14:5554 (1975).

9. D. C. Youvan, S. Ismail, and E. J. Bylina, Chromosomal Deletion and Plasmid Complementation of the Photosynthetic Reaction Center and Light-Harvesting Genes from *Rhodopseudomonas capsulata*, Gene 38:19 (1985).

10. G. A. Francis and W. R. Richards, Localization of Photosynthetic Membrane Components in *Rhodopseudomonas sphaeroides* by a Radioactive Labeling Procedure, Biochemistry 19:5104 (1980).

11. J. D. Houghton, C. L. Honeybourne, K. M. Smith, H. D. Tabba, and O. T. G. Jones, The Use of N-Methylprotoporphyrin Dimethyl Ester to Inhibit Ferrochelatase in *Rhodopseudomonas sphaeroides* and its Effect in Promoting Biosynthesis of Magnesium Tetrapyrroles, Biochem. J. 208:479 (1982).

12. K. K. Wong, 4-Vinyl Protochlorophyllide Excretion by *Rhodopseudomonas gelatinosa* in Nicotinamide-Enriched Medium, Plant Sci. Lett. 13:269 (1978).

13. Y. Shioi, M. Doi, and B. Boddi, Selective Inhibition of Chlorophyll Biosynthesis by Nicotinamide, Arch. Biochem. Biophys. 267:69 (1988).

14. G. H. Kaiser, J. Beck, J. U. von Schütz, and H. C. Wolf, Low Temperature Excitation and Emission Spectroscopy of the Photosynthetic Bacteria *Rhodopseudomonas sphaeroides* "Wild-Type" Strain ATCC 17023, Biochim. Biophys. Acta 634:153 (1981).

15. J. Beck and G. Drews, Tetrapyrrol Derivatives Shown by Fluorescence Emission and Excitation Spectroscopy in Cells of *Rhodopseudomonas capsulata* Adapting to Phototrophic Conditions, Z. Naturforsch. 37c:199 (1982).

16. J. Beck, J. U. von Schütz, and H. C. Wolf, Optically Detected Magnetic Resonance of Porphyrin Complexes in the Bacterium *Rhodopseudomonas sphaeroides*, Z. Naturforsch. 38c:220 (1983).

17. J. Oelze, Regulation of Tetrapyrrole Synthesis by Light in Chemostat Cultures of *Rhodobacter sphaeroides*, J. Bacteriol. 170:4652 (1988).

18. P. A. Reilly and R. A. Niederman, Role of Apparent Membrane Growth Initiation Sites during Photosynthetic Membrane Development in Synchronously Dividing *Rhodopseudomonas sphaeroides*, J. Bacteriol. 167:153 (1986).

19. C. L. Wellington and J. T. Beatty, Promoter Mapping and DNA Sequence of the *bchC* Bacteriochlorophyll Biosynthesis Gene from *Rhodobacter capsulatus*, Gene in press (1989).

THE ANTENNA-REACTION CENTER COMPLEX OF HELIOBACTERIA

P.J.M. van Kan, E.J. van de Meent, F.A.M. Kleinherenbrink, T.J.Aartsma, and J. Amesz

Department of Biophysics
University of Leiden
Leiden, The Netherlands

INTRODUCTION

The discovery of new species and groups of photosynthetic bacteria, as well as the results of 16 S r-RNA sequence analysis have led to a basic revision of the traditional taxonomy of the photosynthetic bacteria (Stackebrandt et al., 1988). At present the following divisions of photosynthetic bacteria are discerned: the purple bacteria, the green sulfur bacteria, the green filamentous (or gliding) bacteria, and the heliobacteria. Of these four groups, the least known are the heliobacteria. This is not surprising, because the first species of this group was discovered only a few years ago, when Gest and Favinger (1983) reported the isolation of Heliobacterium chlorum from a soil sample from the Indiana University campus. Since then, two other species have been found, Heliobacillus mobilis (Beer-Romero and Gest, 1987) and Heliospirillum gestii. The latter two species were isolated from rice paddies in Thailand.

The heliobacteria are strictly anaerobic, photosynthetic, nitrogen fixing organisms. They contain a hitherto unknown bacteriochlorophyll, BChl g (Brockmann and Lipinski, 1983). The photosynthetic apparatus of heliobacteria is contained in the cytoplasmic membrane. Chlorosomes, as in green sulfur and green filamentous bacteria, are absent, and the cytoplasmic membrane does not show invaginations (Gest and Favinger, 1983) as in purple bacteria.

The primary electrondonor of H. chlorum, P-798, is probably a dimer of BChl g (Prince et al., 1985; Brok et al., 1986). Studies of the electron acceptor chain suggested a similarity with green sulfur bacteria and photosystem I of plants. The primary electronacceptor is probably a BChl c or chlorophyll (Chl) a-like pigment, absorbing near 670 nm (Nuijs et al., 1985), and there is some evidence for the involvement of an iron-sulfur center in the acceptor chain (Prince et al, 1985; Brok et al., 1986; Smit et al., 1987).

The present communication concerns studies in our laboratory regarding energy transfer and electron transport at low temperature, and describes the preparation and properties of isolated antenna-reaction center complexes from H. chlorum and Hb. mobilis.

MATERIAL AND METHODS

Heliobacterium chlorum was grown anaerobically on medium No. 112 or 1552, Heliobacillus mobilis on medium No. 1552 of the American Type Culture Collection containing 10 mM ascorbate. Membrane fragments were prepared by sonication followed by centrifugation in a buffer containing 10 mM Tris and 10 mM sodium ascorbate, pH 8.0. For measurements at low temperature 66% (v/v) glycerol was added to prevent crystallisation.

Absorption, fluorescence and fluorescence excitation spectra were recorded on a single beam spectrophotometer (Rijgersberg et al., 1980). Flash-induced absorbance difference changes were measured as described earlier (Nuijs et al., 1985; Smit et al., 1987). Actinic flashes were

Molecular Biology of Membrane-Bound Complexes in Phototrophic Bacteria
Edited by G. Drews and E. A. Dawes
Plenum Press, New York, 1990

261

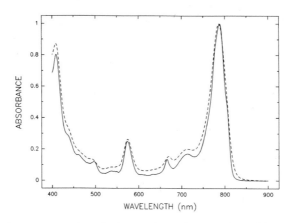

Fig. 1   Absorption spectrum of membranes of *H. chlorum.*
Broken line: room temperature; solid line: 100 K.

provided by the frequency doubled (532 nm) output of a Q-switched or passively mode-locked Nd:YAG laser (15 ns and 25 ps FWHM, respectively). Kinetic data in the ps region were fitted with a convolution of a single- or a bi-exponential decay and an instrument response function represented by a Gaussian of 35 ps FWHM. In some cases we obtained a better fit assuming a finite rise time of 10-20 ps, but the validity of this result is still being investigated.

RESULTS AND DISCUSSION

Absorption Spectra

Fig. 1 shows absorption spectra of isolated membranes of *H. chlorum.* At room temperature the Q_y absorption maximum of BChl g is located at 786 nm, but at low temperature this band can be resolved in at least three components, which were ascribed to different spectral forms of BChl g: BChl g 778, BChl g 793, and BChl g 808 (Van Dorssen et al., 1985). Q_x and Soret bands of BChl g are located at 575 and 410 nm, respectively. The band at 670 nm may, at least partially, be ascribed to a BChl c or Chl a-like pigment. Its amplitude varies with culture conditions, and tends to be larger when air is not rigorously excluded from the culture medium (Prince et al., 1985). Bands of carotenoid (mainly neurosporene; Gest and Favinger, 1983; Van Dorssen et al., 1985) are seen in the region of 450 - 500 nm. Absorption spectra of *Hb. mobilis* membranes were very similar to those of *H. chlorum*, both at room temperature and at low temperature.

Time-resolved Absorption Difference Spectroscopy

Absorbance difference spectra of *H. chlorum* membranes obtained upon excitation with saturating 25 ps laser flashes at 15 K are shown in Fig. 2. The first spectrum (solid circles) was measured at 40 ps after the maximum of the flash. Negative bands are observed near 665, 793, and 812 nm. The first band appears to be superimposed on a broad positive band at 640 - 700 nm. The second spectrum (open circles) was obtained at 350 ps after the flash. The amplitude of the 812 nm band was now strongly reduced, but the band at 793 nm was only slightly smaller than in the 40 ps spectrum.

The bleaching at 793 nm may be attributed to photo-oxidation of P-798. The wavelength of maximum bleaching agrees well with that observed by Smit et al. (1989) in low temperature experiments in the ms region. Kinetics of absorbance changes are shown in Fig. 3. Relatively simple kinetics were observed at low energy density of excitation, where a rapid bleaching occurred to a level which was constant during the first few nanoseconds after the flash. At

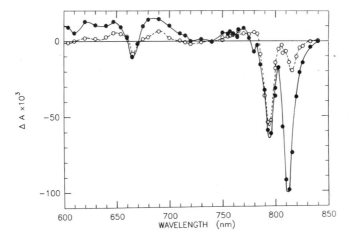

Fig. 2. Absorbance difference spectra of *H. chlorum* membranes, measured at 15 K. Solid circles at 40 ps, open circles at 350 ps after a 25 ps, 532 nm laser flash (3 mJ/cm²). $A_{790} = 0.70$.

about six times higher flash energy the amplitude of the constant component was saturated, and an additional transient was observed after the flash, which decayed with a time constant (1/e) of 30 ps or less. This transient may be ascribed to the formation of excited BChl *g* in the antenna. Comparison of the saturation curves indicated that the efficiency of charge separation was about the same or only slightly lower at 15 K than at room temperature.

It should be noted that the amplitude of the bleaching at 793 nm was about 3 times larger than that of P-798 at room temperature. This factor can only partially be explained by a narrowing of the band of P-798 upon cooling. The cause of this apparent increase in oscillator strength of P-798 is not clear.

The negative band near 665 nm in the difference spectrum may be ascribed to photoreduction of the primary electron acceptor (Nuijs et al., 1985). The kinetics at 668 nm are shown in Fig. 4. The decay could be fitted with a single exponential component with a

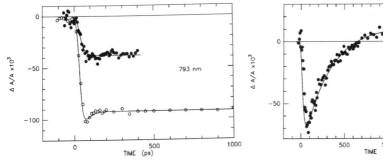

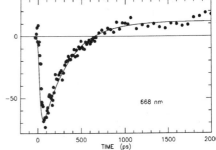

Fig. 3. Kinetics at 793 nm, plotted as the ratio between the absorbance difference and the initial absorbance at the same wavelength. Open circles: flash energy density 3 mJ/cm². Solid circles: flash energy density 0.5 mJ/cm².

Fig. 4. Kinetics at 668 nm. Energy density 3 mJ/cm².

263

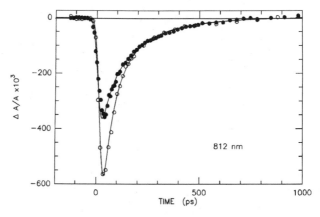

Fig. 5. Kinetics at 812 nm. Open circles: energy density 3 mJ/cm². Solid circles: energy density 0.5 mJ/cm².

time constant of 300 ± 50 ps. At room temperature a decay of 800 ± 50 ps was obtained with the same preparation.

These results support the concept that the charge separation in heliobacteria consists of the transfer of an electron from P-798 to an acceptor absorbing near 670 nm, which is presumably a BChl c or Chl a-like pigment. The decay time of the bleaching at 668 nm is shortened by a factor of 2.5 upon cooling from 300 K to 15 K. Thos would mean that electron transfer to a subsequent acceptor is 2.5 times faster at 15 K than at room temperature.

The bleaching at 812 nm may be attributed to the formation of excited singlet states of the long-wavelength antenna BChl, BChl g 808. The kinetics of this bleaching could be fitted with a bi-exponential decay with components of approximately 50 and 200 ps (Fig. 5). The fast component was relatively large at high excitation energy density, which suggests that it may be caused by excitation annihilation in the antenna.

The maximum amplitude at 812 nm was roughly the same as caused by P-798⁺, but the *relative* bleaching at 812 nm was much larger and amounted to more than 50 % of the initial absorbance at this wavelength. This indicates that ground-state depletion is the main factor which causes saturation of the signal at high flash energies. The wavelength of maximum bleaching at low temperature was significantly red-shifted with respect to the absorption maximum of BChl g 808 (Van Dorssen et al., 1985). Apparently at low temperature the excitations are predominantly located on BChls absorbing at longer wavelengths than the 'bulk' BChl g 808 molecules. Absorbance changes of BChl g 778 and of BChl g 793 were at least an order of magnitude smaller than those of BChl g 808. This indicates that excitation is transferred to BChl g 808 within much less than a few tens of picoseconds at low temperature, in agreement with the lack of short-wave emission in the low-temperature fluorescence spectrum (Van Dorssen et al., 1985).

Isolation and Properties of Antenna-Reactioncenter Complexes

Solubilized antenna-reaction center complexes were obtained by incubation of membranes of *H. chlorum* and *Hb. mobilis* with the detergent sulfobetaine-12 (SB-12, 1.32%), followed by sucrose gradient centrifugation in the presence of 0.1% SB-12 and 0.1% sodium cholate. Similar results were obtained with n-octylglucopyranoside. The solubilized complexes were obtained as a band which settled at approximately 35% sucrose, with a yield of 60-70% for *H. chlorum* and of about 25% for *Hb. mobilis* on a BChl basis.

The absorption spectra of the isolated complexes of the two species (Fig. 6) were very similar to each other, as well as to those of the membranes used as starting material. The low-temperature fluorescence emission and excitation spectra showed efficient energy transfer to BChl g 808 from the other BChls and from neurosporence in the isolated complexes. These observations indicate that the structure of the antenna was conserved during the solubilization and isolation procedure.

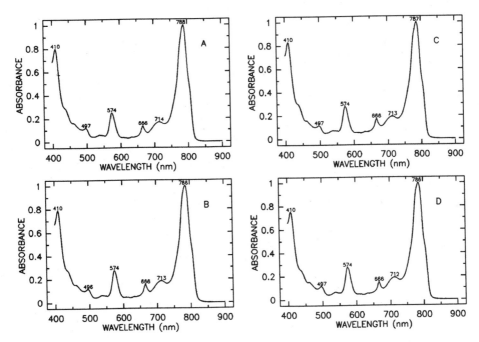

Fig. 6. Absorption spectra at 100 K. A: *H. chlorum* membranes; B: the isolated antenna-reaction center complex of *H. chlorum*; C: *Hb. mobilis* membranes; D: the antenna-reaction center complex of *Hb. mobilis*.

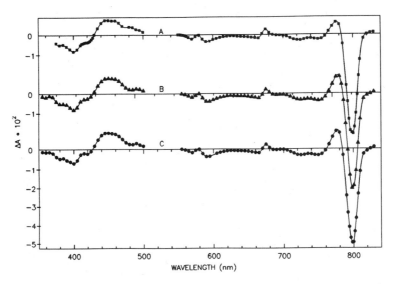

Fig. 7. Absorbance difference spectra of P-798 photo-oxidation measured at room temperature. A: *H.chlorum* membranes; B: the isolated antenna-reaction center complex of *H. chlorum*; C: the antenna-reaction center complex of *Hb. mobilis*. $A_{787} = 1.0$

Measurements of flash-induced absorbance changes showed that the isolated complexes had the same reaction center activity as the starting material. Fig. 7 compares the absorbance difference spectra of P-798 photo-oxidation in saturating flashes for membranes of *H. chlorum* and for the isolated complexes from both species. It can be seen that the shape as well as the amplitude of the spectra were the same for all three preparations, indicating that the structure and the activity of the reaction center was fully retained. The amount of P-798 relative to that of BChl *g* was the same in membranes of *H. chlorum* and of *Hb. mobilis*. Membranes and complexes gave the same saturation curves, which shows that the efficiency of charge separation and of energy transfer to the reaction center was not changed. Re-reduction of P-798$^+$ occurred in the ms time-range; the absence of a fast decay component showed that the electron acceptor chain was at least partially intact in the isolated complexes. Cytochrome *c*-553 photo-oxidation was not observed, indicating that the cytochrome was either lost or inactivated during the isolation.

The complexes were futher purified by size-exclusion HPLC. In this way colorless material was removed, and from the elution volume a molecular weight of approximately 335 kD could be estimated for both complexes. An estimated correction for the weight of the adhering detergent would reduce this number to 310 kD. The height of the protein absorbance in the UV as compared to that of other pigment-protein complexes suggested a BChl *g* content of 10 - 15 % by weight, corresponding to roughly 40 BChl *g* molecules. This would indicate that the complexes contain the antenna complement of one photosynthetic unit, together with the associated reaction center.

SDS-polyacrylamide gel-electrophoresis of the purified complexes of both species revealed the presence of a dominant polypeptide of M_r = 94 kD and weak and variable bands in the range 40-90 kD and above 100 kD. The subunit-structure of the complexes thus appears to deviate considerably from that of the core of photosystem I (Malkin, 1987) and of green sulfur bacteria (Hurt and Hauska, 1984), suggesting that there is no clear structural and evolutionary relationship between these systems and the photosystem of heliobacteria.

ACKNOWLEDGMENTS

The authors wish to thank Mr. R.J.W.Louwe and Ms. M.Nieveen for technical assistance. The investigation was supported by the Netherlands Foundations for Biophysics and for Chemical Research (SON), under auspices of the Netherlands Organization for Scientific Research (NWO).

REFERENCES

Beer-Romero, P., and Gest, H., 1987, *Heliobacillus mobilis*, a peritrichously flagellated anoxyphototroph containing bacteriochlorophyll *g*, <u>FEMS Microbiol Lett.</u>, 41:109.

Brockmann, H., and Lipinski, A., 1983, Bacteriochlorophyll *g*. A new bacteriochlorophyll from *Heliobacterium chlorum.*, <u>Arch. Microbiol.</u>, 136:17.

Brok, M., Vasmel, H., Horikx, J.T.G., and Hoff, A.J., 1986, Electron transport components of *Heliobacterium chlorum* investigated by EPR spectroscopy at 9 and 35 GHz, <u>FEBS Lett.</u>, 194:322.

Gest, H., and Favinger, J.L., 1983, *Heliobacterium chlorum*, an anoxygenic brownish-green photosynthetic bacterium containing a 'new' form of bacteriochlorophyll, <u>Arch. Microbiol.</u>, 136:11.

Hurt, E.C., and Hauska, G., 1984, Purification of membrane-bound cytochromes and a photoactive P840 protein complex of the green sulfur bacterium *Chlorobium limicola* f. *thiosulfatophilum*, <u>FEBS Lett.</u>, 168:149.

Malkin, R., 1987, Photosystem I, in: "Topics in Photosynthesis", Vol. 8, J. Barber, ed., Elsevier, Amsterdam, p. 495.

Nuijs, A.M., van Dorssen, R.J., Duysens, L.N.M., and Amesz, J., 1985, Excited states and primary photochemical reactions in the photosynthetic bacterium *Heliobacterium chlorum*, Proc. Natl. Acad. Sci. USA, 82:6865.

Prince, R.C., Gest, H., and Blankenship, R.E., 1985, Thermodynamic properties of the photochemical reaction center of *Heliobacterium chlorum*, Biochim. Biophys. Acta, 810:377.

Rijgersberg, C.P., van Grondelle, R., and Amesz, J., 1980, Energy transfer and bacterio-chlorophyll fluorescence in purple bacteria at low temperature, Biochim. Biophys. Acta, 592:240.

Smit, H.W.J., Amesz, J., and van der Hoeven, M.F.R., 1987, Electron transport and triplet formation in membranes of the photosynthetic bacterium *Heliobacterium chlorum*, Biochim. Biophys. Acta, 893: 232.

Smit, H.W.J., van Dorssen, R.J., and Amesz, J., 1989, Charge separation and trapping efficiency in membranes of *Heliobacterium chlorum* at low temperature, Biochim. Biophys. Acta, 973:212.

Stackebrandt, E., Embley, M., and Weckesser, J., 1988, Phylogenetic, evolutionary and taxonomic aspects of phototrophic eubacteria, in: "Green Photosynthetic Bacteria", J.M. Olson, J.G. Ormerod, J. Amesz, E. Stackebrandt, and H.C. Trüper, eds., Plenum Press, New York, p. 201.

van Dorssen, R.J., Vasmel, H., and Amesz, J., 1985, Antenna organization and energy transfer in membranes of *Heliobacterium chlorum*, Biochim. Biophys. Acta, 809:199.

IN VITRO SYNTHESIS AND MEMBRANE ASSEMBLY OF PHOTOSYNTHETIC POLYPEPTIDES

FROM RHODOBACTER CAPSULATUS

Dorothee Troschel and Matthias Müller

Biochemisches Institut der Universität Freiburg

F.R. Germany

INTRODUCTION

We have used the facultatively photoheterotrophic bacterium Rhodobacter capsulatus to study the biogenesis of polymeric membrane proteins. Upon lowering the oxygen tension or the light intensity the synthesis of the membrane-located photosynthetic apparatus is induced. This is accompanied by a drastic increase in the surface of the plasma membrane leading to the development of the so-called intracytoplasmic membranes (ICM) (for a recent review see Drews, 1985; Kiley and Kaplan, 1988). The photosynthetic apparatus is organized into two light-harvesting complexes, B870 (LH-I) and B800-850 (LH-II), and a reaction center (RC). Each LH-complex consists of two pigment-binding proteins in a 1:1 stoichiometry: B870 α and ß (M_r = 12 and 7 kDa), and B800-850 α and ß (M_r = 10 and 8 kDa). The B800-850 complex contains in addition the non-pigment-binding protein γ (M_r = 14 kDa). Each of these α and ß peptides has one α-helical transmembrane domain of approximately 20 amino acids with an N-terminus facing the cytoplasm and a C-terminus located in the periplasmic space (Tadros et al. 1984; 1985; 1987). The RC consists of the two pigment-binding proteins L and M (M_r = 20.5 and 24 kDa) each with five membrane-spanning domains, and the non-pigment-binding protein H (M_r = 28 kDa), which is anchored within the membrane by a single hydrophobic stretch. The three RC-proteins are found in the ICM in a 1:1:1 stoichiometry with their N-termini located in the cytoplasm (Tadros et al. 1987).

In order to understand in general the molecular mechanism of protein integration into procaryotic membranes and specifically the assembly of the pigment-containing, polymeric complexes we have developed an in vitro protein synthesis/membrane integration system from Rhodobacter capsulatus. Using a high-speed supernatant of an R. capsulatus cell homogenate we have synthesized in vitro polypeptides of the bacterium's photosynthetic apparatus. When isolated ICM were added cotranslationally, the majority of the de novo synthesized proteins was found to integrate into the lipid bilayer. Moreover, in vitro integrated proteins could be shown to assemble into pigment-containing, native complexes.

Molecular Biology of Membrane-Bound Complexes in Phototrophic Bacteria
Edited by G. Drews and E. A. Dawes
Plenum Press, New York, 1990

RESULTS

In Vitro Synthesis of LH-and RC-Polypeptides

In analogy to protocols developed for E.coli (Müller and Blobel 1984) and Rhodobacter sphaeroides (Chory and Kaplan 1982) we synthesized specific proteins in an R.capsulatus cell extract by a coupled transcription/translation reaction using plasmid pBBC1. This plasmid carries a 9.2 kb insert containing the genes for B870 α and ß, and RC-L and M in addition to several other open reading frames (Klug et al. 1985). Due to the large number of proteins resulting from the in vitro transcription-/translation of plasmid pBBC1, it was necessary to selectively express the individual pigment-binding proteins.

To achieve this, specific plasmids were required encoding only the proteins of interest. We therefore subcloned the puf-genes (B870 α and ß, RC-L and M) present on a 4.8 kb EcoRI-BamHI fragment of pBBC1 into the pSP65 vector rendering them under the control of a heterologous SP6 promoter. The resulting plasmid pSBC57 was transcribed in vitro using SP6 RNA-polymerase, following linearization with StuI. This enzyme has a unique restriction site downstream from the coding sequence for α.

To translate the B870 α and ß-mRNA in vitro, various supernatants were prepared from an R. capsulatus homogenate. An S-30 showed substantial endogenous, i.e. non-B870 mRNA-dependent translation activity, presumably due to polysomal mRNA associated with the high amount of membranes. A 135000 x g supernatant (S-135), however, expressed only two proteins (M_r = 4.7 kDa and 6.7 kDa) dependent on the addition of ßα-mRNA. These in vitro synthesized 4.7 kDa and 6.7 kDa proteins were recognized by antibodies raised against detergent-solubilized B870 proteins purified from ICM. Furthermore the appearance of these two proteins was abolished when transcripts were used which had been interrupted within the coding regions of α or ß. Linearizing pSBC57 DNA by cutting within the coding region of ß separates α from the shared promoter. Neither peptide could be detected in the products of a synthesis reaction using plasmid DNA so linearized. Cutting pSBC57 within the α-gene, however leaves ß intact and, as expected, the 4.7 kDa ß-polypeptide is observed, while the α-polypeptide is not. Clearly, the 4.7- and 6.7 kDa proteins appeared only when the full-length coding regions for α and ß were contained in the transcripts. These results demonstrate that the two in vitro translation products of ßα-mRNA are indeed the B870 polypeptides α and ß.

Integration of In Vitro Synthesized B870 Polypeptides into ICM

In order to study the integration of the in vitro synthesized polypeptides into membrane vesicles of R. capsulatus, the membrane-free S-135 was supplemented with ICM. These were prepared by sucrose gradient purification of a crude membrane pellet derived from phototrophically grown R. capsulatus cells.

To find out whether or not the B870 α and ß proteins became associated with exogenously added ICM in vitro, translation products were separated by centrifugation into soluble and pelletable material. In the absence of ICM, most of α and ß remained soluble, whereas the majority of each protein was converted into a pelletable form upon the addition of membranes.

Alkaline carbonate extraction was employed in order to demonstrate that this membrane association reflected true integration into the lipid bilayer. This treatment, applied to rough ER (Fujiki et al. 1982) and to E. coli plasma membranes (Ahrem et al. 1989; Watanabe et al. 1986), breaks open membrane vesicles and removes and solubilizes loosely attached proteins without influencing lipid-integrated material. The applicability of this method to membrane proteins from R. capsulatus was first confirmed by 0.2 M Na_2CO_3 (pH 11.5) treatment of ICM and analysis of the polypeptide pattern of soluble and pelletable material on Coomassie blue-stained SDS-gels. As anticipated, the in vivo assembled pigment-binding proteins of B870 remained firmly attached to the membranes even after alkaline treatment, while other proteins were solubilized. Most of the in vitro synthesized B870 α and ß which co-sedimented with ICM, also proved to be resistant to alkaline treatment (only 13% of both polypeptides were released by Na_2CO_3)'. Thus it appears that the in vitro synthesized proteins integrate into the lipid bilayer upon addition of ICM. Some of the in vitro synthesized B870 proteins were found in the pellet fraction even in the absence of exogenously added ICM. This probably represented aggregated material. In order to discriminate between this form and true membrane integration, the newly synthesized protein was fractionated by centrifugation through a two-step sucrose gradient and resolved into soluble, membrane-bound and aggregated/pelleted forms.

The majority of protein which previously was found in the pellet fraction was now found associated with the membrane fraction detectable by its strong pigmentation. This association was dependent on the co-translational addition of ICM, i.e. most protein remained soluble in the absence of membranes. The amount of pelleted/aggregated material was independent of the addition of membranes. When ICM were added post-translationally, only small amounts of α and ß were recovered in the membrane fraction. The observed membrane association of de novo synthesized B870 α and ß, therefore, is not dependent on the mere presence of ICM, but on the addition of ICM early during protein synthesis. It is conceivable that α and ß when synthesized to completion prior to the addition of membranes, fold into a stable tertiary structure that is incompatible with membrane integration.

Additional evidence for B870 α and ß being actually integrated into the lipid bilayer was provided by the finding that they could be solubilized by the addition of detergents. The translation reactions were treated with either 1% Triton-X 100 or 1% lauryldimethylamine oxide (LDAO) prior to fractionation. After detergent treatment by either Triton or LDAO the bulk of the α and ß protein was shifted from the membrane fraction to the soluble fraction.

Assembly of de novo Synthesized B870 α and ß into Membrane-Bound, Pigment-Containing Complexes

We examined whether in vitro synthesized B870 α and ß polypeptides would assemble into pigment-containing complexes when integrated into the lipid bilayer of ICM. To this end, pigment-containing protein-complexes were isolated from ICM by mild detergent fractionation using Triton X-100 (Peters et al. 1983). Aliquots of ICM were solubilized by Triton X-100 and fractionated by Triton-PAGE at 4°C in the dark resulting in two pigmented bands. The absorption spectra of these bands (obtained by scanning pigmented parts of the gel with a microspectrophotometer (Mäntele et al.1988)) were characteristic of the RC/B870 and B800-850 complexes, respectively.

If B870 proteins synthesized in vitro in the presence of ICM were analyzed by the above procedure with fluorography, the major part of the radioactively labeled proteins was recovered from the B870-containing band. Some radioactivity was localized in the gel region above the RC/B870 band but none was found in the B800-850 complex. This was expected, since no B800–850 proteins had been synthesized. Furthermore, if no ICM had been added to the in vitro synthesis, all the radioactivity remained in the upper area of the Triton-gel, which did not contain pigment-protein-complexes. These results therefore clearly indicate that a significant part of de novo synthesized B870 α and ß not only integrate into the lipid bilayer of ICM, but also assemble into a pigment-containing RC/B870 complex, whose native structure was retained during isolation as revealed by the authentic absorption profile.

CONCLUSION

We have described the development of a cell-free protein-synthesizing system from R. capsulatus. The two pigment-binding proteins α and ß of the B870 complex were expressed in vitro using a high-speed membrane-free supernatant (S-135) of an R. capsulatus cell extract. The in vitro synthesized proteins were found to integrate into the lipid bilayer of exogenously added ICM. This was demonstrated by: 1) the fact that membrane association of B870 α and ß was resistant to Na_2CO_3 treatment in contrast to peripheral ICM proteins; 2) the detergent-solubility of the membrane-integrated proteins; 3) the finding that membrane integration of the pigment-binding polypeptides was not simply dependent on the presence of ICM, but rather specifically required the addition of membranes cotranslationally for optimum efficiency.

In vitro synthesized B870 α and ß polypeptides not only integrated into exogenously added ICM, but also assembled into pigment-containing complexes. Isolation of native pigment-protein-complexes revealed that a significant amount of de novo synthesized B870 α and ß protein had assembled into a supra-molecular structure characteristic of an RC/B870 complex. Presumably this results from an association of the newly integrated B870 proteins with bacteriochlorophyll and sphaeroidenone present in the ICM. These in vitro assembled light-harvesting complexes appear to further interact with reaction center complexes to form native units as evidenced by the authentic absorption spectrum of an intact RC/B870 unit.

ACKNOWLEDGEMENTS

This work was supported by grant D 29/31–3C from the Deutsche Forschungsgemeinschaft and the Fonds der Chemischen Industrie.

REFERENCES

Ahrem, B., Hoffschulte, H.K. and Müller, M., 1989, In Vitro Membrane Assembly of a Polytopic, Transmembrane Protein Results in an Enzymatically Active Conformation, J. Cell Biol., 108:1637.

Chory, J. and Kaplan, S., 1982, The in Vitro Transcription-Translation of DNA and RNA Templates by Extracts of Rhodopseudomonas sphaeroides, J. Biol. Chem., 257:15110.

Drews, G., 1985, Structure and Functional Organization of Light-Harvesting Complexes and Photochemical Reaction Centers in Membranes of Phototrophic Bacteria, Microbiol. Rev., 49:59.

Fujiki, Y., Hubbard, A.L., Fowler, S. and Lazarow, P.B., 1982, Isolation of Intracellular Membranes by Means of Sodium Carbonate Treatment: Application to Endoplasmic Reticulum, J. Cell Biol., 93:97.

Kiley, P.J. and Kaplan, S., 1988, Molecular Genetics of Photosynthetic Membrane Biosynthesis in Rhodobacter sphaeroides, <u>Microbiol. Rev.</u>, 52:50.

Klug, G., Kaufmann, N. and Drews, G., 1985, Gene Expression of Pigment-Binding Proteins of the Bacterial Photosynthetic Apparatus: Transcription and Assembly in the Membrane of Rhodopseudomonas capsulata, <u>Proc. Natl. Acad. Sci. USA</u>, 82:6485.

Mäntele, W., Steck, K., Becker, A., Wacker, T., Welte, W., Gad'on, N. and Drews, G., 1988, Spectroscopic Studies of Crystallized Pigment-Protein Complexes of R. palustris, <u>in</u>: "The Photosynthetic Bacterial Reaction Center", Breton, <u>J.</u> and Vermeglio, A., eds., Plenum Publishing Corporation, London.

Müller, M. and Blobel, G., 1984, In Vitro Translocation of Bacterial Proteins across the Plasma Membrane of Escherichia coli, <u>Proc. Natl. Acad. Sci. USA</u>, 81:7421.

Peters, J., Takemoto, J. and Drews, G., 1983, Spatial Relationships between the Photochemical Reaction Center and the Light-Harvesting Complexes in the Membrane of Rhodopseudomonas capsulata, <u>Biochemistry</u> 22:5560.

Tadros, M.H., Frank, R., Dörge, B., Gad'on, N., Takemoto, J.Y. and Drews, G., 1987, Orientation of the B800-850, B870, and Reaction Center Polypeptides on the Cytoplasmic and Periplasmic Surfaces of Rhodobacter capsulatus Membranes, <u>Biochemistry</u> 26:7680.

Tadros, M.H., Frank, R., Zuber, H. and Drews, G., 1985, The Complete Amino Acid Sequence of the Large Bacteriochlorophyll-Binding Polypeptide B870 α from the Light-Harvesting Complex B870 of Rhodopseudomonas capsulata, <u>FEBS Lett.</u> 190:41.

Tadros, M.H., Suter, F., Seydewitz, H.H., Witt, I., Zuber, H. and Drews, G., 1984, Isolation and Complete Amino Acid Sequence of the Small Polypeptide from Light-Harvesting Pigment-Protein Complex I (B870) of Rhodopseudomonas capsulata, <u>Eur. J. Biochem.</u> 138:209.

Watanabe, M., Hunt, J.F. and Blobel, G., 1986, In Vitro Synthesized Bacterial Outer Membrane Protein is Integrated into Bacterial Inner Membranes but Translocated across Microsomal Membranes, <u>Nature</u> 323:71.

ASSEMBLY OF CHLOROSOMES DURING PHOTOSYNTHETIC DEVELOPMENT IN

Chloroflexus aurantiacus

T.E. Redlinger, S.J. Theroux, D.L. Driscoll,
S.J. Robinson and R.C. Fuller

Department of Biochemistry and Botany
University of Massachusetts, Amherst, MA   01002 U.S.A.

The chlorosome of the green photosynthetic bacterium, *Chloroflexus aurantiacus*, contains rod elements composed of Bchl c complexed with a 5.6 kDa protein. This protein, termed the Bchl c-binding protein, dimerizes to form subunits which assemble into rods. This Bchl c pigment-protein complex functions in harvesting light and transferring energy with nearly 100 percent efficiency to the Bchl-a located in the chlorosome baseplate, the cytoplasmic membrane (CM), and finally to the reaction center. Surrounding the rod elements is a special envelope which appears in electron micrographs to have characteristics of an unusual lipid protein monolayer. Two proteins, M_r 11,000 and 18,000 KDa, are closely associated with the surface of the chlorosome and appear to be integral components of this envelope (1,2,3,4).

Evidence for chlorosome development in *C. aurantiacus* comes mainly from electron microscopic studies on induced and noninduced cells. Under aerobic/dark growth conditions, chlorosomes are not observed on the CM and bacteriochlorophyll is absent (1,5). The first stage of chlorosome development begins with the formation of membrane attachment sites, visualized as arrays of 5 nm particles on the P face of the CM and suggested to represent clusters of reaction center complexes (1). In the same time frame, the baseplate structure and chlorosome envelope are already visible. The second stage of development involves the filling of the chlorosome core with rod elements, consisting of Bchl and its binding protein.

Chlorosome development has also been studied in cells during adaptation from high to low light intensity (2). Freeze-fracture electron micrographs of these cells indicate that as light intensity is lowered, the chlorosomes grow outward, i.e., they become thicker in appearance, indicating that additional rod elements are added to existing chlorosomes.

Immunochemical analysis of chlorosome development indicate that the major chlorosome polypeptides (M_r 18, 11,and 5.6 kDa) are present as high molecular weight complexes in cells prior to induction of the photosynthetic apparatus. Polyclonal antibodies directed against the individual purified chlorosome polypeptides reacted with two high molecular weight complexes (M$_r$60 and 47 kDa). Lowering the oxygen concentration resulted in disappearance of the high molecular weight species, and concomitant appearance of the component polypeptides of the chlorosome. To explain these results, it was proposed that a single polyprotein precursor was synthesized constitutively, then posttranslationally processed into the individual polypeptides at lowered oxygen concentrations (6,7).

The post-translational polyprotein processing hypothesis predicts that several chlorosome polypeptides should be encoded by one long open reading frame in the *C. aurantiacus* genome. To test this polyprotein hypothesis, we have isolated the gene (*csmA*) encoding for the Bchl *c* binding polypeptide.

We report here that this protein is not encoded as part of a large polyprotein, but rather as a single smaller polypeptide. Sequence analysis of *csmA* indicates that the Bchl *c* binding protein is synthesized with a carboxy terminal extension. We propose that this carboxy terminal extension may have a role in proper incorporation of the Bchl *c* binding protein into the developing chlorosome or in proper association of the binding protein with Bchl *c* during rod assembly.

MATERIALS AND METHODS

Chloroflexus aurantiacus J-10-fl was grown anaerobically with high incident light intensity (1).

To clone the gene for the Bchl *c* binding protein, genomic DNA was purified from *C. aurantiacus* by CsCl density gradient centrifugation (8), then digested with the indicated restriction endonucleases. The resulting restriction fragments were separated by electrophoresis on agarose gels. Southern blots were prepared by transfer of the genomic restriction fragments to nylon membranes, then probed with a labeled oligonucleotide (5'-ATGTTYCAUGGNCAYTGGCAUTGGGT-3') synthesized on the basis of reverse translation of a segment of the protein sequence of the Bchl *c* binding protein (9). Genomic fragments in size range which hybridize with this oligonucleotide were eluted from agarose gels, ligated into a plasmid vector, then used to transform *Escherichia coli* DH5α. Recombinant plasmids were isolated from individual colonies chosen from these size-selected plasmid libraries, then screened by hybridization with labeled oligonucleotide.

Plasmids which contained inserts hybridizing to the oligonucleotide were further analyzed. The sequence of DNA inserts was determined by chain termination reactions on double stranded templates using modified T7 DNA polymerase (10). Sequence data was analyzed using the sequence analysis package (version 5.3) of the University of Wisconsin Genetics Computer Group (11).

Northern blots were prepared using RNA isolated from *C. aurantiacus*, then fractionated on formaldehyde gels (12).

A more detailed description of the methods employed for isolation of *csmA* is presented elsewhere (13).

Nucleotide sequence of csmA, and predited sequence of Bchl c binding
protein

The gene encoding the Bchl c binding protein, which we designate
csmA, was isolated as a 1.1 kb *KpnI* fragment from a size-selected
plasmid library of *C. aurantiacus* genomic DNA. The nucleotide sequence
for this fragment was determined using the strategy depicted in Figure
1. The sequence of *csmA*, with a portion of the upstream and downstream
flanking regions, is presented in Figure 2.

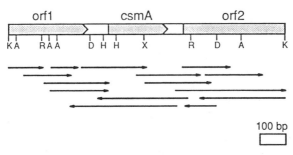

Figure 1. Restriction map and sequencing
strategy for the 1.1 kb *KpnI* fragment
containing *csmA*. Flanking regions contain
putative open reading frames which extend
beyond the boundaries of the 1.1 kb fragment.
Arrows indicate the extent and direction of
the sequencing reactions. Restriction
endonuclease sites indicated on the map are:
K, *KpnI*; A, *AluI*; R, *DdeI*; H, *HinfI*;
X, *XmaI*.

Translation of the *csmA* nucleotide sequence indicates that the BChl
c binding protein is encoded individually, not as a polyprotein.
Separate opening reading frames flank the *csmA* gene both upstream and
downstream, with appropriate translational initiation or termination
signals. We have no evidence to indicate whether these open reading
frames encode other proteins compoments of the chlorosome.

The reading frame of the BChl c protein encodes a protein of 80
amino acids. Comparison with the amino acid sequence data indicates that
the initiator methionine encoded by *csmA* is removed in the mature
protein. The next 50 amino acids align exactly with the sequence of the
BChl c binding protein, at which point translation of the *csmA*
nucleotide sequence indicates a serine residue which was not detected by
peptide sequencing (Fig.3). Most significantly, translation of the *csmA*
nucleotide sequence predicts that the BChl c binding protein is
synthesized with an additional 27 amino acids at the carboxy terminus,
not present in the mature protein isolated from the chlorosome.

```
CTGAGGCAGCCTGAACTGTGGGGTTTCTCGGAGGCTTGCCTCCGGTGACC


TGAATAATGGAGTCCGTTTGATAAGGAGGTGTGTGCATGGCGACGAGAGG
                    ─────              M  A  T  R  G
                       SD                   csmA

CTGGTTCTCGGAGTCGTCGGCGCAGGTGGCGCAAATCGGCGACATCATGT
 W  F  S  E  S  S  A  Q  V  A  Q  I  G  D  I  M  F


TCCAGGGCCACTGGCAATGGGTCTCGAATGCGCTACAGGCCACCGCGGCA
 Q  G  H  W  Q  W  V  S  N  A  L  Q  A  T  A  A


GCGGTTGACAACATCAACCGCAATGCTTACCCGGGCGTGTCCCGGAGCGG
 A  V  D  N  I  N  R  N  A  Y  P  G  V  S  R  S  G


CTCGGGCGAGGGAGCGTTCAGCAGCAGCCCGAGCAACGGCTTCCGTCCGA
 S  G  E  G  A  F  S  S  S  P  S  N  G  F  R  P  K


AGCGCATTCGCTCGCGCTTCAACCGCTAGGCTGGCGGGCAGGCGTGAGGC
  R  I  R  S  R  F  N  R                 ──────────


GTGCGTTCTGTCGCCTCACGCCGATCCCTCGGGAGCGTTATGATTAACCT
►              ◄──────────              M  I  N  L
    IR

GCTCTACTACAGTCATACCACCGTAC
 L  Y  Y  S  H  T  T  V
```

Figure 2. Nucleotide sequence of *csmA* and flanking regions. A Shine-Dalgarno sequence upstream of *csmA* is indicated by dashed underlining; an inverted repeat forming a potentially stable stem-loop downstream of the gene is denoted by underlining with solid arrows. Predicted open reading frames adjacent to *csmA* are also indicated.

1
MetAlaThrArgGlyTrpPheSerGluSerSerAlaGlnValAlaGlnIleGly
  AlaThrArgGlyTrpPheSerGluSerSerAlaGlnValAlaGlnIleGly

AspIleMetPheGlnGlyHisTrpGlnTrpValSerAsnAlaLeuGlnAlaThr
AspIleMetPheGlnGlyHisTrpGlnTrpValSerAsnAlaLeuGlnAlaThr

51
AlaAlaAlaValAspAsnIleAsnArgAsnAlaTyrProGlyVal**Ser**ArgSer
AlaAlaAlaValAspAsnIleAsnArgAsnAlaTyrProGlyValArg

GlySerGlyGluGlyAlaPheSerSerSerProSerAsnGlyPheArgProLys

ArgIleArgSerArgPheAsnArg

Figure 3. Alignment of the Bchl *c* binding protein as determined by peptide sequencing (lower line) with that predicted from csmA gene upper line.) *CsmA* encodes a Bchl *c* binding protein with a carboxy terminal extension of 27 amino acids.

Transcription of *csmA*

The sequence data indicates that *csmA* encodes the BChl c binding
protein. To determine whether this region is transcribed, RNA was
purified from photosynthetic *C. aurantiacus*, grown anaerobically under
high light intensity. Northern blots were prepared by fractionating
this RNA on formaldehyde gels, then transferring the RNA to nylon
membranes. These Northern blots were probed with 1.1 kb *Kpn1* fragment

350 nucleotides

Figure 4. *CsmA* is transcribed into a
detectable message. Total RNA was isolated
from anaerobically induced *C. aurantiacus*,
then separated on an agarose gel in the
presence of formaldehyde. A blot prepared
from this gel was then probed with a labeled
1.1 kb *Kpn1* fragment to detect possible
hybridization with *csmA* and the adjacent
reading frames.

containing *csmA*. A strong signal was detected (Figure 4), indicating
that the mRNA for this gene is 350 nucleotides in length. This is a
length sufficient to span the putative ribosome-binding region for
translation initiation, the *csmA* reading frame, and a downstream
inverted repeat which may act as a transcription termination or
processing signal. The small size of this message is inconsistent with
a polyprotein containing other known components of the chlorosome.

DISCUSSION

We have isolated and characterized the gene (*csmA*) encoding the Bchl *c* binding protein, the rod element subunit, of the *C. aurantiacus* chlorosome. Our sequence data indicate that the *csmA* gene is not large enough to encode more than one known chlorosome polypeptide. *CsmA* encodes a protein of 80 amino acids, rather than the expected 51 amino acids determined by peptide sequencing of the purified Bchl c binding protein. Assuming post-translational removal of the initiator formyl-methionine residue common in prokaryotes, the NH_2-terminal 50 amino acids align exactly with the sequenced protein. At this point, the peptide sequence terminates with Arg, whereas the sequence predicted from *csmA* is SerArg. Introducing a Ser before the terminal Arg increases the predicted mass of the protein to 5695. A value which corresponds closely with the value of M_r = 5698±6 determined by ^{252}Cf plasma desorption mass spectrometry (P. Hojrup, P.O. Gerola, and J.M. Olson; personal communication). It thus appears that this serine at position 50 was not detected by the peptide sequence determination.

The structure of the *csmA* reading frame is consistent neither with a polyprotein containing the other polypeptide constituents of the chlorosome, nor with the simple protein defined by peptide analysis. Rather, nucleotide sequence analysis indicates that the Bchl *c* binding polypeptide is synthesized as a precursor protein, which is then processed into the mature form by removal of a carboxy terminal extension.

We propose three possibilities for the role of this carboxy terminal extension. The first is a modified polyprotein hypothesis: the carboxy terminus of the *csmA* gene product is indeed a second polypeptide, of 27 or fewer amino acids, post-translationally cleaved from the Bchl c binding protein. A second possibility is that the *csmA* carboxy terminal domain may serve as an intramolecular chaperone. This domain stabilizes the Bchl *c* binding protein prior to assembly into the chlorosome, perhaps by preventing incorrect folding of the *csmA* gene product prior to synthesis of Bchl c. The third possibility is that the Bchl *c* binding precursor protein, is processed during transport across the envelope of the developing chlorosome. The carboxy terminal extension may then act as a transit peptide, targeting the *csmA* gene product to the assembling chlorosome.

Signals targeting most bacterial proteins are short stretches of amino acids at or near the NH_2-terminus of the protein. A few eukaryotic membrane-targeting signals have been shown to be located at the carboxy-terminus, e.g., in peroxisomes amd glycosomes (14,15,16). Such carboxy-terminal targeting signals are relatively short hydrophilic peptide regions, rich in hydroxyl and basic amino acids, and devoid of hydrophobic domains (17). These characteristics are shared by the carboxy terminus of the Bchl c binding peptide precursor. Of the 27 amino acids not reported in the sequence of the mature protein, seven are hydroxylated and six are basic. This region is also extremely hydrophilic, especially the domain from residues 68 to 80 (Figure 5). These similarities between the Bchl c binding protein carboxy terminal extension and known peroxisomal targeting domains suggest that the chlorosome envelope is a translocation-competent membrane, across which the Bchl c binding precursor protein is transported.

Possible targeting and chaperon functions for the carboxy terminal extension of the Bchl *c* binding protein may be combined to construct a model of chlorosome rod assembly (Figure 6). The carboxy terminal domain of the *csmA* gene product may be necessary either to stabilize the protein prior to the synthesis of Bchl *c*, or for targeting the protein to the immature chlorosome. In either case, carboxy terminal extension would then be removed during chlorosome maturation. Precedent for involvement of a carboxy terminal extension in assembly of photosynthetic structures is provided by photosystem II in chloroplasts. Here, cleavage of a carboxy terminal extension of the D1 polypeptide facilitates proper assembly of the oxygen-evolving complex, since mutants unable to cleave this extension are unable to assemble correctly (18).

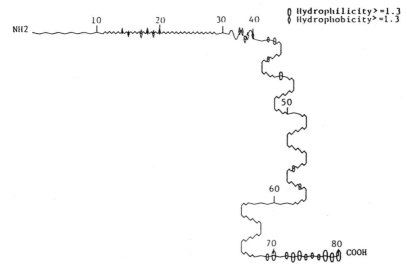

Figure 5. Hydrophilicity predictions by Kyte-Doolittle algorithm indicate that the carboxy terminal domain of the *csmA*-encoded protein is very hydrophilic. The mean hydrophilicity at each residue is determined by averaging structural parameters for surrounding residues and expressing hydrophilicity as an oval symbol proportional to the mean value calculated. Hydrophobic areas are marked with diamonds.

Synthesis of chlorosomes in *C. aurantiacus* does not involve a polyprotein. Rather, predictions of protein sequence based upon translation of the *csmA* gene indicate that the BChl *c* binding protein is synthesized individually, but with a carboxy terminal extension which is not present in the mature chlorosome rod element. We suggest here possible roles for this carboxy terminal extension in assembly of a functional BChl *c* light-harvesting complex within the chlorosome. Isolation of *csmA* (13) provides the opportunity to directly test these possible chaperon or targeting functions.

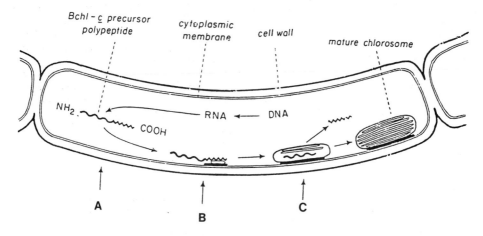

Figure 6. Proposed model of chlorosome rod assembly. A. Synthesis of the *csmA* gene product containing a hydrophilic COOH-terminal extension. B. Targeting of newly synthesized precursor polypeptides to chlorosomes. C. Post-translational processing, association with BChl c, and assembly of mature Bchl c binding polypeptides into rod elements.

REFERENCES

1. Sprague, S.G., L.A. Staehelin, M.J. DiBartolomeis, and R.C. Fuller (1981) Isolation and development of chlorosomes in the green bacterium *Chloroflexus aurantiacus*. J. Bacteriol. **147**: 1021-1031.

2. Sprague, S.G., L.A. Staehelin, and R.C. Fuller (1981) Semiaerobic induction of bacteriochlorophyll synthesis in the green bacterium *Chloroflexus aurantiacus*. J. Bacteriol. **147**: 1032-1039.

3. Feick, R.G., and R.C. Fuller (1984) Topography of the photosynthetic apparatus of *Chloroflexus aurantiacus*. Biochemistry 23: 3693-3700.

4. Betti, J.A.. R.E. Blankenship, L.V. Natagajan, L.C. Dickinson and R.C. Fuller (1982) Antenna organization and evidence for the function of a new antenna pigment species in the green photosynthetic bacterium *Chloroflexus*. BBA 680:194-201.

5. Foster, J., T.E. Redlinger, R. Blankenship, and R.C. Fuller (1986) Oxygen regulation of the photosynthetic membrane system in *Chloroflexus aurantiacus*. J. Bacteriol. **147**: 655-659.

6. Redlinger, T.E., and R.C. Fuller (1985) Protein processing as a regulatory mechanism in the synthesis of the photosynthetic antenna in *Chloroflexus*. Arch. Microbiol. **141**: 344-347.

7. Fuller, R.C., and T.E. Redlinger (1985) Light and oxygen regulation of the development of the photosynthetic apparatus in *Chloroflexus aurantiacus*. In K.E. Steinbeck, S. Bonitz, C.J. Arntzen, and L. Bogorad (ed.), Molecular biology of the photosynthetic apparatus, pp. 155-162. Cold Spring Harbor Laboratory, Cold Spring Harbor, NY.

8. Robinson, S.J. amd T.E. Redlinger (1987) Isolation of genes encoding the photosynthetic apparatus of *Chloroflexus*. In: J. Biggins, ed. Proceedings of VII Internationl Congress on Photosynthesis. Vol. IV: 740-744.

9. Wechsler, T. F. Suter, R.C. Fuller, and H. Zuber (1985) The complete amino acid sequence of the bactriochlorophyll *c* binding polypeptide of the green photosynthetic bacterium *Chloroflexus aurantiacus*. FEBS. Lett. **181**: 173-178.

10. Toneguzzo, F., S. Glynn, E. Levi, S. Mjolness, and A. Hayday (1988) Use of a chemically modified T7 DNA polymerase for manual and automated sequencing of supercoiled DNA. BioTechniquies **6**: 460-469.

11. Devereux, J.R., P. Haeberli, and O. Smithies (1984) A comprehensive set of sequence analysis programs for the VAX. Necl. Acids. Res. **12**: 386-395.

12. Selden, R.F. (1987) Denaturation of RNA ufins formaldehyde. p.4.9.1-4.9.5. *In* F.M. Ausubel, R. Brent, R.E. Kingston, D.D. Moore, J.G.Seidman, J.A. Smith, and K. Struhl (ed.) Current protocols in molecular biology. John Wiley, New York.

13. Theroux, S.J., T.E. Redlinger, R.C. Fuller and S.J. Robinson. Carboxy terminal extension for the bacteriochlorophyll *c* binding protein of *Chloroflexus aurantiacus* predicted from the nucleotide sequence of *csmA*. J. Bact. (in press).

14. Gould, S.J., G.-A. Keller, and S. Subramani (1987) Identification to a peroxisomal targeting signal at the carboxy terminus of firefly luciferase. J. Cell. Biol. **105**: 2923-2931.

15. Small, G.M., L.J. Szabo, and P.B. Lazarow (1988) Acyl-CoA oxidase contains two targeting sequences each of which can mediate protein import into peroxisomes. EMBO J. **7**: 1167-1173.

16. Swindels, B.W., R. Evers, and P. Borst (1988) The topogenic signal of the glycosomal (microbody) phosphoglycerate kinase of *Crithidia fasciculata* resides in a carboxy-terminal extension. EMBO J. **7**: 1159-1165.

17. Verner, K., and G. Schatz (1988) Protein translocation across membranes. Science **241**: 1307-1313.

18. Diner, B.A., D.F. Ries, B.N. Cohen, and J.G. Metz (1988) COOH-terminal processing of polypeptide D1 of the photosystem II reaction center of *Scenedesmus obliquus* is necessary for the assembly of the oxygen-evolving complex. J. Biol. Chem. **263**: 8972-8980.

ACKNOWLEDGEMENTS

This investigation was supported by a Department of Energy grant (DEFG 02-88-ER 13921) to T. E. Redlinger and S. J. Robinson, and by a National Science Foundation grant (DCB 88-03649) to R. C. Fuller.

CAROTENOID ABSORBANCE CHANGES IN LIPOSOMES RECONSTITUTED WITH PIGMENT-PROTEIN COMPLEXES FROM *RHODOBACTER SPHAEROIDES*

Wim Crielaard, Klaas J. Hellingwerf[a] and Wil N. Konings

Department of Microbiology, University of Groningen
Kerklaan 30, 9751 NN Haren, The Netherlands.

[a]Present address: Department of Microbiology, University of Amsterdam
Nieuwe Achtergracht 127, 1018 WS Amsterdam, The Netherlands.

INTRODUCTION

The electrochromic behaviour of carotenoids has been widely used to determine the electrical potential difference ($\Delta\psi$) across photosynthetic membranes, like chromatophores [1] and bacterial cells [2,3]. The carotenoid absorbance change has several advantages over other methods for recording the $\Delta\psi$: (i) the method is non-invasive, (ii) the relationship between the $\Delta\psi$ and the absorbance change is linear and (iii) the method shows a rapid response time. The major disadvantage of the carotenoid band shift as a $\Delta\psi$ indicator is that a calibration of the $\Delta\psi$ dependent bandshifts is not (always) possible in every experimental system [3].

It has been shown that the carotenoids which are responsible for the large linear membrane potential response in photosynthetic bacteria are only those located in the light harvesting complex II (LH_{II}) of the photosynthetic apparatus [4,5]. Exploration of the carotenoid band shift as a $\Delta\psi$ probe in other membrane systems like liposomes and hybrid membranes requires the reconstitution of pigment protein complexes which contain electrochromically active LH_{II}-complexes. The development of such a system is presented in this study.

RESULTS AND DISCUSSION

$RCLH_ILH_{II}$- (reaction center complexes with both antenna complexes still attached) or LH_{II}-complexes were isolated from *Rhodobacter sphaeroides* chromatophores using cholate (1%) or a combination of cholate (1%) and deoxy-cholate (2%) as solibilizing detergents. The complexes were purified by sucrose gradient centrifugation (cf. ref. 6). After reconstitution of these isolated pigment-protein complexes into liposomes made from *Escherichia coli* phospholipids the absorbance change of the reconstituted carotenoids was measured at 503-487 nm upon the induction of a potassium diffusion potential. Fig. 1 shows time courses of absorbance changes in liposomes reconstituted with $RCLH_ILH_{II}$-complexes, loaded with either K^+- or Na^+-buffer. The K^+-loaded liposomes showed, after dilution in Na^+-buffer, an increase in absorbance at 503-487 nm upon the addition of valinomycin. This increase could be abolished by the subsequent addition of nigericin, which leads to uncoupling of the membrane (Fig. 1A). When these liposomes were diluted in K^+-buffer no absorbance changes could be detected upon the addition of valinomycin or nigericin (Fig. 1B). A potential with the opposite polarity, induced by the addition of

Molecular Biology of Membrane-Bound Complexes in Phototrophic Bacteria
Edited by G. Drews and E. A. Dawes
Plenum Press, New York, 1990

285

external KCl to Na^+-loaded liposomes, resulted in a decrease of absorbance at 503-487 nm (Fig. 1C). As in intact photosynthetic membranes [1], the direction of the absorbance change is dependent on the polarity of the applied potential.

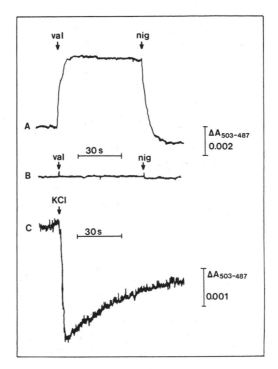

Figure 1. Carotenoid absorbance changes at 503-487 nm induced by K^+-diffusion potentials in liposomes reconstituted with $RCLH_ILH_{II}$-complexes. Liposomes (reconstituted with 70 nmol BChl/mg lipid) loaded with K^+-buffer (1A and 1B) or Na^+-buffer (1C) were diluted 67 fold in 2.0 ml 50 mM Na-phosphate (pH 7.6) 50 mM NaCl (1A and 1C) or 50 mM K-phosphate (pH 7.6) 50 mM KCl (1B). Where indicated valinomycin (val) and nigericin (nig) were added to give final concentrations of 100 nM and 250 nM, respectively. KCl was added from a 3 M stock solution to give a final K^+ concentration of 295 mM. The final BChl concentration in all experiments was 6.6 µM.

Spectra of the absorbance changes

In intact bacterial cells and chromatophores the carotenoid absorbance changes show a characteristic spectrum, which can be explained by a shift (to higher or lower wavelengths) of the absorbance maxima of a part of the carotenoids in all three carotenoid peaks [7,8]. In order find out whether such a characteristic spectrum could be observed in the reconstituted liposomes the spectral changes in the carotenoid region were measured of liposomes which contained $RCLH_ILH_{II}$- or LH_{II}-complexes and were loaded with K^+-buffer. For comparison also the spectral changes in the same region were recorded of *Rb. sphaeroides* chromatophores from which the pigment protein complexes were isolated. Figure 2 shows the result of these analyses. Chromatophores showed a diffusion-potential induced difference spectrum with maxima at 521 nm 488 nm and 458 nm and minima at 506 nm 472 nm and 442 nm (Fig. 2A). Liposomes containing RCLH$_I$LH$_{II}$-complexes showed a similar shaped difference spectrum (Fig 2B), however with different maxima (499 nm and 467 nm) and minima (517 nm, 482 nm and 451 nm). An inverted difference spectrum with respect to the RCLH$_I$LH$_{II}$-liposomes was found in liposomes containing LH$_{II}$-complexes (Fig 2C): maxima at 517 nm, 482 nm, and 451 nm and minima at 499 nm and 467 nm. No changes in the absolute spectra of the different preparations were observed in this spectral region (data not shown).

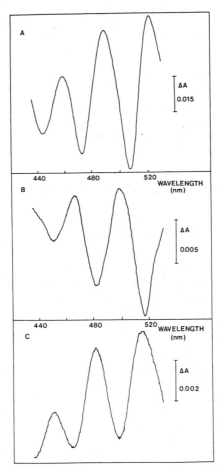

Figure 2. Spectra of carotenoid absorbance changes induced by K$^+$ diffusion potentials in chromatophores of *Rb. sphaeroides*, in liposomes containing RCLH$_I$LH$_{II}$-complexes and in liposomes containing LH$_{II}$-complexes. K$^+$-loaded chromatophores and liposomes were diluted 50 fold into 2 ml 50 mM K$^+$-phosphate (pH 7.6) 50 mM KCl (supplemented with 10% sucrose and 8 mM MgSO$_4$ in the case of chromatophores). The figures show the difference between the spectra before and after the addition of 100 nM valinomycin. A: Chromatophores, final BChl concentration 36 μM, B: RCLH$_I$LH$_{II}$-liposomes (70 nmol BChl/mg lipid), final BChl concentration 8.8 μM, C: LH$_{II}$-liposomes (60 nmol BChl/mg lipid), final BChl concentration 6.4 μM.

It is clear that both RCLH$_I$LH$_{II}$- and LH$_{II}$-liposomes show typical band shift difference spectra (Fig. 2B and 2C), similar to the one obtained for chromatophores (Fig. 2A, see also ref. 1). The most consistent interpretation of the shape of these spectra assumes that solubilisation and/or reconstitution of the pigmented proteins causes a slight blue-shift (appr. 4 to 5 nm) of the field-sensitive carotenoids (i.e. those associated with B800 in LH$_{II}$; see refs. 4 and 5). When this blue-shift is taken into account, it is clear that the carotenoids in chromatophores and in RCLH$_I$LH$_{II}$-complexes give a mirror-response when exposed to a $\Delta\psi$, inside negative. This would be consistent with an opposite orientation in these two samples of the carotenoid associated with B800. Indeed it was found that RCLH$_I$LH$_{II}$-complexes, reconstituted in E. coli lipids, do have their cytochrome c binding site exposed to the external aqueous phase for more than 95% (W.Crielaard and K.J.Hellingwerf, unpublished observations). This is the same orientation as observed for RCLH$_I$LH$_{II}$-complexes from *Rhodopseudomonas palustris* after reconstitution in E. coli phospholipids [6] but opposite to the orientation in chromatophores. Along similar lines it is predicted that LH$_{II}$ will have opposite orientations when reconstituted either as a single complex or in combination with RCLH$_I$ (compare Fig. 2B with Fig. 2C). A small blue-shift (4 nm) has recently been observed in the absorption spectrum of carotenoids in the LH$_{II}$-complex from *Rhodopseudomonas acidophila*, upon incubation with a low concentration of an ionic detergent (lithium dodecyl sulphate) [9]. However, since no differences in the absolute spectra of the different preparations were observed, the actual shift is probably obscured by the carotenoids associated with B850, which do not necessarily have to show an identical shift as the field-sensitive carotenoids. The results described in Fig. 2 make it unlikely that one of the two complexes has been reconstituted in a random orientation. In that case a doubling of the number of minima and maxima in the $\Delta\psi$-induced difference spectrum would be expected.

Relation between the electrical membrane potential and the carotenoid signal

In bacterial systems a linear relationship instead of the theoretically predicted quadratic relationship has been found between the carotenoid absorbance change and the membrane potential [1,10]. It has been suggested that this linear relationship is caused by a permanent electrical field [4,11], resulting from charged amino acids in the neighbourhood of the field-sensitive carotenoids. Purification and/or reconstitution could lead to a change of this permanent field and thereby change the relationship between the electrical field and the induced absorbance change. Figure 3 shows that this is not the case in RCLH$_I$LH$_{II}$-liposomes. The absorbance change at 503-487 nm is proportional to the applied potential in the range from +133 mV (KCl-pulses) to -110 mV (valinomycin pulses).

The internal K$^+$-concentration of the Na$^+$-loaded liposomes after washing (1.25 mM), was calculated by extrapolating the induced absorbance change to zero (where the internal [K$^+$] equals the external [K$^+$]). A correction was made for dilution with KCl (by the addition of an amount of buffer equaling the KCl pulse, after relaxation of the induced potential). From these calculations a calibration factor for the RCLH$_I$LH$_{II}$-liposomes of 6.6 x10^{-6} mV^{-1} x (μM BChl)$^{-1}$ can be determined.

The most attractive feature of the carotenoid absorbance change, the linear relationship with the $\Delta\psi$ [cf ref. 1], has been preserved in the liposomes. Both $\Delta\psi$'s, negative and positive inside, provoke a linear response with a proportionality constant of 6.6 x10^{-6} mV^{-1} x (μM BChl)$^{-1}$. This constant is probably different from the proportionality constant in chromatophores, due to the shifts in maxima and minima. The linear response implies that the postulated local field in the carotenoid region [6,11] has not been abolished by the isolation and reconstitution.

Liposomes containing electrochromically active pigment protein complexes can be used for several purposes. By fusing bacterial membrane vesicles with liposomes containing pigment protein complexes [12] the carotenoid absorbance change can be used as a $\Delta\psi$ probe in non-photosynthetic membranes. Such a system is well-suited to investigate further the discrepancies between the $\Delta\psi$ deduced from distribution measurements of lipophilic ions and from the carotenoid absorbance change [2,3,13]. However, measurements of the band-shift in RCLH$_I$LH$_{II}$-liposomes with a $\Delta\psi$, generated by illumination (cf. ref. 6), have not yet been possible. These measurements are complicated by the spectral overlap between the band-shift and absorbance changes due to the light-dependent changes in the redox state of cytochrome c [6].

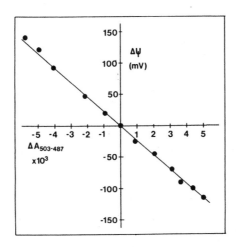

Figure 3. Relation between the carotenoid absorbance changes at 503-487 nm and the membrane potential in RCLH$_I$LH$_{II}$-liposomes. K$^+$- or Na$^+$-loaded liposomes (70 nmol BChl/mg lipid) were diluted 67 fold in 50 mM Na-phosphate (pH 7.6), 50 mM NaCl. Potentials, negative inside (negative $\Delta\psi$), were induced by adding 100 nM valinomycin to the K$^+$-loaded liposomes. The potential was varied by varying the K$^+$-buffer/Na$^+$-buffer ratio. Potentials, inside positive (positive $\Delta\psi$), were induced by the addition of KCl to the Na$^+$-loaded liposomes. KCl was added from a stock solution of 3 M. The $\Delta\psi$ was varied by varying the amount of KCl added. A correction was made for dilution. The $\Delta\psi$'s were calculated after the determination of the internal K$^+$-concentration by extrapolation to zero bandshift. The final BChl concentration was in all cases 6.6 μM.

REFERENCES

1]   Jackson, J.B. and Crofts, A.R. (1969) FEBS Lett. 4, 185-189.
2]   Clark, A.J. and Jackson, J.B. (1981) Biochem. J. 200, 389-397.
3]   Crielaard, W., Cotton, N.P.J., Jackson, J.B., Hellingwerf, K.J. and Konings, W.N. (1988) Biochim. Biophys. Acta 932, 17-25.
4]   Holmes, N.G., Hunter, C.N., Niederman, R.A. and Crofts, A.R. (1980) FEBS Lett. 115, 43-48.
5]   Webster, G.D., Cogdell, R.J. and Lindsay, J.G. (1980) Biochim. Biophys. Acta 591, 321-330.
6]   Molenaar, D., Crielaard, W. and Hellingwerf, K.J. (1988) Biochemistry 27, 2014-2023.
7]   De Grooth, B.G. and Amesz, J. (1977) Biochim. Biophys. Acta 462, 237-246.
8]   De Grooth, B.G. and Amesz, J. (1977) Biochim. Biophys. Acta 462, 247-258.
9]   Robert, B. and Frank, H.A. (1988) Biochim. Biophys. Acta 934, 401-405.
10]   Wraight, C.A., Cogdell, R.J. and Chance (1978) in: "The photosynthetic bacteria" (Clayton R.K. and Sistrom W.R. eds.) pp.471-511, Plenum Press, New York and London.
11]   Kakitani, T., Honig, B. and Crofts, A.R. (1982) Biophys. J. 39, 57-63.
12]   Crielaard, W., Driessen, A.J.M., Molenaar, D., Hellingwerf, K.J. and Konings, W.N. (1988) J. Bacteriol. 170, 1820-1824.
13]   Ferguson, S.J., Jones, O.T.G., Kell, D.B. and Sorgato, M.L. (1979) Biochem. J. 180, 75-85.

PHOSPHORYLATION OF MEMBRANE PROTEINS

IN CONTROL OF EXCITATION ENERGY TRANSFER

John F Allen and Michael A Harrison

Department of Pure and Applied Biology
University of Leeds
Leeds   LS2 9JT
England

INTRODUCTION

In chloroplasts of green plants a broad function for
phosphorylation of light-harvesting polypeptides is relatively
well characterised while the mechanism by which
phosphorylation exerts its effect is mostly a matter for
speculation. The state of knowledge for the photosynthetic
purple bacteria and cyanobacteria is much less satisfactory,
and for the green photosynthetic bacteria it is non-existent.

All photosynthetic prokaryotes have quite different
light-harvesting polypeptides from those of chloroplasts, and
their structure and function cannot therefore be controlled in
precisely the same way. Nevertheless, cyanobacteria show
light-dependent phosphorylation of polypeptides that can be
associated by circumstantial evidence with the light-
harvesting phycobilisome and with photosystem II, and purple
non-sulphur bacteria show light-dependent phosphorylation of
low-molecular-weight chromatophore membrane polypeptides. In
both groups phosphorylation can be demonstrated both in vivo
and in vitro, and correlates with functional changes observed
by chlorophyll fluorescence spectroscopy or chlorophyll
fluorescence induction kinetics.

Here we report on recent data concerning light-induced
phosphorylation specifically of cyanobacterial polypeptides,
purple bacteria having been covered recently elsewhere[1]. In
particular, we have now obtained a partial amino acid sequence
of a cyanobacterial 13 kDa protein previously suggested to be
a light-harvesting protein. Details of the purification
technique and of the sequencing protocol will be published
elsewhere (M A Harrison, J N Keen, J B C Findlay and J F
Allen, submitted). Here we describe this sequence and discuss
the implications of our identification of this protein for
control of excitation energy transfer at a transcriptional as
well as a post-translational level.

Molecular Biology of Membrane-Bound Complexes in Phototrophic Bacteria
Edited by G. Drews and E. A. Dawes
Plenum Press, New York, 1990

PROTEIN PHOSPHORYLATION IN LIGHT-STATE TRANSITIONS

The cyanobacterial thylakoid membrane differs from that of the chloroplast in that LHC II, the principal antenna complex of photosystem II in chloroplasts, is functionally replaced in cyanobacteria by the phycobilisome. The phycobilisome is a macromolecular assembly of water-soluble polypeptides covalently bound to phycobilin chromophores[2].

Chloroplasts and cyanobacteria nevertheless show functional similarities in the way in which they can adapt the light-harvesting capacity of each photosystem in response to the spectral quality of incident light[3,4]. Light preferentially absorbed by photosystem II (light 2) gives rise to a redistribution of excitation energy in favour of photosystem I, while light preferentially absorbed by photosystem I (light 1) gives rise to a redistribution of excitation energy in favour of photosystem II. The states induced by light 1 and light 2 are termed state 1 and state 2 respectively, and the general phenomenon of redistribution between the two states is known as state 1-state 2 transitions. The effect of each transition is to maintain a high quantum yield of photosynthesis despite an change in the spectral composition of incident light. Without the state transition phenomenon, such a change would favour one or other photosystem to the detriment of the balanced utilization of excitation energy that depends on the series connection of the two photosystems for electron transfer[5,6].

The mechanism of state 1-state 2 transitions in chloroplasts involves phosphorylation of LHC II by a protein kinase whose activity is regulated by the redox state of plastoquinone or of another electron transport intermediate also located betwenn the two photosystems[7,8]. Incident light favouring excitation of photosystem II causes reduction of plastoquinone and hence activation of the kinase that catalyzes phosphorylation of LHC II. This results in dissociation of LHC II from photosystem II and in the attendant changes characteristic of the transition to state 2, including increased relative fluorescence yield at 735 nm at liquid nitrogen temperature that suggests reassociation of phospho-LHC II with photosystem I. Incident light favouring photosystem I causes oxidation of plastoquinone and inactivation of the LHC II kinase. Dephosphorylation of LHC II is then catalyzed by a phosphatase assumed to be continually active, and the state 1 transition follows as the LHC II reassociates with PS II.

Initial experiments on protein phosphorylation in cyanobacteria and red algae showed only light-independent effects[9,10]. Subsequent work, however, showed phosphorylation of polypeptides of molecular weight 18.5 kDa, 15 kDa and 13 kDa in Synechococcus 6301 under conditions shown to correspond to state 2[11]. The 18.5 kDa phosphoprotein is water-soluble and co-purifies with the phycobilisome[12]. The 15 kDa polypeptide is either an integral membrane protein or is bound to the thylakoid membrane[11]. The 15 kDa polypeptide is the principal species labelled in vitro with gamma(^{32}P)-ATP[13].   The 13 kDa phosphoprotein can be found to occur in both soluble and membrane phases, but is lost from the latter by washing.   The 18.5 kDa soluble phosphoprotein was proposed as a phycobilisome component and the 15 kDa membrane phosphoprotein as a component of photosystem II[11], proposals explicitly devised to explain their imagined functions in controlling excitation energy transfer from the phycobilisome to PS II[14]. These proposals have not yet been subjected to the crucial test of purification to the level where amino acid sequencing can be performed, though work on this is currently underway in our laboratory.   We have however obtained such results for the 13 kDa protein.

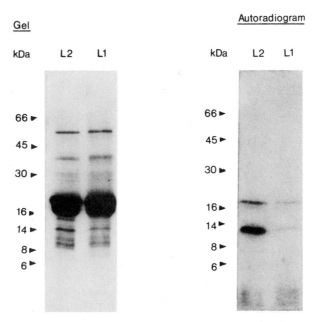

Fig. 1.   Whole cell labelling of polypeptides from the cyanobacterium <u>Synechococcus</u> 6301 (mutant AN112) with ^{32}P-orthophosphate. L1 indicates SDS-PAGE track of sample from cells incubated for 30 minutes in light 1, i.e. photosystem I-absorbed light defined by a red filter transmitting light of wavelength greater than 640 nm (50% transmittance)[21]. L2 indicates SDS-PAGE track of sample from cells incubated for 30 minutes in light 2, i.e. photosystem II-absorbed light defined by an orange filter transmitting light between 560 nm and 620 nm (50% transmittance)[21]. "Gel" shows bands stained with Coomassie Brilliant Blue, "Autoradiogram" shows ^{32}P-labelling of the same gel, with molecular weights of standards in kDa as indicated.

Fig. 1 shows effects of 30 minutes' incubation under light absorbed preferentially by photosystem I and photosystem II on ^{32}P-labelling of proteins <u>in vivo</u> in the AN112 mutant[10] of <u>Synechococcus</u> 6301. In the cyanobacteria it is possible to define rigorously wavelength bands specific to photosystems I and II because of the distinctive phycobilin light-harvesting system of photosystem II with an absorption band complementary to that of the chlorophyll light-harvesting system of photosystem I. Fig. 1 shows ^{32}P-labelling of polypeptides running at apparent relative molecular masses of 18.5 kDa and 13 kDa.

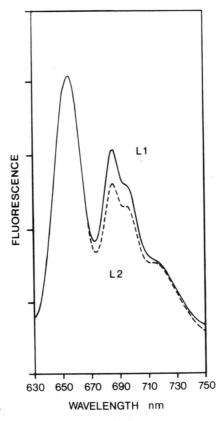

Fig. 2. Fluorescence emission spectra of cells of <u>Synechococcus</u> 6301 (mutant AN112) at 77 K, at 10 <u>microg</u> ml[-1] chlorophyll concentration with excitation wavelength 600 nm, using a Perkin-Elmer LS5 fluorescence spectrometer. L1 and L2 indicate cells pre-illuminated as in Fig. 1., showing a change characteristic of that between light-state 1 and light-state 2[4,11,21]. Spectra were normalized to the phycocyanin emission maximum at 650 nm.

The effects of the illumination conditions of Fig. 1 on distribution of absorbed excitation energy between the pigment beds of photosystems I and II were confirmed by fluorescence emission spectroscopy, as shown in Fig. 2. Fluorescence emission from photosystem I (at 720 nm) relative to that from photosystem II (at 685 to 695 nm) is increased by illumination, prior to freezing, with light absorbed by photosystem II. In contrast, pre-illumination with light absorbed by photosystem I causes a decrease in low-temperature fluorescence emission from photosystem I relative to that from photosystem II. These effects indicate altered excitation energy distribution between photosystem I and photosystem II, a phenomenon resulting from imbalance in electron transport rates between the two photosystems.

We have developed a procedure for purification of the 13 kDa protein. The procedure gave a single band as detected by staining with Coomassie Brilliant Blue. The ^{32}P-labelled and purified cyanobacterial 13 kDa protein was subjected to acid hydrolysis and the hydrolysate to high-voltage paper electrophoresis. The radioactivity migrated with the tyrosine standard, indicating tyrosine as the the labelled amino acid in the modified 13 kDa polypeptide seen in Fig. 1.

The purified 13 kDa protein was subjected to solid-phase N-terminal sequencing as described. The sequence of 30 amino acids obtained is as follows, together with the sequence of greatest similarity from the OWL protein sequence database of the Leeds-Birkbeck ISIS integrated sequence/integrated structure data resource[15].

```
MKXIEAIIRPFKLDEVKIALVNAGIVGMTV    Synechococcus 13 kDa
MKKIDAIIKPFKLDDVRERLAEVGITGMTV    E.coli P_II
```

With reference to the <u>Synechococcus</u> sequence, an amino acid identity is represented by # and a conservative substitution by * in the <u>E. coli</u> sequence in the following comparison.

```
MKXIEAIIRPFKLDEVKIALVNAGIVGMTV    Synechococcus 13 kDa
## #*####*#######*#*  # * ## ####    E.coli P_II
```

It is seen that the closest match was obtained between the cyanobacterial 13 kDa protein and the P_{II} protein encoded by the glnB gene of <u>E. coli</u>[16], the two proteins having 63% of amino acids in common at each position. The second closest match for amino acid sequence identity with the 13 kDa cyanobacterial protein was for another protein related to <u>E. coli</u> P_{II} whose sequence has been deduced from the sequence of a gene from <u>Rhizobium leguminosarum</u> (not shown). This showed 60% of amino acids identical to those of the 13 kDa protein.

PROTEIN MODIFICATION IN TRANSCRIPTIONAL REGULATION

It is likely that the ^{32}P-labelling of the 13 kDa cyanobacterial P_{II} protein (Figs. 1) results from uridylylation of tyrosine, as in E. coli[17]. In E. coli the P_{II} protein is uridylylated in response to a decrease in ratio of glutamine to 2-ketoglutarate in the cell[17]. The uridylylated form of P_{II} promotes activation of the enzyme glutamine synthetase by deadenylylation. Uridylylation of P_{II} also results in activation of transcription of the glnA structural gene for glutamine synthetase. This activation of transcription results from a removal of the unmodified P_{II} that is required as a cofactor for the protein phosphatase activity of a combined kinase-phosphatase termed NR_{II}. The substrate of NR_{II} therefore becomes phosphorylated as a result of uridylylation of P_{II}. The substrate of NR_{II} is an activator of transcription of several E. coli genes, including glnA, and it activates transcription in its phosphorylated form. This transcriptional activator is the protein NR_I. Phosphorylation of NR_I is likely to regulate directly binding of RNA polymerase at promoters, since the NR_I protein itself shows the helix-turn-helix motif that is characteristic of DNA-binding proteins and also binds the sigma^{54} RNA polymerase[17].

Our discovery of the P_{II} protein in a cyanobacterium suggests the existence in photosynthetic organisms of a combined cascade control of enzyme activity and transcription analogous to that involved in regulation of ammonia assimilation in E. coli. Control of modification of P_{II} by photosynthetic light-harvesting and electron transport (Figs. 1 and 2.) may indicate a coupling between the photochemical reactions of photosynthesis and assimilatory nitrogen metabolism.

TRANSCRIPTIONAL CONTROL OF EXCITATION ENERGY TRANSFER

Light-modified transcriptional regulators such as P_{II} may contribute to feedback control of the synthesis and assembly of the photosynthetic apparatus. For example, in both cyanobacteria[18] and eukaryotic organisms[19] alteration of the stoichiometry of photosystems I and II occurs, with the effect of maintaining equal rates of light utilisation in a process analogous to the short-term redistribution of light-harvesting complexes during state 1-state 2 transitions. The redox state of the electron carrier plastoquinone acts as a sensor of imbalance in excitation of the two photosystems[7] and it has been suggested that changes in plastoquinone redox state initiate changes in gene expression that give rise to altered photosystem stoichiometry[20,21]. Redox control of P_{II} modification at the level of a photosynthetic electron transport component situated between photosystems I and II is indicated by the results in Fig. 1. The possibility therefore arises that short-term light-state transitions[3-7] and photosynthetic control of gene expression share common components and respond to the environmental changes via the same trigger represented by perturbation of redox poise.

Control of light-harvesting phycobiliprotein stoichiometry by complementary chromatic adaptation[22] could also involve redox control of gene expression by such a route.

A feedback control by which components of the photosynthetic apparatus control their own synthesis and assembly may provide an alternative to specific photoreceptors such as phytochrome, whose primary role might then be in non-photosynthetic tissue otherwise incapable of responding to environmental light.

AKNOWLEDGEMENTS

We thank Dr A N Glazer for the gift of AN112 cells, Drs J B C Findlay and J Keen for protein sequencing, and the UK SERC for research grants to JFA and for support (to Dr J B C Findlay) of the Leeds protein sequencing unit.

REFERENCES

1. J. F. Allen, M. A. Harrison and N. G. Holmes, Protein phosphorylation and control of excitation energy transfer in photosynthetic purple bacteria and cyanobacteria, Biochimie in press.
2. A. N. Glazer, Light harvesting by phycobilisomes, Annu. Rev. Biophys. Biophys. Chem., 14:47 (1985).
3. C. Bonaventura and J. Myers, Fluorescence and oxygen evolution from Chlorella pyrenoidosa, Biochim. Biophys. Acta, 189:366 (1969).
4. N. Murata, Control of excitation transfer in photosynthesis. I. Light-induced changes of chlorophyll a fluorescence in Porphyridium cruentum, Biochim. Biophys. Acta, 172:242 (1969).
5. J. Myers, Enhancement studies in photosynthesis, Annu. Rev. Plant Physiol., 22:289 (1971).
6. W. P. Williams and J. F. Allen, State 1/state 2 changes in higher plants and algae, Photosynth. Res., 13:19 (1987).
7. J. F. Allen, J. Bennett, K. E. Steinback and C. J. Arntzen, Chloroplast protein phosphorylation couples plastoquinone redox state to distribution of excitation energy between photosystems, Nature, 291:25 (1981).
8. P. Horton, J. F. Allen, M. T. Black and J. Bennett, Regulation of phosphorylation of chloroplast membrane polypeptides by the redox state of plastoquinone, FEBS Lett., 125:193 (1981).
9. G. Schuster, G. C. Owens, Y. Cohen and I. Ohad, Light-independent phosphorylation of the chlorophyll a/b protein complex in thylakoids of the prokaryote Prochloron, Biochim. Biophys. Acta, 767:596 (1984).
10. J. Biggins, C. L. Campbell and D. Bruce, Mechanism of the light state transition in photosynthesis: II. Analysis of phosphorylated polypeptides in the red alga Porphyridium cruentum, Biochim. Biophys. Acta, 806:230 (1984).
11. J. F. Allen, C. E. Sanders and N. G. Holmes, Correlation of membrane protein phosphorylation with excitation energy distribution in the cyanobacterium Synechococcus 6301, FEBS Lett., 193:271 (1985).

12. C. E. Sanders and J. F. Allen, The 18.5 kDa phosphoprotein of the cyanobacterium Synechococcus 6301: a component of the phycobilisome, in: "Progress in Photosynthesis Research. Vol. II," J. Biggins, ed., Martinus Nijhoff, Dordrecht (1987).

13. C. E. Sanders and J. F. Allen, Effects of divalent cations on 77K fluorescence emission and on membrane protein phosphorylation in isolated thylakoids of the cyanobacterium Synechococcus 6301, Biochim. Biophys. Acta, 934:87 (1988).

14. J. F. Allen and N. G. Holmes, A general model for regulation of photosynthetic unit function by protein phosphorylation, FEBS Lett., 202:175 (1986).

15. D. Akrigg, A. J. Bleasby, N. I. M. Dix, J. B. C. Findlay, D. Parry-Smith, J. C. Wootton, T. L. Blundell, S. P. Gardner, F. Hayes, S. Islam, M. J. E. Sternberg, J. M. Thornton and I. J. Tickle, A protein sequence/structure database, Nature 335: 745 (1988).

16. H. S. Son and S. G. Rhee, Cascade control of Escherichia coli glutamine synthetase. Purification and properties of P_{II} protein and nucleotide sequence of its structural gene, J. Biol. Chem. 262:8690 (1987).

17. B. Magasanik, Reversible phosphorylation of an enhancer binding protein regulates transcription of bacterial nitrogen utilization genes, Trends Biochem. Sci. 13:475 (1988).

18. M. Kawamura, M. Mimuro and Y. Fujita, Quantitative relationship between two reaction centres in the photosynthetic system of blue green algae, Plant Cell Physiol. 20:697 (1979).

19. A. Melis and J. S. Brown, Stoichiometry of system I and system II reaction centers and of plastoquinone in different photosynthetic membranes, Proc. Natl. Acad. Sci. U.S.A. 77:4712 (1980).

20. Y. Fujita, A. Murakami and K. Ohki, Regulation of photosystem composition in the cyanobacterial photosynthetic system: the regulation occurs in response to the redox state of the electron pool located between the two photosystems, Plant Cell Physiol. 28:283 (1987).

21. A. Melis, C. W. Mullineaux and J. F. Allen, Acclimation of the photosynthetic apparatus to photosystem I or photosystem II light: evidence from quantum yield measurements and fluorescence spectroscopy of cyanobacterial cells, Z. Naturforsch. 44c:109 (1989).

22. N. Tandeau de Marsac, Phycobilisomes and complementary chromatic adaptation in cyanobacteria, Bulletin l'Institut Pasteur, 81:201 (1983).

LIMITED PROTEOLYSIS AND ITS INFLUENCE ON THERMAL STABILITY OF THE

PHOTOSYNTHETIC REACTION CENTER FROM *Chloroflexus aurantiacus*

Michail A. Kutuzov, Nellie B. Levina,
Najmoutin G. Abdulaev and Alexander S. Zolotarev

Shemyakin Institute of Bioorganic Chemistry
USSR Academy of Sciences
Ul. Miklukho-Maklaya, I6/IO
II787I GSP Moscow, V-437 USSR

SUMMARY

Limited proteolysis with proteases of different specificity has been used to obtain a number of preparations of the reaction center (RC) from the thermophilic green bacterium *Chloroflexus aurantiacus*. The absorption spectrum of the native RC are preserved in these preparations, even after the shortening of the N-termini of both RC subunits to different extent. Considerable part of the hydrophilic N-terminus of the L-subunit has been shown to be unaccessible to the proteases. AlaL3-ArgL25 region contribute to the thermal stability of the *Chloroflexus* RC.

INTRODUCTION

Thermophilic green bacterium *Chloroflexus aurantiacus* is a member of one of the most ancient branches of eubacteria[1]. Its photosynthetic reaction center (RC) is the simplest one of the "quinone type" known at present[2]. It consists of only two subunits and may serve as the simplest model of photosystem II. Recently the primary structures of L and M subunits of *Chloroflexus* RC have been determined and are very similar to the RC's of the purple bacteria[3-5]. The increased thermal stability of the *Chloroflexus* RC renders it as a unique object for studies the mechanisms involved in stabilizing the tertiary structures of proteins at higher temperature. In this work the accessibility of the N-terminal regions of both subunits to proteases treatment and their contribution to the thermal stability of the RC have been studied.

RESULTS AND DISCUSSION

Together with similarities, there are also essential differences between the RC's of *Chloroflexus* and purple bacteria. The most prominent are: the absence of H subunit and replacement of one of the bacterio-chlorophylls by bacteriopheophytin in *Choloroflexus* RC[2]. Moreover, there is almost no homology between the N-terminal sequences of both *Chloroflexus* subunits and corresponding regions of the purple bacteria RC's; furthermore, the N-terminus of the *Chloroflexus* L subunit is about 30 amino acid residues longer than that in the purple bacteria.

Molecular Biology of Membrane-Bound Complexes in Phototrophic Bacteria
Edited by G. Drews and E. A. Dawes
Plenum Press, New York, 1990

In order to elucidate the possible role of the N-terminal suquences we used limited proteilysis of the RC with enzymes of different specificity followed by the identification of the cleavage sites. We tested the apectral properties, thermal stability and electron transfer of some of the preparations (to be descrebed elsewhere). Trypsin, clostripain, proteinase from *St. aureus* V8, chymotrypsin, thermolysin and pronase E have been used in this assay. Initially we attempted to digest RC reconstituted into azolection liposomes using cholate dialysis method. This method yielded proteoliposomes with 30-40% RC's oriented their acceptor side (i.e. N-termini) out, as determined according to[6]. Unfortunately, the rate of digestion was very low even at rather high enzyme to protein ratios (from I:I up to IO:I (w/w)). Due to the poor yields of cleaved products even after more than 24 h incubation, we decided to use another approach. Enzymatic cleavage was carried out in 0.I% lauryldimethylamine N-oxide (LDAO). At this condition the rate of proteolysis was considerably higher. Similar results have been obtained in the case of halorhodopsin[7]: membrane-associated protein appeared to be unaccessible to trypsin attack, while detergent-solubilized one was cleaved in several sites. The authors have suggested that detergent-solubilized and membrane-bound proteins adopt different conformations, the latter being more tightly structured than the former.

Proteolysis was carried out in the dark either at room temperature or at 37°C. The progress of the reaction was followed by SDS-PAGE; RC absorption spectra were also controlled. All the enzymes except chymotrypsin yielded relatively stable cleavage products with electrophoretic mobilities somewhat lower than those of the noncleaved RC subunits (Fig.I). These products had almost unchanged absorption spectra except that the absorption of accessory bacteriochlorophyll (8I2 nm) was reduced up to 3-5% or 5-IO% after cleavage at room temperature or 37°C, respectively. However, stability of the cleaved RC preparations was noticeably decreased. The cleaved preparations were stable only for I-2 days, while freshly isolated RC can be stores at 4°C in 0.1% LDAO, pH 9.0 several weeks without loss of its spectral properties. Prolonged incubation of RC with proteases (up to 24 h at 37°C) except clostripain resulted in gradual disappearance of the bands shown on fig.I and of the

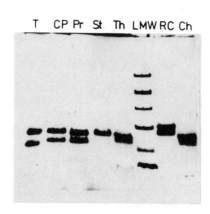

Fig. I. SDS-PAGE of the partially digested *Chloroflexus* RC. The hydrolyses have been carried out at the same conditions: I.5 h, 37°C, pH 9.0. T, trypsin (enzyme to protein ratio I:5); CP, clostripain (I:30); Pr, pronase E (I:40); St, proteinase from *St. aureus* V8 (I:IO); Th, thermolysin (I:7); LMW, molecular weight standards 94, 67, 43, 30, 20 and I4 kd; RC, entire RC; Ch, chymotrypsin (I:2).

typical absorption spectrum of RC. In the case of V8-staphylococcal protease hydrolysis two additional weak bands I9 and 20 kd transiently appeared. Probably, prolonged digestion of the detergent-sulubilized RC with proteases leads to the cleavage of the loops connecting transmembrane helices followed by rapid protein denaturation.

Cleavage sites have been identified by sequencing on gas-phase sequenator either the mixture of produced membrane fragments after their precipitation with ethanol or individual fragments electroblotted onto PVDF-membrane after SDS-PAGE separation. Identified sites of the deepest cleavage are shown on fig.2. Site of the chymotryptic cleavage of L subunit was not identified because of the formation of the mixture of derivatives.

It is shown (fig.2), that the whole hydrophilic N-terminal region of the M subunit is accessible for proteolysis. These data fairly fit well the earlier suggested folding model *Chloroflexus* RC in the membrane[3,4]. This indicates that the N-terminus of the M subunit is easily accessible on the surface, flexible and not highly structured. As far as hydrophilic N-terminus of L subunit is concerned only its distal half is accessible to the proteases. We suggest that the proximal part of N-terminus of L subunit is highly structured or buried inside the protein.

In order to investigate the possible contribution of the N-termini in stabilization of the RC tertiary structure at elevated temperature we have tested the thermal stability of two RC preparations obtained by clostripain hydrolysis. This enzyme has been chosen because it has no potential cleavage sites in the C-termini of RC and yields a limited number of easily identifiable intermediate products.

The effect of temperature on native and cleaved samples were examined by incubating them (A_{812} 0.I–0.3) in 50 mM Tris-HCl buffer (pH 9.0) containing 0.I% (w/v) lauryldimethylamine N-oxide at given temperature for I5 min. The samples were immediately cooled down to IO°C on water bath and absorption spectra were recorded at room temperature. The extent of

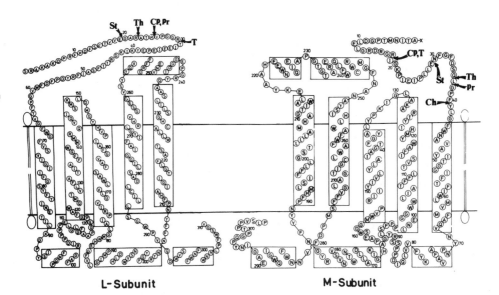

L-Subunit   **M-Subunit**

Fig. 2. Model of polypeptide chain folding of the *Chloroflexus* RC subunits. Arrows indicate the identified sites of enzymatic cleavage. Proteases designated as on fig. I.

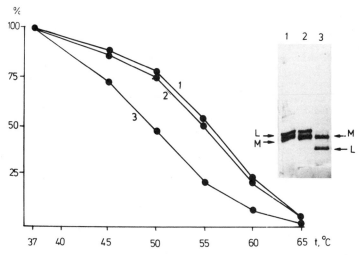

Fig. 3. Heat denaturation curves of the entire RC (I); the intermediate
(2) and the final (3) products of its partial clostripain digest.
Insert: SDS-PAGE, 5-20% gradient gel, 0.3% SDS. Lane I, entire
RC; lane 2, intermediate product; lane 3, final product.

denaturation was appreciated as decrease of absorption at 8I2 nm. Heat
denaturation curves of the entire RC, the intermediate and final products
of its partial clostripain digest are shown on fig.3. The final product
lacks I8 N-terminal amino acid residues of M subunit and 25 N-terminal
amino acid residues of L subunit. The intermediate product lacks the
same I8 amino acid residues of the M subunit but only two amino acid
residues of L subunit. As shown on fig.3 heat denaturation curves of the
entire RC and intermediate products are nearly identical, while the
thermal stability of final product is markedly decreased. Thus, cleavage
from the L and M subunits of 2 and I8 N-terminal amino acid residues,
respectively, has almost no effect on thermal stability of the RC.
Additional cleavage of 23 amino acid residues (AlaL3-ArgL25) doesn't change
the absorption spectrum of RC but decreases its thermal stability.

One of the more interesting and apparently significant factors in
thermal stability involves solvent-accessible arginyl residues. Comparative
studies of the same enzymes from thermophilic and mesophilic organisms
have shown positive correlations between thermal stability and increased
arginine content[8]. Moreover, guanidination and amidination of a number of
soluble proteins have been shown to increase their thermal stability at
low level of modification[9]. Comparing the primary structures of the
RC's from *Chloroflexus* and purple bacteria one can see that content of
arginyl residues indeed is higher in N-terminal regions of *Chloroflexus*
RC. However, it is obviously not possible to explain the increased thermal
stability of this protein only by higher content of arginyl residues. The
removal of four of them as in the case of intermediate product doesn't
change the thermal stability. But further shorting of only another three
arginyl residues within the region AlaL3-ArgL25 decreases it noticeably.
This region has high density of charged amino acid residues. It appears
that this could stabilize RC structure by forming salt bridges with some
other part of the molecule. Stabilization of the soluble protein from
thermophiles due to the increased number of salt bridges has been previously
described[10].

It will be of great interest to further investigate how the region AlaL3-ArgL25 stabilizes the *Chloroflexus* RC at higher temperature. In any case determination of its tertiary structure is essential to answer this question.

REFERENCES

I. C. R. Woese, Bacterial evolution, Microbiol. rev. 5I: 22I (I987)

2. B. K. Pierson, J. P. Thornber, and R. E. B. Seftor, Partial purification, subunit structure and thermal stability of the photochemical reaction center of the thermophilic green bacterium *Chloroflexus aurantiacus*, Biochim. Biophys. Acta 723: 322 (I983)

3. Yu. A. Ovchinnikov, N. G. Abdulaev, A. S. Zolotarev, B. E. Shmukler, A. A. Zargarov, M. A. Kutuzov, I. N. Telezhinskaya, and N. B. Levina, Photosynthetic reaction centre of *Chloroflexus aurantiacus*. I. Primary structure of L-subunit, FEBS Lett. 23I: 237 (I988)

4. Yu. A. Ovchinnikov, N. G. Abdulaev, B. E. Shmuckler, A. A. Zargarov, M. A. Kutuzov, I. N. Telezhinskaya, N. B. Levina, and A. S. Zolotarev, Photosynthetic reaction centre of *Chloroflexus aurantiacus*. Primary structure of M-subunit, FEBS Lett. 232: 364 (I988)

5. J. A. Shiozawa, F. Lottspeich, D. Oesterhelt, and R. Feick, The primary structure of the *Chloroflexus aurantiacus* reaction-center polypeptides, Eur. J. Biochem. I80: 75 (I989)

6. M. D. Mamedov, N. I. Zacharova, A. A. Kondrashin, and A. Yu. Semenov, The effect of o-phenanthroline on electrogenesis under primary charge separation in the reaction center of *Rhodopseudomonas sphaeroides*, Biol. Membrany (Russ) 3: 5I3 (I986)

7. B. Schobert, J. K. Lanyi, and D. Oesterhelt, Structure and orientation of halorhodopsin in the membrane: a proteolytic fragmentation study, EMBO J. 7: 905 (I988)

8. D. J. Merkler, C. K. Farrington, and F. C. Wedler, Protein thermostability, Int. J. Pept. Protein Res. I8: 430 (I98I)

9. F. S. Qaw and J. M. Brewer, Arginyl residues and thermal stability of proteins, Mol. Cell. Biochem. 7I: I2I (I986)

IO. M. F. Perutz and H. Ridt, Stereochemical bases of heat stability in bacterial ferredoxins and hemoglobin A2, Nature 255: 256 (I975).

EXAMINATION OF THE *RHODOBACTER CAPSULATUS* SPECIAL PAIR IN WILD-TYPE AND HETERODIMER-CONTAINING REACTION CENTERS BY TIME-RESOLVED OPTICALLY DETECTED MAGNETIC RESONANCE

Stephen V. Kolaczkowski[1,3], Edward J. Bylina[2,4], Douglas C. Youvan[2] and James R. Norris[1]

[1]Chemistry Division, Argonne National Laboratory, Argonne, IL 60439; [2]Department of Chemistry, Massachusetts Institute of Technology, Cambridge, MA 02139; [3]current address: Department of Chemistry, University of Arizona, Tucson, AZ 85721; [4]current address: Biotechnology Program, Pacific Biomedical Research Center, University of Hawaii at Manoa, Honolulu, HI 96822

One of the exciting new techniques available to photosynthetic researchers is the ability to perform site-directed mutagenesis on the photosynthetic apparatus. The X-ray crystal structures of reaction centers (RCs) can be used to identify important amino acid residues (Michel et al., 1986; Chang et al., 1986; Yeates et al., 1988). Genetic manipulations that alter such residues allow for unprecedented experimental possibilities. *Rb. capsulatus* possesses the best-developed genetic system for the study of the photosynthetic apparatus (Scolnick and Marrs, 1987). A system of deletion strains (Youvan et al., 1985) and complementing plasmids (Bylina et al., 1986, 1989) has been used to generate a diverse collection of reaction center mutations. When the histidine at M200 was replaced with either leucine or phenylalanine, the (Bchl)$_2$ primary donor is replaced with a Bchl-Bphe heterodimer (Bylina and Youvan, 1988). The isolation of heterodimer-containing RCs is one of the first true tests of the importance of a special pair type structure for efficient photosynthetic charge separation.

His$_{M200 \rightarrow}$Leu RCs have the following properties: (1) the absorption spectrum of these RCs is drastically altered relative to wild-type RCs (Bylina and Youvan, 1988); (2) the efficiency of the initial charge separation step is reduced 50% (Kirmaier et al., 1988); (3) Linear dichroism experiments indicate that the pigment organization in these RCs is similar to wild-type RCs (Breton, et al., 1989); (4) EPR experiments indicate that both the cation radical and the triplet state of the primary donor in these RCs are highly asymmetric and monomeric in nature (Bylina, et al., submitted); and (5) the triplet yield in these heterodimer-containing RCs is reduced relative to the triplet yield found in wild-type RCs (Bylina et al., submitted).

These EPR experiments also identified the action of the radical pair mechanism (RPM) as the mechanism of triplet formation in heterodimer-containing RCs. The polarization of the triplet EPR spectra of heterodimer-containing RCs showed that this triplet state is T_0 polarized. The role of the RPM in the formation of the T_0 polarized triplet states was first proposed for and shown to be correct in (Bchl)$_2$-containing RCs from *Rb. sphaeroides* R-26 (Thurnauer et al., 1975; Bowman et al., 1981; Norris et al., 1982).

The RPM has been well characterized in photosynthetic systems (Hoff, 1981; Norris, et al., 1982; Boxer et al., 1983). A simple description of the RPM begins with the reaction sequence found in figure 1. Optical excitation of the SP populates the first excited state of the SP. Electron transfer from this excited state to the photoactive Bphe results in the formation of a singlet correlated radical pair, ^{1}(SP+I-). When electron transfer to the primary quinone is blocked, the state

Molecular Biology of Membrane-Bound Complexes in Phototrophic Bacteria
Edited by G. Drews and E. A. Dawes
Plenum Press, New York, 1990

305

$^1(SP+I^-)$ can (1) recombine to reform the excited state of the SP with rate k_s', which then decays to the special pair ground state; (2) recombine and directly return to the ground state with rate k_s''; or (3) undergo Radical Pair-Intersystem Crossing (RP-ISC) to form the triplet correlated radical pair state $^3(SP+I^-)$ with rate ω. The state $^3(SP+I^-)$ may (1) revert back to $^1(SP+I^-)$ with rate ω by further action of RP-ISC; or (2) decay by charge recombination to form $^3(SP)$ with rate k_t. The kinetic scheme can be simplified by combining k_s' and k_s'' into a single rate constant k_s. The possible fates of $^1(SP+I^-)$ are determined by the rate constants k_s, k_t and ω. In this case, the quantum mechanical coherence of the two electron spins is greatly simplified by treating the coherence as two identical classical rate constants. These simplifications of the RPM allow the radical pair to be modeled by four simultaneous differential equations (Budil, 1986; Budil et al., 1987):

$$d[^3(SP+I^-)]/dt = \omega [\, ^1(SP+I^-)] - (\omega + k_t) [\, ^3(SP+I^-)] \tag{1}$$

$$d[^1(SP+I^-)]/dt = -(\omega + k_s) [\, ^1(SP+I^-)] \tag{2}$$

$$d[^3(SP)]/dt = k_t [\, ^3(SP+I^-)] \tag{3}$$

$$d(SP_G)/dt = k_s [\, ^1(SP+I^-)] \tag{4}$$

These equations can be used to obtain an expression for the triplet yield in terms of only k_s, k_t and ω:

$$\Phi_T = \omega k_t \,/\, [\omega (k_s + k_t) + k_s k_t] \tag{5}$$

This is not a rigorously quantum mechanical analysis and the assumptions made in the derivation must be applied with caution. However, this analysis provides a simple but firm starting point for understanding the processes that underlie $^3(SP)$ formation.

In RP-ISC, the two electrons of the radical pair reside on two separate molecules. In general, these two molecules have different magnetic environments that effect the two electrons to differing degrees (Haberkorn and Michel-Bayerle, 1979). In low magnetic fields, the major influence on the two spins are the nuclear hyperfine moments of the hydrogen atoms of the SP and photoactive Bphe molecules (Boxer et al., 1983). Application of small magnetic fields, typically under 100 gauss for photosynthetic systems, can either enhance or inhibit RP-ISC (Budil, 1986; Brettel et al.; 1987). By changing the applied magnetic field and measuring the $^3(SP)$ yields, values for k_s, k_t and ω can be obtained (Budil, 1986; Budil et al., 1987; Kolaczkowski et al., 1987). In addition, the energy difference between the $^1(SP+I^-)$ and $^3(SP+I^-)$ radical pair states can be determined. This energy splitting is equal to the 2J value for the radical pair (Norris et al., 1982). The J value of a radical pair measures the strength of interaction between the two members of the radical pair, and as such is dependent on both the distance between the molecules of the radical pair and their orientation. The J value can be determined by measuring the magnetic field where the maximum triplet yield is

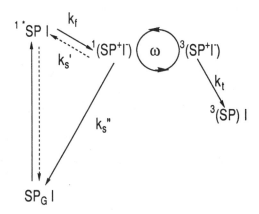

Figure 1. Schematic of the radical pair mechanism

observed. At this field strength, either the T_+ or T_- state of the radical pair is degenerate with the singlet radical pair state.

In this paper, we describe a series of optically detected magnetic resonance experiments that probes the source(s) of the unusually low triplet yield observed in heterodimer-containing RCs. Since carotenoidless RCs facilitate the direct observation of the triplet state of the primary donor by eliminating triplet energy transfer to the carotenoid, a carotenoidless (blue-green) *Rb. capsulatus* *puf* operon deletion strain was isolated. This deletion strain was then used to produce the carotenoidless wild-type and His^{M200}→Leu RCs used in this study.

METHODS

A frameshift mutation in the *pufX* gene was generated as previously described (Bylina and Youvan, 1988). An M13mp18 derivative containing the unique *KpnI*-*SacI* fragment of pU29 (Bylina et al., 1986) was used for the construction of the mutation. DNA inserts were shuttled into pU29 derivatives as described (Bylina and Youvan, 1987). The shuttling of the mutation into pU29 was verified by the presence of the *BamHI* site generated by the mutation. The mutation was introduced into plasmid pU2922 as described (Bylina et al., 1989). Plasmid pU2922 derivatives bearing *puf* operon mutations were conjugated into *Rb. capsulatus* deletion strains as described (Bylina et al., 1989). Photosynthetic growth assays were performed as previously described (Bylina et al., 1989). Near-infrared fluorescence was measured using either high-speed infra-red photography (Youvan et al., 1983) or a video-based digital imaging spectrometer (Yang and Youvan, 1988).

Reaction centers were purified from cultures grown semiaerobically (Yen and Marrs, 1977) in RCV+ containing 30 µg/ml kanamycin by a modified DEAE chromatography method (Prince and Youvan, 1987; Bylina and Youvan, 1988). Purified RC samples (1 ml of OD800 ~ 10 in 10 mM Tris-HCl pH 8.0, 25% ethylene glycol) were reduced by either the addition of a small amount of sodium dithionite or the addition of 1 µmole sodium ascorbate and illuminating while freezing. The samples were held in a quartz cuvette (2 mm path) sealed by a gas tight stopcock. The samples were placed in a N_2 flowing gas dewar and maintained at -30° C. The dewar was held in a pair of Helmholtz coils, which provided a static magnetic field along the axis of the probe light when energized. The magnetic field was controlled by a feedback circuit in the power supply, Lambda model # LFG341A FM, and swept by a digital output from the controlling computer. Further experimental details are provided in Wasielewski et al. (1984). *Rb. sphaeroides* R-26 RCs were used to check the instrumentation, especially before and after experiments using His ^{M200}→Leu RCs.

RESULTS

Isolation of a Carotenoid-Minus Deletion Strain

Oligonucleotide mutagenesis was used to introduce a frameshift into the sequence of the *pufX* gene by inserting a single base (C) after its 23rd codon, resulting in the introduction of a *BamHI* site at this position in the gene. A pU2922 derivative bearing this mutation was introduced into *Rb. capsulatus* deletion strain U43. The resulting mutant is photosynthetically competent and possesses an enhanced fluorescence phenotype. This is consistent with previous experiments that suggest that the *pufX* gene is required for normal ratios of antenna complexes (Klug and Cohen, 1988). When this mutant is spotted onto solid media and incubated under photosynthetic conditions in high light, blue-green sectors appear on the edges of the spots. These sectors remained stable during purification and were used to isolate a blue-green *puf* operon deletion strain.

A carotenoidless deletion strain was isolated as follows: (1) One of the repurified blue-green sectors isolated during photosynthetic growth was subcultured repeatedly in media lacking antibiotics. Plasmid pU2922 is slowly lost from *Rb. capsulatus* strains when grown without antibiotic selection (Bylina et al., 1989); (2) The subcultures were diluted and spread onto antibiotic-free plates, in order to obtain plates each containing 50-100 colonies; (3) The plates were analyzed with a video-based digital imaging spectrometer (Yang and Youvan, 1988). Colonies which were no longer fluorescent were characterized further and found to be lacking the plasmid pU2922. In all cases, loss of the plasmid did not restore carotenoid biosynthesis.

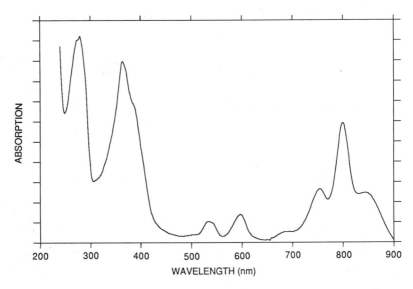

Figure 2. Absorption spectrum of carotenoidless L228BamHI (wild-type) RCs.

Plasmid pU2922 derivatives bearing either the wild-type *puf* operon or the HisM200→Leu mutation were reintroduced to one of these plasmid-free strains (U43b) via conjugation. L228BamHI (wild-type) and HisM200→Leu RCs purified from these transconjugants lacked carotenoids, as demonstrated by the wild-type RC absorption spectrum in figure 2.

Transient Absorbance and Magnetic Field Effect Measurements

Figure 3 shows the transient absorbance traces measured at 860 and 545 nm for carotenoidless and carotenoid containing RCs [both L228BamHI (wild-type) and HisM200→Leu] from *Rb. capsulatus*. The carotenoidless L228BamHI RCs show the transient formation of (SP+I-) followed by electron back transfer to form ^{3}(SP) and (SP)$_G$. The carotenoid-containing wild-type RCs show the transient formation of (SP+I-) followed by electron back transfer and formation of the carotenoid triplet, ^{3}C. ^{3}C is formed by rapid triplet energy transfer from ^{3}(SP) to the ground state carotenoid. The (SP+I-) decay rate is 14 ns, which is in good agreement with the rates measured for both carotenoidless and carotenoid-containing RCs from *Rb. sphaeroides* (Schenck et al., 1982; Cogdell et al., 1975).

The HisM200→Leu RCs show unusual transient absorbance traces. In both the carotenoidless and carotenoid-containing cases, a short lived transient is observed for the decay of (SP+I-). This transient is not well resolved and appears within the ~ 5 ns pulse width of the excitation laser. The instrument function of the system is approximately 2 ns (Wasielewski et al., 1984). This palces an upper limit on the lifetime of the (SP+I-) state, in quinone reduced RCs, at 2 ns. In the case of the carotenoidless HisM200→Leu RCs, no residual bleaching is observed after the decay of (SP+I-). This is consistent with the low ^{3}(SP) yield observed by EPR (Bylina et al., submitted). The HisM200→Leu trace, recorded in the region where ^{3}C absorbs, detects some ^{3}C formed, although the amount is almost negligible. Since the extinction coefficient for the carotenoid $T_1 \rightarrow T_n$ transition is much larger than the SP→^{1}SP in both wild-type and heterodimer-containing RCs, the amount of ^{3}(SP) formed can be observed as ^{3}C after triplet energy transfer to the carotenoid (Norris et al., 1985).

The magnetic field effect data on ^{3}(SP) yields are shown in figure 4 for both carotenoidless wild-type and carotenoid-containing HisM200→Leu RCs. The triplet yield data are normalized to one at zero field. The error bars represent ± 2 standard deviations in the data. These data show a slight increase in triplet yield with increasing magnetic fields, followed by a strong decrease in triplet

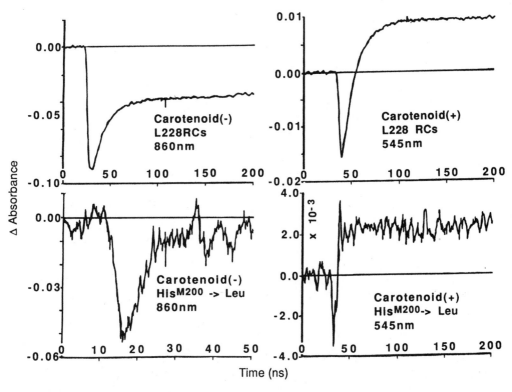

Figure 3. Transient absorption traces for wild-type and His[M200]→Leu RCs.

yield with larger magnetic fields. The maximum triplet yield occurs at 10 ± 5 gauss, yielding a J value of 5 ± 3 gauss. Detection of a magnetic field effect in carotenoidless His[M200]→Leu RCs was prevented by the inability to measure any ^{3}(SP) or magnetic field induced changes in the rate of (SP+I⁻) decay, since all decays were within the pulse width of the excitation laser. A very slight magnetic field effect is measured for the carotenoid-containing His[M200]→Leu RCs. The error bars equal ± 2 standard deviations in the data. The noise in this data attests to the low ^{3}(SP) yields and the poor signal to noise in the transient absorbance change traces. A speculative view of the data might observe a J value of zero for these His[M200]→Leu RCs.

The results of time-resolved optically detected magnetic resonance experiments on both wild-type and His[M200]→Leu RCs are summarized below:

Species :	Rb. sphaeroides	Rb. capsulatus	Rb. capsulatus
Primary Donor :	(Bchl a- Bchl a)	(Bchl a- Bchl a)	(Bchl a-Bphe a)
2J value :	14 ± 2 gauss[a]	10 ± 5 gauss[c]	~0 gauss[c]
k_s :	4 ± 0.5 x 10^7/sec[a]	3.7 ± 0.5 x 10^7/sec[c]	≥5 x 10^8/sec[c]
k_t :	5.0 ± 0.5 x 10^8/sec[a]	5.0 ± 0.5 x 10^8/sec[c]	nd
ω :	~ 4 x 10^7/sec[a]	~4 x 10^7/sec[c]	nd
(SP+I⁻) lifetime :	13 ± 1 ns[a,b]	14 ± 1 ns[c]	≤ 2 ns[c]

[a]Budil, 1986; Budil et al., 1987; Kolaszkowski et al., 1987 and repeated in this work
[b] Schenck et al., 1982   [c] this work   nd = not determined

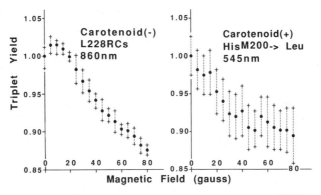

Figure 4. Plots of triplet yield vs. magnetic field for wild-type and HisM200→Leu RCs.

DISCUSSION

The transient absorbance traces and magnetic field effect data for the wild-type RCs from *Rb. capsulatus* show the same behavior of the radical pair state as is observed in *Rb. sphaeroides* RCs. The maximum triplet yield is observed at 10 ± 5 gauss for the wild-type RCs from *Rb. capsulatus* in comparison to the 14 ± 2 gauss observed for *Rb. sphaeroides* (Budil, 1986; Kolaczkowski et al., 1987). The slight decrease in the 2J value for *Rb. capsulatus* RCs may originate from the same source as the increase in radical cation linewidth and triplet ZFS parameters for isolated *Rb. capsulatus* RCs as compared to the measurements of the same signals in chromatophore membranes (Prince and Youvan, 1987). It has also been observed that the isolated wild-type RCs from *Rb. capsulatus* show a shift in the optical absorbance of the special pair band from the expected value of 865 nm (the value observed in *Rb. sphaeroides*) to ~ 850 nm (Prince and Youvan, 1987). The decrease in J may be due to the same source as the blue shift in the absorbance spectrum of the special pair.

The HisM200→Leu RCs have a (SP+I-) lifetime considerably shorter than the wild-type RCs. We interpret this result as an overall increase in the rate of ^{1}(SP+I-) recombination or k_S. An increase in k_S will tend to decrease the overal ^{3}(SP) yield, if k_t and ω remain relatively unchanged (Eqn 5). This increase in k_S is certainly the source of the low ^{3}(SP) yield observed in HisM200→Leu RCs.

Because we were unable to fully resolve the actual (SP+I-) recombination rate and obtain high quality magnetic field effect data for HisM200→Leu RCs, we cannot precisely identify the source for the increase in k_S. The possible decrease in J observed in the magnetic field effect data is compatible with the slowing of the forward charge separation step observed previously (Kirmaier et al., 1988). The large difference between the forward charge separation state (2.8 x 10^{11}/sec) and the reverse charge recombination rate (7.7 x 10^7/sec) in wild-type RCs is reduced in HisM200→Leu RCs. This reduced difference in the forward and backward rates of the initial electron transfer step in these RCs is reminiscent of artificial electron-transfer model systems (Wasielewski, 1988), where there is little difference in the forward and backward rates of electron transfer. These results suggest that the 'one-way switch' which functions during electron transfer [i.e. conformational transition(s) in the RC] may be modified in HisM200→Leu RCs.

ACKNOWLEDGEMENTS

This work was supported (D.C.Y) by the National Science Foundation (DMB-8609614). Work at ANL was supported by the U.S. Department of Energy, Office of Basic Energy Sciences, Division of Chemical Sciences under contract W-31-109-Eng-38.

REFERENCES

Bowman, M.K., Budil, D.E., Closs, G.L., Kostka, A.G., Wraight, C.A. & Norris, J.R. (1981) Magnetic resonance spectroscopy of the primary state P^f, of bacterial photosynthesis, Proc. Natl. Acad. Sci. USA 78, 3305-3307.

Boxer, S.G., Chidsey, C.E.D. & Roelofs, M.G. (1983) Magnetic field effects in the solid state: An example from photosynthetic reaction centers, Ann. Rev. Phys. Chem. 34, 389-398.

Breton, J., Bylina, E.J. & Youvan, D.C. (1989) Pigment organization in genetically modified reaction centers of *Rhodobacter capsulatus*, Biochemistry 28, 6423-6430

Brettel, K. & Setif, P. (1987) Magnetic effects on Primary reactions of Photosystem I, Biochim. Biophys. Acta 893, 109-114.

Budil, D.E. (1986) Magnetic characterization of the primary radical pair state of bacterial photosynthesis, Thesis Dissertation, University of Chicago, Chicago, Illinois.

Budil, D.E., Kolaczkowski, S.V., & Norris, J.R. (1987) Temperature dependence of electron in "Progress in Photosynthesis Research" (Biggins, J., Ed.) Vol 1, pp 125-128, Martinus Nijhoff, Dordrect, The Netherlands.

Bylina, E.J., Ismail, I. & Youvan, D.C. (1986) Plasmid pU29, a vehicle for mutagenesis of the photosynthetic *puf* operon in *Rhodopseudomonas capsulata,* Plasmid 16, 175-181.

Bylina, E.J., Kirmaier, C., Mc Dowell, L., Holten, D. & Youvan, D.C. (1988) Influence of an amino acid residue on the optical properties of a photosynthetic reaction center complex, Nature 336, 182-184.

Bylina, E.J. & Youvan, D.C. (1987) Genetic engineering of herbicide resistance: saturation mutagenesis isoleucine 229 of the reaction center L subunit., Z. Naturforsch. 42c: 769-774.

Bylina, E.J. & Youvan, D.C. (1988) Directed mutations affecting spectroscopic and electron transfer properties of the primary donor in the photosynthetic reaction center, Proc. Natl. Acad. Sci. USA 85, 7226-7230.

Bylina, E.J., Jovine, R.V.M. & Youvan, D.C. (1989) A genetic system for rapidly assessing herbicides that compete for the quinone binding site of photosynthetic reaction centers, Bio/Technology 7, 69-74.

Bylina, E.J., Kolaczkowski, S.V., Norris, J.R., & Youvan, D.C. EPR characterization of genetically modified reaction centers of *Rhodobacter capsulatus,* submitted.

Chang, C.-H., Tiede, D., Tang, J., Smith, U., Norris, J.R., & Schiffer, M. (1986) Structure of the *Rhodopseudomonas sphaeroides* R-26 reaction center, FEBS Lett. 205, 82-86.

Cogdell, R.J., Monger, T.J. & Parson, W.W. (1975) Carotenoid triplet states in reaction centers from *Rhodopseudomonas sphaeroides* and *Rhodospirillium rubrum,* Biochim. Biophys. Acta 408, 198-199.

Hoff, A.J. (1981) Magnetic field effects on photosynthetic reaction centers, Quart. Rev. Biophys. 14, 599-665.

Kirmaier, C., Holten, D., Bylina, E.J. & Youvan, D.C. (1988) Electron transfer in a genetically modified bacterial reaction center containing a heterodimer, Proc. Natl. Acad. Sci. USA 85, 7562-7566.

Klug, G. & Cohen, S.N. (1988) Pleiotropic effects of localized *Rhodobacter capsulatus puf* operon deletions on production of light-absorbing pigment-protein complexes, J.Bacteriol. 170, 5814-5821.

Kolaczkowski, S.V., Budil, D.E. & Norris, J.R. (1987) 3($P+I^-$) lifetime as measured by B1 field dependent RYDMR triplet yield, in "Progress in Photosynthesis Research" (Biggins, J., Ed.) Vol 1, pp 1213-1216, Martinus Nijhoff, Dordrect, The Netherlands.

Kolaczkowski, S.V. (1989) On the mechanism of triplet energy transfer from the triplet primary donor to spheroidene in photosynthetic reaction centers from *Rhodobacter sphaeroides* 2.4.1, Thesis Dissertation, Brown University, Providence, Rhode Island.

Michel, H., Epp, O. & Deisenhofer, J. (1986) Pigment-protein interactions in the photosynthetic reaction center from *Rhodopseudomonas viridis,* EMBO J. 5,2445-2451.

Norris, J.R., Bowman, M.K., Budil, D.E., Tang, J., Wraight, C.A. & Closs, G.L. (1982) Magnetic characterization of the primary state of bacterial photosynthesis, Proc. Natl. Acad. Sci. USA 79,5532-5536.

Norris, J. R., Budil, D.E., Kolaczkowski, S.V., Tang, J.H. & Bowman, M.K. (1985) Photoinduced charge separation in bacterial reaction centers investigated by triplets and radical pairs, in: Antennas and Reaction Centers of Photosynthetic Bacteria (Michel-Beyerle, M.E., Ed.) pp 190-197, Springer-Verlag, New York.

Prince, R.C. & Youvan, D.C. (1987) Isolation and spectroscopic properties of photochemical reaction centers from *Rhodobacter capsulatus*, Biochim. Biophys. Acta 890, 286-291.

Schenck, C.C., Blankenship, R.E. & Parson, W.W. (1982) Radical pair decay kinetics, triplet yields and delayed fluorescence from bacterial reaction centers, Biochim. Biophys. Acta 680, 44-59.

Scolnick, P.A. & Marrs, B.L. (1987) Genetic research with photosynthetic bacteria, Ann. Rev. Microbiol. 41, 703-726.

Thurnauer, M.C., Katz, J.J. & Norris, J.R. (1977) The triplet state in bacterial photosynthesis: possible mechanisms of the primary photoact, Proc. Natl. Acad. Sci. USA 75, 3270-3274.

Wasielewski, M.R. (1988) Synthetic models for photosynthesis, Photochem. Photobiol., 47, 923-929.

Wasielewski, M.R., Norris, J.R. & Bowman, M.K. (1984) Magnetic resonance studies of short-lived radical pairs in solution, Faraday Disc. Chem. Soc. 78, 279-288.

Yang, M.M. & Youvan, D.C. (1988) Applications of imaging spectroscopy in molecular biology: I. screening photosynthetic bacteria, Bio/Technology 6, 939-942.

Yeates, T.O., Komiya, H., Chirino, A., Rees, D.C., Allen, J.P., & Feher, G. (1988) Structure of the reaction center from *Rhodobacter sphaeroides* R-26 and 2.4.1: protein-cofacter (bacteriochlorophyll, bacteriopheophytin and carotenoid) interactions, Proc. Natl. Acad. Sci. USA 85, 7993-7997.

Youvan, D.C., Hearst, J.E. & Marrs, B.L. (1983) Isolation and characterization of enhanced fluorescence mutants of *Rhodopseudomonas capsulata*, J. Bacteriol. 154, 748-755.

Youvan, D.C., Ismail, S. & Bylina, E.J. (1985) Chromosomal deletion and plasmid complementation of the photosynthetic reaction center and light harvesting genes from *Rhodopseudomonas capsulata*, Gene 38, 19-30.

THE BINDING AND INTERACTION OF PIGMENTS AND QUINONES IN

BACTERIAL REACTION CENTERS STUDIED BY INFRARED SPECTROSCOPY

Werner Mäntele[*], Monika Leonhard[*], Michael Bauscher[*]
Eliane Nabedryk[$], Gerard Berger[$], Jacques Breton[$]

[*]Institut für Biophysik und Strahlenbiologie der Universität Freiburg,
Albertstraße 23, D-7800 Freiburg, FRG
[$]Service de Biophysique, Département de Biologie, CEN Saclay
F-91191 Gif-sur-Yvette, France

INTRODUCTION

The primary photochemistry in bacterial photosynthesis involves a charge separation, stable for milliseconds to seconds, between specialized pigments acting as the primary electron donor and the intermediary electron acceptor, and quinones acting as primary and secondary electron acceptors. In bacterial reaction centers, the primary electron donor (P) is a bacteriochlorophyll (BChl) a or b dimer, the intermediary acceptor (H) is a bacteriopheophytin (BPheo) a or b monomer, and ubiquinones or menaquinones have been identified as electron acceptors (Q). The specifity, efficiency and stability of this charge separation relies on the arrangement and specific interaction of the pigments and redox components in the protein matrix. X-ray structures available for bacterial RC [1,2] provide a static picture of the quiescent state and suggest specific interactions of the pigments and quinones with their host site. However, additional information on the mechanisms and dynamics of primary electron transfer and on the concomitant change of interactions and protein conformation is required. Time-resolved optical spectroscopy (for a review, see [3]) as well as resonance Raman spectroscopy (for a review, see [4]) have provided such information.

Infrared difference spectroscopy has been established as an appropriate tool to study the molecular processes concomitant with primary electron transfer [5-7]. Using time-resolved or Fourier-transform infrared (FTIR) techniques, a sensitivity can be obtained which is high enough to detect the alteration of individual bonds against the large protein background. Highly detailed light-induced difference spectra between the quiescent RC state (PHQ) and the reduced acceptor state (PH$^-$Q) [6,7] or the charge separated state (P$^+$HQ$^-$) [5,7] were obtained in the mid-infrared (1800 cm^{-1}-1000 cm^{-1}). In these difference spectra, the cofactors are expected to contribute. Infrared spectroscopy, however, is nonselective; thus, absorbance changes from protein constituents, lipids, and water might contribute as well. For this reason, and due to the complexity of the IR difference spectra, a comparison with model compound spectra of chlorophylls and quinones is needed for an interpretation on the level of individual bonds.

Infrared spectra of the neutral cofactors have been useful for the study of the ligation properties of these molecules. In the reaction center, however, these cofactors form cation radicals, anion radicals or dianions. The neutral cofactors are thus poor models for the binding and interaction properties in vivo. Using a combination of electrochemical and IR-spectroscopic techniques [8], we have recorded IR spectra of the isolated cofactors in their neutral and in their radical state. We present here a comparison

Molecular Biology of Membrane-Bound Complexes in Phototrophic Bacteria
Edited by G. Drews and E. A. Dawes
Plenum Press, New York, 1990

313

of these model spectra with the light-induced IR difference spectra, which allows to deduce information on the binding and interaction of chlorophylls and quinones in vivo.

TECHNICAL ASPECTS

Infrared difference spectroscopy of reaction centers

Vibrational spectroscopy of isolated chlorophylls or quinones in organic solvents has been used to investigate their molecular structure, their aggregation and ligation with each other and with solvent molecules [9,10]. For the investigation of the cofactors of the reaction centre in their native environment a different strategy, that of "*reaction-modulated*" infrared difference spectroscopy, has to be employed. The two main concepts that have found successful application are that of "*time-resolved*" IR difference spectroscopy described to some detail in [11] and more recently in [12], as well as that of Fourier-transform infrared (FTIR) difference spectroscopy [13]. The following schemes illustrate these concepts.

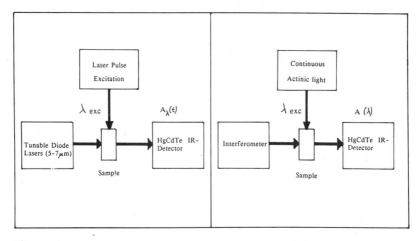

Fig. 1. Time-resolved IR spectroscopy   FTIR difference spectroscopy

In the reaction centers, the IR background absorption through the protein constituents, lipids and water is as high as to completely mask the cofactor absorption. If, however, electron transfer reactions are induced by additional continous illumination or a flash, molecular changes concomitant with primary electron transfer are selectively detected in the light-induced difference spectra, whereas the absorption of the main part of the protein, which is not involved in the reaction, cancels.

Sample preparations for RC and membranes

Water absorbs strongly in the mid-infrared. For the investigation of functionally intact reaction centers, a compromise must be found between the water content necessary for unperturbed electron transport reactions and the water content tolerable in the IR. Careful drying of detergent-solubilized RC or of RC reconstituted in lipid vesicles onto CaF_2 IR windows followed by hydration in an atmosphere of controlled humidity has been utilized successfully. Whenever pH, ionic strength or redox potential are critical parameters, or artificial donors/acceptors have to be present, the corresponding buffers can be added and the films then be covered with a second window separated by a 10-20 μm spacer to form UV, optical and IR transparent microcells as described in [6]. However, strict controls of the electron transport functions as in [5], [6] or [14] are essential.

Preparation and purification of chlorophylls for model studies.

Chlorophylls can be prepared by a number of well-known classical techniques.

However, without recrystallization, these preparations may contain contaminants that can catalyze pigment degradation by light, as well as large amounts of lipids. In order to avoid this, we have combined mild extraction procedures with a reverse-phase HPLC purification to obtain pigments of high spectral purity and very low contamination [15]. The BPheos are prepared according to procedures described in [16]. Quinones used for model studies were used in commercial purity or after sublimation.

Electrochemical and spectroelectrochemical techniques

The electrochemical techniques used to characterize the formation of radical ions of BChls, BPheos and their corresponding analogs, as well as the formation of quinone mono- and dianions are essentially described in [16-18]. The thin-layer electrochemical cells that allow to combine UV/VIS/IR spectroscopic investigations with these electrochemical techniques are described in detail in [8,18]. Briefly, a transparent gold or platinium grid placed in the beam is used as a working electrode to apply an electric potential, a Pt electrode placed outside the beam area serves as a counter electrode, and an Ag/AgCl couple as an internal [8] or external [18] reference electrode in the corresponding organic solvent. In view of the high sensitivity of the pigments, cells are used strictly anaerobic.

RESULTS AND DISCUSSION

Carbonyl groups as markers for pigment bonding and interaction

A total of four carbonyl groups surround the BChl molecule (for the structure of BChl a, see Fig.2). The acetyl C=O group at the 2a position (IUPAC 3^1), the keto C=O group at the 9 position (IUPAC 13^1) and the two ester C=O groups at the 10a and 7c (IUPAC 13^3 and 17^2) differ considerably in their bond strength and their electronic coupling to the π-electron system. Consequently, they are expected to absorb at different positions within the carbonyl frequency range, coarsely determined by the character of the group, and fine-tuned by the frequency-lowering effect of hydrogen-bonding. Strong hydrogen-bonding, in addition, may lead to a weaker absorption. Within the C=O absorbance range from $1760 \, cm^{-1}$ to $1620 \, cm^{-1}$, the ester C=O groups are expected to absorb at $1760 \, cm^{-1}$ to $1720 \, cm^{-1}$, the keto C=O group at $1720 \, cm^{-1}$ to $1655 \, cm^{-1}$, and the acetyl C=O from $1660 \, cm^{-1}$ to $1620 \, cm^{-1}$.

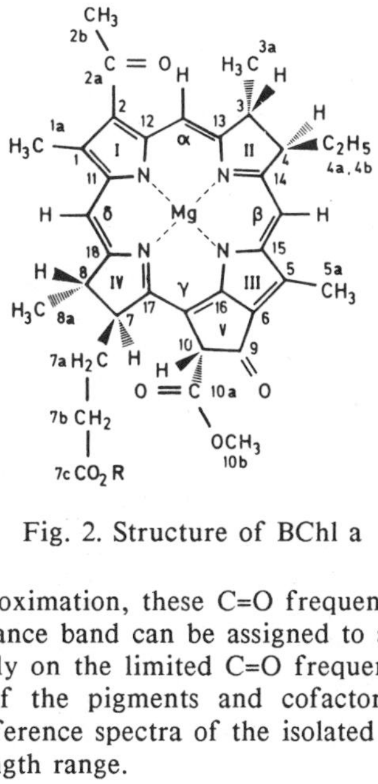

Fig. 2. Structure of BChl a

Within reasonable approximation, these C=O frequencies can be regarded as group frequencies, i.e. an IR absorbance band can be assigned to a single vibrator. We will thus focus for the discussion mainly on the limited C=O frequency range, although the light-induced difference spectra of the pigments and cofactors in the RC as well as the electrochemically-induced difference spectra of the isolated pigments provide information in a very much larger wavelength range.

A light-induced FTIR difference spectrum obtained from a thin film of *Rb. sphaeroides* RC is shown in fig. 3b, together with the IR absorbance spectrum (fig 3a). The difference spectrum represents the molecular processes concomitant with primary charge separation, i.e. the difference spectrum between the PQ and the P^+Q^- state. The bands corresponding to the charge-separated state are chosen to appear positive, those corresponding to the neutral state appear negative.

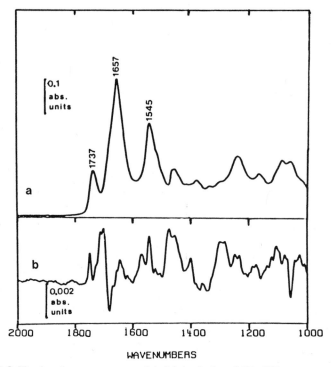

Fig.3. RC IR **absorbance** spectrum (a); Light-induced IR **difference** spectrum (b)

With respect to the absorbance spectrum, the light-induced difference spectrum is displayed on a 50 times enlarged absorbance scale. Thus, only a very small number of bonds in the RC exhibit changes of geometry or bond strength upon primary charge separation. The main contribution to the IR absorbance spectrum (fig 3a) arises from the peptide groups which give rise to the amide I band around 1650 cm^{-1} and the amide II band around 1550 cm^{-1}. The absence of large bands in these spectral regions in the difference spectrum (fig 3b) allows to exclude conformational changes of the peptide backbone involving more than one or two peptide bonds.

By far the largest bands in the IR difference spectrum (fig 3b) are observed in the C=O range (approx, 1760 cm^{-1} to 1620 cm^{-1}), indicating changes of bonding and interaction of the C=O groups of the pigments. Using the concept of time-resolved IR difference spectroscopy, we have recorded time-resolved IR absorbance changes in this frequency range. Figure 4a (taken from [12]) shows the absorbance change at 1716 cm^{-1} (corresponding to the main positive peak in the IR difference spectrum of fig 3b) following a nsec laser flash. A very rapid increase of absorption is observed, unresolved with the time resolution of 500 nsec presently available [12]. This absorbance increase decays biphasic, with kinetic components (see fit parameters) according to charge recombination from Q_A and Q_B from different RC populations. When difference spectra (4b) are calculated from time-resolved signals taken in the 1700 cm^{-1} to 1760 cm^{-1} range, they confirm those obtained by FTIR spectroscopy as shown by chosing the same wavelength region (4c).

The inset of fig. 4a shows the rise of the signal at 1716 cm^{-1} at higher time resolution. The lack of a 200 μsec kinetic component (which would be typical for electron transfer from Q_A to Q_B) excludes contributions from the quinone C=O in this spectral region. Time-resolved IR difference spectra thus provide additional information on the dynamic parameters of the processes and the reaction mechanisms involved.

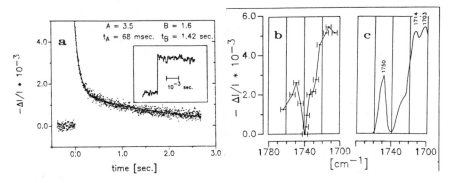

Fig 4. Time-resolved IR signals (a); IR difference spectra (b,c)

In addition to the IR difference spectra of the light-induced charge separated state (P^+Q^-/PQ), that of the reduced intermediary electron acceptor state (H^-/H) can be obtained using the technique of "photochemical trapping" with chemically prereduced samples. H^-/H difference spectra were obtained from RC with bound cytochromes such as *Rp. viridis* [6] and *C. vinosum* [20] and show highly detailed and reproducible difference bands upon the photochemical reduction of BPheo a and b.

IR spectra of electrochemically generated pigment radicals

In order to discern in these light-induced difference spectra the contribution of the cofactors from contributions eventually arising from the protein constituents, lipids or water, comparisons with pigment and quinone (see below) model spectra are essential. Quantitative and reversible cation formation of BChl a and b as well as anion formation of BPheo a and b have been obtained in thin-layer spectroelectrochemical cells [7,8,16]. As an example, the absorbance spectra of neutral and cationic BChl a in tetrahydrofuran, together with the difference spectrum (cation-minus-neutral), are shown in fig. 5.

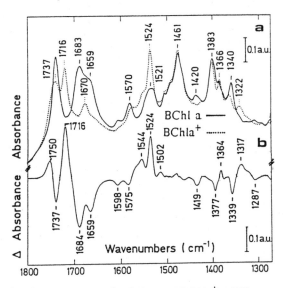

Fig. 5: IR absorbance spectra of BChl a and BChla$^{+\cdot}$, difference spectrum

The IR spectra of pigment radicals differ considerably from those of the neutral species. Several effects can account for the spectral shifts and intensity changes upon radical formation:

- partial change of bond order of individual bonds of the conjugated system due to the addition or abstraction of an electron;
- inductive coulombic effects of the modified charge distribution in the conjugated system on bonds not involved in conjugation;
- inductive effects on hydrogen-bonds from the pigment to the solvent molecules, or, _in vivo_, to the protein.

Implications for the binding of the pigments

In the carbonyl frequency range, the prominent band shift from 1684 cm^{-1} to 1716 cm^{-1} can be assigned to the 9 keto C=O group, the one from 1737 cm^{-1} to 1750 cm^{-1} to the 10a ester C=O group since it is absent in pyrobacteriochlorophyll [16]. The 7c ester C=O remains unaltered upon radical formation. The negative band at 1659 cm^{-1} can be assigned to the acetyl C=O of neutral BChl a. In a similar way, assignment is obtained for BChl b and its cation, as well as for BPheo a and BPheo b and their respective anions [16,21].

We have discussed above the influence of hydrogen bonding on the position and absorption strength of the C=O groups. The IR spectra of pigment-radicals, generated electrochemically in solvents of different polarity and different ability to form hydrogen bonds, can now serve to probe the local environment of the carbonyl groups of the primary donor BChls and the intermediary acceptor. A comparison of the primary donor cation radical formation in BChl a/b -containing RC with the formation of BChl a/b$^+$ _in vitro_ in a hydrogen-bonding and a non-hydrogen-bonding solvent suggests H-bonding of the 10a ester C=O group in the neutral state, but an ester group which is essentially free in the cation state (for a detailed discussion of this comparison, see [7]). For the 9-keto C=O group, the same comparison indicates a non-interacting group in the neutral as well as in the radical state.

When H$^-$/H difference spectra are compared with anion-minus-neutral difference spectra of BPheo a or b, a marked difference appears. A single negative band in the ester C=O frequency region appears in the model spectra, which, using chemically modified pheophytins, can be assigned to the 10a ester C=O. Thus, the absorbance of the 10a C=O decreases upon reduction. BPheo anion formation in RC, however, always gives rise to a difference spectrum with two negative bands in the ester C=O range [6,7]. Moreover, this double band structure is common to the reduction of all intermediary acceptors in plant and bacterial RC [22,23] and thus points to a general process. There is considerable evidence that the negative band at higher frequency (at 1747 cm^{-1} in _Rp. viridis_ and _C. vinosum_, at 1739 cm^{-1} in photosystem II) may be assigned to a protein side chain group protonated in the neutral state of the intermediary acceptor H. In _Rp. viridis_, this group may be associated to the glutamic acid residue GLU L104 [1] and appears accessible for ^{1}H-^{2}H exchange as indicated by a decrease and shift in the IR difference spectra.

Contributions from quinone reduction

Having thus established that, in the C=O frequency region, the strongest bands in the light-induced IR difference spectra arise from the C=O groups of the pigments, we now turn to the question of the quinone contribution. The quinone C=O vibration is expected in the 1600-1700cm^{-1} spectral region, at an absorption strength comparable to that of the pigment carbonyls [10]. Bands in the light-induced IR difference spectra from _Rb. sphaeroides_ and _Rp. viridis_ in the 1640-1660 cm^{-1} are thus possible candidates for the quinone C=O, but the assignment in [7] remained speculative.

In order to determine the absorbance range and extinction coefficients of quinone C=O and C=C vibrations, the IR spectra of quinone model compounds were recorded using the spectroelectrochemical techniques described above. Fig. 6 shows the IR absorbance spectrum of neutral ubiquinone (dashed line). The main peaks at 1660 cm^{-1} and at 1606 cm^{-1} can be assigned to the C=O and to the C=C modes, respectively. Upon reduction to the monoanion, both peaks fully disappear and a new peak at 1500 cm^{-1} appears which can be assigned to the C-O modes of the anion. Dianion formation shifts this peak further to 1473 cm^{-1} (full line).

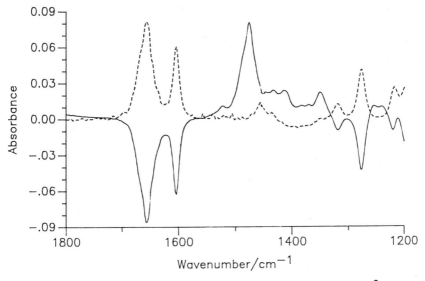

Fig. 6: IR absorption (Q,---) and redox-induced difference spectrum (Q^{2-}/Q,——) of ubiquinone

The redox-induced IR difference spectra of the quinones have allowed us to estimate the contribution from quinone reduction in the RC to the light-induced difference spectra (fig. 3b). For "normal" binding of the quinone carbonyls of Q_A or Q_B, a negative band in the difference spectrum (fig 3 b), of a strength comparable to the highest frequency band assigned to the BChl 10a ester C=O mode, would be expected around 1660 cm^{-1} for the disappearing C=O mode. Correspondingly, a negative band around 1606 cm^{-1} and a positive band around 1490 cm^{-1} should be observed due to the disappearing C=C of Q and the appearing C-O mode of the $Q^{\cdot-}$ and Q^{2-}.

While the quinone model spectra can give limits for the contribution of the quinone(s), a clear assignment is not possible. In addition, model spectra taken in solvents of different H-bonding character have shown a remarkably small influence of H-bonding, in contrast to solvent polarity, on the absorption maximum and extinction coefficient of the C=O bond. In order to assign the C=O and the C=C modes of the quinones, Bagley et al. [24] have used *Rb. sphaeroides* RC reconstituted with ^{13}C- and ^{18}O-labelled quinones. Indeed, P^+Q^-/PQ difference spectra of such Rc have indicated small but reproducible features in the 1600 cm^{-1} to 1680 cm^{-1} and the 1400 cm^{-1} to 1500 cm^{-1} spectral region. The intensities of these bands, however, are much smaller than what is suggested from the model spectra presented here. Nevertheless, a line in the P^+Q^- difference spectra at 1604 cm^{-1} [24], closely coinciding with the C=C frequency in the model spectra at 1606 cm^{-1}, can be assigned to the ethylenic mode of Q_B. Although very strong in model compounds, no conclusive assignment can be given for the C=O mode of the quinones in the RC due to the absence of an isotopic shift [24]. The only conclusive explanation for the absence of a strong C=O band in vivo is "unnormal", possible heterogeneous, binding or a distorted geometry of the quinone resulting in an extreme line broadening or weakening of the C=O mode in the native environment.

CONCLUSIONS:

A possible way the charge-separated state in the RC can be stabilized might be by conformational changes of the protein to a state which energetically favours the cationic and anionic radical state of the cofactors and thus prevents recombination. The light-induced IR difference spectra have demonstrated, through the absence of large signals in the amide I and II frequency range, that conformational changes are absent or at least very small upon primary charge separation or intermediary electron acceptor reduction. Instead, it appears that generation of an cationic or anionic (radical) state of a cofactor modifies its hydrogen bonding pattern to the protein and thus leads to stabilization. For the RC of *Rp. viridis* and *Rb. sphaeroides* the available X-ray structures [1,2] serve as a reference. In general, reasonable agreement is obtained on the bonding of the primary donor BChl's and BPheo's in the neutral state. However, in addition to X-ray crystallography, where specific interactions are suggested for the quiescent state of the RC, the light-induced IR difference spectra provide information on the bonding in the charge-separated state and thus on the processes in the pigment environment that may govern the efficiency of charge separation and stabilization. Furtheron, the necessary prerequisite for an investigation of cofactor bonding and interaction by IR difference spectroscopy are much less stringent than for crystallographic investigations. Bonding and interaction can even be investigated for RC in intact membranes. This aspect seems to be important for the structure-function relationships, since it has been frequently argued that structural details derived from crystalline proteins may represent energetically favorable "frozen" states of the protein, which differ from the average state observed in vivo. Furthermore, the ionic conditions or pH of a successful crystallization may differ significantly from in vivo conditions. Finally, the introduction of time-resolved techniques in IR spectroscopy allows the study of dynamic processes in the protein during charge separation.

ACKNOWLEDGEMENTS

The authors would like to thank S. Andrianambinintsoa, J. Kleo and S. Chaudhuri for their technical assistance and R. Hienerwadel, Dr. K.A. Bagley, Dr. D.A. Moss and Prof. W. Kreutz for helpful discussions. Part of this work was supported by the Deutsche Forschungsgemeinschaft (Ma 1054/2-1) and the European Community (ST 2J-0118-D).

REFERENCES

[1]   H. Michel, O. Epp and J. Deisenhofer; Pigment-Protein Interactions in the Photosynthetic Reaction Centre from Rhodopseudomonas Viridis; The EMBO Journal 5:2445 (1986)

[2]   J.P. Allen, G. Feher, T.O. Yeates and D.C. Rees; Structure of the Reaction Center from Rhodobacter Sphaeroides R-26: The Cofactors; Proc. Natl. Acad. Sci. USA 84:5730 (1988)

[3]   W.W. Parson; Photosynthetic Bacterial Reaction Centers; Ann. Rev. Biophys. Bioeng. 11:57 (1982)

[4]   M. Lutz; Resonance Raman Studies in Photosynthesis; in: Advances in Infrared and Raman Spectroscopy (Clark, R.H.J and Hester, R.E., eds.) Wiley Heyden 11:211 (1984)

[5]   W. Mäntele, E. Nabedryk, B.A. Tavitian, W. Kreutz, and J. Breton; Light Induced Fourier Transform Infrared (FTIR) Spectroscopic Studies of the Primary Donor Oxidation in Bacterial Photosynthesis; FEBS Letters 187:227 (1985)

[6]   E. Nabedryk, W. Mäntele, B.A. Tavitian and J. Breton; Light-Induced Fourier-Transform Infrared (FTIR) Spectroscopic Investigations of the Intermediary Acceptor Reduction in Bacterial Photosynthesis; Photochem. Photobiol. 43:461 (1986)

[7]   W. Mäntele, A.M. Wollenweber, E. Nabedryk, J. Breton; Infrared Spectroelectrochemistry of Bacteriochlorophylls and Bacteriopheophytins: Implications for the Binding of the Pigments in the Reaction Center from Photosynthetic Bacteria; Proc. Natl. Acad. Sci. USA 85:8468 (1988)

[8]   W. Mäntele, A.M. Wollenweber, F. Rashwan, J. Heinze, E. Nabedryk, G. Berger, J. Breton; Fourier-Transform Infrared Spectroelectrochemistry of the Bacteriochlorophyll Anion Radical; Photochem. Photobiol. 47:451 (1988)

[9]   K. Ballschmiter, J.J. Katz; An Infrared Study of Chlorophyll-Chlorophyll and Chlorophyll-Water Interactions; J. Am. Chem. Soc. 91:2661 (1969)

[10]   B.R. Clark and D.H. Evans; Infrared Studies of Quinone Radical Anions and Dianions Generated by Flow-Cell Electrolysis J. Electroanal. Chem. 69:181 (1976)

[11]   W. Mäntele, F. Siebert and W. Kreutz; Kinetic Properties of Rhodopsin and Bacteriorhodopsin Measured by Kinetic Infrared Spectroscopy (KIS) Methods in Enzymology 88:729 (1982)

[12]   R. Hienerwadel, W. Kreutz and W. Mäntele; Time-Resolved Infrared Spectroscopy using Tunable Diode Lasers: Characterization of Intermediates in Light-Induced Electron Transfer of Photosynthesis; Procedings of the III European Conference on the Spectroscopy of Biological Molecules, Rimini (1989), in the press

[13]   M.S. Braiman and K.J. Rothschild; Fourier-Transform Infrared Techniques for Probing Membrane Protein Structure; Ann. Rev. Biophys. Biophys. Chem. 17:541 (1988)

[14]   R. Hienerwadel, W. Kreutz and W. Mäntele; A Period-Four Infrared Signal from Active Water-Splitting Complex; Proceedings of the VIIIth International Congress on Photosynthesis; Stockholm 1989, in the press

[15]   G. Berger, A.M. Wollenweber, J. Kleo, S. Andrianambinintsoa, W. Mäntele; A Rapid Preparative Method for Purification of Bacteriochlorophyll A and B; J. Liq. Chrom. 10:1519 (1987)

[16]   M. Leonhard, A.M. Wollenweber, G. Berger, J. Kleo, E. Nabedryk, J. Breton and W. Mäntele; Infrared spectroscopy and Electrochemistry of Chlorophylls: Model Compound Studies on the Interaction in their Native Environment; In: Techniques and New Developments in Photosynthesis; (Barber,J. ed.) in the press

[17]   J. Heinze; Cyclic Voltammetry -"Electrochemical Spectroscopy"; Angew. Chemie (Int. Ed. Engl.) 23:831 (1984)

[18]   M. Bauscher, K. Bagley, E. Nabedryk, J. Breton, W. Mäntele; Models for Ubiquinones and their Anions Involved in Photosynthetic Electron Transfer, Characterized by Thin-Layer Electrochemistry and FTIR/UV/VIS Spectroscopy; Proceedings of the VIIIth International Congress on Photosynthesis; Stockholm (1989), in the press

[19]   R.C. Prince, D.M. Tiede, J. P. Thornber and P.L. Dutton; Spectroscopic Properties of the Intermediary Electron Carrier in the Reaction Center of *Rhodopseudomonas viridis*; Biochim. Biophys. Acta 462:467 (1977)

[20]   E. Nabedryk, S. Andrianambinintsoa, W. Mäntele, J. Breton; FTIR Investigations of the Intermediary Electron Acceptor Photoreduction in Purple Photosynthetic bacteria and Green Plants; in: The Photosynthetic Bacterial Reaction Center - Structure and Dynamics (J. Breton & A. Vermeglio, eds) NATO ASI Series (1988), Plenum Press

[21]   M. Leonhard, E. Nabedryk, G. Berger, J. Breton, W. Mäntele; Model Compound Studies of Pigments Involved in Photosynthetic Energy Conversion: Infrared (IR) -Spectro-Electrochemistry of Chlorophylls and Pheophytins; Proceedings of the VIIIth International Congress on Photosynthesis,Stockholm (1989), in the press

[22]   B.A. Tavitian, E. Nabedryk, W. Mäntele, J. Breton; Light-Induced Fourier-Transfrom Infrared (FTIR) Spectroscopic investigations of Primary Reactions in Photosystem I and Photosystem II; FEBS Lett 201:151 (1986)

[23]   E. Nabedryk, S. Andrianambinintsoa, G. Berger, M. Leonhard, W. Mäntele, J. Breton; Characterization of Bonding Interactions of the Intermediary Electron Acceptor in the Reaction Center of Photosystem II by FTIR Spectroscopy; Manuscript submitted to BBA (July 1989)

[24]   K.A. Bagley, E. Abresch, M. Okamura, G. Feher, M. Bauscher, W. Mäntele, E. Nabedryk, J. Breton; FTIR studies of the $D^+Q_A^-$ and $D^+Q_B^-$ States in Reaction Centers from *Rb. sphaeroides*; Proceedings of the VIIIth International Congress on Photosynthesis; Stockholm (1989), in the press

SPECTROSCOPIC AND STRUCTURAL STUDIES OF CRYSTALLIZED REACTION

CENTRES FROM WILD TYPE *RHODOBACTER SPHAEROIDES* Y

F. Reiss-Husson[1], B. Arnoux[2], A. Ducruix[2], K. Steck[3],
W. Mäntele[3], M. Schiffer[4] and C. H. Chang[4]

[1]UPR 407 and [2]ICSN, CNRS, 91198 Gif sur Yvette, France
[3]Institut für Biophysik und Strahlenbiologie, 7800 Freiburg, FRG
[4]Biology Division, Argonne National Lab., Argonne, Il 60439, USA

INTRODUCTION

Spectroscopic studies on reaction centre (RC) crystals are a powerful tool to answer a number of questions. First one may ask if the pigments have been altered or not during the crystallization process. Second, photoinduced electron tranfer may be studied in the crystalline state and kinetics may be compared to those observed in solution. Finally, when the three-dimensionnal structure of the reaction center has been solved, the polarized absorption spectra may be analyzed to assign the individual absorption bands to oriented transition moments of each chromophore [1,2].

For this purpose, we present here spectroscopic data on the polarized absorption and the charge separation in crystallized RC from wild type *Rhodobacter* (*Rb.*) *sphaeroides* Y. A peculiar feature of this Y strain is that the metal interacting with the quinones in the RC is easily exchanged *in vivo*; this allows to isolate [3] and to crystallize [4] RCs containing predominantly Mn^{2+} instead of Fe^{2+}. Furthermore, this RC contains a bound spheroidene molecule, which has been shown by resonance Raman spectroscopy to be a cis-isomer [5,6].

We have recently solved the structure of this RC at 3 Å resolution, and refinement of the data is in progress (B. Arnoux, A. Ducruix, F. Reiss-Husson, M. Schiffer ,J. Norris and C. H. Chang, unpublished results). We have already published polarized absorption spectra of orthorhombic RC crystals [7]. High dichroism was observed for the bacteriopheophytin (BPheo) Q_x and Q_y transitions, as well as for the Q_y transitions of the dimeric and of the accessory bacteriochlorophylls (BChls). Low absorption and dichroism of the carotenoid transition in the plane of the crystal was reported. We also demonstrated photoactivity of the crystals under steady and modulated light illumination. In this report, we present an extension of this work. Photoactivity of the crystals has now been studied after single flash excitation, and the kinetic parameters of the charge recombination have been measured. Polarized absorption spectra have been obtained at different angular positions of the polarization with respect to the morphological axes of the crystal, and at various tilt angles of the crystal plane with respect to the beam axis. These polarized absorption spectra will be interpreted later on the basis of crystallographic data recently obtained for the structure of the chromophores in Y RC.

EXPERIMENTAL

Crystallization of *Rb. sphaeroides* Y RC was performed at 18°C by microdialysis in the presence of 0.8% octylglucoside and 0.22 M NaCl, with PEG 4000 as precipitant [8]. Orthorhombic crystals grew mainly as long prisms with a diamond shaped cross-section.

Molecular Biology of Membrane-Bound Complexes in Phototrophic Bacteria
Edited by G. Drews and E. A. Dawes
Plenum Press, New York, 1990

323

Space group is $P2_12_12_1$, and the unit cell parameters are a= 143.7 Å, b= 139.8 Å and c= 78.65 Å.

Very thin crystals (<20 μm thick) were selected for microspectrophotometry, and transferred from the mother liquor to a microcell formed by two cover slides separated by a spacer of 20 μm thickness. Due to their dimensions (up to 300 μm x 100 μm x 20 μm), the platelets were aligned parallel to the plane of the microcell.

Absorption and linear dichroism spectra were recorded on a single beam microspectrophotometer described in [2], in the spectral range from 450 nm to 1000 nm, with a resolution of 1 nm. The measuring light intensity was varied with neutral density filters in order to avoid actinic effects. The flash-induced photochemistry was initiated by a dye laser (Lambda Physik DL 1000, pumped by a Lambda Physik EMG 53 MSC excimer laser) using Rhodamine 6G as a dye at 590 nm. Pulses were of approx. 15 nsec duration. The actinic light was sent through a plastic monofiber optical guide onto the crystal, colinearly with the measuring beam. Care was taken to homogeneously illuminate the crystal. Saturated excitation was obtained with a pulse energy of approx. 0.5 mJ, measured at the entrance of the fiber.

The detector signal corresponding to the optical transients was further amplified, appropriately filtered and stored in a Nicolet 2090 transient recorder interfaced to an IBM computer, where signal averaging was performed using software developed in our laboratory. Standard algorithms were used to fit the transient absorbance changes to a sum of exponential decay functions.

RESULTS

Linear dichroism of the crystals

In the orthorhombic platelets which were used for the optical measurements, a set of three orthogonal axes is clearly defined (Fig.1). The x axis, which is parallel to the largest dimension of the platelet, has been shown by X-ray cristallography to coincide with the a* axis of the cell; y and z axes however are not parallel, respectively, with the crystallographic b* and c* axes.

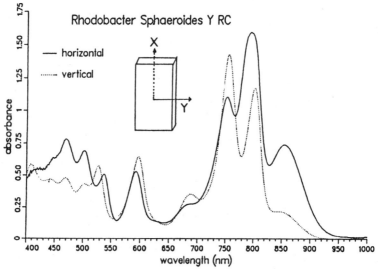

Fig. 1. Polarized absorbance spectra of a single RC crystal

On a number of such crystals, series of spectra were measured with the polarizer rotated from the x to the y direction. These spectra were generally similar to those already published (see Fig.1a in [7]), with a very high dichroism for the Q_y transitions of the BChl dimer, the monomeric BChls and BPheos. However, two spectral features were eventually variable in some crystals. First, the Q_y transition of the dimer was sometimes abnormally weak, even under very low intensity of the measuring light. It was increased to a normal value when ascorbate was added to the mother liquor. This observation indicated that some chemical, reversible oxidation of the BChl dimer occured during the lengthy crystallization process. In order to avoid this, we have modified our crystallization protocol: the microdialysis was performed in vials which were flushed with argon and then tightly closed. Crystals produced under these new conditions reproducibly contained a high proportion of the dimer in the reduced state (see Fig.1).

A more surprising observation concerned the carotenoid visible absorption bands, whose intensity was variable from one crystal to the other, even when these crystals were grown in the same capillary. Apparently in some of these small crystals the carotenoid was altered or lost. This effect occured even when the crystallization was done under an inert atmosphere, and did not depend on the time of growth. We still do not have an explanation for this variability which is all the more puzzling as spheroidene is present in the larger crystals used for X-ray diffraction.

A series of spectra with varying angles of polarization was taken for a crystal presenting a "normal" carotenoid absorption. The spectra taken at 0° and at 90° for the measuring beam perpendicular to the plane of the crystal are shown in fig. 1. In order to record the out-of-the-plane transition moments, spectra were also recorded at 0° and 90° polarisation angles for various tilt angles (up to ± 40°) of the crystal plane with respect to the beam axis.

A convenient way to represent the anisotropic absorption in the crystals is to plot the absorbance strength in the different planes of the crystals as ellipses, or, in a 3D plot, as ellipsoids. As defined in the inset of fig. 1, the x- and y-axis span the plane of the crystal perpendicular to the beam. Transition moments in the z-axis (the direction of the measuring light) have to be extrapolated from spectra obtained from samples tilted by up to ± 40°. In order to avoid errors due to the increasing path-length through the crystal upon tilting, a normalization procedure described in [2] is applied. Figure 2 shows the ellipses representing the absorption strength of the primary donor absorption at 865 nm (Qy 865) and of the bacteriopheophytin absorption at 750 nm (Qy Bph) for the x-y and the z-y planes.

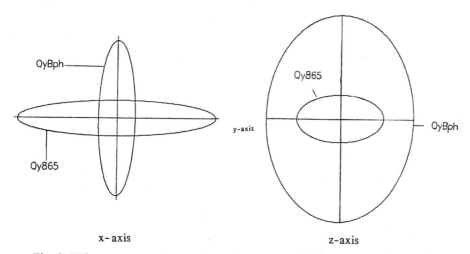

Fig. 2. Ellipses representing the absorption strength in the x-y and the z-y planes.

These plots indicate that the linear dichroism is highest in the x-y plane for the special pair absorption as well as for the BPheo absorption. The extrapolated linear dichroism in the z-y plane, however, is much smaller for the special pair. Almost no dichroism is observed for the BPheos in the z-y plane.

Photochemical activity of the crystals

Excitation of the RC crystals by a saturating laser flash resulted in transmittance changes which were analyzed in the Q_y band of the dimer. The measuring light intensity was systematically reduced using neutral density filters in order to avoid its bleaching effects on those RCs which contain Q_B [7]. Furthermore, the transients were measured at various polarisation angles, and at different wavelengths throughout this band.

Figure 3 shows examples of light-induced time-resolved absorbance changes measured at 865nm with the measuring light polarized along x, and at 950 nm with the measuring light polarized along y. The kinetic data were analyzed according to a sum of two exponential rate expressions. The two components had life times (= k^{-1}) of 1.13 ± 0.09 sec and 131 ±13 msec, respectively. These values may be compared to lifetimes of $P^+Q_A^-$ and $P^+Q_B^-$ states of the Y RCs in detergent [9]. The ratios of the (slow-to-fast) terms is 3.5. This indicates that in about 75% of the RCs the secondary quinone is still bound and functional.

As a control, a crystal was soaked in its mother liquor during one night in the presence of an excess of terbutryn. After this treatment, the kinetic traces following light-induced charge separation could be fitted to a single exponential decay function, with a much faster lifetime of 60ms. Obviously terbutryn was able to bind to the Q_B site in the crystal and to inhibit the Q_A to Q_B electron transfer.

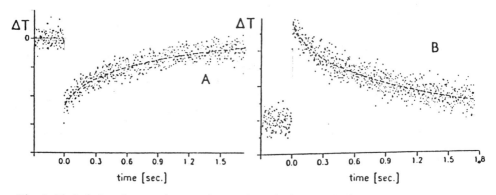

Fig. 3. Flash-induced transmittance changes in a single crystal of *Rb. sphaeroides* Y RC. Measuring light-intensity was reduced with neutral density filters to 10% of the value used to record the spectra in fig 1 in order to avoid actinic effects.
(a) 865 nm detection with polarization along the x-axis (see Fig. 1).
(b) 960 nm detection with polarization along the y-axis. Traces were obtained by averaging 20 (a) and 40 (b) sweeps. Full lines represent theoretical curves obtained by fits to two-exponential decay functions (see text).

DISCUSSION

The kinetics of charge recombination after single flash excitation confirm that in *Rb. sphaeroides* Y RC crystals photoinduced electron transfer takes place from the primary electron donor to the secondary quinone electron acceptor. The presence of this secondary quinone thus determined agrees well with the crystallographic analysis. Indeed, the cycle and the isoprenoid chain of both quinones Q_A and Q_B are well-defined in the electron density map (B. Arnoux, A. Ducruix, F. Reiss-Husson, M. Schiffer, J. Norris and C. H. Chang, to be published). Both quinones are also present in crystals from *Rb. sphaeroides* R26 RC [10,11] and in those of wild type 2.4.1. strain [11]. However, other authors have described crystals lacking Q_B activity [12].

It is interesting to note that, after excitation with single flashes, the rates of charge recombination of the states $P^+Q_A^-$ and $P^+Q_B^-$ is not modified in the crystal state as compared to the detergent solution. This indicates that the geometry of the chromophore array and the interactions between the chromophores and the surrounding amino-acid residues are not changed upon crystallization. Furthermore, in spite of the ordered packing of the RCs in the unit cell, interactions between neighboring arrays of chromophores do not seem to modify the kinetics as compared to the isolated state in solution.

The situation is different if charge recombination from the photoaccumulated states ($P^+Q_A^-$ and $P^+Q_B^-$) is measured as described in [2]. The life-times measured with steady or modulated illumination were comparable for the decay from $P^+Q_A^-$ (108 msec instead of 131 msec), but considerably different for the decay from $P^+Q_B^-$ (0.375 sec instead of 1.13 sec). In addition, the lifetimes were dependent on the measuring wavelength. We have attributed this dependence to the different fractions of the RCs in constant turnover through the measuring light [2]. In addition, at the high concentration of RC in the crystal, the interaction of the oxidizing and the reducing side of the RC with the small volume of the aqueous phase (only about 70 % of the total volume) should not be neglected. The single-turnover measurements of the photochemical activity of the crystals with short intense flashes, however, should not be affected.

CONCLUSIONS

Measurements of polarized absorption spectra proved to be of great value for controlling the integrity of the pigments in the crystals. Autooxidation of the primary donor, which was observed in our first experiments [7], could be prevented by avoiding exposure to O_2 during crystallization. Variability in carotenoid absorption, however, remains unexplained and could not be prevented in the small and thin crystals used for spectroscopic investigations. In contrast, in the larger crystals used for X-ray crystallography spheroidene could be localized and its conformation be described (B. Arnoux et al., to be published). At the present stage of refinement of the structure, the spheroidene molecule was found to adopt a quite asymmetrical conformation, with a central 15-15' -cis bond located at close proximity of ring I of the accessory BChl on the M side. The electron density corresponding to the two extremities of the molecule, however, was not well defined, and only 21 out of the 40 carbon atoms could be fitted in the density map. This might be indicative of some disorder, which would lead to low dichroism and explain partially the spectroscopic experiments. Another explanation would be oxidative bleaching of spheroidene during crystallization; this would affect differently crystals growing at various times or rates. Obviously, further explanation is required on this point.

Finally, as refinement of the structure of the *Rb. sphaeroides* Y RC at 3 Å is underway, it will soon be possible to analyze the polarized absorption spectra (Fig.1) in terms of the contribution of each pigment as for the *Rhodopseudomonas viridis* RC [1].

REFERENCES

[1]   Knapp, E. W., Fischer, S. F., Zinth, W., Sander, M., Kaiser, W., Deisenhofer, J. and Michel, H. (1985); Proc. Natl. Acad. Sci. USA **82**, 8463-8467.

[2]   Mäntele, W., Steck, K. Becker, A., Wacker, T., Welte, W., Gad'on, N., and Drews, G. (1988); in: The Photosynthetic Bacterial Reaction Center: Structure and Dynamics (Breton, J. and Vermeglio, A. eds.) NATO ASI Series Vol. **149**, pp. 33-39.

[3]   Rutherford, A. W., Agalidis, I. and Reiss-Husson, F. (1985); FEBS Lett. **182**, 151-157.

[4]   Ducruix, A. and Reiss-Husson, F. (1987); J. Mol. Biol. **193**, 419-421.
[5]   Lutz, M., Agalidis, I., Hervo, G., Cogdell, R. and Reiss-Husson, F.(1978);
        Biochim. Biophys. Acta **503**, 287-303.
[6]   Lutz, M., Szponarski, W., Berger, G., Robert, B. and Neumann, J. M. (1987);
        Biochim. Biophys. Acta **894**, 423-433.
[7]   Reiss-Husson, F. and Mäntele, W. (1988); FEBS Lett. **239**, 78-82
[8]   Ducruix, A., Arnoux, B. and Reiss-Husson, F. (1988); in: The Photo-
        synthetic Bacterial Reaction Center: Structure and Dynamics (Breton, J.
        and Vermeglio, A., eds.) NATO ASI Series, Vol. **149**, 419-421
[9]   Agalidis, I. (1987); Eur. J. Biochem. **166**, 235-239
[10]   Chang, C. H., Tiede, D. M., Tang, J., Smith, U., Norris, J. and Schiffer, M.
        (1986); FEBS Lett. **205**, 82-86
[11]   Yeates, T. O., Komiya, H., Chirino, A., Rees, D. C., Allen, J. P. and
        Feher, G. (1988); Proc. Natl. Acad. Sci. USA **85**, 7995-7997
[12]   Taremi, S. S., Violette, C. A. and Frank, H.A. (1989);
        Biochim. Biophys. Acta **973**, 86-92

THE ELECTROGENIC EVENT ASSOCIATED WITH THE REDUCTION OF THE

SECONDARY QUINONE ACCEPTOR IN RHODOBACTER SPHAEROIDES

REACTION CENTERS

A. Yu. Semenov, M.D. Mamedov, V.P. Shinkarev,
M.I. Verkhovsky, and N.I. Zakharova

A.N. Belozersky Laboratory and Biological
   Department
Moscow State University
Moscow 119899
USSR

An electrometric method was used to investigate flash-induced electrogenic stages in proteoliposomes containing photosynthetic RCs of Rhodobacter sphaeroides. Besides the very fast electrogenic step associated with primary dipole formation, an additional electrogenic stage with a rise-time of 0.25 ms (pH 7.5) appeared to be induced by even numbered flashes. The maximal amplitude of this stage contributes ~0.3 to the fast phase, associated with the charge separation between P870 and Q_A. The similarity of the rise-time of this phase, the time of the disproportioning reaction of semiquinones Q_A and Q_B and the rate of proton uptake by RCs, its appearance only after even-numbered flashes, the sensitivity to o-phenanthroline as well as the increase of its rise-time with pH indicate that the additional electrogenic phase arises from the reaction:

$$Q_A^- (H^+) Q_B^- + H^+ \longrightarrow Q_A Q_B H_2$$

A kinetic model is suggested describing the flash number dependence of the observed changes of the amplitude of this phase.

Abbreviations: RC, reaction center; P870, reaction center bacteriochlorophyll dimer; Q_A, Q_B, primary and secondary quinine acceptors; $\Delta \Psi$, transmembrane electric potential difference; TMPD, N.N.N',N', tetramethyl-p-phenilenediamine.

Molecular Biology of Membrane-Bound Complexes in Phototrophic Bacteria
Edited by G. Drews and E. A. Dawes
Plenum Press, New York, 1990

329

1. INTRODUCTION

There are many works that have been devoted to the electrogenicity of the reactions in RC of purple bacteria (1-6). By a variety of methods it has been demonstrated that the reactions: $P870\ Q_A \longrightarrow P870^{\ddagger}\ Q_A^-$ and $cyt\ c_2\ P870^{\ddagger} \longrightarrow cyt\ c_2^+\ P870$ give a major contribution to the electrogenesis of RC [1-6]. An additional o-phenanthroline-sensitive electrogenic phase due to the electrogenic events in the quinone acceptor complex was observed by direct electrometry assay [5]. The phase with a rise-time of 2.5 ms, which was associated with the reduction of Q_B^- to a doubly reduced ubiquinone species, was observed after the second flash in the RC complex of Rb. sphaeroides, R-26, incorporated into the bilayer [7]. The time of this phase was longer than that observed in Rhodospirillum rubrum [8] and Rb. sphaeroides [9,10] chromatophores and in reaction center proteoliposomes of Rhodopseudomonas viridis [11] in later investigations. In the work reported here we investigate the electrogenic reaction $Q_A^-(H^+)Q_B^- + H^+ \longrightarrow Q_A Q_B H_2$ in reaction center proteoliposomes of Rhodobacter sphaeroides.

2. MATERIALS AND METHODS

RC complexes were isolated from chromatophores as described previously [12]. Proteoliposomes were prepared by cholate dialysis as in Refs [13,14] TMPD, ubiquinone (Q-10), asolectin wre from Sigma, buffers Mes, Mops, Hepes and Tris being obtained from Serva. Kinetic measurements of generation were described in [8,15]. Saturating light pulses were delivered from a LOMO OGM-40 ruby laser. A home made single-beam spectrophotometer was used to measure fast absorption kinetics [16]. The data storage and processing system included a DL-1080 (Data Lab) transient recorder interfaced to a Nova-3D mini-computer (Data General). Q_B function in collodion film-associated proteoliposomes was reconstituted by adding Q-10 (20 mg/ml) to the solution of asolectin in decane, which was used to impregnate the collodion film into the measuring cell as described in (8).

3. RESULTS

The formation of the ubisemiquinone anion of secondary quinone acceptor measured by absorption increase at 450 nm is observed to oscillate with flash number displaying a periodicity of two in RC preparations [17,18]. Q_B^- is formed in response to odd flashes and disappeared in response to even flashes. Factors that influence the binary oscillations are an actinic light intensity, time interval between flashes, rate of Q_B oxidation by the redox mediator, the extent of $P870^+$ reduction during the time between flashes, the quinone concentration in the RC preparation, the equilibrium constant of electron transport between Q_A and Q_B, the amount of P870 reduced by Q_B but not by the mediator [19,20]. Only with all these factors being optimal, can one clearly see binary oscillations of Q_B formation. In order to analyse the electrogenic phases of quinone acceptors, we investigated the photoinduced $\Delta \Psi$ generation in RC proteoliposomes under specific conditions, defined from spectral behaviour of the semiquinone species.

Fig. 1A shows photoelectric responses induced by the first (curve 1) and second (curve 2) flashes and their difference (curve 3). One can see that curves 1 and 2 largely differ. Following the first flash, one observes only a fast phase of generation (τ <0.1 μ s), which is associated with charge separation between P870 and Q_A [6-10]. An additional phase of $\Delta \Psi$ generation with a rise-time of ~ 0.25 ms at pH 7.5 (curve 2) appears after the second flash.

This phase is sensitive to o-phenanthroline, an inhibitor of electron transfer from Q_A to Q_B (not shown). Its sensitivity to o-phenanthroline is explained by both charge recombination between P870$^+$ and Q_A^- and a faster oxidation of Q_A^- by the mediator than the oxidation of Q_B [21]. Therefore, some of RCs are in a state in which Q_A is oxidized before the second flash. That causes a decrease of the amplitude of the second flash-induced "quinone" phase. The amplitude of the additional "quinone" phase decreases with increasing time interval between the flashes and with decreasing the actinic light intensity [8,9].

As it was mentioned above, the "quinone" electrogenic phase is a result of the second electron transfer to the semiquinone anion Q_B^- followed by the practically simultaneous proton binding from the aqueous phase. It is not clear, whether the correlation exists between the kinetics of the electrogenic process and the electron and proton transfer reactions? To clarify this problem we measured the rate of the disproportioning semiquinones Q_A^- and Q_B^- at 450 nm. Fig 1B shows that a first flash induces the increase of absorption related to the semiquinones Q_B^- formation which is stable in the second time scale. After the second flash the

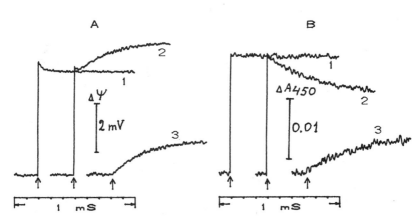

Fig. 1A. Photoelectric responses in proteoliposomes of Rhodobacter sphaeroides RC-complexes induced by the first (1) and the second (2) flash.3 - difference between the 2nd and the 1st flash kinetics. B. Semiquinone Q_B^- formation and disappearance in response to the 1st (1) and 2nd (2) flashes. RC concentration 2.3 μM. 3 - difference between the 1st and the 2nd flash-induced kinetics.

initial absorption increase follows the slower decrease due to the semiquinone Q_B^- disappearance. The first minus second flash-induced absorption change difference represents the kinetics of the disproportioning semiquinones Q_A^- and Q_B^-. From the results presented in fig 1A and 1B it is seen that a good correlation is observed between the rate of the rise of the "quinone" electrogenic phase and the rate of disproportioning semiquinones Q_A^- and Q_B^-.

It is well known, that the flash-induced proton binding by the RCs preparations could be observed by the absorption change registration of some pH-indicator dyes [18,22]. For this purpose we measured the flash-induced absorption changes of cresol red dye in the isosbestic point of pigment absorption change (Δ_A 583 nm). Fig. 2 shows the flash-induced absorption increase of cresol red in the Rb. sphaeroides RCs suspension due to the proton binding from the aqueous phase. The kinetics of the second electron transfer to the Q_B^- semiquinone anion is presented on the same fig. 2 for the comparison. A good correlation between the rates of electron and proton transfer in the quinone acceptor complex of RCs is observed.

An additional evidence of the interconnection between the all three processes is the coincidence of the pH-dependences of their rate constants, presented in fig. 3.

Fig. 4 shows the dependence of the $\Delta\Psi$ amplitude on flash number. The curve drawn via experimental points can mathematically be described by a formula : $\Psi(n) = A(1(1- \gamma)^{n-1})$ (1), where (Ψ n) is the amplitude of the "quinone" phase of $\Delta\Psi$ observed after the n-th flash; A is a constant; γ is a parameter which characterises the sum of the quantum yields of electron transfer to Q_B after the first and the second flash [19]. The fit of model and experimental points indicate that the binary oscillations are really due to Q_B functioning. It is important to emphasize that the amount of Q_B, stabilized in RC before the second flash depends essentially on the extent of Q_B^- oxidation during the time between the flashes and on the rate of P870$^+$ reduction.

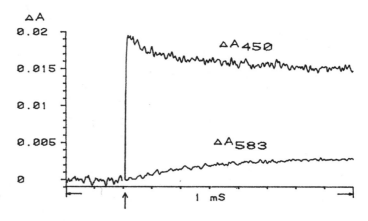

Fig. 2 Comparison of the flash-induced proton binding kinetics and the rate of the disproportioning reaction of semiquinones Q_A^- and Q_B^-. Incubation medium: 100 mM KCl, 0.4 mM DAD, pH=7.2. Each curve is derived by averaging of 90 curves. For ΔA_{583} measurements cresol red, 50μM, was added.

In our experiments, a major factor that accounts for the decrease of the amplitude of the quinone phase is the oxidation of Q_B^-. In 10 s after the flash (the time interval between the ruby laser flashes used by us), the oxidation of Q_B^- by the mediator is 20-30 percent. Hence it follows that the amplitude of the "quinone" phase of $\Delta\Psi$ must be increased by 1.3 times. Thus the amplitude of the electrogenic "quinone" phase becomes as large as $\sim 30\%$ of that associated with the charge separation between P870 and Q_A.

4. DISCUSSION

An investigation on RC-proteoliposome preparations provides convincing evidence that the reaction: $Q_A^-(H+)Q_B^- + H^+ \longrightarrow Q_A Q_B H_2$, is electrogenic. The conclusion that the differnce in the amplitudes of the photoelectric responses induced by the second and the first flash is due to this reaction is made from the specific binary oscillations of $\Delta\Psi$ (fig.1A and Fig.4), from the dependency of the amplitude induced by the second flash on the time interval between flashes, from the intensity of the flash dependence and from its sensitivity to o-phenanthroline. The pH dependence of the rise-time of this phase (fig. 3) also provides evidence in favor of this conclusion. Moreover, a good correlation has been observed between the rate of this phase, the rate of disproportioning semiquinones Q_A^- and Q_B^- (fig. 1.A and B), and the rate of proton binding by the RCs (fig. 2).

To explain the relatively small pH dependence of rise-time of "quinone" phase of $\Delta\Psi$ generation (fig. 3) one can assume that protonation of doubly reduced secondary quinone occur from protonated amino acid residues of the RC protein (22). At high pH, when residues are deprotonated, the value of τ is determined by proton diffusion from medium.

A similar electrogenic phase, associated with the

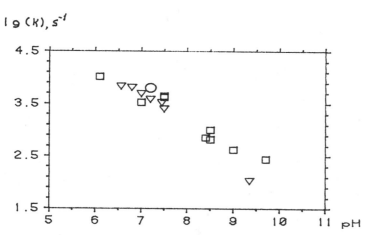

Fig. 3. The rate constants ($k=1/\tau$) for submillisecond phase of $\Delta\Psi$ generation induced by the second flash ($\square$) and for semiquinones disproportioning, measured at 450 nm ($\triangledown$), as a function of pH. Conditions as in fig. 1 except for varying buffers. Buffers used for varying pH: Mes, Mops, Hepes, Tris and Ches at 20 mM. A circle (0) indicates the rate of proton binding; the data derived from fig. 2

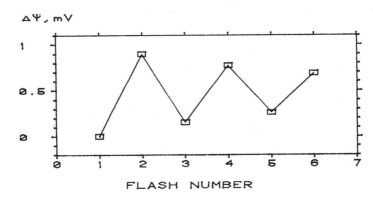

Fig. 4. Flash number dependence of the amplitude of additive submillisecond phase of $\Delta \Psi$, measured in 1 ms after flash (□). Solid line represents the theoretical function described by Eqn. (1) with γ = 1.83.

$Q_B{}^{2-}$formation has been observed by Feher and Okamura using RC complexes, incorporated into bilayer [7]. However, the time of the "quinone" phase observed by them after the second flash was 2.5 ms, that is an order of magnitude longer than ours results. The cause for this discrepancy is unclear.

REFERENCES

1. Jackson, J. B. and Dutton, P. L. (1973) Biochim. Biophys. Acta 325, 102-113.
2. Blatt, Y., Gopher, A., Montal, M. and Feher, G. (1983) Biophys. J. (Abstr.) 41, 121a.
3. Packham, N. K., Dutton, P. L. and Mueller, P. (1982) Biophys. J. 37, 465-473.
4. Tiede, D. M., Mueller, P. and Dutton, P. L. (1982) Biochim. Biophys. Acta 681, 191-201.
5. Drachev, L. A., Semenov, A. Yu., Skulachev, V. P., Smirnova, I. A., Chamorovsky, S. K., Kononenko, A. A., Rubin, A. B. and Uspenskaya, N. Ya. (1981) Eur. J. Biochem. 117, 483-489.
6. Drachev, L. A., Semenov, A. Yu. and Skulachev, V. P. (1979) Doklady AN SSSR 245, 991-994.
7. Feher, G. and Okamura, M. Y. (1984) in: Advances in Photosynthesis Research (Sybesma, C., ed.) Vol II, pp. 155-164, Martinus Nijhoff/Dr. W. Junk Publishers, Dordrecht, The Netherlands.
8. Kaminskaya, O. P., Drachev, L. A., Konstantinov, A. A., Semenov, A. Yu. and Skulachev, V. P. (1986) Biol. Membranes (USSR) 3, 557-562.
9. Semenov, A. Yu., Mamedov, M. D., Mineev, A. P., Chamorovsky, S. K. and Grishanova, N. P. (1986) Biol. Membranes (USSR) 3, 1011-1019.
10. Drachev, L. A., Mamedov, M. D., Mulkidjanian, A. Ya., Semenov, A. Yu., Shinkarev, V. P. and Verkhovsky, M. I. (1988) FEBS Lett. 233, 315-318.
11. Dracheva, S. M., Drachev, L. A., Konstantinov, A. A., Semenov, A. Yu., Skulachev, V. P., Arutjunian, A. M., Shuvalov, V. A. and Zaberezhnaya, S. M. (1988) Eur. J. Biochem. 171, 253-264.

12. Zakharova, N. I., Fabian, M., Uspenskaya, N. Ya., Kononenko, A. A. and Rubin, A. B. (1981) Biokhimiya (USSR) 46, 1703-1711.
13. Drachev, L. A., Zakharova, N. I., Karagulyan, A. K., Kondrashin, A. A. and Semenov, A. Yu. (1984) Doklady AN SSSR 275, 193-198.
14. Racker, E. (1972) J. Membrane Biol. 10, 221-235.
15. Skulachev, V. P. (1982) Meth. Enzymol. 88, 35-45.
16. Drachev, L. A., Kaurov, B. S., Mamedov, M. D., Mulkidjanian, A. Ya., Semenov, A. Yu., Shinkarev, V. P., Skulachev, V. P. and Verkhovsky, M. I. (1988) Biochim. Biophys. Acta, in press.
17. Vermeglio, A. (1977) Biochim. Biophys. Acta 459, 516-524.
18. Wraight, C. A. (1977) Biochim. Biophys. Acta 459. 525-531.
19. Shinkarev, V. P., Verkhovsky, M. I., Kaurov, B. S. and Rubin, A. B. (1981) Mol. Biol. (USSR) 15, 1069-1082.
20. Rubin, A. B. and Shinkarev, V. P. (1984) Electron Transport in Biological Systems, Nauka, Moscow (in Russian).
21. Agalidis, I. and Velthuys, B. R. (1986) FEBS Lett. 197., 263-266.
22. Wraight, C. A. (1979) Biochim. Biophys. Acta 548, 309-327.

COMPETITION BETWEEN TRAPPING AND ANNIHILATION IN *RPS. VIRIDIS* PROBED BY FAST PHOTOVOLTAGE MEASUREMENTS

H.-W. Trissl, J. Deprez[*], A. Dobek[%], W. Leibl, G. Paillotin[*], J. Breton[*]

[*] Abt. Biophysik, FB Biologie/Chemie, University, D-4500 Osnabrück
[*] Dept. Biologie, Centre d'Etudes Nucléaires de Saclay, F-91191 Gif-Sur-Yvette, France
[%] Institute of Physics, A. Mickiewicz University, 60-780 Poznan, Poland

INTRODUCTION

The study of exciton transfer, exciton-exciton interaction, and primary charge separation in the photosynthetic membrane requires excitation by picosecond flashes. If a significant fraction of the reaction centers (RCs) is closed by a single flash, the excitation density reaches a level where several excitons reside in the pool of antenna pigments at the same time. Then excitons can be lost by singlet-singlet annihilation before they are trapped by the primary photochemistry in the RC. Furthermore, annihilation leads to an apparent acceleration of all other reactions connected with the exciton dynamics. As will be shown, the quantitative treatment of this competitive deactivation path allows to determine molecular parameters that characterize a given antenna system.

We have studied exciton annihilation and trapping in *Rps. viridis* by probing the photochemical path into the RC with a photoelectric method based on the light-gradient effect[1,2]. The time resolution was sufficient to resolve the two electrogenic steps of the primary charge separation and to determine the relative dielectric distances between the primary donor, P, the pheophytin intermediary acceptor, H, and the first quinone acceptor, Q_A. The comparison of the photovoltage evoked by long and short flashes (i.e. without and with annihilation) as well as time-resolved measurements together with a theory on exciton dynamics[3] allowed us to evaluate (i) the bimolecular rate constant for annihilation, (ii) a parameter that describes the competition between trapping and annihilation, and (iii) the rate constant of trapping in the low energy limit and at higher energies.

THEORY

Exciton interactions are described by an overall bimolecular annihilation rate constant $\gamma = \gamma_1 + 2\gamma_2$ which accounts for the two singlet-singlet reactions:

Molecular Biology of Membrane-Bound Complexes in Phototrophic Bacteria
Edited by G. Drews and E. A. Dawes
Plenum Press, New York, 1990

337

$$S_1 + S_1 \xrightarrow{\gamma_1} S_1 + S_0$$

$$S_1 + S_1 \xrightarrow{\gamma_2} S_0 + S_0$$

Under the assumption of a fast equilibration of the excitation energy and free energy exchange between neighbouring photosynthetic units, the system can be described by two states, which are characterized by quenching rate constants k_o and k_c for open and closed RCs, respectively:

$$k_o = k_t + k_l \qquad (1)$$

$$k_c = k_q + k_l \qquad (2)$$

where k_t denotes the rate constant of trapping, k_q the quenching rate constant of P^+, and k_l the rate constant of all other loss processes. Then the time-dependent exciton density, $n(t)$, and the fraction of closed RCs, $q_c(t)$, is given by the set of coupled differential equations:

$$dn(t)/dt = -k_o \cdot q_o(t) \cdot n(t) - k_c \cdot q_c(t) \cdot n(t) - 1/2\, \gamma \cdot n(t)^2 \qquad (3)$$

$$dq_c(t)/dt = \Gamma \cdot k_o \cdot q_o(t) \cdot n(t) \qquad (4)$$

where q_o is the fraction of open RCs ($q_o + q_c = 1$) and Γ the quantum yield of the primary photochemistry. An analytical solution for $q_o(t)$ is given elsewhere[3]. It contains a parameter α which describes the competition between annihilation and trapping according to $\alpha = \gamma/2\Gamma k_o$.

Introducing an electrogenicity factor A_1, the time course $A_1 \cdot q_c(t)$ represents a displacement current which yields by integration the first phase of the photovoltage kinetics. It describes the trapping process which is defined as the appearance of the first charge separated state P^+H^-. The process of charge stabilization on Q_A, with the electrogenicity factor A_2, yields a second phase of the photovoltage kinetics.

MATERIALS AND METHODS

The experiments were made with whole cells at an OD of 0.3 in a 0.1 mm cuvette (1 064 nm). In all experiments 30 μM phenazin metosulfate was added. An absorption cross section of bacteriochlorophyll b in the membranes of $\sigma = 8.6 \cdot 10^{-17}$ cm^2 was used for the calculations[4].

Photovoltage measurements were made with a micro-coaxial cell[5] connected to broad band amplifiers of 10 GHz and 8 GHz bandwidth, respectively, and recorded with a 7 GHz transient digitizing oscilloscope (7250, Tektronix). At the end of each sweep a marker signal from a picosecond photodiode was added, which had a fixed time delay to the excitation flash, and which served to shift the individual traces to a common origin before averaging on a personal computer. This procedure avoids loss of time resolution due to the jitter between different single-shot traces. The time resolution and apparative response function was obtained by means of the ultra-fast charge separation occurring in purple membranes[6,7]. The data analysis included the specific light-gradient effects[8].

The excitation source was a Nd-YAG laser delivering flashes of 12 ns or 30 ps duration at 1 064 nm. Flash energies were measured with an energy meter (RJ 7200, Laser Precision). The repetition rate was 0.1 Hz. Fluorescence was detected by a Ge-Avalanche picosecond photodiode (AR-G20, Antel) using a long pass filter RG 835. These signals were amplified 10-times (B.& H. Electronics 3 GHz-amplifier, type AC-3010) and recorded on a 1 GHz oscilloscope.

RESULTS

The time course of the light-gradient photovoltage from whole cells of *Rps. viridis* with oxidized and reduced Q_A is shown in Fig. 1 *a* and *b*, respectively. The fit parameters obtained by convolution of the displacement current according to a two-step consecutive reaction scheme with the response function of the apparatus are given in the insets. (Note, that after the normalization, $A_1 = 1$, three fit parameters are left.) The trapping time of 30 to 55 ps is close to our present time resolution, but simulations with 20 ps or less give significantly worse fits. In the oxidized case the reduction of Q_A occurs with 140 ps and a relative electrogenicity factor of $A_2 = 1.5$. In the reduced case a charge recombination occurs with a rate constant of $k_{-1} = (2.4 \text{ ns})^{-1}$ ($A_2 = -1$).

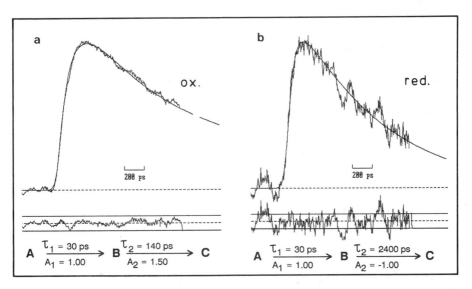

Fig. 1. Kinetics of the photovoltage from dark-adapted *Rps. viridis* cells with oxidized Q_A (a) and reduced Q_A (b). Amplitudes normalized to equal heights. Excitation at 1 064 nm with a 30 ps flash. The solid lines represent the best fits (indicated parameters) according to a convolution of a biexponential displacement current with the response function of the apparatus. The lower traces are residuals with lines indicating ±5 % deviations.

The dependence of the photovoltage amplitude on the excitation energy for the oxidized and reduced case, as well as the dependence of the apparent trapping time on the excitation energy, E, is shown in Fig. 2. The excitation energy is expressed in hits per RC, $z=\sigma NE$, taking an antenna size of $N=24$. All these dependencies are predicted by the above theory (solid lines). The corresponding best fit parameters are listed in Table 1. The ratio of quenching rate constant k_0/k_c is obtained from fluorescence measurements[4]. The quantum yield of the primary photochemistry was found to be $\Gamma=0.95\pm.05$. The electrogenicity factor A_2 is evaluated twice, from the two-step forward charge separation and from the comparison of the photovoltage amplitude between oxidized and reduced case (dashed line in Fig. 2). As seen from the difference of the photovoltage amplitude evoked by non-annihilating and annihilating flashes significant annihilation losses occur for exciton densities higher than about 0.1 hits per trap.

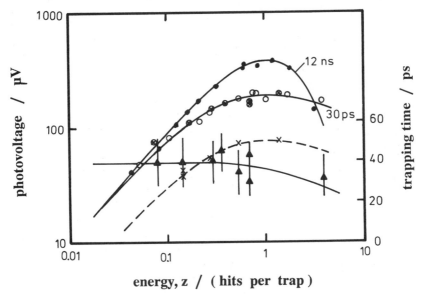

energy, z / (hits per trap)

Fig. 2. Double-logarithmic plot of the energy dependence of the photovoltage amplitudes (left scale) from *Rps. viridis* cells with oxidized Q_A (rund symbols) evoked by non-annihilating (12 ns) and annihilating flashes (30 ps). Photovoltage from cells with reduced Q_A evoked by 30 ps flashes (x). The dashed line is derived from the fit to the data of the oxidized case with ps excitation (solid line) by multiplication with 0.4. Semi-logarithmic plot of the energy dependence of the trapping time in open RCs (▲; right scale). The solid lines result from a global fit of eqs. 3-4 to the data with the parameters listed in Table 1.

The photoelectrically detected 2.4 ns backreaction kinetics when Q_A is reduced, is also found in fluorescence measurements in the sub-nanosecond time range (Fig. 3). These experiments were made with 30 ps flashes from a Ruby laser ($\lambda=694$ nm). The fluorescence kinetics from dark-adapted *Rps. viridis* showed a fast transient that decayed close to the base line within the photodiode's response time (Fig. 3a). The reduction of Q_A by dithionite induced an additional slower decaying fluorescence phase with an exponential time constant of 2.4 ± 0.3 ns (Fig. 3b). The fluorescence yield obtained by deconvolution of the two-phases yielded a ratio of the fluorescence yield for oxidized and reduced Q_A of 1.45. Since the slow fluorescence phase correlates well with the electrically measured back-reaction, it can be assigned to the charge recombination of P^+H^- that repopulates the excited state.

The not normalized data showed that the fast fluorescence transient had the same size and shape as in the oxidized case, except for the additional contribution of the slow phase (accuracy $\pm 10\%$). This latter observation indicates that the trapping kinetics are not much affected by Q_A-reduction.

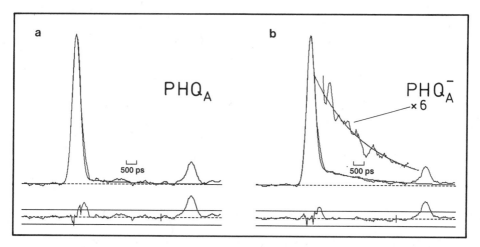

Fig. 3: a) Time course of the fluorescence evoked by a 30 ps flash at 694 nm from whole *Rps. viridis* cells with open RCs. b) Time course of the fluorescence evoked by a 30 ps flash at 694 nm from *Rps. viridis* cells with reduced RCs. Inset: 6-times digitally amplified decaying phase together with a calculated curve (smooth line) reflecting an exponential decay with a time constant of 2.4 ns. The small peaks at the right end of the traces are due to reflections in the cable. Recording bandwidth: 1 GHz.

DISCUSSION

The parameters that follow from the present analysis are listed in Table 1. The trapping time of approximately 40 ps appears to be remarkably fast in view of the unfavourable energetics between the antenna pigments, absorbing maximally at 1 015 nm, and the primary donor, absorbing at 965 nm. Compared to other antenna systems[3], also the rate of annihilation is very fast. A straight forward explanation is given by the small photosynthetic unit size of only 24 pigments, which leads to higher exciton densities in small photosynthetic units than in photosynthetic units with larger antenna sizes. (This is a consequence of the normalization of the excitation energy in hits per trap.) Due to the fast trapping time, the competition

Table 1. Best fit parameters of photovoltage (1 064 nm excitation) and fluorescence measurements from *Rps. viridis* whole cells.

k_o/k_c	A_2/A_1	k_o^{-1} / ps	k_{-1}^{-1} / ns	γ^{-1} / ps	α
1.4	1.5	40 ± 10	2.4	9	2.6

parameter of $\alpha = 2.6$ is not so high (according to its definition by the ratio of the rates of annihilation and trapping). The value is between those found for *Rb. sphaeroides* and *Rb. rubrum*[10].

The relative dielectric position of H with respect to P and Q_A was found by two independent experiments to be $A_2/A_1 = 1.5 \pm 0.05$. From the crystallographic data of the isolated RC[9] the corresponding relative geometric distance, d_2/d_1, can be estimated to be $d_2/d_1 = 0.9$. From the comparison of both ratios it can be concluded that the protein environment around P-H is more polar than around H-Q_A. This more polar surrounding might be essential for supporting the primary charge separation step.

The dielectrically weighted distance in purple bacteria[4,10] may be compared to that of photosystem II (PS II) of higher plants, for which a ratio of $A_2/A_1 = 0.9 \pm 0.1$ was reported[11,12]. Hence, H and Q_A in PS II lie dielectrically closer together than in purple bacteria and, consequently, the influence of the negative charge on Q_A on the standard free energy (=Gibbs function) of the radical pair P$^+$H$^-$, ΔG, is expected to be more pronounced in PS II than in purple bacteria.

For the photosynthetic unit of *Rps. viridis* the ΔG for the decay of the excited state into P$^+$H$^-$, when Q_A is reduced, can be calculated from our experimentally determined parameters to be $\Delta G(N=24) = -R \cdot T \cdot \ln(k_o/k_{-1}) = -103$ meV at 20 °C. The free energy change of the isolated RC can be calculated according to:

$$\Delta G(N') = -R \cdot T \cdot \ln [k_1/k_{-1}] + R \cdot T \cdot \ln N' = -R \cdot T \cdot \ln [(k_1/N')/k_{-1}] \tag{5}$$

This formula relates the free energy change of a reaction center connected to antenna pigments to the free energy change of an isolated reaction center by adding the entropy term connected with the degeneracy of the excited state on the antenna pigments and on P*. N' is an effective antenna size which has to be calculated as sum over states:

$$N' = \sum_i N_i \cdot \exp[-(E_i - E_o)/(R \cdot T)] \tag{6}$$

where N_i is the number of pigments in a given pigment pool and E_i the energy of their relaxed excited states. E_o is the energy of the lowest excited state involved. Taking the absorption maxima of 1 015 nm and 965 nm and the Stokes shifts of

240 cm^{-1} and 500 cm^{-1} for the antenna pigments and the primary donor, respectively[13,14], as well as 24 antenna pigments (N=24) and 1 special pair bacteriochlorophyll b, one obtains N'=24.3 and ΔG(N=24) = -103 meV. This is sufficient negative to justify an analysis by an irreversible reaction scheme. Extrapolation to the case of isolated reaction center (N=0; N'=1) with reduced Q_A yields ΔG(RC)= -184 meV. This number agrees well with values reported for oxidized as well as reduced isolated RCs from $Rb.$ $sphaeroides$ and $R.$ $rubrum$[13,15]. Hence, there is only a minor effect of the negative charge on Q_A on the change of the free energy of the radical pair in $Rps.$ $viridis$, which is in contrast to the $\Delta\Delta G(Q_A/Q_A^-)$ of +50 meV or +70 meV reported for PS II[16,12]. It is suggestive to ascribe this difference to the larger dielectric distance between H and Q_A in $Rps.$ $viridis$ and $Rb.$ $sphaeroides$[10] as compared to the one in PS II[11,12].

REFERENCES

1. Fowler, C. F., and Kok, B., 1974, Direct observation of a light-induced electric field in chloroplasts, Biochim. Biophys. Acta, 357:308.
2. Witt, H.T. and Zickler, A., 1973, Electrical evidence for the field indicating absorption change in bioenergetic membranes, FEBS Lett., 37:307.
3. Deprez, J., Paillotin, G., Dobek, A., Leibl, W., Trissl, H.-W., and Breton, J., 1989, Competition between energy trapping and exciton annihilation in the lake model of the photosynthetic membrane of purple bacteria, Biochim. Biophys. Acta, xxx: in the press.
4. Trissl, H.-W., Breton, J., Deprez, J., Dobek, A., and Leibl, W., 1989, Trapping kinetics, annihilation, and quantum yield in the photosynthetic purple bacterium $Rps.$ $viridis$ as revealed by electric measurment of the primary charge separation, Biochim. Biophys. Acta, xxx: in the press.
5. Trissl, H.-W., Leibl, W., Deprez, J., Dobek, A., and Breton, J., 1987, Trapping and annihilation in the antenna system of photosystem I, Biochim. Biophys. Acta, 893:320.
6. Groma, G., Szabo, J., and Varo, Gy., 1984, Direct measurement of picosecond charge separation in bacteriorhodopsin. Nature 308:557.
7. Trissl, H.-W., Gärtner, W., and Leibl, W., 1989, Reversed picosecond charge displacement from the photoproduct K of bacteriorhodopsin demonstrated photoelectrically, Chem. Phys. Lett., 158:515.
8. Leibl, W., and Trissl, H.-W., 1989, Relationship between the fraction of closed photosynthetic reaction centers and the amplitude of the photovoltage from light-gradient experiments, Biochim. Biophys. Acta, xxx: in the press.
9. Deisenhofer, J., Epp, O., Miki, R., Huber, R., and Michel, H., 1984, X-ray structure analysis of a membrane protein complex, J. Mol. Biol., 180:395.
10. Dobek, A., Deprez, J., Paillotin, G., Leibl, W., Trissl, H.-W., and Breton, J., 1989, Excitation trapping efficiency and kinetics in $Rb.$ $shaeroides$ R26 whole cells probed by photovoltage measurements in the picosecond time scale, Biochim. Biophys. Acta, xxx: in the press.
11. Trissl, H.-W., and Leibl, W., 1989, Primary charge separation in photosystem II involves two electrogenic steps, FEBS Lett., 244:85.
12. Leibl, W., Breton, J., Deprez, J., and Trissl, H.-W., 1989, Photoelectric study on the kinetics of trapping and charge stabilization in oriented PS II membranes, Photosynth. Res., xxx: in the press.
13. Carithers, R. P., and Parson, W. W., 1975, Delayed fluorescence from $Rhodopseudomonas$ $viridis$ following single flashes, Biochim. Biophys. Acta, 387:194.

14. Scherer, P. O. J., Fischer, S. F., Hörber, J. K. H., and Michel-Beyerle, M. E., 1986, On the temperature dependence of the long wavelength fluorescence and absorption of *Rhodopseudomonas viridis* reaction centers, in "Antennas and reaction centers of photosynthetic bacteria," M. E. Michel-Beyerle, ed., Springer Verlag, Berlin.

15. Woodbury, N. W., and Parson, W. W., 1986, Nanosecond fluorescence from chromatophores of *Rb. sphaeroides* and *R. rubrum*, Biochim. Biophys. Acta, 850:197.

16. Schatz, G. H., Brock, H., and Holzwarth, A. R., 1988, Kinetik and energetic model for the primary processes in photosystem II, Biophys. J., 54:397.

Spectroscopic properties of pigment-protein complexes from photosynthetic purple bacteria in relation to their structure and function

F. van Mourik[a], R. W. Visschers[a], M. C. Chang[b], R. J. Cogdell[d],
V. Sundström[c] and R. van Grondelle[a]

[a]Dept. of Biophysics, Physics Laboratory of the Free University,
Amsterdam, The Netherlands
[b]Dept. of Biochemistry, North Western University, Evanston, USA
[c]Dept. of Physical Chemistry, University of Umeå, Umeå, Sweden
[d]Dept. of Botany, University of Glasgow, Glasgow, Scotland

Introduction

In photosynthesis the light-energy, necessary to drive the electrochemical processes occurring in the reaction center, is collected by a light-harvesting antenna. Many (bacterio)chlorophyll-protein complexes cooperate in the absorption of the energy from the sun and the transport of the excitation energy to the reaction center.

The study of the bacterial light-harvesting antenna complexes has arrived at the point that good crystals have been grown, and high-resolution X-ray diffraction studies are possible. In this respect a detailed understanding of the spectroscopic properties of the pigment-protein complexes may have to wait until this exact structural data becomes available. On the other hand, it is of interest to discuss whether it is possible to correlate the observed spectral properties of the pigment-protein complexes with a (unique) model for the organization of the pigments. The 'history' of the bacterial reaction center is in this respect revealing, since detailed ideas existed about the organization of the pigments in the RC, long before the crystal structure was obtained [1,2], and some of these models turned out to be quite accurate [3].

Although from a biochemical point of view the components that constitute the light-harvesting antenna are far less complex than those of the RC, their capability to form large, apparently well organized, aggregates complicates the structural and spectroscopic analysis. The light-harvesting complexes can be isolated using detergents, and are obtained in a relatively pure form, while retaining their *in vivo* spectroscopic properties. Unfortunately, the isolated (detergent) pigment-protein complexes are still composed of large aggregates of the functional/theoretical basic unit [4,5].

In general, the antenna-proteins are divided into two main classes: the core antenna or LH-1 and the peripheral antenna or LH-2. LH-1 is the antenna component that surrounds the reaction center, and it is produced in a fixed stoichiometry to the reaction center. In most species LH-2 is the variable component, whose synthesis is enhanced in low light intensities [6]. The LH-2 complexes which occur in e.g. *Rhodobacter sphaeroides* and *Rhodopseudomonas palustris* [7] usually absorb in the 800-850 nm region, while the LH-1 antenna absorbs around 870-880 nm in most purple bacteria; the spectra are strongly red-shifted compared to monomeric Bchl a in organic solvents. In all cases well-

Molecular Biology of Membrane-Bound Complexes in Phototrophic Bacteria
Edited by G. Drews and E. A. Dawes
Plenum Press, New York, 1990

defined CD-spectra are observed: for LH-2 these signals are intense. For LH-1 some species show intense CD around 870-880 nm, while for other species the CD is weak. Fluorescence polarization of the near-IR bands of LH-1 and the B850 component of LH-2 is low, suggesting very rapid energy transfer among a group of spectroscopically more or less identical but differently oriented pigments. Linear dichroism experiments, however, indicate a large degree of organization. All the Q_y transitions of the Bchl molecules appear to be oriented parallel to the plane of the membrane, which is in good agreement with the results from the polarized fluorescence measurements [8,9].

Recently, we have followed two different approaches to come to a futher understanding of the spectroscopic properties of the antenna of purple bacteria:

(1) In 1987 Miller *et al* [10-12] reported the preparation of a subunit form of the LH-1 antenna of *Rhodospirillum rubrum*, which showed a narrow absorption peak at about 820 nm and a CD spectrum indicative of an exciton pair with a high-energy component at about 780 nm. A similar pigment-protein complex can be prepared from LH-1 of *Rb. sphaeroides* and *R. capsulata* [13,14].

This complex may serve as a model system for the basic 'minimal unit' of the LH-1 antenna. Gel-filtration and sedimentation measurements indicate that the B820 complex contains approx. 4 Bchl molecules [10,15,16]. Depending on the detergent concentration, reversible transitions occur between this B820 form, the 'intact' LH-1 form, B873, and a dissociated species, B777.

We have addressed the question of how the pigments are organized in the B820 particle: is the reassociated B873 form spectroscopically distinct from the 'in-vivo' LH-1 antenna, and is aggregation the only cause for the different spectral forms ?

(2) When grown under 'low-light' conditions *R. palustris* produces an LH-2 antenna that is quite different from its 'high-light' type of LH-2 [7]. In the 'high-light' type the 850 nm band is more intense than the 800 nm band, while in the 'low-light' type this ratio is reversed.

We have decided to look for homologies between this apparently very different type of LH-2 complex and the 'high-light' B800-850 from *R. palustris*, which in its turn is rather similar to the *Rb. sphaeroides* B800-850.

In addition, one might wonder how in the low-light grown species the overall antenna organization has been changed so as to let the organism be optimally adopted to the altered light conditions. This question is even more interesting in view of the fact that in the 'normal' type of LH-2 the Bchl-850 components play a prominent role in the energy-transport to the reaction center, whereas in these 'low-light' complexes the energy-transfer appears to proceed with at least equal efficiency with a relatively low amount of Bchl-850.

In this manuscript we will present some of the spectroscopic properties of both these low-light *R. palustris* LH-2 complexes and of the B820 LH-1 subunit-form and discuss them in the light of the currently available models for light-harvesting antenna pigment organization.

Materials and Methods

Fluorescence, CD and LD spectra were recorded on a home-built spectrophotometer which will be described elsewhere. Absorption spectra were recorded on a Cary 219 spectrophotometer. Low temperature samples were prepared in 50 % glycerol, in acrylic 1cm fluorescence cuvettes. For the 77K CD spectra 2 mm quartz cuvettes with a removable coverplate were used. LD spectra were measured by biaxial orientation in 15 % acrylamide gels as in [17].

Spectral simulations were performed on a SUN 4/280, using a home developed program similar to the program described in [19].

Absorbance 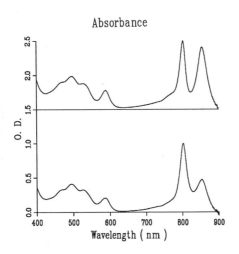 Circular Dichroism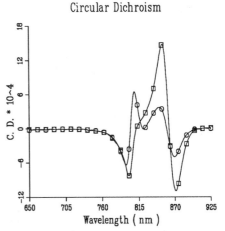

Fig. 1. Room temperature spectra of high-light (upper trace, offset by 1.5) and low-light B800-850 complexes.(squares)

Fig. 2. Room temperature CD spectra of low light B800-850 (circles) and high-light B800-850 complexes. Spectra were normalised to OD_{800} = 1.0

Results: 1. The Low-Light B800-850 complex from *Rhodopseudomonas palustris*

Fig. 1 clearly shows the most obvious difference between the high-light and low-light forms of the *Rps. palustris* B800-850 complexes: in the low-light complex the 800 nm band is far more intense than the 850 nm band. Both complexes show similar carotenoid absorption, with peaks at 465, 495 and 528 nm.

The CD spectra of the two types of B800-850 complexes are shown in Fig. 2. The high-light complex has a CD spectrum that is almost identical to the type-I B800-850 complex of *Rps. acidophila* [18]. The 860-900 nm region of the low-light B800-850 CD spectrum is very similar to that of the high-light B800-850, and the magnitude of this part of the spectrum appears to be proportional to the amount of Bchl 850. The part of the spectrum around 800 nm is not proportional to the intensity of the 800 nm absorption band, suggesting that the difference between the two forms is not only an increase of the amount of Bchl800 in the low-light form. The most likely explanation for the more complex 800 nm CD in the low-light B800-850 is that the Bchl800 are more densely packed in the low-light complex. The organization of the Bchl850's is not influenced by this change in packing.

The 77K absorption spectra of low-light membranes and B800-850 complexes are shown in fig. 3. A closer look at the near-IR B850 band shows that it is non-symmetric, and there appears to be a minor component between the B800 and B850 bands located at approx. 820 nm.. The existence of this additional band is more clearly demonstrated in the 77K CD spectrum (fig. 4) of the isolated low-light B800-850 complex, which shows 5 distinct near-IR CD-peakes.The CD bands at 880 and 850 nm are due to the B850, the other three bands, at 798, 807 and 820 nm, appear to be associated with the B800 'system'. This notion is supported by low-temperature fluorescence measurements (results not shown): the residual fluorescence that comes from the B800 is emitted by the long-wavelength component. Also, picosecond absorption measurements show that upon excitation between 780 and 820 nm, the excitation density is located on the red-most component within the time-resolution of the apparatus, i.e. << 1 ps at 77K (results not shown). These observations suggest that the three bands of the B800 'system' are excitonic transitions.

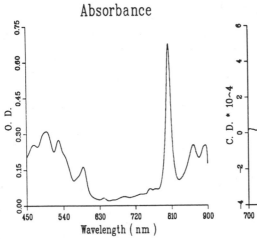

Absorbance

Fig. 3. 77K absorption spectrum of low-light low-light membranes

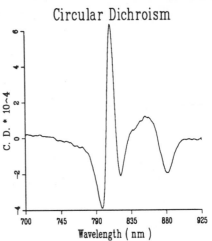

Circular Dichroism

Fig. 4. 77K CD spectrum of low-light B800-850, $OD_{800}=1$.

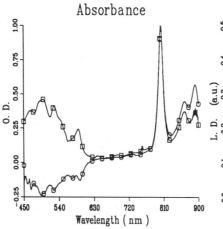

Absorbance

Fig. 5. 77K absorption (squares) and LD (circles, arb. units) of low-light membranes from *R. palustris*.

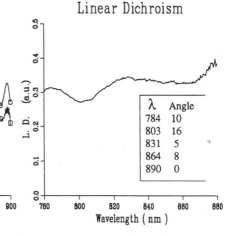

Linear Dichroism

λ	Angle
784	10
803	16
831	5
864	8
890	0

Fig. 6. LD/A spectrum of the spectra in Fig 5. Inset: calculated angles with the plane of the membrane.

Fig. 5 shows the 77K LD spectrum of low-light membranes from *Rps. palustris* measured in a squeezed polyacrylamide gel, together with the 77K absorption spectrum.

The 'parallel' orientation of all the Q_y transition moments, and the overall 'perpendicular' orientation of the Q_x transitions and the carotenoids is clear.

The 77K LD/A spectrum of the low-light membranes is shown in fig. 6. The angles displayed in the inset were calculated using the formula of Ganago for disc-shaped particles [17].

We note that the B875 IR-transition makes an angle of nearly 90 degrees with the membrane normal. B850 is slightly less ordered, while the B800 LD reflects a significant tilting of some of the Bchl Q_y transitions, away from the membrane plane (see below). The complicated nature of the B800 absorption band is clearly demonstrated by the LD/A spectrum between 780 and 840 nm, where at least three different components can be distinguished. The peak positions roughly correlate with those observed in the CD-spectrum.

Apart from the contribution from the B875, the LD spectra from the membranes were almost identical to those of the complexes, indicating that the complexes form discoid aggregates in (detergent)solutions. In the Q_x region at least three Bchl LD bands can be distinguished, a negative band at 600 nm, a positive band at 590 nm and a negative band at 580 nm. (results not shown)

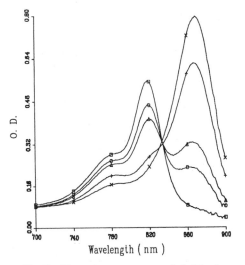

Fig. 7. Absorption spectra, recorded 20 min. x-x, 1 hour +-+, 2 hours (triangles), 3 hours (circles) and 5 hours after adding 0.8 % OG to LH-1 of *Rs. rubrum*.

Fig. 8. 77K absorption spectra of B820 and (reassociated) B873 (squares) of *Rs. rubrum* normalized to their resp. absorption maxima.

2. The LH-1 subunit form

In this section we shall describe the spectroscopic features of the LH-1 antenna and its B820 'sub-unit form', as studied by low-temperature polarized fluorescence, CD and OD.

Fig. 5 follows in time the changes that occur to the absorption spectrum the carotenoid-extracted LH-1 complex of *Rs. rubrum* after adding OG to a final concentration of 0.8% . The transitions (in this case 875 to 820 nm) induced by OG are fully reversible, and moreover, the different forms can also be produced by reconstituting the particles from the α and β polypeptides and Bchl [11], indicating that the differences between the three forms represent different aggregational forms of the basic components. For *Rs. rubrum* the absorbance maxima of the three components are at 873, 820 and 777 nm. From *Rb. sphaeroides* LH-1 similar particles can be obtained, with absorption maxima at 873, 815 and 777 nm.

Fig. 6 shows the 77K absorption spectra of the B820 and the (reassociated) B873 forms. In the B820 spectrum the shoulder at approx. 780 nm in the RT spectrum now appears as a distinct band. As the excitation spectra will show most of this 780 nm band is not coupled to the main 820 nm band and it probably represents a small amount of the 777 form, in equilibrium with the B820 form. Also, at slightly lower OG concentrations, mixed B820 and B873 forms can be obtained which do not show energy transfer from B820 to B873 (results not shown) so the 'mixed' spectra represent mixtures of distinct 'pure' particles.

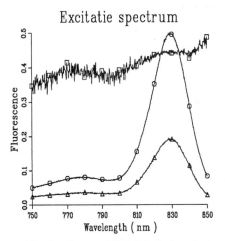

Fig. 9.   Corrected 77K Polarized excitation spectra of *Rs. rubrum* B820, detection wavelength 850 nm. Circles: vertical detection, Triangles: horizontal detection, Squares: Polarization spectrum

Fig. 10 Idem, for the reassociated B875-form, detection wavelength 920 nm.

The excitation spectra shown in fig. 7, from a sample with an absorption spectrum as in fig 6 but with a maximal absorption at 825 nm of 0.05, clearly demonstrate the absence of the 780 nm shoulder in the excitation spectrum of the B820 form. The maximum of the emission spectrum of the B820 form is at 830 nm at 77K. The fluorescence polarization spectrum (Fig. 7) shows that the polarization of the 820 nm band is constant (apart from the red edge, which is contaminated with scattered light), P=0.43. However, there is a distinct dip in the curve around 790 nm, which we tentatively ascribe to a high energy exciton component of the B820 form.

Assuming that we are dealing with a Bchl dimer (which the spectra strongly suggest) the contribution of this high-energy component to the OD spectrum can be estimated from the excitation spectra, since its transition dipole must be perpendicular to the low-energy component. It is found to be in the order of 2%, and not more than 5% .

Note that the polarization of the 820 nm band is very high compared to similar measurements on other pigment-protein complexes. For monomeric Bchl *a* in 2% OG at 77K a P-value close to 0.5 was obtained (results not shown).

Fig. 8 shows the polarized excitation spectra of the reassociated B873 form (reassociated from *Rs. rubrum* B820), the polarization value of approx. 0.1 is in good agreement with the values found both in membranes and in isolated LH-1 complexes [8,9,20]. We obtain similar spectra for the reassociated B873 form of *Rb. sphaeroides*, the only difference being that for the *Rb. sphearoides* B873 the increase in polarization upon excitation in the red wing of the absorption band of was more pronounced, in agreement with earlier studies on the intact *Rb. sphaeroides* LH-1 [9]. The value of 0.1 found for the polarization indicates energy transfer among a number of non-parallel pigments with their transition dipoles parallel to one common plane. Thus, upon formation of the

B873 form the B820's probably aggregate into larger complexes. This agrees with studies showing that the Bchl molecules are more exposed in the B820 form than in the B873 form [14].

Circular Dichroism

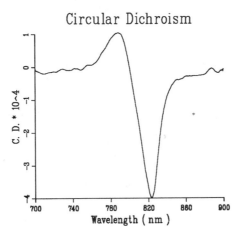

Figure 11. 77K CD spectrum of B820 subunit of LH1 from *Rs. rubrum*. Normalized to $OD_{820} = 1$

The 77K CD spectrum of the B820 form is shown in Fig. 11. As for the OD spectra, the bands of the CD spectrum sharpen at lower temperature, causing slight band-shifts of the CD spectrum, but no additional bands appear. The positive lobe, located at approx. 780 nm at room temperature, shifts to approx. 790 nm, due to the decreased influence of the negative 820 nm band. The true position of the positive band may be even more to the red, which would agree with the position of the high-energy component in the polarized excitation spectra.

Models for the pigment organization in antenna pigment-proteins

Several models have been proposed for the light-harvesting antenna proteins using both the polypeptide-sequence data [21] and the spectroscopic properties[22,23]. Although the biochemical (sequence) data for the light-harvesting polypeptides is almost complete, a consensus on the organization of the chromophores has not been reached [12,24]. One of the unresolved questions is which of the LH-1 polypeptides bind(s) the Bchl875: α or β with two Bchl-molecules on one polypeptide [12,10,11], or both α and β with each carrying one Bchl molecule[24].
The model in which one of the polypeptides binds two Bchl-molecules was proposed by Loach et al. [12] and is consistent with their successful reconstitution of the B820 form using only the β-polypeptide [11]. However, this arrangement is troublesome when it comes to explaining the hyperchromism and red-shift, as discussed in [25], and would give rise to an inverted LD-signal.
The model proposed in 1984 by Kramer *et al* for the B800-850 complex from *Rb. sphaeroides* [22] was mainly based on the polarized fluorescence and LD properties of this LH-2 complex. The red-shift of the pigments was not explicitly considered and it was assumed that either exciton-interactions or pigment-protein interactions might be responsible for the observed absorption spectra. Moreover, the CD of the complex was not explicitly calculated (see below). The Kramer-model includes the interaction of at least 2 Bchl800, 4 Bchl850 and 3 carotenoid molecules in the LH-2 complex of purple bacteria, and proposes very rapid (≤ 1 ps) energy transfer within such a complex. Pico-

second absorption spectroscopy later confirmed the proposed rapid initial depolarization of the Bchl850 excited state [26] and the ps transfer from Bchl800 to Bchl850 [27].

Alternatively, Scherz and Parson, after having demonstrated that the spectroscopic properties of light-harvesting complexes can be simulated/accounted for by dimers of Bchl [28,19], proposed a model for the B850 part of the LH-2 antenna, that attempted to explain the hyperchromism, the red-shift of the absorption bands, and the CD-signal [25]. In this model the red-shift is proposed to be solely due to a strongly interacting Bchl-dimer, while the CD-signal arises from the weaker interaction between different dimers, placed at relatively large distances from each other. This model did not attempt to account for the observed polarized fluorescence and LD spectra.

We shall illustrate the fundamental reasons why at least two different 'types' of exciton interaction are required in order to explain the near-IR absorption and CD spectra of the B850 band of LH-2. The CD spectrum shows a negative lobe at about 870 nm, and a positive lobe at 850 nm (77K CD spectrum in Fig. 7A in [1]). Thus, the CD spectrum is indicative of a *blue*-shifted dimer. On the other hand, absorption-recovery experiments [29,30] show that the effect of exciting (i.e. bleaching) one of the Bchl850 molecules is a significant blue-shift of the whole absorption spectrum. These observations indicate that the IR-CD and IR-OD spectra have a different excitonic origin. Moreover, it suggests that the interactions, which dominate the red-shift, originate from several pigments, probably more than two [29].

In modeling these spectra there are two major problems. The first is the choice of the amount of red-shift caused by non-excitonic pigment-protein interactions. The second is that probably no unique model exists.

One feature that we wish to introduce in our modeling is that all fluorescence and LD-measurements show that the Q_y transitions (or better near-IR transitions, since they represent mixed states) of B875 and B850 are oriented in the plane of the membrane. To obtain the CD as proposed in the Scherz and Parson model, the vector connecting the two dimers can not lie in this same plane. However, this is not very likely, since several lines of evidence show that the Bchl850's are bound to conserved histidines on the α and β polypeptides. In our calculations we will simulate the spectroscopic properties of the B820, and use these as a starting point for the calculation of the LH-1 spectra. The CD is in that case an intrinsic property of the basic unit and it is not due to long-range interactions.

Simulation of the B820 subunit form

The very high value we found for the polarization of the B820 fluorescence imposes severe restrictions on the sort of model that one can construct, in order to describe the particle in terms of interacting Bchl molecules. Both the high polarization and the low dipolar strength of the high energy exciton component indicate that the angle between the Q_y transition dipoles must be small. The fact that the exciton-interactions induce both red-shift and (some) hyperchromism indicates that the most likely organization of the Q_y transitions is head to tail. Using this information, and similar procedures as in [19], we constructed the model depicted in fig. 12, which produces the correct CD and absorption spectra for the B820 form.

The dipole strengths of the $Q_{x,y}$ and $B_{x,y}$ were taken from [6]. The monomer Q_y transition was chosen at 805 nm, as suggested by the 77K CD and fluorescence data. The spectra were calculated as in [19], with mixing of non-degenerate states. The spectra were constructed from the calculated transition dipoles and rotational strengths using Gaussian band shapes with a bandwidth as suggested by the 77K OD spectrum.

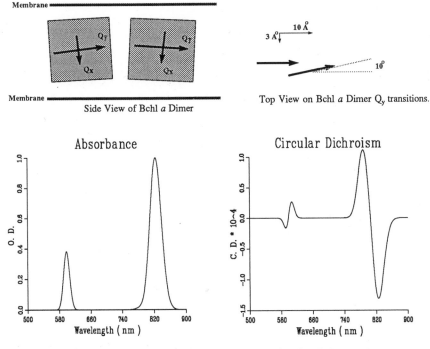

Fig. 12. Model for the B820 complex (top), with corresponding OD (left) and CD (right) spectra.

The CD spectrum is clearly less non-conservative as the spectrum in Fig. 11, also the spectrum in Fig. 11 looks different because the bandwidths of the two bands are not identical.

To calculate the spectra in Fig. 13, twelve hypothetical 'B820-dimers' were placed in a circular arrangement with a radius of 3.5 nm.

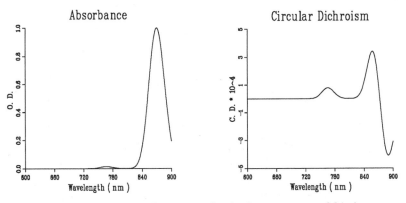

Fig 13. OD (left) and CD spectra of a circular arrangement of 24 pigments.

In the calculation of the corresponding spectra only the Q_y transitions were used, to save computer-time. This results in conservative CD spectra. The introduction of non-degenerate transitions and groundstate-perturbation into the calculations would also give rise to a more drastic red-shift (and hyperchromicity) since it will effectively increase the dipole-strength of the IR transitions, thereby increasing the interaction between the pigments. To compensate for this we increased the interactions by 60 %. Contrary to the simulations in [31] our circular arrangement does give rise to CD-bands. This is due to the fact that we have made the arrangement less symmetric. Note that in order to produce CD we have given the transitions a small angle with the plane of the membrane, another way of producing CD would be to displace one of the monomers slightly along the membrane normal (i.e. placing the pigments in two parallel planes).

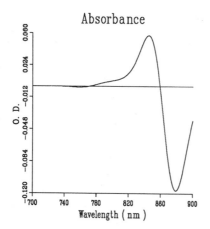

Fig. 14. The effect of bleaching one of the pigments in Fig. 13.

In Fig. 14 the effect of exciting (bleaching) one of the pigments in the model of Fig. 13 is simulated. The spectrum shown represents the difference spectrum of the spectrum of the 24-mer and that of a 23-mer. It is clear that the the bleaching of one of the pigments does not result in the appearance of a monomer absorptionband. Instead, the whole spectrum is partially bleached and *blue*-shifted.

Discussion

The main conclusion from our measurements on the low-light *Rps. palustris* B800-850 complex is that despite its very different OD and CD spectra the particle is very similar to the B800-850 complex from *Rb. sphaeroides*.
As found for the *Rb. sphaeroides* B800-850 complex [22], the B850 band appears in the LD/A spectrum as a homogeneous band, and has a conservative CD spectrum.
The B800 band of the low-light *Rps. palustris* B800-850 complex looks very different from its *Rb. sphaeroides* equivalent, however, the B800-part of the CD spectrum of the *Rb. sphaeroides* B800-850 complex is rather variable (compare [22],[18]), and in some preparations a three-banded B800 CD spectrum is actually observed. The LD/A signal of the low-light B800 band is not homogeneous, which we ascribe to exciton splitting. In order to explain the CD and LD/A spectra in the 800 nm region we must assume that the organization of the pigments is less symmetric than that of the B850 pigments. In the Kramer-model one of the B800 pigments makes a larger angle with the plane of the membrane than the other one and a similar arrangement, but containing more B800 pigments would describe the low-light *Rps. palustris* B800-850 as well.
The CD spectra of the low-light and high-light *Rps. palustris* B800-850 suggest that the main difference between them is an increase of the amount of B800. Studies of the protein contents of these complexes indicate that the pigment to protein ratio is higher in the low-light form [32], which makes these complexes relatively 'cheap' and explains why they are produced at low-light intensity. The polypeptides have now been se-

quenced [33] and the high-light complex has been cristallized [34], from these new data we should be able to produce a model for the pigment organization in the low-light B800-850.

The model we present for the B820 complex is not a unique model. Surely different solutions exist that comply with our experimental data. However, it clearly demonstrates that the OD and CD spectra of the B820 form can be described in terms of a dimer of interacting Bchl *a* molecules. The resemblance of the calculated CD spectrum with that presented in Fig. 11 is far from perfect. One of the weak points of the methods for the calculation of the spectra is that it treats transition dipoles as point-dipoles. For a dimer arranged as in Fig 12 , where part of the porphyrin-rings are 'overlapping', the use of the point-monopoles would be more appropriate. Also, the calculations do not take the band-shapes into account, while Fig. 11 clearly demonstrates that the band-shapes of the B820 CD-spectrum are not identical.

The calculated bleaching-spectrum bears a strong resemblance with the spectra observed in pico-second absorption measurements. It might also explain the measurements of Rafferty [35].

The actual shape of the 'induced absorptionband' of the bleaching-spectrum depends on how the origin of the red-shift is distributed over monomer-monomer and dimer-dimer interactions. A monomer-like band will only appear in the induced-absorptionband if the monomer-monomer interaction is significantly larger than the total interaction of the dimer with its neigbours.

Shuvalov and Parson measured the T-S spectrum of the B850 band of the *Rb. sphaeroides* R26 B800-850 complex [36]. The discussion above shines a new light on their results. The induced absorption band in fig. 5B in [36] appears to be composed of two bands, one around approx. 800 nm which could well be the monomer-band, while the other one could be part of the band-shift spectrum.

We have very recently measured the T-S spectrum of the B820 complex. The near-IR region of the spectrum we obtained could be described by only two components, a bleached band around 820 nm with the shape of the dimer-spectrum, and a band around 805 nm which we think represents the monomer band.

Acknowledgments

This work was supported by the Dutch Foundation for Biophysics, RJC and RvG were supported by a grant from the EEC Grant nr. SC1-0004-C.
We thank Drs A.M. Hawthornthwaite and C.A. Vonk for the use of their data on the *Rps palustris* particlès.

References

1) Deisenhofer J., Epp O., Miki K., Huber R. and Michel H. (1984) *J. Mol. Biol.* 180, 385-398

2) Deisenhofer J., Epp O., Miki K., Huber R. and Michel H. (1985) *Nature* 318, 618-624

3) Shuvalov V.A., Asadov A.A. (1979) *Biochim. Biophys. Acta* 545, 296-308

4) van Grondelle R., Hunter C.N., Bakker J.G.C., Kramer J.M. (1983) *Biochim. Biophys. Acta* 723, 30-36

5) Hunter C.N., Pennoyer J.D., Sturgis J.N. Farrelly D., Niederman R.A. (1988) *Biochem.* 27, 3459-3467

6) Aagard J., Sistrom W.R. (1972) *Photochem. Photobiol.* 15, 209-225

7) Firsow N. N., Drews G. *Arch. Microbiol.* (1977) 115, 299-306

8) Breton j., Farkas D.L., and Parson W.W. (1985) *Biochim. Biophys. Acta* 808, 421-427

9) Kramer H.J.M., Pennoyer J.D., van Grondelle R., Westerhuis W.H.J., Niederman R.A. and Amesz J. (1984) *Biochim. Biophys. Acta* 767, 335-344

10) Miller J.F., Hinchigeri S.B., Parkes-Loach P.S., Callahan P.M., Sprinkle J.R., Riccobono J.R., Loach P.A. (1987) *Biochem.* 26, 5055-5062

11) Loach, P.S., Sprinkle J.R., Loach P.A. (1988) *Biochem.* 27, 2718-2727

12) Loach P.A., Parkes P.S., Miller J.F., Hinchigeri S.B., Callahan P.M. (1985) in *Molecular Biology of the Photosynthetic Apparatus* editors: Arntzen C., Bogorad L., Bonitz S. & Steinback K. pp 197-209

13) Heller B.A., Loach (1989) Submitted to *Photochem. Photobiol.*

14) Chang M.C., Callahan P.M., Parkes-Loach P.S., Cotton T.M. and Loach P.A. (1989), Submitted to *Biochemistry.*

15) Ghosh R., Hauser H., Bachofen R. (1988) *Biochem.* 27, 1004-1014

16) Ghosh, R., Rosatzin, T., Bachofen, R. in *Photosynthetic Light-Harvesting Systems Organization and Function*, editor Scheer, Schneider, pp 93-102

17) Ganago A. O., Fok M. V., Abdurakhmanov I. A., Solov'ev A. A., Erokhin E. *Molekulyarnaya Biologiya* (1980) 14, 381-389

18) Cogdell R.J., Scheer H. (1985) *Photochem. Photobiol. 42, 669-678*

19) Scherz A., Parson W. W. *Biochim Biophys. Acta* (1984) 766, 666-678

20) Bergström H., Westerhuis W.H.J., Sundström V., van Grondelle R., Niederman R.A. and Gillbro T. (1988) *FEBS Lett.* 233, 12-16

21) Zuber H. (1986) *Trends in Biochem. Sci.* 11, 414-419

22) Kramer J. M., van Grondelle R., Hunter C. N., Westerhuis W. H. J., Amesz J. (1984) *Biochim. Biophys. Acta* 765:156-165

23) Breton j., Vermeglio A., Garrios M., Paillotin G. in *Photosynthesis III Structure of the Photosynthetic Apparatus* editor George Akoyunoglou (1981) 445-459

24) Picorel R., L'Ecuyer A., Potier M., Gingras G. (1986) *J. Biol. Chem.* 261, 3020-3024

25) Scherz A., Parson W. W. *Photosynthesis Research* (1986) 9, 21-32

26) Bergström H., Sundström V., van Grondelle R., Gillbro T., Cogdell R. (1988) *Biochim. Biophys. Acta* 936, 90-98

27) Bergström H., Sundström V., van Grondelle R., Åkesson E., Gillbro T. (1986) *Biochim. Biophys. Acta* 852, 279-287

28) Scherz A., Parson W. W. *Biochim. Biophys. Acta* (1984) 766, 653-655

29) Nuys A.M., van Grondelle R., Joppe H.L.P., Bchove A.C., Duysens N.M. (1985) *Biochim. Biophys. Acta* 810, 94-105

30) van Grondelle R., Bergström H., Sundström V., Gillbro T. (1987) *Biochim. Biophys. Acta* 894, 313-326

31) Pearlstein R.M., Zuber H. in *Antennas and Reaction centers of photosynthetic Bacteria* (1985) pp. 53-61, editor Michel-Beyerle.

32) Hayashi H., Nakano M., Morita s. *J. Biochem.* (1982) 92 1805-1811

33) Evans M.B., Hawthornthwaite A.M. and Cogdell R.J. (1989) Submitted to *Biochim. Biophys. Acta.*

34) Mäntele W., Steck K., Becker A., Wacker T., Welte N., Gad'on N. and Drews G. (1988) in *Structure of Bacterial Reaction Centers: X-Ray Crystallography and Optical Spectroscopy with Polarized Light* editors Breton j., Vermeglio A.

35) Rafferty C.N., Bolt J., Sauer K. and Clayton R.K. (1979) *Proc. Natl. Acad. Sci. USA* 76, 4429-4432

36) Shuvalov V.A., Parson W.W. (1981) *Biochim. Biophys. Acta* 638, 50-59

EXCITATION ENERGY TRANSFER IN PHOTOSYNTHETIC SYSTEMS:

PIGMENT OLIGOMERIZATION EFFECT

Zoya Fetisova

A.N. Belozersky Laboratory of Molecular Biology and
Bioorganic Chemistry, Moscow State University, Moscow
119 899, USSR

STATEMENT OF THE PROBLEM

Our previous theoretical analysis has shown that the structure of
a photosynthetic unit (PSU) should be strictly optimized in vivo to ensure the high quantum yield values (~90%) found experimentally for the
primary photochemistry (Fetisova and Fok, 1984). We have already studied
the basic principles of the structural organization of an optimal model
light-converting systems with certain simple types of their lattices
(Fetisova, Fok and Shibaeva, 1985; Fetisova and Shibaeva, 1987). However,
in all the known photosynthetic organisms, the three-dimensional array
of pigment-protein complexes form the cluster structure of the PSU lattice: the distances between molecules within a single complex are ~10 Å,
while for the nearest neighbour molecules which belong to adjacent complexes the distances are 20 - 30 Å.

Here we deal with the problem: does a cluster structure of a PSU
lattice in vivo reflect an optimal space distribution for a given number
of pigment molecules within a fixed area occupied by a PSU? In other
words, can a clustering of the PSU pigment molecules in vivo be considered as one of the optimizing structural factors ensuring the high efficiency of excitation energy transfer from an antenna to a reaction center
(RC)?

This problem is examined here by mathematical simulation of the
light-harvesting process in model systems.

The impact of pigment clustering on the rate of energy transfer from
antenna to RC was considered earlier by Knox (1977) for a case of weak
interactions between all the molecules of a model two-dimensional PSU
which contains 120 antenna molecules per RC, packed in clusters of 5, with
the minimal intercluster (R) and minimal intracluster (r) distances between molecules such that $R/r = \sqrt{2}$. Comparison of this model with the model of a uniform isotropic PSU (with square lattice) of the same size and
area showed that the time of Förster's inductive-resonance energy transfer
from antenna to RC is practically the same in the two models. However, it
had later been found that the distance ratio R/r varies from 2 to 3 in vivo, and the extrapolation of the Knox's calculations to these higher R/r
values is not necessarily valid. Besides, as the intracluster distances r
decrease (and hence the intercluster distances R increase), the
interactions between the nearest molecules of adjacent clusters become even
weaker; while the strong intermolecular interactions within each cluster
may, in principle, arise and should be then taken into account. Moreover,

Molecular Biology of Membrane-Bound Complexes in Phototrophic Bacteria
Edited by G. Drews and E. A. Dawes
Plenum Press, New York, 1990

357

there are numerous experimental evidences that interactions between the molecules within each cluster of an in vivo PSU are substantially stronger than interactions between the molecules of adjacent clusters (for review see Hanson, 1988; Scherz and Rosenbach-Belkin, 1989).

However, in order not to limit the generality of our problem consideration, it is necessary to examine the both weak and strong intracluster interactions. So, we have studied the impact of a cluster structure of a model PSU lattice on the efficiency of energy transfer from antenna to RC for the two limits:

(1) weak intermolecular interactions between all the pigments of a PSU, which implies Förster-type energy transfer between all the pigment molecules in a PSU; and

(2) strong intermolecular interactions within each cluster and weak interactions between molecules of adjacent clusters, which implies that the time of energy transfer within a cluster can be neglected compared to the time of inductive-resonance energy transfer between adjacent clusters; here each cluster is considered as a single "supermolecule".

MODELS AND METHODS

All the model PSUs studied here have regular lattices only (Fetisova, Freiberg and Timpmann, 1988). We have examined two-dimensional aggregates of elementary PSUs that comprise infinite lattices with translational symmetry. In each model, elementary PSUs contain either

(1) N = 48 antenna molecules plus 1 RC (Fig. 1, models A, B, C). Here the clusters (models B and C) were of minimal size, i.e. contained n = 2 molecules, n being the number of pigment molecules in a cluster, or

(2) N = 120 antenna molecules plus 1 RC (Fig. 1, models D, E, F). In these PSUs (models E and F), the clusters contained n = 5 molecules (except for the corner ones in model E).

Efficiency of energy transfer from antenna to RC for a model PSU with regularly clustered pigment molecules has been compared with that for a corresponding uniform isotropic model PSU with the same number of antenna molecules arrayed in a square lattice (models A and D in Fig. 1). In all the computations, the area occupied by an elementary PSU has been taken constant for a given number of antenna molecules. The R/r ratio of distances has been changed up to 3. For the case of weak intermolecular interactions (models A, B, D, and E) dipole-dipole approximation has been used to describe energy transfer between all the molecules of a PSU. For the case of strong intermolecular interactions within a cluster in a PSU, each cluster has been considered as a single "supermolecule" (models C and F), and dipole-dipole approximation has been used for inductive-resonance energy transfer between these "supermolecules".

In all the computations performed, the random orientation of dipole moment vectors of all the PSU "molecules" has been assumed.

To mimic absorption spectra of PSUs, we have used absorption spectra of the two light-harvesting pigments of bacterial PSUs, namely, bacteriochlorophyll a (BChl a) and bacteriochlorophyll c (BChl c). The following three combinations of spectra of these pigments have been used:

(1) BChl a in vitro (Scherz and Rosenbach-Belkin, 1989):

    (i) BChl a in formamide/H_2O/TX-100 (monomeric BChl a, absorption maximum is at 780 nm);

    (ii) BChl a in formamide/H_2O (oligomeric BChl a, absorption maximum is at 860 nm);

(2) BChl a in vivo (Picorel et al., 1986):

    (i) BChl 868 - monomer (absorption maximum is at 868 nm) derived from the intact dimeric B 880-holochrome of purple bacterium Rhodospirillum rubrum, mutant strain F24 lacking RC;

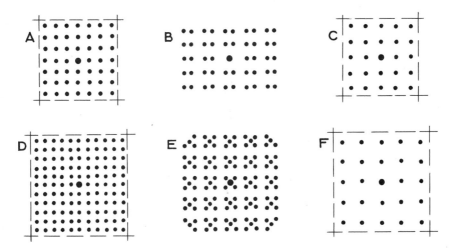

Fig. 1. Elementary fragments of infinite two-dimensional uniform models
of multicentral photosynthetic units. In the upper row, each
fragment of macroscopic PSUs contains N = 48 antenna molecules
(small circles) per one reaction center (large circles). In models A and B, each element is monomer. In model B, the antenna
molecules are packed in clusters of two weakly bound molecules.
Each element of model C is dimer, i.e. represents the cluster
of two strongly bound molecules. In the lower row, each fragment
of macroscopic PSUs contains N = 120 antenna molecules per one
reaction center. In models D and E, each element is monomer. In
model E, the molecules are packed in clusters of five weakly
bound molecules. Each element of model F is oligomer, i.e. represents the cluster of five strongly bound molecules.

       (ii) B 880 - dimer (absorption maximum is at 880 nm) of the intact
           B 880-holochrome of R. rubrum, strain F24;
   (3) BChl c in vitro (Olson et al., 1987):
       (i) BChl c in CCl_4 (monomeric BChl c, absorption maximum is at
           667 nm);
       (ii) BChl c in CCl_4 (dimeric BChl c, absorption maximum is at
           706 nm).

Thus, the absorption spectra of monomeric pigments listed above have
been used to simulate the spectra of model PSUs comprised of monomeric pigments (models A, B, D, and E). The shape of absorption spectra of oligomeric pigments has been used to simulate the spectra of model PSUs composed
of the clusters containing n strongly coupled pigment molecules, i.e.
either dimeric (model C, n = 2) or oligomeric pigment clusters (model F,
n = 5). Besides, the hyperchromism of the Q_y band for an oligomer has been
assumed to be the same as for a dimer.

We have further assumed that the fluorescence spectra are mirror symmetric to the absorption spectra. The Stokes shift values (ΔSt) for oligomeric and monomeric pigments have been taken so that $\Delta St_{ol} / \Delta St_m = 2$
in accordance with available experimental data.

Efficiency of energy transfer from antenna to RC is determined by the
time of this transfer. The time of excitation trapping by RC has been computed for diffusion-type exciton motion in the antenna at low excitation
levels. The method of the computation has been described earlier (Fetisova
and Fok, 1984). We have computed the time t needed for excitation quen-

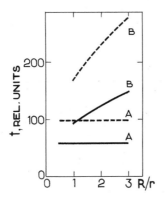

Fig. 2. Dependence of the t time of excitation energy trapping by reaction centers for PSU model B (N = 48, n = 2) on the ratio R/r of distances between nearest molecules which belong to adjacent clusters (R) and to the same cluster (r). Curves B represent this dependence t(R/r) for two cases corresponding to irreversible excitation trapping by reaction centers (solid curve) and to reversible one (broken curve). Horizontal lines A show the trapping time value for the corresponding uniform isotropic model A, again, for both irreversible trapping (solid line) and reversible one (broken line).

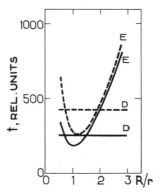

Fig. 3. Dependence of the t time of excitation energy trapping by reaction centers for PSU model E (N = 120, n = 5) on the ratio R/r of the distances between nearest molecules of adjacent clusters (R) and of the same cluster (r). Curves E represent this dependence t(R/r) for both irreversible (solid curve) and reversible excitation trapping by reaction centers (broken curve). Horizontal lines D show the trapping time value for the corresponding uniform isotropic model D, again, for both irreversible trapping (solid line) and reversible one (broken line).

ching to a 1/10 of its initial value due to excitation trapping by RCs, in other words, during this time 90% of excitation energy is captured by RCs. The trapping time t is expressed in arbitrary units identical for all model PSUs of the same size (i.e. containing the same number of antenna molecules in an elementary PSU). In all the computations performed, a uniform initial excitation distribution among all the lattice sites has been assumed. All computations have been made for both irreversible ($\varphi_t = 1.0$) and reversible excitation energy trapping by RC; in the latter case we have chosen $\varphi_t = 0.5$.

EFFICIENCY OF LIGHT HARVESTING IN PSUs WITH DIFFERENT ARRANGEMENTS

Limit of Weak Interactions between All the Pigment Molecules

Fig. 2 shows dependence of the trapping time t on the ratio R/r of distances between nearest molecules which belong to adjacent clusters (R) and to the same cluster (r). The curves B represent this dependence for cluster model B for both irreversible (solid curve) and reversible trapping of excitation energy by RC (broken curve). The horizontal lines A show the trapping time values for a corresponding uniform isotropic model A, again, for irreversible (solid line) and reversible trapping (broken line).

It is noteworthy that anisotropy of model B at R/r = 1 makes this model relatively inefficient compared to the corresponding uniform isotropic model A (where R/r = 1 also).

As can be seen from Fig. 2, the trapping time increases upon the increase in R/r ratio, in other words, formation of clusters leads to a significant decrease in efficiency of a PSU with inductive-resonance energy transfer among all its molecules. E.g., for $\varphi_t = 1.0$ and R/r = 3 the trapping time for model B is 1.6-fold greater than it is at R/r = 1, and 2.7-fold greater as compared to the corresponding uniform isotropic model A. The same effect is observed if an RC is not an absolute trap.

Fig. 3 shows results of computations made for larger PSUs with larger clusters. As can be seen from this figure, the trapping time for cluster model E increases when the distance between clusters of pigment molecules grows.

At R/r < 1.5, model E is slightly more efficient than model D. At R/r = 1.5 the efficiences of these models are equal. It should be noted that our calculations agree well with those reported earlier by Knox (1977) for the models D and E with a fixed ratio R/r = $\sqrt{2}$ and $\varphi_t = 1.0$. At R/r > 1.5, the trapping time is longer for the cluster model E as compared to the corresponding uniform isotropic model D. So, at R/r = 3 and $\varphi_t = 1.0$, t is 3.8-fold greater for E (see solid curve E) than for D (horizontal solid line D). At $\varphi_t = 0.5$ the difference is smaller (2.3-fold) yet significant.

Therefore, cluster formation in a PSU has a negative effect on the rate of energy transfer from antenna to RC in systems where the interaction among all the pigment molecules being weak lead to inductive-resonance energy transfer over all the molecules of PSU. At R/r values close to those in vivo, cluster formation decreases the rate of energy transfer from antenna to RC several times compared to a corresponding uniform isotropic model PSU of the same size and area.

Limit of Weak Intercluster and Strong Intracluster Interactions

The situation described above changes drastically if the intermolecular interactions within a cluster is strong enough that each cluster behaves as a single "supermolecule" and inductive-resonance energy transfer occurs between these "supermolecules".

Table 1 comprises the normalized values of trapping time t computed for the models A and C. The trapping time t for each dimeric model C has

Table 1. Normalized Values of the Time of Excitation Trapping by Reaction Centers in PSU Models A (Monomeric) and C (Dimeric) for Both Irreversible and Reversible Trapping

Antenna Pigment Species	PSU Models		
	dimeric[a] (C)	monomeric (A)	
		$\varphi_t = 1.0$	$\varphi_t = 0.5$
BChl a in vivo	1	3.4	3.3
BChl a in vitro	1	3.2	3.1
BChl c in vitro	1	2.2	2.1

[a]The time value for each dimeric model has been taken to be equal to 1 unit.

Table 2. Normalized Values of the Time of Excitation Trapping by Reaction Centers in PSU Models D (Monomeric) and F (Oligomeric) for Both Irreversible and Reversible Trapping

Antenna Pigment Species	PSU Models		
	oligomeric[a] (F)	monomeric (D)	
		$\varphi_t = 1.0$	$\varphi_t = 0.5$
BChl a in vivo	1	3.9	3.6
BChl a in vitro	1	3.7	3.4
BChl c in vitro	1	2.5	2.3

[a]The time value for each oligomeric model has been taken to be equal to 1 unit.

been taken to be equal to 1 unit. Table 1 shows that dimeric model C is more effective than monomeric one of the same size and area. Really, the trapping time for dimeric model is 2.1 - 3.4-fold smaller than that time for the corresponding monomeric isotropic model A.

The results of analogous computations made for larger model PSUs are shown in Table 2. Here the trapping time t for each oligomeric model F has been taken also to be equal to 1 unit. All conclusions made in the case of smaller PSUs with clusters of minimal size are valid for greater PSUs with larger clusters. The rate of energy transfer from antenna to RC in oligomeric model is 2.3 - 3.9-fold higher than in the corresponding monomeric isotropic model D.

Therefore, the oligomer or dimer formation in photosynthetic systems in vivo can be considered as one of the optimizing structural factors ensuring the high efficiency of excitation energy delivery to RC. The effect of other optimizing factors reported by us earlier (Fetisova, Fok, and Shibaeva, 1985; Fetisova and Shibaeva, 1987) can be extended to PSUs with such a clustered arrangement, assuming that N, having been earlier the number of PSU molecules, denotes now the number of clusters (i.e. oligomers). For PSUs of such a type, all quantitative results published by us previously are also valid.

CONCLUSIONS

The analysis performed enables us to conclude that:

(i) In the case of weak interactions between all the PSU molecules defining the inductive-resonance-type of energy transfer, the molecular cluster formation in PSU plays a negative role in the delivery of the excitation energy from the excited antenna molecules to the reaction center, i.e. slows it down.

(ii) In the case of strong interactions within each PSU cluster (i.e. upon oligomerization of molecules in the cluster allowing to consider each oligomer as a single "supermolecule"), and weak interactions between adjacent clusters, the molecular cluster formation in PSU accelerates the delivery of the excitation from the antenna to the reaction center and hence increases the efficiency of such a PSU as compared to that of the corresponding monomeric uniform isotropic PSU of the same size and area.

In accordance with available data, this is the situation that is probably realized in an vivo PSUs. Therefore, the molecular cluster formation in natural photosynthetic systems being a structural factor optimizing the excitation energy delivery from the antenna to the reaction center is biologically expedient.

REFERENCES

Fetisova, Z.G., and Fok, M.V., 1984, The ways of optimization of light energy conversion in primary steps of photosynthesis. I. Necessity of photosynthetic unit structure optimization and the method of its efficiency calculation (Engl. Transl.), Molek. Biol., 18 : 1354

Fetisova, Z.G., Fok, M.V., and Shibaeva, L.V., 1985, The ways of optimization of light energy conversion in primary steps of photosynthesis (Engl. Transl.), Molek. Biol., 19 : 802; 19 : 809; 19 : 1202; 19 : 1212

Fetisova, Z.G., and Shibaeva, L.V., 1987, Principles for designing optimal artificial light-harvesting molecular system, in: "Proceedings of the 1986 International Congress on Renewable Energy Sources", S. Terol, ed., C.S.I.C. Publisher, Madrid

Fetisova, Z.G., Freiberg, A.M., and Timpmann, K.E., 1988, Long-range molecular order as an efficient strategy for light harvesting in photosynthesis, Nature, 334 : 633

Hanson, L.K., 1988, Theoretical calculations of photosynthetic pigments,
Photochem. Photobiol., 47: 903

Knox, R.S., 1977, Photosynthetic efficiency and exciton transfer and trapping, in: "Topics in Photosynthesis", J. Barber, ed., Elsevier,
Amsterdam

Olson, J.M., Van Brakel, G.H., Gerola, P.D., and Pedersen, J.P., 1987, The
bacteriochlorophyll c dimer in carbon tetrachloride, in: "Progress in Photosynthesis Research", J. Biggins, ed., Martinus
Nijhoff, Dordrecht

Picorel, R., L'Ecuyer, A., Potier, M., and Gingras, G., 1986, Structure of
the B880 holochrome of Rhodospirillum rubrum as studied by the
radiation inactivation method, J. Biol. Chem., 261: 3020

Scherz, A., and Rosenbach-Belkin, V., 1989, Comparative study of optical
absorption and circular dichroism of bacteriochlorophyll oligomers in Triton X-100, the antenna pigment B850, and the primary
donor P-860 of photosynthetic bacteria indicates that all are
similar dimers of bacteriochlorophyll a, Proc. Natl. Acad. Sci.
USA, 86: 1505

EXCITATION ENERGY TRANSFER IN PURPLE PHOTOSYNTHETIC BACTERIA: ANALYSIS BY THE TIME-RESOLVED FLUORESCENCE SPECTROSCOPY

Mamoru Mimuro[1], Keizo Shimada[2], Naoto Tamai[3,4] and Iwao Yamazaki[3,4]

[1]National Institute for Basic Biology, Myodaiji, Okazaki, Aichi 444, [2]Department of Biology, Faculty of Science, Tokyo Metropolitan University, Fukazawa, Setagaya, Tokyo 164 and [3]Institute for Molecular Science, Myodaiji, Okazaki, Aichi 444, (Japan)

INTRODUCTION

Absorption of light is the primary event to drive photosynthesis in photosynthetic bacteria. The light energy is then transferred among antenna pigments and finally delivered to reaction center (RC) where the photochemical charge separation takes place. The RC polypeptides are known to be highly conserved through many kinds of photosynthetic organisms [1]; RC II in higher plants, "quinone type RC", is similar to RC of purple photosynthetic bacteria, whereas RC I, "Fe-S type RC", to that of (strictly) anaerobic green bacteria (*Chlorobium limicola* or *Heliobacterium chlorum*). On the other hand, the polypeptides of antenna pigment protein complex are divergent; only a partial similarity is suggested [2]. Physical basis to sustain the function of pigment protein complex, however, might be similar to each other, which is not clearly elucidated.

In purple photosynthetic bacteria, two types of pigment protein complexes are known; B800-B850 (or B800-B820) (so called LH2), B875 and B890 (LH1) whose presence and relative content depend on the species and growth conditions [2]. Recently, a new type of component was found spectroscopically (B896 [3,4,5,6] or B905 [7]), even though its chemical nature is not elucidated yet. The antenna complex consists of two types of polypeptides; α and β. Minimum functional unit is proposed to be $(\alpha\beta)_2$ [4] or $(\alpha\beta)_3$ [8]. This unit forms a higher order structure, giving rise to hexamer or dodecamer as a functional unit *in vivo* [4]. In the cells, the pigment protein complexes interconnect each other to yield a large domain where up to 3000 Bchls can function as common antenna to every RC [3,9,10].

For the analysis of energy transfer process, the critical point is the assumption of presence of components and their decay kinetics [11]. On the latter, two decay kinetics have been introduced to describe the time behaviour of components [12]; exponential decay or the decay proportional to the square root of time. Application of either of two kinetics depends on the structure of pigment systems and the interaction between the pigment molecules. In general, exponential decay is applicable to "trap-limited random walk process" and the decay proportional to the square root of time, to "diffusion-limited trapping process". Selection of decay kinetics is the critical point for the analysis. On the other hand, the estimation of number of components is not always easy based on the decay curves. Even if the global analysis in adopted, the

Present address: [4]Department of Chemical Process Engineering, Faculty of Engineering, Hokkaido University, Sapporo 060 (Japan).
Abbreviations used: Bchl, bacteriochlorophyll; RC, reaction center

Molecular Biology of Membrane-Bound Complexes in Phototrophic Bacteria
Edited by G. Drews and E. A. Dawes
Plenum Press, New York, 1990

365

number of components is sometimes to be assumed *a priori* [13]. Contrary to this, the time-resolved spectra will give clear changes in the spectrum without the assumption of decay kinetics. This method is more suitable for the multi-component system, like as the case of whole cells or pigment protein complexes.

In this study, we presented the energy flow process in purple photosynthetic bacteria analyzed mainly with the time-resolved fluorescence spectroscopy. We used two species of purple bacteria; *Rhodobacter sphaeroides* and an aerobic photosynthetic bacterium *Erythrobacter* sp. OCh 114 [14]. Compared with the former, the latter lacks one antenna component which corresponds to B850. Despite of the difference in the pigment composition, common features in the energy flow were clearly detected; fast energy transfer from LH2 to LH1 and the presence of the longer-wavelength antenna. Based on these results, the overall kinetics of energy transfer in the purple photosynthetic bacteria is discussed.

MATERIALS AND METHODS

Culture of bacteria: *Rhodobacter sphaeroides* 2.4.1 was grown photoheterotrophically at 30°C in a medium composed of 0.5 % (w/v) polypeptone, 0.1 % yeast extract and 0.4 % sodium lactate (pH 7.0). *Erythrobacter* sp. OCh 114 was grown heterotrophically at 25°C under the sufficiently aerated condition [15]. Cells at the late log-growth phase were collected, suspended in the buffer (10 mM MOPS (pH 7.0) for *Rb. sphaeroides* or 50 mM MOPS (pH 7.5) containing 0.34 M NaCl for *Erythrobacter* sp.) and used for measurements.

Time-resolved fluorescence spectroscopy and data analysis: Time-resolved fluorescence spectrum in the ps time range was measured with the apparatus reported previously [16,17]; in principle, the time-correlated single photon counting method [16]. The feature of our system is a detector, a microchannel-plate photomultiplier with the photocathode of so-called "S-1 type" (R-1564U-05, Hamamatsu Photonics, Japan). The excitation pulse was 6 ps-width (fwhm, 800 kHz) and its pulse intensity was in a range of 10^8 to 10^9 photons/cm^2, which is low enough to avoid singlet-singlet annihilation process. The time-resolution of this optical set-up was 6 ps. All the measurements were carried out at 22°C. Fluorescence spectrum was deconvoluted into component bands with the assumption of Gaussian-band shape and lifetimes of the resolved components were estimated by the convolution calculation with reference to the excitation pulse profile [17,18].

RESULTS

I. Excitation energy flow in *Rb. sphaeroides*

The pigment content of *Rb. sphaeroides* under our culture condition was estimated by the absorption spectrum shown in Fig. 1A; the ratio of B800-B850 to B875 was

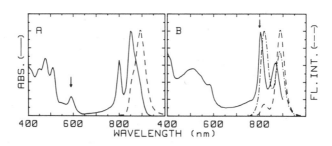

Fig. 1. Absorption and fluorescence spectra of *Rb. sphaeroides* (A) and *Erythrobacter* sp. OCh 114 (B) at room temperature at the steady-state. Full lines show the absorption spectra and the broken lines, fluorescence spectra. Arrows show the excitation wavelength. In (B), the fluorescence spectrum after the addition of Triton X-100 (0.1 %) was also shown by an altenate line.

about 2.5 [19]. Thus the excitation light at 590 nm was estimated to be absorbed by B800-B850 complex by more than 65 %, even if the difference in the extinction coefficients of two complexes at 590 nm was taken into account [4,20]. The time-resolved fluorescence spectra (Fig. 2A) show that the main fluorescence component just after the excitation pulse was located at 889 nm with a clear shoulder around 860 nm; no emission was observed in the wavelength region from 800 to 850 nm. It is readily to be interpreted that the 889 and 860 nm components arise from B875 and B850, respectively. A higher intensity from B875 than that from B850 even in the initial time range clearly indicates fast energy transfer from B850 to B875. The transfer time is shorter than the time resolution of the apparatus (6 ps). This fast transfer has been suggested [21,22], and resolved by Freiberg et al. [23] as the 8-ps component. Changes in the spectra with time is small; compared with the spectrum 335 ps after the excitation pulse (shown by dotted lines in Fig. 2A), two differences are clear; one is the decrease in the intensity of 860 nm band and the other, a clear blue-shifted main fluorescence band in the initial time range. The latter suggests the presence of a new component band in the longer wavelength region of the main fluorescence component. Within 100 ps, the spectrum became essentially the same as that in a later time range, indicating fast equilibration between components.

Fluorescence spectra were resolved into components by the following procedure. Difference in the spectra at two different times (Fig. 2B) indicates the presence of the components around 910 nm, in addition to the emissions from B850 and B875. With the assumption of these three components, the time-resolved fluorescence spectrum was deconvoluted into components. The components, thus resolved, were F859, F889, F909 and F955. The last one is most probably the vibrational band. The former three can be assigned to be B850, B875 and a new component corresponding to B896 [3,4,5].

The B896 was named according to the estimated locations at low temperature, contrary to other components (B800, B850 and B875), the observed locations at physiological temperature. Thus we estimated the location of B896 at physiological temperature by using the Stepanov equation [24]. Corresponding to the observed fluorescence maxima at 859, 889 and 909 nm, the calculated peaks for B850, B875 and a longer-wavelength antenna were 851, 879 and 894 nm, respectively. As for the measurements at −196°C, 875-, 911- and 925-nm fluorescence maxima correspond to 855-, 892- and 902-nm absorption bands, respectively (data not shown, see ref. 18). The locations of 851 and 879 nm at 22°C, and 855 nm at −196°C agree with the

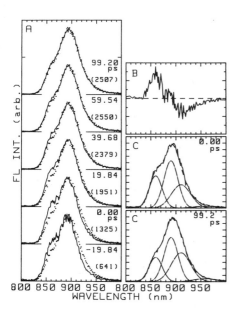

Fig. 2. Time-resolved fluorescence spectra of *Rb. sphaeroides* at 22°C under the excitation at 590 nm (A) and deconvolution of those spectra (C). In (A), each spectrum was normalized to the maximum intensity and numbers in parentheses indicate the maximum number of photons in the respective spectrum. Dotted lines show the spectrum 335 ps after the excitation pulse. (B) Difference spectra at two different times (0 ps *minus* 133.9 ps). (C) Deconvolution of the spectra with the assumption of Gaussian band shape as a function of wavenumber. Actual calcaulation was carried out with 1 nm interval. (———), observed spectrum; (— · —), component band and (— — —) sum of component bands.

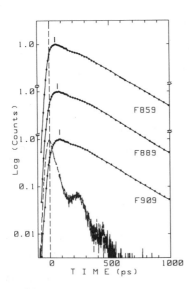

Fig. 3. Rise and decay curves of individual fluorescence components in *Rb. sphaeroides* at 22°C. Each point was obtained by the relative intensity of the component band and actual count of fluorescence. Broken line shows the pulse profile. Small bars over the kinetics indicate the time for the maximum intensity. For the lifetimes, see Table 1.

previous reports within the difference of 1 nm [25,26]. However our estimate 892 nm for B875 at −196°C are longer by 4 nm than the other report (888 nm) [26]. Thus we applied the same magnitude of blue-shift for the location of the longer-wavelength antenna and deduced its location to be at 890 nm. Hereafter, that component is called B890 as the physiological form in *Rb. sphaeroides*.

The rise and decay curves of individual fluorescence components were obtained based on the deconvoluted spectra (Fig. 3). The kinetics of three components are similar to each other; a difference was found in the time for the maximum intensity. A sequential shift was clearly observed in the order of F859, F889 and F909, indicating the sequential energy flow among these components.

Based on the observed decay kinetics, lifetimes of fluorescence components were estimated by convolution calculation using multi-exponential decay function with reference to the excitation pulse profile (Table 1). Note that the decay curve is pure in terms of component. The lifetime of the main decay component is 255 ± 10 ps, in agreement with the previous reports for that with the closed RC [3,6,7,21]. The rise term was not resolved in the kinetics of F859, however it was clearly found in the kinetics of F889 and F909; those were 20.0 and 35.0 ps, respectively. These rise terms can be interpreted as a generating process of the excited population in the case of closed trap, that is, the equilibration process among antenna molecules. The equilibration time between B890 and RC was not obtained, simply because the fluorescence from RC could not be detected in the cells.

Table 1. Lifetimes of fluorescence components in *Rb. sphaeroides* at 22°C.

Component	Rise term τ (ps)	Decay terms τ_1 (ps)	A_1	τ_2 (ps)	A_2
(B800)	−	−		−	
F859 (B850)	−	12.7	0.66	255	0.34
F889 (B875)	20.0	255	0.99	1017	0.01
F909 (B890)	35.0	255	0.98	1020	0.02

The equilibration time is the sum of rate constants for forward and backward transfer. As proposed by Zankel [27], the ratio of forward transfer time to backward transfer time betweem B850 and B875 is half. Thus 20-ps equilibration time gives rise to about 6 ps transfer time from B850 to B875. On the other hand, dominance of the F889 even at 0 ps indicates that the transfer time is shorter than 6 ps. This fast transfer is the main cause for kinetic mismatch between the decay of B850 and

368

the rise of the B875. The 6 ps is the boundary of time resolution. More accurate estimation of transfer time is required to clarify this point. On the transfer from B875 to B890, the 35 ps rise term can give about 10 ps transfer time with the same assumption of the ratio of energy transfer for forward and backward transfer. This estimation is essentially agreeable with the 15 ps transfer time measured at −196°C [28]. At this experimental stage, the transfer time from B890 to RC is not clear, however it could be in a range up to 20 ps (cf. ref. 18). Therefore, the overall transfer time can be estimated to be shorter than 50 ps, which is significantly shorter than the estimation by Hunter et al. [29].

The relative amount of B890 can be estimated by the method propoesed by Zankel [27] based on the observed fluorescence intensity in the equilibrium state. The ratio of integrated areas under each component fluorescence was 1.00:1.94:1.70 for F850:F889:F909 (Fig. 2C). Corresponding energy levels of individual components were expressed by their locations at 851, 878 and 890 nm, thus the Bchl population giving the above ratio of fluorescence yield is estimated to be 1.00:0.33:0.14 on Bchl basis. In the *Rb. sphaeroides* cells we used, the relative content of B800-B850 to B875(+B890) complex was about 2.5 [19]. In the B800-B850 complex, B850/B800 was 2 on Bchl basis [20]. In RC-B875(+B890) complex, the molar ratio of Bchl to RC is estimated to be 28, of which 4 belongs to RC and the remaining 24, to B875 plus B890 [30]. This stoichiometry gives a value of 0.51 for (B875+B890)/B850 in the cells, which is in good agreement with the value of 0.47 calculated from the deconvoluted spectra (Fig. 4). Based on the above results, the content of B890 per RC is estimated to be 7±1 Bchl with 17±1 Bchl being for B875. This B890 content is very close to the estimation by Sundstrom and van Grondelle [6], who reported the content to be 6 Bchl per RC.

II. Excitation energy flow in *Erythrobacter* sp. OCh 114

The feature of pigment system in *Erythrobacter* sp. OCh 114 is shown in Fig. 1B. Two prevailing absorption maxima were found at 806 nm and 870 nm; those are separated as the complexes, B806-complex and B870-RC complex. The relative content was estimated to be almost 1 to 1 on Bchl *a* basis. Bchl corresponding to B850, a common Bchl in purple photosynthetic bacteria, is missing in this species. Two complexes seems to interact loosely, because an addition of 0.1 % Triton X-100 induced the uncoupling of energy transfer (Fig. 1B). This feature leads to the idea that the transfer time from B806 to B870 is slow due to a lower spectral overlap

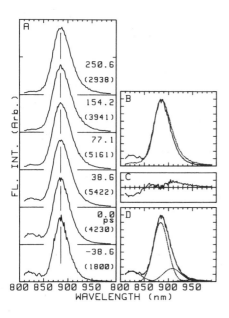

Fig. 4. Time-resolved fluorescence spectra of *Erythrobacter* sp. OCh 114 at 22°C under the excitation at 800 nm (A) and deconvolution pattern with a longer-wavelength antenna (D). (B) shows the spectra at two different times (——— , 0 ps) and (—··— , 250.6 ps), and (C) the difference spectrum between them.

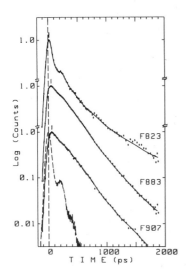

Fig. 5. Rise and decay curves of individual fluorescence components in *Erythrobacter* sp. OCh 114 at 22°C. Broken line shows the pulse profile. For the lifetimes, see Table 2. For details, see Fig. 3.

between the components. Thus energy flow in this species was carefully examined by the same method as for *Rb. sphaeroides*.

The time-resolved fluorescence spectra at 22°C with the preferential excitation of B806 by the 800-nm light pulse were shown in Fig. 4A. The main fluorescence was observed at 884 nm with a small band around 822 nm at 0 ps. The former originates from B870 and the latter, B806. It is remarkable that the emission from B870 was dominant even in the initial time range. This fact indicates the fast energy transfer from B806 to B870 in this species, and is the same as the observation in *Rb. sphaeroides*, despite of the difference in the pigment system. Time-dependent changes in the spectra is small. The main fluorescence is located at 884 nm up to 2.0 ns (data not shown). Changes in the spectra were detected only in the initial time range; after 80 ps, the spectra were almost the same, indicating the establishment of equilibrium between the components.

The time-resolved spectra were resolved into component bands by deconvolution. The bandwidth of the spectrum at 0 ps (Fig. 4B, full lines) is clearly narrower than that at 250.6 ps (altenate line). The difference spectrum between them (Fig. 4C) clearly indicates the presence of a new component around 910 nm. Thus, the spectrum could be simulated by three components with one vibrational band (Fig. 4D). Those were called F823, F883, F907 and F946. The F907, a longer wavelength antenna, corresponds to F909 in *Rb. sphaeroides*. The location of absorption maximum of F907 was estimated by the Stepanov equation [24] to be located at 888 nm, close to the case of *Rb. sphaeroides* (890 nm).

Rise and decay curves of individual components were shown in Fig. 5. The curve for F823 is a typical of a short decay component with an additional long-lived component. On the other hand, the decay kinetics of the F883 and F907 were similar to each other. A shift of the times for the maximum intesities were clearly observed in the order of F823, F883 and F907, indicating the sequentual energy flow among these components.

Table 2. Lifetimes of fluorescence components in *Erythrobacter* sp. OCh 114 at 22°C.

Components	Rise term			Decay terms			
	τ (ps)	τ_1 (ps)	A_1	τ_2 (ps)	A_2	τ_3 (ps)	A_3
F823 (B806)	–	19	0.95	210	0.045	910	0.005
F883 (B870)	–	191	0.902	385	0.098		
F907 (B888)	9	195	0.91	405	0.09		

Fluorescence lifetimes of individual components were estimated with the assumption of exponential decay (Table 2). Under B806 excitation condition, the rise of the F823 was not resolved; instead three decay components are necessary to describe the decay kinetics (19, 210 and 910 ps). Out of three, the 19-ps component might responsible for fast energy flow to B883. However, corresponding rise term was not resolved in the kinetics of the F883. Inconsistency of the decay kinetics between B806 and B870 and the absence of the rise term in the acceptor molecules (B870) clearly indicate that the energy transfer between B806 and B870 occurs within the time resolution of the apparatus (6 ps). The 19-ps decay time of the B806, thus, may not reflect the main energy flow but the residual part. The decay of F883 is biphasic; 191 ps and 385 ps. The major part most probably corresponds to the average lifetime of the excited molecules with the closed trap, as estimated in other photosynthetic bacteria (200 ps for *Rb. sphaeroides* and *R. rubrum*) [3,6,7,21]. In the decay of the F907, a rise term was resolved with the lifetime of 9 ps and the main decay was also 195 ps. The 9-ps component should be regarded as the shift of the equilibrium between F883 and F907, thus actual transfer time could be much shorter, probably in a range up to 3 ps.

The content of B888 per RC was estimated by the same method as for *Rb. sphaeroides* [27]. The ratio of fluorescenec yield of between B870 and B888 were estimated to be 3.1±0.2. Provided that the locations of B870 and B888 are to be at 870 and 888 nm, respectively, the ratio of number of molecules was calculated to be 10.0±0.7. The B870 content per RC was 26±2 Bchl *a* (Shimada et al., unpublished data). The molar ratio of B888, thus, was estimated to about 2, at most 3 per RC. This number is significantly smaller than that in *Rb. sphaeroides* (7±1) [18] or *R. rubrum* (6) [5,6]. This smaller number is responsible for lack of the red shift of the fluorescence maximum even in a later time range (Fig. 4A).

DISCUSSION

Two common features became clear in the energy transfer processes in two species of purple photosynthetic bacteria, despite of the difference in the constitution of pigment systems; fast energy transfer from LH2 to LH1 and the presence of a longer-wavelength antenna. We, therefore, discuss the reason for the above results, mainly based on the structure of the pigment systems (Fig. 6).

Fast energy transfer from LH2 to LH1

In *Rb. sphaeroides*, B800 is known to localize near the cytoplasmic side of the membranes, on the other hand, B850 and B875, near the periplasmic side where the special pair of RC is also localized [2,29]. B850 is the dimer, which is known by a clear couplet type of CD spectrum. The absorbed photon energy by B800 is transferred to the pigment in the other side of membrane (B850) and transferred through pigments in the periplasmic side (B875 and B890) to the RC. On the other hand, molecular topology of pigments is not clear in *Erythrobacter* sp. OCh 114. Spectroscopic properties and polypeptide composition of B870 and RC in *Erythrobacter* sp. OCh 114 are very similar, respectively, to the corresponding complexes of *Rb. sphaeroides* [15,31], and the LH1 content per RC (26 Bchl *a*) is almost the same as that of *Rb. sphaeroides* (24 Bchl *a*). The polypeptide for B806 is only one type as far as we studied [15] and its primary structure is not known yet. However this B806 shows a strong CD spectrum, suggesting the dimer structure [31], as similar to the B850 in *Rb. sphaeroides* [3]. A direct comparison of B806 with B850 is not straightforward, however fast energy flow from B806 to B870 strongly suggests that the B806 is equivalent to B850, not to B800, even if the location of the absorption maximum is close to that of B800. Thus it is reasonably proposed that the energy transfer in *Erythrobacter* sp. OCh 114 occurs only in the periplasmic side of the membranes.

Absence of the B850 in *Erythrobacter* sp. OCh 114 did not induce substantial slow down of energy transfer time. Due to the limitation of time resolution of the

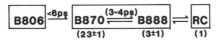

Rb.sphaeroides

| B800 | <1ps | B850 | <6ps | B875 | (~10ps) | B890 | ⇌ | RC |

B875 (17±1) B890 (7±1) RC (1)

Erythrobacter sp. OCh 114

| B806 | <6ps | B870 | (3~4ps) | B888 | ⇌ | RC |

B870 (23±1) B888 (3±1) RC (1)

Fig. 6. Schematic model for energy flow among pigment protein complexes in *Rb. sphaeroides* and *Erythrobacter* sp. OCh 114. Numbers under the components show the relative contents of Bchl *a* per RC.

apparatus (6 ps), we could not identify the real difference in the two bacterial systems. The exact transfer time should be measured to elucidate the funciton of B850 in the energy transfer sequence. One possible explanation for the fast energy flow in the *Erythrobacter* sp. OCh 114 is a tight coupling between LH1 and LH 2; the output of LH2 is specifically interacts with the input of LH 1, which can compensate the absence of B850.

Presence of a longer-wavelength antenna

In both species, a longer-wavelength antenna component was clearly detected by the time-resolved fluorescence spectra; B890 in *Rb. sphaeroides* and B888 in *Erythrobacter* sp. OCh 114. Its presence is also known in *R. rubrum* [5,6,18] and *R. palustris* [18]. Thus the presence of a longer-wavelength antenna can be ascribed to a common feature in the pigment system of purple photosynthetic bacteria. The locations of absorption maximum of this special pigment is temperature dependent as similar to other antenna pigments; at −196°C, those are estimated to be located at 898 nm for *Rb. sphaeroides* and 890 nm for *Erythrobacter* sp. (Mimuro et al., unpublished data).

The chemical nature of the longer-wavelength antenna is not known simply because it has not been isolated yet. It is supposed that this pigment is to be formed only in the complex, or by a specific interaction between LH1 and RC complexes. Difference in the number of molecules would indicate that difference in the interacting site determines the number of molecules; in *Erythrobacter* sp. OCh 114, only one side of the RC is the site responsible for it, whereas in *Rb. sphaeroides*, both side of RC are responsible for the B890.

The function of a longer-wavelength antenna might not necessarily be the same in both species; when it is abundant, its function can be assumed to concentrate the excitation energy around RC when it is closed. However when a longer-wavelength antenna is not abundant, the above function will become minor. Instead, we tends to propose an additional function of the longer-wavelength antenna; a structural factor to keep the LH1-RC complex or a functional factor to give a supplemental conditions necessary for the charge separation and/or stablization process. The exact function of the longer-wavelength antenna is not clear at this experimental stage, thus it should be resolved from the structural and also functional point of view.

Decay kinetics in bacterial antenna system

As shown by rise and decay kinetics of fluorescence components (Figs. 3 and 5), exponential decay is most probable to describe the kinetics in bacterial antenna systems. This is remarkable difference from the kinetics in phycobilin-chl *a* system [16]. The decay kinetics depend on the structure of pigment system and also on the interaction between pigments. The transfer time between individual Bchl molecules is fast, probably in the order of 1 ps, and energy migration to RC takes place within 50 ps after the excitation of the pigments in the highest energy level. In RC, the charge separation takes place in about 3 ps [32] and the charge will be stabilized by the electron transfer to quinone (Q_A) with the rate constant of about 200 ps [32]. The average decay time of fluorescence with the closed RC (200 ps) [3,6,21] might reflect this process. Contrary to this, when the RC is open, the average decay time is known to be 60 ps [3,6,21] which originates from antenna Bchl but not from RC itself. This rate constant is close to our estimation of overall transfer time to RC. Freiberg et al. [23] reported the same overall transfer time with the estimation of the transfer time from B875 to B890 to be 25 ps. Thus, if the overall energy transfer is the rate-limiting step, this 60 ps lifetime reasonably explain the time behaviour of the fluorescence components. This situation should be called "diffusion-limited" process, not the "trap-limited" process. This interpretation leads to the idea that the decay kinetics strongly depend on the measuring conidition with reference to the open/closed state of RC. In general, it is known that the diffusion-limited process can be described by the kinetics proportional to the square root of time [12]. This discrepancy should be deeply considered.

Acknowledgement

The authors thank the Instrument Center, Institute for Molecular Science for the operation of picosecond spectroscopy. This work is supported in part by the Grant-in-Aid for the Scientific Research from the Ministry of Education, Science and Culture, Japan to MM (62540520) and KS (60304007).

REFERENCES

1. Blankenship, R.E., Brune, D.C., Freeman, J.M., King, G.H., McManus, J,H., Nozawa, T., Trost, J.T. and Wittmershaus, B.P. (1988) in Green Photosynthetic Bacteria, (Olson, J.M., Ormerod, J.G., Amesz, J., Stackebrandt, E. and Truper, H.G. eds.), pp. 57-68, Plenum Press, New York.

2. Zuber, H. (1987) in The Light Reactions (Barber, J. ed.), pp. 197-259, Elsevier, Amsterdam.

3. van Grondelle, R. (1985) Biochim. Biophys. Acta 811, 147-195.

4. Kramer, H.J.M., Pennoyer, J.D., van Grondelle, R., Westerhuis, W.H.J., Niederman, R.A. and Amesz, J. (1984) Biochim. Biophys. Acta 767, 335-344.

5. van Grondelle, R. and Sundstrom, V. (1988) in Photosynthetic Light-Harvesting Systems; Organization and Function, (Scheer, H. and Schneider, W. eds.), pp. 403-438, Walter de Gryuter, Berlin.

6. Sundstrom, V., van Grondelle, R., Bergstrom, H., Akesson, E. and Gillbro, T. (l986) Biochim. Biophys. Acta 851, 431-446.

7. van Grondelle, R., Hunter, C.N., Bakker, J.G.C. and Kramer, H.J.M. (1983) Biochim. Biophys. Acta 723, 30-36.

8. Borisov, A.Yu., Gadonas, R.A., Danielius, R.V., Piskarskas, A.S. and Razjivin, A.P. (1982) FEBS Lett., 138, 25-28.

9. Hunter, C. N., Kramer, H.J.M. and van Grondelle, R. (1985) Biochim. Biophys. Acta 807, 44-51.

10. Vos, M., van Dorssen, R.J., Amesz, J., van Grondelle, R. and Hunter, C.N. (1988) Biochim. Biophys. Acta 933, 132-140.

11. Mimuro, M. (1988) in Photosynthetic Light-Harvesting Systems; Organization and Function, (Scheer, H. and Schneider, W. eds.), pp. 589-600, Walter de Gruyter, Berlin.

12. General Discussion in Chlorophyll Organization and Energy Transfer in Photosynthesis (1978) Ciba-Foundation Symposium 61, pp. 341-364, Excerpta Medica, Amsterdam.

13. Beechem, J.M. and Brand, L. (1986) Photochem. Photobiol., 44, 323-329.

14. Harashima, K., Shiba, T. and Murata, N. (1989) in Aerobic Phototsynthetic Bacteria, Japan Scientific Societies Press, Tokyo.

15. Shimada, K., Hayashi, H. and Tasumi, M. (1985) Arch. Microbiol., 143, 244-247.

16. Yamazaki, I., Mimuro, M., Murao, T., Yamazaki, T., Yoshihara, K. and Fujita, Y. (1984) Photochem. Photobiol. 39, 233-240.

17. Mimuro, M., Yamazaki, I., Itoh, S., Tamai, N. and Satoh, K. (1988) Biochim. Biophys. Acta 933, 478-486.

18. Shimada, K., Mimuro, M., Tamai, N. and Yamazaki, I. (1989) Biochim. Biophys. Acta, 975, 72-79.

19. Itoh, M., Matsuura, K., Shimada, K. and Satoh, T. (1988) Biochim. Biophys. Acta, 936, 332-338.

20. Kramer, H.J.M., van Grondelle, R., Hunter, C.N., Westerhuis, W.H.J. and Amesz, J. (1984) Biochim. Biophys. Acta 765, 156- 165.

21. Sebban, P., Jolchine, G. and Moya, I. (1984) Photochem. Photobiol. 39, 247-253.

22. Borisov, A.Yu., Freiberg, A., Godik, V.I. Rebane, K.K. and Timpmann, K.E. (1985) Biochim. Biophys. Acta 807, 221-229.

23. Freiberg, A., Godik, V.I., Pullertis, T. and Timpman, K. (1989) Biochim. Biophys. Acta, 973, 93-104.

24. Stepanov, B.I. (1957) Dokl. Acad. Nauk USSR 112, 839-841.

25. Goedheer, J.C. (1972) Biochim. Biophys. Acta, 275, 169-176.

26. Sebban, P., Robert, B. and Jolchine, G. (1985) Photochem. Photobiol., 42, 573-578.

27. Zankel, K.L. (1978) in The Photosynthetic Bacteria (Clayton, R.K. and Sistrom, W.R., eds.), pp. 341-347, Plenum Press, New York.

28. Bergstrom, H., Westerhuis, W.H.J., Sundstrom, V., van Grondelle, R., Niederman, R.A. and Gillbro, T. (1988) FEBS Letters, 233, 12-16.

29. Hunter, C.N., van Grondelle, R., and Olsen, J.D. (1989) Trends Biochem. Sci., 14, 72-76.

30. Matsuura, K. and Shimada, K. (1986) Biochim. Biophys. Acta 852, 9-18.

31. Hayashi, H., Shimada, K., Tasumi, M. Nozawa, T. and Hatano, M. (1986) Photobiochem. Photobiophys., 10, 223-231.

32. Ke, B. and Schuvalov, V.A. (1987) in The Light Reaction, (Barber, J. ed.), pp. 31-93, Elsevier, Amsterdam.

BIOCHEMICAL EVIDENCE FOR CHROMOPHORE–CHROMOPHORE INTERACTIONS AS THE MAIN ORGANIZATIONAL PRINCIPLE IN CHLOROSOMES OF CHLOROFLEXUS AURANTIACUS

Kai Griebenow and Alfred R. Holzwarth

Max–Planck–Institut für Strahlenchemie
Stiftstraße 34 – 36
D–4330 Mülheim/Ruhr, F.R.G.

INTRODUCTION

Chlorosomes are the main light–harvesting antennae of Chlorobiaceae and Chloroflexaceae. Chlorosomes from C. aurantiacus have been prepared in the past only by methods based on density gradient centrifugation (SDGC) (1–3) without detergents (2) or with miranol (1) as detergent. These chlorosomes always contain BChl a which was thought to be a part of the so–called baseplate (1) which attaches the chlorosomes to the cytoplasmic membrane (CM). Studies by PAGE of these chlorosomes showed the occurance of at least four proteins in the chlorosome band isolated from the gradient. Different values have been reported for the M_r of these proteins. Schmidt et al. (4) resolved only two proteins with M_r of 10 and 15 kD while Sprague et al. (3) reported three proteins with M_r of 13.7, 12.4 and < 12.4 kD (approx. 10 kD from the Fig. given). Feick et al. (5) resolved four proteins with M_r of 18, 11, 5.8 and 3.7 kD. The very different values reported particularly for the M_r of the smallest protein may be explained by the different gel–systems and marker proteins used but could also indicate differences between these preparations.

The model for chlorosomes developed by Wechsler et al. (6) is based on the study by Feick et al. (5) where it was suggested from results obtained by various techniques that their 3.7 kD protein should be the BChl c–binding protein in chlorosomes. A BChl c/5.6 kD protein ratio of 5 – 8 was calculated. Wechsler et al. (6) sequenced this protein and found an absolute molecular weight of 5592 Dalton. The protein was thus renamed to 5.6 kD protein. New spectroscopic results from artificial BChl c aggregates in organic solvents (7,8) are difficult to reconcile with this protein–chromophore complex model (6) but would seem to fit better to a chromophore–chromophore interaction model (9,10).

The present work therefore aims at elucidating the structure and composition of chlorosomes in order to distinguish between these two models. The key to this problem consists in the analysis of the protein patterns of chlorosomes isolated in different ways.

Molecular Biology of Membrane-Bound Complexes in Phototrophic Bacteria
Edited by G. Drews and E. A. Dawes
Plenum Press, New York, 1990

375

C. aurantiacus was grown and chlorosomes free from BChl a_{790} were prepared as described before (11). Membranes were prepared according to (11) and further purified by centrifugation. The resulting pellet was resuspended in Tris–HCl–ascorbate buffer 20 mM Tris, 2 mM ascorbate, pH 8.0 (TAB) and used for the isolation of chlorosomes by SDGC. Chlorosomes with attached BChl a_{790} were prepared by separation on a continuous sucrose gradient after incubation of the isolated membranes with various detergents. The detergents used were miranol, deriphat–160 (lauryl β– iminodipropionic acid), LDAO (lauryl dimethylamine– N –oxide) and DDM (dodecyl β– D –maltoside). Membranes were incubated with the detergent at $4^{o}C$ in the dark under slow stirring and layered on a gradient. The chlorosome band was collected with a Pasteur pipette. A second SDGC purification step normally followed using a 20 – 40 % sucrose gradient (w/v). Optimal detergent/BChl c ratios were 0.054 µl miranol per nmol BChl c and molar ratios of 13:1 for deriphat, 98:1 for LDAO and 9:1 for DDM.

Polyacrylamide gel electrophoresis (PAGE) was carried out both on the Pharmacia PhastSystem^{SM} using 0.1 mm thick gels and on a conventional vertical system from Atlanta (SE 600). With the PhastSystem we used the SDS buffer strips and 8 – 25% commercial gels from Pharmacia. The gels were run as described in the instruction manual. Silver staining was performed as described by Heukeshoven et al. (12) and resulted in a very high sensitivity. The sensitivity was tested using marker proteins from Sigma (α–Lactalbumin 14.2 kD, Trypsin Inhibitor 20.1 kD, and Albumin 66 kD) for different regions of the gels. The tested sensitivity of a single protein band on the gel was calculated to be minimal 1 ng per band but in most cases 0.5 ng can be detected easily. The 5.6 kD protein, which was kindly provided by Prof. Zuber, ETH Zürich, was also detectable at a concentration of 1 ng per band. Incubation of the sample was performed by boiling for 10 minutes in a water bath. The incubation buffer contained 5% LDS, 250 mM Tris–HCl, pH 8.8, 2% dithiothreitol, 20% sucrose, and a few drops of bromophenolic blue.

With the conventional Atlanta system we used 1 mm thick gels (18 x 16 cm). The gel system was a modified Laemmli system (13) as described by Feick et al. (1). Staining was carried out either by coomassie staining or the more sensitive silver staining. The silver staining was an improved procedure according to refs. (12,14,15) with modifications accounting for the special problems encountered in this case. In detail the procedure consisted of the following steps:
i) fixation in 10% EtOH, 5% HAc, over night;
ii) 1% glutaraldehyde, 45 min;
iii) 10% EtOH, 5% HAc, 30 min;
iv) 3 times destilled water for 15 min;
v) 0.2% $AgNO_3$, 25 min;
vi) washing once with destilled water ;
vii) washing with 1.25% Na_2CO_3, 371 µl formaldehyde per liter for a very short time (washing away silver from the surface);
viii) developing for 30 min as in vii);
ix) reducing background with 1.5 g $K_3[Fe(CN)_6]$, 3.0 g $Na_2S_2O_3$, 0.5 g Na_2CO_3 per liter time dependent on background, reducing very carefully.
x) 3 times destilled water (10 min);
xi) 5% glycerol, 30 min;

For membranes both staining methods were tested. Silver staining both increases the intensity of each band stained as compared to staining by coomassie and also additional bands appear. Thus all proteins appear to be stained by the silver method. Relative molecular weights (M_r) were calculated according to Weber et al. (16) using the following calibration proteins obtained from Sigma: α–lactalbumin 14.2 kD, trypsin inhibitor 20.1 kD, trypsinogen 24 kD, carbonic anhydrase 29 kD, glycerinealdehyde–3–phosphate dehydrogenase 36 kD, albumin from egg 45 kD and bovine serum albumin 66 kD. For calculation of M_r for proteins with low molecular weights we have additionally used the following marker proteins obtained from Serva: carbonic anhydrase 29 kD, trypsin inhibitor from soy bean 21 kD, cytochrome c 12.5 kD and trypsin inhibitor from lung 6.5 kD. Before PAGE all chlorosome samples were adjusted to the same BChl c concentration. This procedure should simplify the identification of the BChl c–binding protein in different preparations.

Calculation of the pigment content, absorption– and fluorescence spectroscopy was carried out as described by us earlier (11).

<u>Sensitivity Estimates for Silver Staining and the PhastSystem</u>

The previously reported BChl c:5.6 kD protein ratio is 5 – 8:1. Thus at a concentration of 5 – 8 nmol/ml BChl c the sample would contain 1 nmol/ml protein or 5.6 µg/ml for the 5.6 kD protein. On the PhastSystem we use 1 µl of sample per lane corresponding to 5.6 ng of the 5.6 kD protein. At a staining sensitivity of 1 ng protein/band the minimal concentration of BChl c which must be present in the incubated sample must therefore be approx. 1 – 2 nmol/ml. The pure 5.6 kD protein (donated by Prof. Dr. Zuber) is clearly detectable at an amount of 1 ng on the gel. Other tested proteins with different molecular weights are also in all cases detectable with a minimum sensitivity of 1 ng per band.

RESULTS AND DISCUSSION

Fig. 1 shows the main proteins which can be found in different chlorosome preparations isolated by SDGC. At least three proteins can be identified in all preparations shown in lanes 1 – 5 of Fig. 1. Calculations of the relative molecular weights (M_r) using the Sigma proteins gave values of 11500 (s=360), 14000 (s=300), 14500 (s=100) and 17000 (s=400) Dalton for the proteins 1 to 4. Calculations with Serva proteins gave values 4 – 5 kD, 12.5 kD, 13 kD and 17 kD for these proteins. The results using Sigma markers thus correspond more to the results reported by Schmidt et al. (4) and Sprague et al. (3), while results obtained relative to Serva marker proteins agree more with the results of Feick et al. (5). Using 8 – 25% Pharmacia PhastGels we have calculated a M_r of 8.5 kD for the "BChl c–binding" 5.6 kD protein. Because of the different M_r values obtained with different gel systems we have also compared the 5.6 kD protein on the different gel–systems as an internal standard. This protein always represents to the smallest protein found on the gels (calculated M_r of 4 – 5, 8.5, 11 kD) which is identical to protein 1.

The vertical contaminations (lanes 1 – 5, Fig. 1; lanes 1 – 6, Fig. 3) are caused by a minimal background reduction. Stronger background reduction (Fig. 1, lanes 6 – 10) decreases the sensitivity of the staining substantially. The 5.6 kD protein, which is not the most concentrated protein, is not always detectable after stronger background reduction. The various results can be collected as follows:

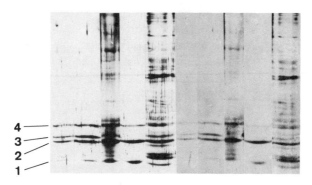

Fig. 1. PAGE (conventional 1 mm gel, 15%) of chlorosomes obtained by SDGC using different detergents after the first purification step. Lanes 1 and 2: Miranol, lane 3: Deriphat, lane 4: LDAO, lane 5: membranes. Lanes 6 – 10 correspond to lanes 1 – 5 after stronger background reduction. All samples were adjusted to 10 nmol BChl c. 50 µl were used for each lane.

i) The 5.6 kD protein (protein 1) is always easily detectable in membrane fractions and also in fractions of chlorosomes with high CM–contaminations (Fig. 1, lanes 5 and 10; Fig. 2, lane 1; Fig. 3, lane 7; Fig. 4, lanes 1 and 3). A first purification step by SDGC and various detergents often decreases the concentration of the 5.6 kD protein (Fig. 1, lanes 1,2,6,7; Fig. 2, lane 3) drastically as compared to membranes. Since all samples were adjusted to the same BChl c concentration the BChl c–binding protein should be found in membranes, CM contaminated chlorosome fractions, and purified chlorosomes in approx. the same concentration.

ii) Further purification of chlorosomes and chlorosomes with CM contaminations (bands 2 and 3 from the first gradient) reduces the concentration of all proteins (Fig. 2). This also

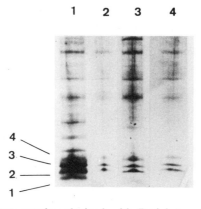

Fig. 2. PAGE of different samples obtained with Deriphat under optimal conditions on a 8 – 25% gradient gel (0.1 mm, PhastSystem). All samples were adjusted to 5 nmol/ml BChl c. Lane 1 shows band 3 (chlorosomes and CM contamination) and lane 2 further purified band 3, lanes 3 and 4 chlorosomes (enriched and further purification of band 2).

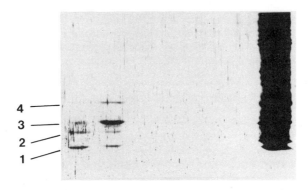

Fig. 3. PAGE (conventional 1 mm gel, 15%) of chlorosomes using a combination of Miranol and DDM treatment in comparison to GEF–chlorosomes. Lane 1: further purification with DDM, lane 2: Miranol, lanes 3 – 6: GEF–preparations, lane 7: membranes. Samples in lanes 1 – 5 were adjusted to 25 nmol BChl c/ml, the sample of lane 6 to 45 nmol BChl c/ml, membranes to 20 nmol BChl c/ml. 50 µl sample were used for each lane.

applies to the 5.6 kD protein. All samples were again adjusted to the same Bchl c concentration.

   iii) The 5.6 kD protein is drastically reduced and nearly undetectable in BChl a–free GEF–chlorosomes (Fig. 3, lanes 3 – 6; Fig. 4, lanes 2,4 – 7). In Fig. 3 the GEF samples are adjusted to the same Bchl c concentration as the SDGC preparation. In Fig. 4 the 5.6 kD protein was present in lane 8 with 1 ng. Based on these results and taking into account our

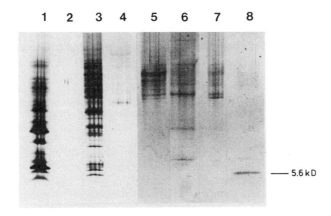

Fig. 4. PAGE of GEF–chlorosomes (lanes 2 and 4 – 7) and chlorosome enriched samples obtained with DDM (lane 1) or Deriphat (lane 3) on a 8 – 25% gradient gel (PhastSystem, 0.1 mm). For comparison the 5.6 kD protein is presented in lane 8 with a concentration of about 1 ng. All samples were adjusted to 12.5 nmol BChl c/ml before PAGE. As clearly seen GEF–chlorosomes do not contain detectable proteins with molecular weights below 20 kD.

theoretical estimates given above we can calculate from the PhastGels (8 – 25%) were we have used 10 nmol BChl c/ml for the PAGE (sensitivity 1 ng/band) a ratio of BChl c/5.6 kD protein which must be at least a factor of five higher than estimated by Feick et al. (5), i. e. minimal 25 – 40 BChl c molecules per 5.6 kD protein. From conventional 1 mm gels we can calculate a similar factor of minimal 4.5. The 5.6 kD protein was always detectable in SDGC preparations and membranes adjusted to 10 nmol BChl c/ml before PAGE. GEF preparations adjusted to 25 nmol BChl c/ml before PAGE never showed this protein (factor 2.5) and even with a preparation adjusted to 45 nmol BChl c/ml the protein was not detectable (factor 4.5).

No other proteins can be found in GEF–preparations either which could serve the function for BChl c–binding with respect to the model of Wechsler et al. (6). Sometimes different amounts of high molecular weight proteins can be found but the patterns of these contaminations vary between different preparations. Thus these proteins are contaminations which were not totally removed by the GEF–procedure.

Our results show that it is possible to reduce and even remove the so–called BChl c–binding protein by various techniques from chlorosomes with and without BChl a_{790}. All or most of the protein is removed by the GEF–procedure. The protein concentration in this case is almost below the sensitivity of the staining procedure what is never the case for "normal" preparations and membranes. It is essential that in all cases where the 5.6 kD protein is present in slightly reduced concentration or even undetectable, the normal spectroscopic properties of the chlorosomes are fully contained, except for the absence of BChl a in GEF–chlorosomes. Thus the 5.6 kD protein can be excluded to be the BChl c–binding protein in the sense of Feick et al. (5) and Wechsler et al. (6). No other protein can be found either which can serve alternatively the function of the BChl c–binding protein. Thus our results present strong biochemical evidence for a model where chromophore–chromophore interactions play the key role for the pigment organization. The function of the 5.6 kD protein is not clear.

Chlorosomes of C. aurantiacus and perhaps also those of other organisms to our knowledge present the first photosynthetic antenna system whose organizational principle is not based preferentially on chromophore–protein interaction. Our results demand for a new model of the chlorosome structure which is based preferentially on the interaction of large aggregates of BChl c–chromophores. The small amount of protein probably present in chlorosomes might fulfill the function of ordering these chromophore aggregates relative to each other and may also serve as terminal elements in these aggregates. In vitro formation of BChl c aggregates with spectroscopic properties similar to those of the BChl c part of chlorosomes has been demonstrated by several authors (9,10). Our results present the first evidence showing that such aggregates are indeed present and functional in intact chlorosomes.

ACKNOWLEDGEMENTS

We acknowledge Mrs. B. Kalka, Mrs. A. Keil and Mr. S. Emunds for valuable technical support. We also should like to thank Prof. K. Schaffner for his interest and support of this work. Some results are part of a diploma work (K.G.) accepted by the Philipps–Universität Marburg, FRG.

REFERENCES

1. Feick,R.G., Fitzpatrick,M. and Fuller,R.C. J. Bacteriol. 150, 905–915 (1982).

2. Schmidt.K. Arch. Microbiol. 124, 21–31 (1980).

3. Sprague,S.G., Staehelin,A., DiBartolomeis,M.J. and Fuller,R.C. J. Bacteriobiol. 147, 1021–1031 (1981).

4. Schmidt,K., Maarzahl,M. and Mayer,F. Arch. Microbiol. 127, 87–97 (1980).

5. Feick,R.G. and Fuller,R.C. Biochemistry 23, 3693–3700 (1984).

6. Wechsler,T., Suter,F., Fuller,R.C. and Zuber,H. FEBS Lett. 181, 173–178 (1985).

7. Blankenship,R.E., Brune,D.C., Freeman,J.M., Trost,J.T., King,G.H., McManus,J.H., Nozawa,T. and Wittmershaus,B.P. in Green Photosynthetic Bacteria (eds: Olson, J.M., Ormerod, J.G., Amesz, J., Stackebrandt, E. and Trüper, H.G.) 57–68 (Plenum Press, New York 1988).

8. Lutz,M. and van Brakel,G. in Green Photosynthetic Bacteria (eds: Olson, J.M., Ormerod, J.G., Amesz, J., Stackebrandt, E. and Trüper, H.G.) 23–34 (Plenum Press, New York 1988).

9. Bystrova,M.I., Mal'gosheva,I.N. and Kranovskii,A.A. Mol. Biol. 13, 440–451 (1979).

10. Smith,K.M. and Kehrs,K.A. J. Am. Chem. Soc. 105, 1387–1389 (1983).

11. Griebenow,K. and Holzwarth,A.R. Biochim. Biophys. Acta 973, 235–240 (1989).

12. Heukeshoven,J. and Dernick,R. Electrophoresis 9, 28–32 (1988).

13. Laemmli,U.K. Nature 227, 680–685 (1970).

14. Heukeshoven,J. and Dernick,R. Electrophoresis 6, 103–112 (1985).

15. Heukeshoven,J. and Dernick,R. Elektrophorese Forum 86, Diskussionstagung Okt. 1986: 22–27.

16. Weber,K. and Osborn,M. J. Biol. Chem. 244, 4406–4412 (1969).

PICOSECOND ENERGY TRANSFER KINETICS BETWEEN DIFFERENT PIGMENT POOLS IN CHLOROSOMES FROM THE GREEN BACTERIUM CHLOROFLEXUS AURANTIACUS

Kai Griebenow, Marc G. Müller and Alfred R. Holzwarth

Max–Planck–Institut für Strahlenchemie
Stiftstr. 34 – 36
D–4330 Mülheim/Ruhr, F.R.G.

INTRODUCTION

Chloroflexus aurantiacus, a thermophilic green bacterium, contains at least four different bacteriochlorophyll (BChl)–complexes which are coupled in a specific way to optimize the energy transfer from the main antenna, the so–called chlorosome, to the reaction center (1–4). Chlorosomes contain about 1000 – 16000 BChl c molecules (5) which are believed to be organized in rod–like substructures (6). Two BChl a–protein complexes function as intermediate pigment pools. The BChl a_{790} complex is believed to be the first and the B 806–866 complex the second one in the energy transfer chain from the chlorosome to the reaction center (1,2). Energy transfer studies have been carried out in the past using steady state measurements (1,2), as well as picosecond absorption (7), and picosecond fluorescence measurements (2,8). The samples used were either membranes and whole cells or isolated chlorosomes containing BChl a_{790}. We have recently reported on the preparation of chlorosomes free from bound BChl a_{790} (9). In this report we compare the energy transfer kinetics in both types of chlorosomes.

MATERIAL AND METHODS

C. aurantiacus was grown and membranes were isolated as described before (9). Chlorosomes free from BChl a (GEF–chlorosomes) were prepared as described (9). Chlorosomes containing BChl a were prepared according to Feick et al. (10) with the following modifications: Miranol was used with an detergent:BChl c ratio of 0.054 µl miranol per nmol BChl c. Alternatively we used the detergent deriphat–160 for preparations of chlorosomes with an optimal molar ratio deriphat/BChl c of 13:1. Detergent incubation of the membranes was carried out for one hour at $4°C$ under slow stirring in the dark. Before incubation the membranes were further purified by pelleting and resuspending them. Chlorosomes were separated by sucrose density gradient centrifugation (SDGC) on a 10 – 40% sucrose gradient (w/v). The isolated chlorosome band was subjected to an

Molecular Biology of Membrane-Bound Complexes in Phototrophic Bacteria
Edited by G. Drews and E. A. Dawes
Plenum Press, New York, 1990

383

additional purification step by a second detergent incubation followed by a second SDGC. Chlorosomes obtained with deriphat as detergent were similar to those obtained with miranol.

Picosecond fluorescence decays were measured with a single–photon–timing apparatus (11). The decays were recorded at several emission wavelengths and analyzed by a global analysis procedure (11). The wavelength dependence of the corresponding amplitudes were plotted as decay–associated spectra (DAS).

RESULTS AND DISCUSSION

Fig. 1 shows the typical DAS of chlorosomes without BChl a. Four exponentials are required to fit the fluorescence decays over the whole wavelength range measured in each case.

BChl a–Free Chlorosomes

In BChl a–free GEF–chlorosomes the shortest lifetime is approx. 5 ps and the DAS shows both a positive and a negative amplitude. This finding is typical for an energy transfer between two different pigment pools or for a transition between two excited states. A second component with prominent amplitude has a lifetime of about 11 – 16 ps. Two longer–lived components with low amplitudes have lifetimes of about 30 – 50 ps and 450 ps. Two interpretations for these results are possible:

i) The ultrafast 5 ps component describes an energy transfer between two distinct BChl c pigment pools. Further evidence for this interpretation is provided by the temperature

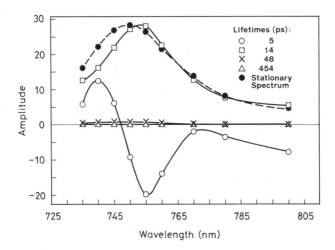

Fig. 1. Decay–associated spectra of BChl a–free chlorosomes from C. aurantiacus as obtained by global analysis of the fluorescence decays measured at room temperature (λ_{exc} = 721 nm).

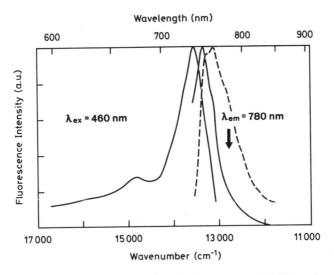

Fig. 2. Corrected fluorescence emission and excitation spectra of BChl a–free chlorosomes. The T = 85 K emission spectrum of GEF chlorosomes is shown in dashed line.

dependence of the fluorescence (Fig. 2) and some results obtained with chlorosomes containing BChl a (see below). The temperature dependence of the fluorescence emission spectrum shows the existence of two emission bands at 750 and 760 nm. Lowering the temperature shifts the emission intensity maximum continuously from 750 to 760 nm. Chlorosome isolates containing BChl a sometimes only show the fluorescence band at 760 nm and not that at 750 nm.

ii) The ultrafast 5 ps component reflects the relaxation between different states of excitonically coupled BChl c pigments. It is known that the BChl c molecules in chlorosomes are strongly coupled both from measurements of fluorescence annihilation (7) and from CD–spectra (12) which show typical shapes of excitationally coupled pigments.

At present we can not exclude any of these possibilities or even a combination of the two as an interpretation for the observed picosecond signals.

BChl a–Containing Chlorosomes

BChl a–containing chlorosomes (Fig. 3) show a short lifetime of about 11 ps which spectrally differs from the 5 ps component of GEF–chlorosomes. The positive part of the DAS has a similar shape as the BChl c emission, the negative part is a mirror image of the BChl a emission. Therefore this component must be attributed to the BChl c → BChl a_{790} energy transfer. The other components with lifetimes of approx. 27, 210 and 420 ps are attributed to more or less efficiently quenched excited states of either BChl c and/or Bchl a. A short lifetime component with 5 ps can not be easily resolved in BChl a–containing chlorosomes. By addition of a fifth component also a lifetime of 3 – 5 ps can be found in these chlorosomes with a spectral shape similar to the one found in GEF–chlorosomes. The characteristic zero crossing was near 745 nm so that we attribute this component to the same process as in GEF–chlorosomes. The error in the corresponding

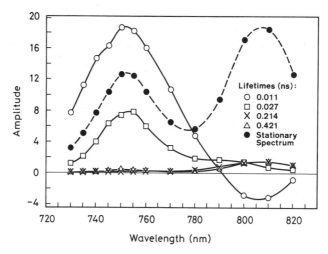

Fig. 3. Decay–associated spectra of BChl a–containing chlorosomes prepared with miranol from C. aurantiacus as obtained by global analysis of the fluorescence decays measured at room temperature (λ_{exc} = 721 nm). A similar result has also been obtained with deriphat–isolated chlorosomes.

DAS is high, however, due to difficulties of resolving five components with three close–lying short lifetimes.

Kinetic Model

We propose the minimal kinetic scheme for the excited state kinetics in chlorosomes as described in Fig. 4. The $(BChl\ c_I)^*$ and $(BChl\ c_{II})^*$ states either present spectrally distinct BChl c pools or the upper and lower states of excitonically coupled BChl c aggregates. Nonradiative quenching of BChl c and/or of BChl a by an unknown quencher occurs with a wide range of rates. This leads to lifetimes of 300 to 500 ps for the terminal pigments.

Relaxation processes within the Bchl c pool have not been resolved so far. Also the BChl c → BChl a transfer in chlorosomes has not been identified so far by a risetime in the BChl a* state. The direct measurement is demonstrated here for the first time by the resolution of a rise–time (negative amplitude) component in the BChl a emission region. The transfer time found here is shorter than the time resolution of the apparatus used by Brune et al. (2). The latter authors set an upper limit of 30 ps for the BChl c→BChl a_{790} transfer in C. aurantiacus chlorosomes.

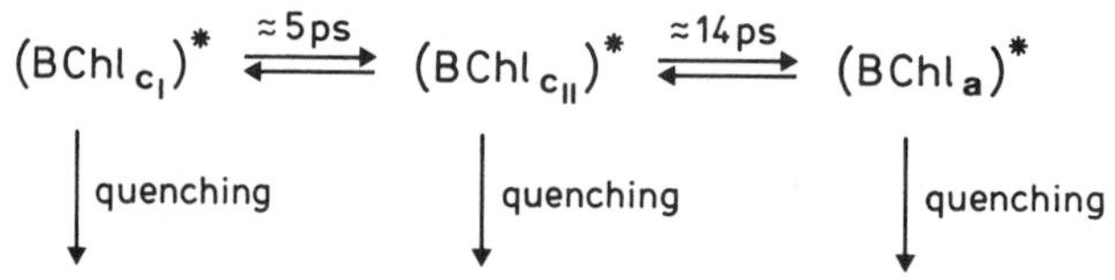

$$(BChl\ c_I)^* \xrightleftharpoons{\approx 5\,ps} (BChl\ c_{II})^* \xrightleftharpoons{\approx 14\,ps} (BChl\ a)^*$$

quenching   quenching   quenching

Fig. 4. Proposed kinetic scheme for energy flow in isolated chlorosomes of C. aurantiacus.

ACKNOWLEDGEMENTS

We acknowledge Mrs. B. Kalka, Mrs. A. Keil and Mr. S. Emunds for valuable technical support. We also should like to thank Prof. K. Schaffner for his interest and support of this work. Partial financial support was provided by the Deutsche Forschungsgemeinschaft. Some results presented here are part of a diploma work (K.G.) accepted by the Philipps–Universität Marburg, FRG.

REFERENCES

(1)   Betti,J.A., Blankenship,R.E., Natarajan,L.V., Fuller,R.C.
      and Dickinson,L.C. Biochim. Biophys. Acta 680, 194–201 (1982).

(2)   Brune,D.C., King,G.H., Infosino,A., Steiner,T., Thewalt,M.L.W.
      and Blankenship,R.E. Biochemistry 26, 8652–8658 (1987).

(3)   Van Dorssen,R.J., Vasmel,H. and Amesz,J. Photosynth. Res. 9, 33–45 (1986).

(4)   van Dorssen,R.J., Vos,M., and Amesz,J. in Photosynthetic Light–Harvesting
      Systems, (eds. H. Scheer and S. Schneider) 531–541
      (de Gruyter, Berlin, New York 1988).

(5)   Golecki,J.R. and Oelze,J. Arch. Microbiol. 148, 236–241 (1987).

(6)   Staehelin,L.A., Golecki,J.R., Fuller,R.C. and Drews,G.
      Arch. Mikrobiol. 119, 269–277 (1978).

(7)   Vos,M., Nuijs,A.M., van Grondelle,R., van Dorssen,R.J., Gerola,P.D. and Amesz,J.
      Biochim. Biophys. Acta 891, 275–285 (1987).

(8)   Fetisova,Z.G., Freiberg,A.M. and Timpmann,K.E. Nature 334, 633–634 (1988).

(9)   Griebenow,K. and Holzwarth,A.R. Biochim. Biophys. Acta 973, 235–240 (1989).

(10) Feick,R.G., Fitzpatrick,M. and Fuller,R.C. J. Bacteriol. 150, 905–915 (1982).

(11) Holzwarth,A.R., Wendler,J. and Suter,G.W. Biophys. J. 51, 1–12 (1987).

(12) Blankenship,R.E., Brune,D.C., Freeman,J.M., Trost,J.T.,
      King,G.H., McManus,J.H., Nozawa,T., and Wittmershaus,B.P.,
      in Green Photosynthetic Bacteria
      (eds. Olson, J., Ormerod, J.G., Amesz, J., Stackebrandt,
      E., and Trüper, H.G.) 57–68 (Plenum Press, New York 1988).

AN INTRODUCTION AND OVERVIEW TO THE SECTION ON ELECTROCHEMICAL GRADIENTS ACROSS MEMBRANES

J.B. Jackson

School of Biochemistry
University of Birmingham
P O Box 363, Birmingham B15 2TT UK

INTRODUCTION

Many if not most of the proteins associated with the cytoplasmic membranes of phototrophic bacteria are involved in the generation or the utilization of the proton electrochemical gradient (Δp). It is now widely recognized that Δp has a central role in the energy economy of the cell and while a number of unexplained observations remain, it is generally thought that it serves to couple the "energy-generating" reactions of electron transport with the "energy-consuming" reactions of ATP synthesis, nicotinamide nucleotide transhydrogenase and certain kinds of solute translocation . An outstanding problem in biochemistry is to try to understand the mechanisms by which some membrane proteins operate as proton translocators and thus use the energy of Δp to drive chemical conversions and other transport processes. This section is devoted to work in which this problem is being tackled. Many other contributions to the Symposium, notably the articles of F. Daldal, H.W. Trissl, A.Y. Semenov, and D.B. Knaff should also be consulted for relevant discussions.

Generation of Δp by photosynthetic electron transport

In phototrophic bacteria the major generators of Δp are the respiratory and the photosynthetic electron transport systems. Some species lack respiratory chains. The cyclic photosynthetic electron transport system, typified by that in Rhodobacter sphaeroides, comprises a photosynthetic reaction centre complex, a cytochrome bc_1 complex, ubiquinone and cytochrome c_2. The conventional view is that the reaction centre serves to oxidise cytochrome c_2 and reduce ubiquinone whereas the bc_1 complex reduces cytochrome c_2 and oxidises the quinone. Electrogenic charge separation across the membrane takes place in both the reaction centre and the bc_1 complex. Proton binding on the cytoplasmic side of the membrane accompanies the reduction of bound quinone at the Q_B site in the reaction centre and at the Q_C site in the bc_1 complex. Proton release on the periplasmic side of the membrane takes place during oxidation of quinol at the Q_z site in the bc_1 complex. Thus the transport of two reducing equivalents through the cyclic electron transport chain effectively leads to the translocation of four H^+ outwards across the bacterial membrane and hence to the generation of Δp.

Molecular Biology of Membrane-Bound Complexes in Phototrophic Bacteria
Edited by G. Drews and E. A. Dawes
Plenum Press, New York, 1990

W. Crielaard, K.J. Hellingwerf and W.N. Konings in their contribution to the Symposium described an artificial cyclic electron transport system in liposomes containing just photosynthetic reaction centres and cytochrome c. The need for a bc_1 complex was circumvented by including in the reconstitution system the water soluble, UQ_0 , which is able slowly but directly to reduce the external cytochrome c. A light-driven Δp across the vesicle membranes was demonstrated. An interesting development described by these authors is the successful reconstitution into liposomes of light-harvesting complex II from Rb. sphaeroides. The sensitivity of the absorption spectrum of carotenoids in the complex to external electric fields is retained. Thus the possibility arises in a reconstituted system to use the carotenoid "voltmeter" to characterise the electrogenic reactions of other membrane proteins. Importantly, the electrochromic response in both natural and reconstituted membranes is linear with the applied membrane potential.

Recently there has been a re-kindling of interest in photosynthetic processes in other more exotic species of phototrophic bacteria. The approach is to try to learn more about the essentials of photosynthesis by studying diversity in distantly related organisms. The photosynthetic reaction centre of Chloroflexus aurantiacus has a simpler peptide composition than that found in Rhodobacter sp. and a different pigment composition. Moreover this bacterium employs menaquinone, not ubiquinone, at the quinone binding sites in the reaction centre and in the quinone pool communicating with the bc_1 complex. These factors prompted G. Venturoli, R. Feick, M. Trotta and D. Zannoni to examine properties of menaquinone bound to the reaction centre. It transpires that although these menaquinones operate at a slightly lower potential than those in the purple bacteria similar thermodynamic constraints (such as the E_m/pH relationships) seem to apply.

Generation of Δp by respiratory electron transport

The respiratory electron transport chains of the phototrophic bacteria are less well characterised than the photosynthetic systems. In Rhodobacter sp., the best studied group, the respiratory chain is supplied by reducing equivalents from NADH-dehydrogenase, succinate dehydrogenase or hydrogenase feeding into the ubiquinone pool. The output of reducing equivalents proceeds from the Q-pool either to the bc_1 complex, cytochrome c_0 and cytochrome c_2 oxidise or through the so-called "alternative" oxidase. There may be a "sharing" of the ubiquinone pool, the bc_1 complex and cytochrome c_2 between the photosynthetic and the respiratory electron flow pathways. The electrogenic and protolytic reaction of the respiratory chain are not well characterised and this was the subject of the paper by J. Varela and J.M. Ramirez using phototrophically grown cells of Rhodospirillum rubrum. The respiratory pathways in this organism are analogous to those in the Rhodobacter species but there is an additional "photo-oxidase" pathway emanating from the bound quinones in the reaction centre and proceeding by way of rhodoquinone to molecular oxygen. Varela and Ramirez monitored pH changes in darkened intact cell suspensions of Rhs. rubrum after oxygen pulses. They attributed the burst of acidification to outward proton translocation catalysed by the respiratory pathways. Through the use of specific inhibitors and respiratory mutants they were able to calculate the ratio of protons translocated per O atom reduced in each of the three pathways. Respiratory electron flow through the cytochrome c_2 oxidase had the largest H^+/O ratio, then electron flow through the alternative pathway and finally electron flux via the rhodoquinone dependent pathway.

Over the last decade it has become clear that some species of the photosynthetic bacteria are capable of "anaerobic respiration" using NO_3^-, N_2O, trimethylamine-N-oxide (TMAO) and dimethyl sulphoxide (DMSO) as terminal oxidants. In the paper by A.G. McEwan, D.J. Richardson, J.B. Jackson and S.J. Ferguson it was explained how the use of mutants and specific inhibitors has led to elucidation of the electron transport pathways to these oxidants in Rb. capsulatus. The reduction of NO_3^-, TMAO and DMSO proceeds through the ubiquinone pool but quite independently of the cytochrome bc_1 complex and cytochrome c_2. However the reduction of N_2O in large measure does proceed through bc_1 and cytochrome c_2. Although in some circumstances these oxidants can support growth by anaerobic respiration in the dark it was proposed that their main function is in photosynthesis. Thus several kinds of experiments revealed that they have a role as "auxiliary oxidants" for the disposal of excess reducing equivalents during phototrophic growth.

Consumption of Δp

The major consumer of Δp under many physiological conditions is the ATP synthase. However chromatophore membranes from purple bacteria can have activities of light-driven nicotinamide nucleotide transhydrogenase approaching the rates of photophosphorylation. J.B. Jackson, N.P.J. Cotton, T.M. Lever, I.J. Cunningham, T. Palmer and M.R. Jones described the first successful solubilization and purification procedure for transhydrogenase from photosynthetic bacteria. Structually it is much simpler than ATP synthase and, together with the availability of a real-time assay and the fact that the equilibrium constant in de-energised membranes is close to unity, transhydrogenase is a valuable model for the study of protonmotive enzymes. The way in which membrane potential (Δψ) might drive the reaction was investigated by observing the dependence on bulk phase pH of the relation between the rate of transhydrogenase and the value of Δψ. The possibilities were considered that the reaction might be accelerated by Δψ either by increasing the occupancy by H^+ at a site at the bottom of a proton well in the enzyme or by increasing the rate at which H^+ is driven across an energy barrier. The data show that there are features of both models that might be involved in the process of energy transduction.

During growth it is expected that the operation of some solute transport sytems will lead to a heavy demand on Δp. Rhodobacter sp. are commonly grown in the laboratory using dicarboxylic acids. It is therefore surprising that little information is available on the nature of the proteins responsible for the transport of these compounds. D.K. Kelly, M.J. Hamblin and J.G. Shaw presented an interesting paper on the preliminary characterisation of the physiology and genetics of C4-dicarboxylate transport in Rb. capsulatus. They isolated Tn5 mutants which were unable to grow on malate in darkened cultures. Unexpectedly they found that these mutants were able to grow, albeit at a low rate, under photosynthetic conditions with malate as C-source. Their experiments suggest that there may be two dicarboxylate transport systems, one operating under aerobic conditions, the other specifically under photosynthetic conditions. The former system has high affinity for malate, also works with succinate and fumarate, and possibly employs a periplasmic binding protein.

In this Symposium the subject of sensory signalling has been considered in the context of the proton electrochemical gradient. This is appropriate since bacteria are expected to respond chemotactically to enable them to locate a move favourable (or less hostile) environment and, in some circumstances, this will be influenced by their energy metabolism. However, in the paper by J.P. Armitage, P.S. Poole , W.A. Havelka and S. Brown it was argued that there was no direct involvement in Rb. sphaeroides of the the proton electrochemical gradient or of electron transport in the chemotactic response. Transport of an effector is required for chemotaxis and, at least in the case of ammonium, some metabolism is necessary. Thus it was considered that the stopping frequency of the flagellum is controlled by the changing concentration of several specific classes of metabolites.

LIGHT-INDUCED ELECTRON TRANSFER AND ELECTROGENIC REACTIONS IN THE bc

COMPLEX OF PHOTOSYNTHETIC PURPLE BACTERIA RHODOBACTER SPHAEROIDES

*Shinkarev V.P. ,Drachev A.L., Drachev L.A., Mamedov M.D.,
*Mulkidjanian A.Ya., Semenov A.Yu., Verkhovsky M.I.

*Biophysics Section, Department of Biology and A.N.Belozersky
Laboratory of Molecular Biology and Bioorganic Chemistry,
Moscow State University, Moscow 119899,USSR

INTRODUCTION

In chromatophores of non-sulphur purple bacteria, the cyclic electron transport chain consist of a photosynthetic reaction center (RC) and bc complex (ubiquinol: cytochrome c_2 oxidoreductase). Following light exposure, the former reduces the ubiquinone and oxidizes the cytochrome c_2 as the latter subsequently oxidizes the ubiquinol and reduces the cytochrome c_2. During functioning of this cycle, the electron transport is coupled to the generation of electrochemical potential difference of hydrogen ions needed to drive ATP synthesis [1-4].

Ubiquinol molecule formed in RC after the second turnover is transferred to the ubiquinol-oxidizing center Z of bc complex where one electron returns via the high-potential path, including the Rieske iron-sulphur protein and cytochromes c_1 and c_2 to P, and other is passed via hemes b-566 and b-561 to the ubiquinone-reducing center C to reduce an ubiquinone molecule which originates from the ubiquinone pool [3-6]. It has been suggested that the semiquinone formed after the first turnover of bc complex remains bound to center C and may be reduced by cytochrome b-561 after the second turnover. In the presence of antimycin A which inhibits all processes in the center C of bc complex the photo-induced electron transfer is limited by mainly one turnover of bc complex leading to reduction of cytochrome b-561 and carriers of high-potential path [1,3,7].

Our knowledge of electrogenic events within the RC and bc complex of the purple bacteria is derived essentially from measurements of carotenoid electrochromic absorption changes, which accompany the generation of $\Delta\Psi$ [8-10]. It has been shown that electrochromic changes of carotenoids are proportional to $\Delta\Psi$[8]. Therefore it is possible to estimate relative electric distances between different redox centers in native chromatophore membrane [11,12],as well as the distances of electrogenic steps associated with proton transfer [13,14]. The flash-induced electrochromic changes of carotenoids in chromatophores consist of three phases. The phases I and II

Abbreviations: RC, reaction center; P, reaction center bacteriochlorophyll dimer; Q_A, Q_B, primary and secondary quinone acceptors of RC; $\Delta\Psi$, transmembrane electric potential difference; TMPD, N,N,N',N'-tetrametyl-p-phenylenediamine; bc complex, ubiquinol:cytochrome c_2 oxidoreductase; b-566 and b-561, low and high potential cytochrome b hemes ; FeS, Rieske iron-sulphur center; F-electrogenic phase of bc complex with rise time 7 ms which is insensitive to antimycin A, but is suppressed by myxothiazol.

Molecular Biology of Membrane-Bound Complexes in Phototrophic Bacteria   393
Edited by G. Drews and E. A. Dawes
Plenum Press, New York, 1990

were related with reactions of RC and phase III - with functioning of bc complex [6-12].

The phase III of carotenoid bandshift is attributed to electrogenic events in bc complex and may be subdivided into a few phases on the basis of their sensitivity to antimycin A and myxothiazol. The kinetics of this phases are dependent on the way by which electron transfer in bc complex is induced [7,10]. At redox potential below ~150 mV, when redox reactions in bc complex are initiated by photo-induced oxidation of high-potential components of bc complex, the faster kinetics, with $\tau \sim 2$ ms, are observed. At redox potential higher than ~200 mV when redox reactions in bc complex are induced by ubiquinol produced in RC the slower kinetics of carotenoid bandshift, with $\tau > 10$ ms are observed. The part of the phase III sensitive to antimycin A is associated with electrogenic reactions at center C of bc complex and its secondary turnovers. Myxothiazol also inhibits phase III of the carotenoid changes [6,11]. It has been established, that the antimycin-insensitive, myxothiazol-sensitive phase of $\Delta\Psi$ is mainly associated with electrogenic electron transfer between cytochrome b hemes [6,11,12].

Robertson and Dutton [12] have tryed to investigate the topography of electron transfer in bc complex, using the total amplitude of phases I and II as internal calibration of $\Delta\Psi$. But existence of a few electrogenic phases in the time domain about 0.1 ms within RC [13-15] the relative amplitudes of which are depended on redox potential, pH of medium, flash number and so on, as well as low signal-to-noise ratio of the carotenoid bandshift did not allow to measure accurately phase I and II kinetics and to determine electric distances of electrogenic processes in bc complex.

In the present work with aid of the electrometric technique which has been developed earlier [13-16] we investigated the kinetics and amplitudes of electrogenic phase(s) due to functioning of bc complex in the presence of antimycin A .This method allowed us to distingwish electrogenic phases, which not possible to determine by carotenoid bandshift method and to normalize all electrogenic events to phase associated with charge separation between P and Q_A.

MATERIALS AND METHODS

Cells of Rhodobacter sphaeroides (wild type, strain R-1) were grown

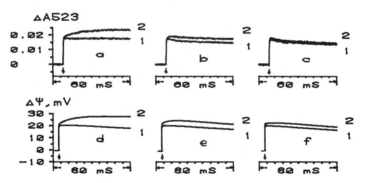

Fig. 1. Electric responses of Rb.sphaeroides chromatophores induced by the first (1) and second (2) flashes measured spectrometrically, by monitoring electrochromic changes of carotenoids at 523 nm (top) and electrometrically (bottom). a,d, with no inhibitors; b,e, in the presence of 4 μM antimycin A; c,f, in the presence of 4 μM antimycin A and 5 μM myxothiazol. Incubation medium: 30 mM HEPES (pH 7.5), 50 μM TMPD, 2 mM potassium ferrocyanide; E_h = 300 mV. The concentration of RC and bc complexes in chromatophores, used for monitoring electrochromic changes, are 0.57 μM and 0.26 μM, respectively. Arrows show laser flashes spased 1 sec apart.

and chromatophores isolated by French-press treatment as described elsewhere [14]. Association of chromatophores with a phospholipid-impregnated collodion film was obtained by 1 h incubation with 20 mM $CaCl_2$.
Kinetic measurements of $\Delta\Psi$ generation were conducted as described in [14]. Saturating flashes were given from a QUANTEL Nd laser.

Absorption changes were measured with a home-made single-beam spectrophotometer. A shutter was used to prevent the dark-adapted samples (dark adaptation time, 5 min) from incidence of measuring light. The data storage and processing system included a DL-1080 (Data Lab) transient recorder interfaced to a Nova-3D mini-computer (Data General). Kinetic curves were analyzed using a modification of the DISCRETE program [17].

Ubiquinone function in collodion film-associated chromatophores was reconstituted by adding Q-10 (20 mg/ml) to the solution of asolectin in decane used to impregnate the collodion film in the measuring cell.

RESULTS

Comparison of the data obtained by two different methods

After dark adaptation, binary oscillation of the ubiquinol formation could be observed under certain conditions in Rb. sphaeroides chromatophores. Ubiquinone produced after odd-numbered flashes stay within the RC, whereas ubiquinol molecules produced by even-numbered flashes are rapidly substituted with ubiquinone molecules from the membrane pool. The released ubiquinol may be oxidized by the bc complex. When the quinone pool has been oxidized prior to the flash, the extent of reduction of cytochrome b-561 and the transmembrane electric potential oscillate both with flash number [18,19].

Fig.1 represents photoelectric responses of Rb. sphaeroides chromatophores induced by a first (1) and second (2) flash, as measured spectrophotometrically (top) and electrometrically (bottom). As seen, the data obtained by electrometry agree fairly well with those derived by monitoring the carotenoid bandshift. After the first flash, the $\Delta\Psi$ increases rapidly (τ <0.1 ms) which is attributed to the separation of charges between P and Q_A in RC [20]. The following slow decay is due to the passive discharging of the membrane or partially by reverse electron transfer from quinone acceptors to P.

The additional slower electrogenic phases appeared under the second flash. To elucidate the origin of the phases, bc complex inhibitors were used. The addition of antimycin A and myxothiazol causes a decrease of the amplitude of the slow phase, induced by the second flash. In the presence of both inhibitors the photoelectric responses after the second flash are some larger than after the first flash. This is due to the fast electrogenic phase, associated with the protonation of reduced Q_B.

The obvious influence of the inhibitors of bc complex on 523 nm absorbance changes indicate that some of the bc complexes are activated even after the first flash, despite the dark adaptation period 5 min. The probable cause of this may be the effect of measuring light, which convert the secondary quinones in some of the RCs into the semiquinone form. In the case of electrometry, where no measuring beam is used, Q_B molecules are fully oxidized before the first flash, which is seen from the lack of the influence of the bc complex inhibitors on the $\Delta\Psi$ kinetics induced by the first flash (fig.1, bottom).

The results obtained herein by the two methods clearly indicate that the data obtained are qualitatively identical, but the electrometric method is more suitable because of the much larger signal/noise ratio and the absence of the measuring light beam effects.

To obtain more detailed kinetic information on $\Delta\Psi$ generation, the differences in the photoelectric potentials induced by the second and first flash (bottom at Fig.1) are presented on two different timescales (Fig.2). As follows from the comparison of curves in Fig.2, antimycin A and myxothiazol have virtually no effect on the fast $\Delta\Psi$ phase ($\tau \sim 0.15$ ms) associated with protonation of Q_B^{2-}, but cause a decrease of the slow phase $\Delta\Psi$ induced by the second flash. In the absence of inhibitors (curve a), the slow phase can be approximated by a single exponential curve with $\tau \sim 20$ ms. Its rise-time reduces to 7 ms after the addition of antimycin A (curve b). Subtracting curve c from b in Fig.2 gives the electrogenic phase F with rise-time 7 ms which is insensitive to antimycin A, but is suppressed by myxothiazol.

The main purpose of this work is the investigation of the kinetics and amplitude of slow phase of $\Delta\Psi$ generation by bc complex in the presence of antimycin A.

The kinetics of $\Delta\Psi$ formation in the presence of antimycin A

The amplitude of the 0.15 ms phase is a proportional to the amount of second-flash-produced ubiquinol. The amplitude of the antimycin-insensitive, but myxothiazol-sensitive phase of $\Delta\Psi$ is proportional to the population of reacting bc complex. Thus, the behavior pattern of this single kinetic curve reflects events occuring in both the RC and bc complex.

In case of a bimolecular character of interaction between the ubiquinol expelled from the RC and bc complex, the rate of $\Delta\Psi$ formation by the bc complex would slow essentially with decreasing the amount of ubiquinol.

The formation of the ubiquinol in RC was varied in two ways, by decreasing the intensity of the second flash (using light filters) or by increasing the dark time lag between the first and second flash (to cause Q_B^- oxidation in the dark by a mediator).

Fig.3 shows the rise-time (a) and amplitude (b) of the electrogenic phase of $\Delta\Psi$ related to the bc complex in the presence of antimycin A as a function of the amplitude of the fast ($\tau \sim 0.15$ ms) electrogenic phase. It is seen that a tenfold decrease in the amount of ubiquinol produced causes a negligible increase in the time of $\Delta\Psi$ generation, but considerable decrease in the amplitude.

In the case of bimolecular pattern of the reaction between the ubiquinol and bc complex , its rate would decrease with lowering both the ubiquinol available and the amount of active bc complexes. The study was continued by measuring the kinetics of the flash-inducing reactions of cytochrome b-561 in the presence of antimycin A. As the intensity of the flashes is decreased or the amount of active bc complex was varied by myxothiazol, the amount of reduced cytochrome b-561 decreases, but its reduction time remaining practically unchanged (see also [21]).

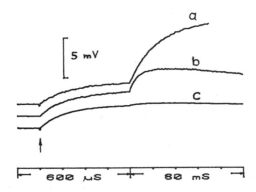

Fig. 2. The difference in photoelectric responses induced by the second and first flashes (bottom at Fig.1), plotted on two timescales a, no inhibitors; b, 4 µM antimycin A; c, in the presence of 4 µM antimycin A and 5 µM myxothiazol.

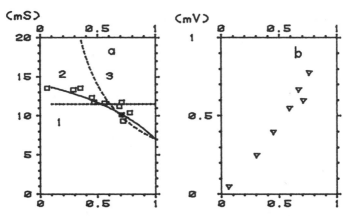

Fig. 3. The rise-time (a) and amplitude (b) of the antimycin-insensitive and myxothiazol-sensitive phase of $\Delta\Psi$ formation as a function of photoelectric response associated with electrogenic protonation of the secondary quinone after second flash. The incubation medium contained: 30 mM MES (pH 6.1), 2 mM ferrocyanide, 50 µM TMPD, 4 µM antimycin A, 1 µM methylene blue. Theoretical curves (a): bimolecular interaction between ubiquinol and bc complex (curve 3); supercomplex formed by the RC and bc complex (curve 1); interaction between the RC and bc complex via a local quinone poole (curve 2).

Amplitude of phase III in the presence of antimycin A

The time of $\Delta\Psi$ generation by bc complex is much larger than that generated by RC. Thus one must take into acccount the passive decay of the membrane potential by leakage of charges through the membrane. It is obvious that the slower decay of membrane potential is, the more precisely the amplitude of $\Delta\Psi$ phase may be determined. The main factor which accelerates the dark relaxation of $\Delta\Psi$ in chromatophores is the nature and concentration of redox mediators, used to poise the redox potential [22]. In this work we used 2 mM ferrocyanide and 30-50 µM TMPD, which do not considerably accelerate the decay of membrane potential. Another factor which is essential only in electrometry, is the time needed for the chromatophores to associate with the collodion film. The latter must be taken lesser 1 h instead of 4 h, used earlier by us in registrating $\Delta\Psi$ in RC [13-16]. In this case the total $\Delta\Psi$ amplitude is lesser, but slower than 200-400 ms. To illustrate the influence of rate of membrane potential decay on the amplitudes of the electrogenic phases related to the functioning of bc complex, the theoretical curves of generation after the second flash are shown in Fig.4.

One can see that increasing the rate of $\Delta\Psi$ relaxation and decreasing the rate of $\Delta\Psi$ generation causes a considerable distortion of both the amplitude of $\Delta\Psi$ generation by bc complex and its observed rise-time.

Using experimental data, presented in Fig.2, one can estimate that the amplitude of antimycin insensitive, mixothiazol-sensitive phase constitutes as much as ~15 per cent of the amplitude of the phase, arising from the charge separation between P and Q_A.

DISCUSSION

Applying the electrometric method developed in [14-16,20] to analysis of flash-induced electrogenic phase insensitive to antimycin A, but sensitive to myxothiazol have revealed unusual kinetic properties of this phase, as well as its very small amplitude.

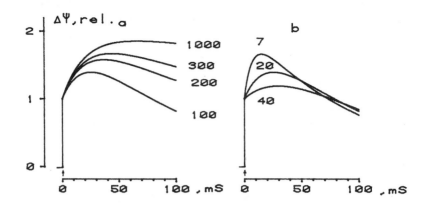

Fig. 4. Theoretical curves of $\Delta\Psi$ generation induced by the second flash for different time (t_2) of $\Delta\Psi$ decay due to the pasive discharging of chromatophore membrane (a) and different time (t_1) of the electrogenic phase $\Delta\Psi$ generation by bc complex (b). Each curve is described by the following Eqn:

$$\Delta\Psi(t) = A \exp(-t/t_2) + B(\exp(-t/t_2) - \exp(-t/t_1))t_2/(t_2 - t_1)$$

there A,B- relative amplitudes of $\Delta\Psi$ due to the reactions in RC and bc complex, respectively. A=1; B=1; t_1=20 ms (a), t_2=100 ms (b). Figures at the curves are the time (t_2) of $\Delta\Psi$ decay (a) and time (t_1) of $\Delta\Psi$ generation by bc complex (b) in millisecond.

The nature of interaction between RC and bc complexes

A bimolecular mechanism of interaction between ubiquinol and bc complex is now generally accepted [7,23]. Evidence for this is the increase of the rate of photo-induced reduction of cytochrome b-561 with increasing the amount of ubiquinol reduced in equillibrium (E_h=200 mV) or produced under oxidizing conditions (E_h=220 mV) following two (not one) successive flashes spaced at 0.7 ms [7]. It is important that in the experiments mentioned the ubiquinol concentration was higher than that of the bc complexes. The foregoing observations of the rate of cytochrome b-561 reduction and $\Delta\Psi$ generation provide convincing arguments that the reaction may be monomolecular when the ubiquinol concentration is no more than the concentration of bc complexes. The momomolecular character of the reaction, when $[QH_2]/[bc]<1$ (this work) and bimolecular, when $[QH_2]/[bc]>1$ [7,23], may be explained by the lack of the fast quinone exchange. The ubiquinol produced in the RC is kept in a local pool and is oxidized by its own (presumably neighbouring) bc complex. Since the number of ubiquinol molecules delivered to the pool at a time may be two, one or none (the RC:bc stoicheometry is 2:1) the rate of b-561 reduction induced by ubiqinol can be change at most twice.

Our data (Fig.3) are in disagreement with a model in which the rate of the reaction is proportional to the product of the concentrations of QH_2 and bc complexes (Fig.3a). In terms of our model, the bimolecular pattern of reaction between the ubiquinol molecule available at equilibrium and the bc complex may be explained by a change in the population of ubiquinol in the local pool. The data obtained cannot be explained in terms of the general point of view, according to which quinones form a pool common to all the RCs and bc complexes in chromatophores. They are in keeping with the presented concept of local pools between which the exchange occurs more slowly than the reduction of cytochrome b-561.

Thus the slow rate of cytochrome b-561 reduction at high redox-potential may be explained by bimolecular reaction between RC-produced ubiquinol and bc complex, as well as by the competition between reduced and oxidized form of quinones in this local pool.

The calculated amplitude of the antimycin-insensitive, myxothiazol-sensitive phase of $\Delta\Psi$(phase F) accounts for as much as ~15 per cent of the phase due to charge separation between P and Q_A (see Fig.1, bottom and Fig.2). The concentration of bc complex is 2 times lower than that of RC. Hence the amplitude of this phase must be increase two times, becoming equal to ~30% of that seen in RC.

As seen from Fig.3b the amplitude of phase F is equal to amplitude of $\Delta\Psi$, caused by electrogenic protonation of secondary quinone acceptor, which constitutes no more than 0.3 of that, due to P^+Q_A formation [28]. Thus the amplitude of phase F of bc complex also form no more than 0.3 of that associated with charge separation between P and Q_A in RC. This amplitude is consistent with estimate, obtained from data of Fig.1 and Fig.2.

The distance between cytochrome b hemes [24,25] is almost the same as that between P and Q_A [26]. Thus the amplitude of this phase must be almost the same as that of phase $\Delta\Psi$(PA) related with charge separation between P and Q_A. The contradiction may be due to the different dielectric permeabilities of RC and bc complex or - to decreasing of phase $\Delta\Psi$(Qb), arising from electron transfer $Q_2^-H \longrightarrow$ b-566 $\longrightarrow$ b-561, by phase $\Delta\Psi$(Qc$_1$), caused by electron transfer $Q_zH_2 \longrightarrow$ FeS $\longrightarrow c_1$ in the opposite direction, i.e. $F = \Delta\Psi$(Qb)$- \Delta\Psi$(Qc$_1$). The latter hypothesis is in accordance as with the known topography of RC and bc complex as well as with the slow proton release during the oxidation of Q_zH_2 [27] which occur with a time of about 30 ms. This is longer than 7 ms of phase F. Proton release cannot therefore interfere with electron transfer.

The electrical distance r(PA) between P and Q_A is approximately equal the distance between P and nonheme iron in RC which as well as the distance r(bb) between cytochrome b hemes may be estimated from the distance between histidines which are their ligands. In this case the amplitude of phase $\Delta\Psi$(PA) (16 amino acid residues between histidines [26]), must be about 1.2 times more than amplitude of phase $\Delta\Psi$(bb), is accounted for electron transfer between cytochrome b hemes (13 residues between histidines [24,25]). Assuming equal dielectric permeabilities in RC and in bc complex and taking into account that the ratio of concentrations of RCs and bc complexes in chromatophores is about 2, we can write:

$$F/\Delta\Psi(PA) \cong (+r(Qb_1)+r(bb)-r(Qc_1))/2r(PA)$$

where r(Qb$_1$), r(Qc$_1$) is electrical distances between Q_z and b-566, Q_z and c$_1$, respectively. If Q_z is closer to c-side of membrane surface than b-566 heme, the term r(Qb$_1$) must be taken with plus and vice versa. Using the latter relation and taking into account that r(PA)$\approx$ 27 Å [26] and r(bb)$\approx$ 0.83·r(PA)$\approx$22 Å, one can obtain the next approximation: r(Qc$_1$)+ r(Qb$_1$)$\approx$ r(bb)-0.3·r(PA)$\approx$ 14 - 15 Å. If Q_z is closer to c-side of membrane surface than b-566 heme, the electrical distance between Q_z and c$_1$ is more than 15 Å.

CONCLUSIONS

In contrast with common point of view which suggest simple bimolecular reaction between ubiquinol and bc complex at high redox potential, we have discovered that when reaction in bc complex induced by ubiquinol formed in RC, the time of reduction of cytochrome b-561 and antimycin-insensitive, myxothiazol-sensitive phase of $\Delta\Psi$ are practically independent from concentration of ubiquinol delivering from RC. The amplitude of antimycin-insensitive, myxothiazol-sensitive phase derived with aid of electrometric method, constitutes as much as 15 % of the amplitude of the phase arising from the charge separation between P and Q_A. From the small amplitude of this phase it is concluded that it is due to the difference of electrogenic electron transfer through the low-potential pathway of bc complex

$(Q_zH \longrightarrow b\text{-}566 \longrightarrow b\text{-}561)$ and in the opposite direction in the high potential pathway $(Q_zH_2 \longrightarrow FeS \longrightarrow c_1)$.

REFERENCES

1. Crofts, A.R. & Wraight, C.A., 1983, Biochim. Biophys. Acta, 726:149.
2. Rubin, A.B. & Shinkarev, V.P., 1984, "Electron transport in Biological Systems", Nauka,Moscow (in Russian).
3. Crofts, A.R., 1985, in: "Enzymes of Biological Membranes", Martonosi, N. ed., pp.347-382, Plenum Press, New York.
4. Dutton, P.L., 1986, in: "Encyclopedia of Plant Physiology", v.19, Staehelin, A. & Arntzen,C.J., eds.,pp.197-237. Springer-Verlag, West Berlin.
5. Robertson, D.E., Giangiacomo, K.M., De Vries, S., Moser,C.C. & Dutton, P.L., 1984, FEBS Lett., 178:343.
6. Glaser, E.G. and Crofts, A.R., 1984, Biochim. Biophys. Acta, 766:322.
7. Crofts, A.R., Meinhardt, S.W., Jones, K.R. & Snozzi, M., 1983, Biochim. Biophys. Acta, 723:202.
8. Jackson, J.B. & Crofts, A.R., 1971, Eur. J. Biochem., 18:120.
9. Jackson, J.B. & Dutton, P.L., 1973, Biochim. Biophys. Acta, 325:102.
10. Matsuura, K., O'Keefe, D.P. & Dutton, P.L., 1983, Biochim. Biophys. Acta,722:12.
11. Glaser, E. & Crofts, A.R., 1988, in: "Cytochrome Systems: Molecular Biology and Bioenergetics", Papa, S.,Chance, B., Ernster, L. & Jaz,J., eds., pp.625-631, Plenum Press, New York.
12. Robertson, D.E. & Dutton, P.L., 1988, Biochim.Biophys. Acta, 935:273.
13. Kaminskaja, O.P., Drachev, L.A., Konstantinov, A.A., Semenov, A.Yu. & Skulachev, V.P., 1986, FEBS Lett., 202:224.
14. Drachev, L.A., Kaurov, B.S., Mamedov, M.D., Mulkidjanian, A.Ja., Semenov, A.Yu, Skulachev, V.P., Shinkarev, V.P. & Verkhovsky, M.I., 1989, Biochim. Biophys. Acta, 973:189.
15. Drachev, L.A., Mamedov, M.D., Mulkidjanian, A.Y., Semenov, A.Y., Shinkarev, V.P., & Verkhovsky, M.I., 1989, FEBS Lett., in press.
16. Skulachev, V.P., 1982, in: "Methods Enzymology", 88:35.
17. Provencher, S.V., 1976, Biophys. J., 16:27.
18. Bowyer, J.R., Tierney, G.V., Crofts, A.R., 1979, FEBS Lett., 101:201.
19. De Grooth, B.G., Van Grondelle, R., Romijn, J.C. & Pulles, M.P.J., 1978, Biochim. Biophys. Acta, 503:480.
20. Drachev, L.A.,Semenov, A.Y. & Skulachev, V.P., 1979, Dokl. Acad. Nauk SSSR (USSR),245:991.
21. Drachev, L.A., Mamedov, M.D., Mulkidjanian, A.Y., Semenov, A.Y., Shinkarev, V.P., & Verkhovsky, M.I., 1989, FEBS Lett., 245:43.
22. Van den Berg, W.H., Bonner, W.D. & Dutton, P.L., 1983, Arch. Biochem. Biophys., 222:299.
23. Ventruolli, G., Fernandez-Velasco, J.G., Crofts, A.R. & Melandri, B.A., 1986, Biochim. Biophys. Acta, 851:340.
24. Saraste, M.,1984, FEBS Lett., 166:367.
25. Widger, W.R., Cramer, W.A., Herrmann, R.G. & Trebst, A., 1984, Proc. Natl. Acad. Sci. USA, 81:674.
26. Allen, J.P., Feher, G., Yeates, T.O., Komiya, H. & Rees, D.C., 1987, Proc. Natl. Acad. Sci. USA, 84:5730.
27. Taylor, M.A. & Jackson, J.B., 1985, FEBS Lett., 180:145.
28. Drachev, L.A., Mamedov, M.D., Mulkidjanian, A.Y., Semenov, A.Y., Shinkarev, V.P., & Verkhovsky, M.I., 1988, FEBS Lett., 233:315.

THE RHODOSPIRILLUM RUBRUM CYTOCHROME bc_1 COMPLEX

David B. Knaff

Department of Chemistry and Biochemistry
Texas Tech University
Lubbock, Texas 79409-1061
U.S.A.

INTRODUCTION

The cytochrome bc_1 complex was first characterized as one of the membrane-bound, multipeptide complexes of the mitochondrial electron transport chain. Subsequently, functionally equivalent cytochrome bc_1 complexes were discovered in the photosynthetic and respiratory chains of bacteria. These complexes, and the related cytochrome b_6f complexes of oxygenic photosynthetic organisms, play a central role in energy transduction associated with photosynthetic and respiratory electron transfer, as the electron flow from quinol to cytochrome c catalyzed by the complexes is coupled to the formation of a proton gradient and an electrical potential across the membrane(1-3). The cytochrome bc_1 complexes of photosynthetic purple non-sulfur bacteria have proven particularly useful for studying the pathway of electron movement for the following reasons:1. The ability to perform kinetic measurements after electron flow is initiated by a short, single-turnover light flash has provided a detailed sequence of electron transfer steps in Rhodobacter sphaeroides and Rhodobacter capsulatus(4,5) and, to a lesser extent, in Rhodospirillum rubrum (6); 2. While the electron-carrying chromophore content of the complexes isolated from photosynthetic bacteria and mitochondria are identical, the peptide compositions of the former are much simpler than those of mitochondria(i.e., three or four vs. ten subunits-Ref. 1,3,7); 3. The amino acid sequences, deduced from the nucleotide sequences of the corresponding genes, are known for all three chromophore-containing subunits of the Rb. capsulatus complex(8,9). Sequences should be available soon for the Rb. sphaeroides and Rhodopseudomonas viridis complexes (A.R. Crofts and D. Oesterhelt, peronal communications) and sequence work is beginning on the R. rubrum complex(See below). The R. rubrum cytochrome bc_1 complex(10,11) has several additional advantages, including its unambiguous three peptide subunit composition, its stability and the fact that it contains a high affinity binding site for its electron acceptor, cytochrome c_2(12).

We have recently developed an improved protocol for purifying the R. rubrum cytochrome bc_1 complex(11) and used this preparation for Resonance

Molecular Biology of Membrane-Bound Complexes in Phototrophic Bacteria
Edited by G. Drews and E. A. Dawes
Plenum Press, New York, 1990

Raman and Electron Spin Echo Envelope Modulation (ESEEM) measurements on the heme and Rieske iron-sulfur cluster chromophores, respectively (13,14). We have also determined the midpoint oxidation-reduction potential (E_m) values of the three hemes in the complex(11) and examined some aspects of binding to the complex by its two substrates, ubiquinone(11) and cytochrome c_2(12,15). Finally, using probes prepared from the genes for cytochrome c_1, cytochrome b and the Rieske iron-sulfur protein from Rb. capsulatus(9), we have located the genes for the three corresponding R. rubrum proteins.

METHODS

The cytochrome bc_1 complex was solubilized from membranes of photosynthetically-grown R. rubrum(Strain S1) using the detergent dodecylmaltoside and purified by anion exchange chromatography on DEAE-Biogel A and DEAE-Sepharose 6B according to a modification(11) of the procedure of Ljungdahl et al.(7). Activity assays, spectroscopic measurements, polyacrylamide gel electrophoresis, Western blots and oxidation-reduction titrations were all performed using standard techniques. Photoaffinity labeling of the complex with UV-photolyzed [^{3}H]3-azido-2-methyl-5-methoxy-6-(3,7-dimethyloctyl)-1,4-benzoquinone (Azido-Q) was carried out according to Yu et al.(16). The preparation of the lysine-modified equine cytochrome c and R. rubrum cytochrome c_2 derivatives used in the kinetic studies has been described previously(15). Differential chemical modification of R. rubrum cytochrome c_2 by acetic anhydride in the presence or absence of the R. rubrum cytochrome bc_1 complex has also been described previously(12), as has been the synthesis of 2,3-dimethoxy-5-methyl-6-(10-bromodecyl)-1,4-benzoquinol($Q_0C_{10}BrH_2$-Ref. 17).

RESULTS

Figure 1 shows both the absolute and the reduced minus oxidized difference spectra of the purified R. rubrum cytochrome bc_1 complex. Most preparations were free of bacteriochlorophyll a(BChl a) as judged either by the absence of any detectable absorbance at 881 nm or the absence of any absorbance at 773 nm in a 7:2 acetone/methanol extract(18). Although preparations occasionally did contain small amounts of residual BChl a(See Table 1), the BChl a:cytochrome c_1 ratio never exceeded 0.03(11). The ascorbate-reduced minus ferricyanide-oxidized difference spectrum of the R. rubrum complex shows α-band and β-band maxima at 553 and 525 nm, respectively, due to the presence of reduced cytochrome c_1(1,7,10,19). The fourth derivative of this difference spectrum shows only a single component centered at 553 nm, indicating that cytochrome c_1 is the only heme-containing component reduced by ascorbate. The dithionite-reduced minus ferricyanide-oxidized difference spectrum contains, in addition to features attributable to cytochrome c_1, α-(560 nm) and β-(532) bands arising from reduced cytochrome b(1,7,10,11). Heme analysis (Table 1) shows that the R. rubrum complex contains 2 cytochrome b hemes(i.e., protoheme) per cytochrome c_1(heme c), as has previously been found for other cytochrome bc_1 complexes(1-3,5,10). The fourth derivative of the dithionite minus ascorbate difference spectrum(not shown) revealed that two different spectral forms of cytochrome b are present in the R. rubrum complex. Evidence for two different spectral forms of cytochrome b has previously been obtained with R. rubrum membranes (20) and with other isolated cytochrome bc_1 complexes(1-3,5,10). The non-heme iron and acid-labile sulfide content of the complex (Table 1) are consistent with the presence of one [2Fe-2S]-containing Rieske iron-sulfur protein(1-3,5,10) per cytochrome c_1. We had previously demonstrated the presence of a Rieske protein in the R. rubrum complex by electron paramagnetic resonance(EPR) spectroscopy (10) and have confirmed this earlier observation using ESEEM spectroscopy and Western blots (See below). The complex also contains one ubiquinone per cytochrome c_1(10).

The R. rubrum complex catalyzed electron flow from either duroquinol(10) or $Q_0C_{10}BrH_2$(11) to either equine cytochrome c or R. rubrum cytochrome c_2. The turnover number for electron flow from $Q_0C_{10}BrH_2$ to equine cytochrome c was 75 s^{-1}(11). Electron flow was inhibited (10,11) by either of two specific inhibitors(1,2,4,5) of cytochrome bc_1 complexes:Antimycin A(87% inhibition at 0.8μM) and myxothiazol(96% inhibition at 1μM).

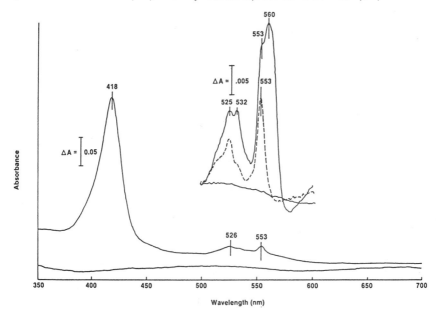

Figure 1. Absorbance spectra of the R. rubrum cytochrome bc_1 complex.
The solid line extending from 350 nm to 700 nm represents the spectrum of the complex as isolated. The two inserts, which were obtained at 10-fold higher sensitivity than the absolute spectrum, show the ascorbate-reduced minus ferricyanide-oxidized
(---) and dithionite-reduced minus ferricyanide-oxidized difference spectra in the α- and β-band regions(From Ref. 11).

Oxidation-reduction titrations (Fig. 2A and B) of the R. rubrum cytochrome bc_1 complex allowed determination of the E_m values for cytochrome c_1(E_m=+320mV at pH 7.4) and confirmed the presence of two different forms of cytochrome b, with E_m values of -33 and -90 mV, respectively, at pH 7.4. The fact that the two cytochrome b hemes have different E_m values has been observed previously with other isolated cytochrome bc_1 complexes(1-3,5) and in R. rubrum membranes(20). The differences between the E_m values for the three hemes in the R. rubrum complex are large enough to allow spectroscopic measurements to be performed on samples in which the hemes are in well-defined redox states. In particular, the 350 mV difference in E_m values between cytochrome c_1 and the higher potential cytochrome b heme makes it easy to poise the R. rubrum complex in a state where cytochrome c_1 is fully reduced and the two cytochrome b hemes are fully oxidized.

Table 1. Prosthetic group content of the R. rubrum cytochrome bc_1 complex(From Ref. 11).

Component	Content (nmol/mg protein)
Cytochrome c_1(heme c)	6.1
Cytochrome b(protoheme)	12.1
Acid-Labile Sulfide	12.4
Non-Heme Iron	14.0
Ubiquinone	6.8
Bacteriochlorophyll	0.18

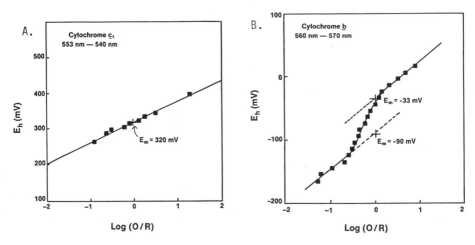

Figure 2. Oxidation-reduction titrations of the R. rubrum cytochrome bc_1 complex.
   A. Cytochrome c_1(553-540 nm). B. Cytochrome b(560-570 nm). The solid lines are least-squares best fits to the n=1 Nernst equation
   (From Ref. 11).

This redox poising can be illustrated using recently obtained Resonance Raman spectra of the R. rubrum cytochrome bc_1 complex(13). Resonance Raman spectroscopy, which allows detection of vibrational modes of the heme chromophores of the complex, can give considerable information about the environment of these heme prosthetic groups. Stoichiometric addition of sodium dithionite under strictly anaerobic conditions to the fully oxidized R. rubrum cytochrome bc_1 complex allowed Resonance Raman spectra to be obtained for defined oxidation states of the complex. Raman vibrational bands of the two heme types can be further distinguished by taking advantage of differences in electronic transitions of the b- and c-type hemes.

Laser excitation at 550 nm will enhance primarily the reduced heme c-chromophore, while 560 nm excitation will lead to enhancement of the protoheme prosthetic groups of reduced cytochrome b.

Figure 3 shows Resonance Raman spectra(13) of the fully reduced R. rubrum cytochrome bc_1 complex using laser excitation that preferentially enhances vibrational modes associated with either:(a) cytochrome c,(550 nm excitation) or (b) cytochrome b(560 nm excitation). Using 550 nm excitation, v_{10} can be seen at 1624 cm^{-1}, a position similar to that found for other c-type cytochromes. Other vibrational features consistent with the presence of ferrous cytochrome c can be seen at 1589 cm^{-1} and 1542 cm^{-1}. The latter two bands shift to 1585 and 1536 cm^{-1} when 560 nm excitation is used(v_{10} appears to be obscured under these conditions), clearly indicating that the two types of hemes can be selectively excited.

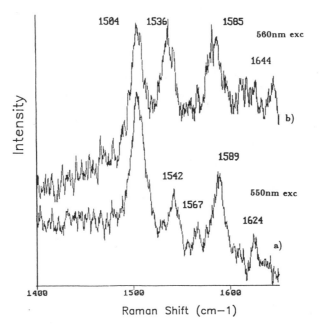

Figure 3. Resonance Raman spectra of the fully reduced R. rubrum cytochrome bc_1 complex. (a)550 nm excitation. (b)560 nm excitation. The complex was judged to be fully reduced by following the size of the cytochrome b α-band during the addition of sodium dithionite under anaerobic conditions(From Ref. 13).

Figure 4 shows the effect of progressive additions of reductant to the fully oxidized R. rubrum cytochrome bc_1 complex on the Resonance Raman spectra of the complex produced by excitation at a single wavelength in the cytochrome Soret region. Shifts in v_4 from 1375 cm^{-1} in the fully oxidized complex to 1360 cm^{-1} in the four electron-reduced complex are indicative of progressive reduction of the hemes. Addition of one electron equivalent(in terms of cytochrome $\underline{c}_1$) of dithionite produces a mode at 1590 cm^{-1}, consistent with that observed for most ferrous, $\underline{c}$-type cytochromes. It should be pointed out that because the Rieske iron-sulfur protein and cytochrome $\underline{c}_1$ are likely to be approximately isopotential(1-3,5), addition of one electron equivalent to the complex would be expected to reduce only about one half of the total cytochrome $\underline{c}_1$. Further additions of dithionite result in the appearance of a v_2 mode at 1582 cm^{-1} as cytochrome $\underline{b}$ becomes reduced, but a shoulder at 1590 cm^{-1} can still be observed. Similarly, v_{10} for ferrous cytochrome $\underline{c}_1$ can be observed at 1623 cm^{-1} in the Resonance Raman spectrum of the one electron-reduced complex and further reduction results in the appearance of the v_{10} mode for ferrous cytochrome $\underline{b}$ at 1617 cm^{-1}. The latter value for v_{10} is similar to that reported for cytochrome $\underline{b}_5$ and is consistent with the idea that the two protohemes of cytochrome $\underline{b}$ in the complex both have two histidine axial ligands(21). These preliminary studies have yielded the first Resonance Raman spectra of an isolated cytochrome $\underline{bc}_1$ complex from a photosynthetic bacterium and demonstrate that the two different heme chromophores in the $\underline{R}$. rubrum complex can be spectrally isolated and characterized. Analyses of these spectra and of spectra taken in other frequency regions are currently underway, as are comparisons of the Resonance Raman spectra of the $\underline{R}$. rubrum cytochrome $\underline{bc}_1$ comlex with those of the spinach chloroplast cytochrome $\underline{b}_6f$ complex.

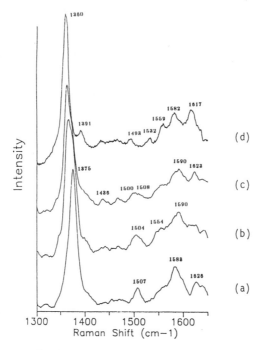

Figure 4. Resonance Raman spectra of the $\underline{R}$. rubrum cytochrome $\underline{bc}_1$ complex at different states of reduction. Excitation wavelength=410 nm. (a)The complex was pre-oxidized with potassium ferricyanide. (b)One electron-reduced complex. (c)Three electron-reduced complex. (d)Four electron-reduced complex. A calibrated solution of sodium dithionite was used to progressively reduce the complex(From Ref. 13).

It has been proposed that the iron-sulfur cluster of the Rieske protein in cytochrome $\underline{bc_1}$ complexes, like the clusters in the Rieske protein isolated from the thermophilic bacterium Thermus thermophilus(22,23) and in phthalate dioxgenase isolated from Pseudomonas cepacia(23,24), contains nitrogen ligands to at least one of the cluster irons. ENDOR(Electron-Nuclear Double Resonance) measurements on the yeast mitochondrial Rieske iron-sulfur protein (25) demonstrated the presence of at least one nitrogen ligand to the iron-sulfur cluster, the first demonstration of such ligation in an iron-sulfur protein component of a membrane-bound electron transfer chain. Electron Spin Echo Envelope Modulation(ESEEM) is a pulsed EPR technique that can be used to determine the transition frequencies of paramagnetic nuclei magnetically coupled to electron spins(26) and is thus useful for examining the Rieske iron-sulfur proteins of cytochrome $\underline{bc_1}$ complexes for nitrogen ligands. Fig. 5 shows the cosine Fourier transform of the three pulse ESEEM pattern arising from the reduced Rieske iron-sulfur center of the solubilized, purified R. rubrum cytochrome $\underline{bc_1}$ complex. The broad features between 3 and 4 MHz and the pair of peaks at 6.32 and 7.27 MHz arise from two inequivalent ^{14}N ligands to the Rieske iron-sulfur cluster. The hyperfine coupling constants for the two ^{14}N ligands are A_1=ca. 4.6 MHz and A_2=ca. 3.8 MHz. Similar ESEEM results were obtained for the Rieske iron-sulfur centers in a purified, detergent-solubilized cytochrome $\underline{bc_1}$ complex isolated from Rb. sphaeroides and a cytochrome $\underline{b_6f}$ complex isolated from spinach chloroplast membranes(14). In contrast to the ESEEM spectra of these three Rieske proteins, the ESEEM spectrum of reduced spinach ferredoxin(a protein known to have only sulfur ligands to the cluster irons) contained no features in the frequency region above 5.0 MHz(14).

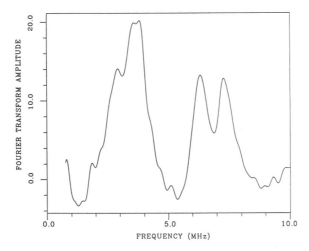

Figure 5. Cosine Fourier transform of the three pulse ESEEM pattern obtained at the g=1.92 maximum of the R. rubrum Rieske iron-sulfur protein. The R.rubrum cytochrome $\underline{bc_1}$ complex was present in buffer containing 33% glycerol(v/v). The time between microwave pulses I and II was 170 ns and that between pulses II and III was varied from 80 to 3000 ns in 10 ns steps. Data were recorded at a temperature of 4.2 K, a microwave frequency of 9.2277 GHz and a magnetic field of 3.463 kG(From Ref. 14).

The above results indicate that the Rieske iron-sulfur proteins of these two photosynthetic bacteria and chloroplasts all probably contain two histidines coordinated to one of the cluster irons, as has been established for the P. cepacia phthalate dioxygenase using ^{15}N ENDOR (24). There are four conserved cysteines in the C-terminal portion of the Rieske iron-sulfur protein of Rb. capsulatus, two of which presumably supply sulfur ligands to the cluster(8,9). There are also two conserved histidines in this region of the protein(8,9) that could provide nitrogen ligands to the cluster. We have recently located the gene for the R. rubrum Rieske iron-sulfur protein in an EcoRI digest of R. rubrum DNA(Shanker, S., Knaff, D. B. and Harman, J. G., unpublished observations) using a probe constructed from the Rb. capsulatus Rieske protein gene(kindly provided by Prof. F. Daldal) and plan to sequence the gene in order to determine whether these histidines are also present in the R. rubrum protein.

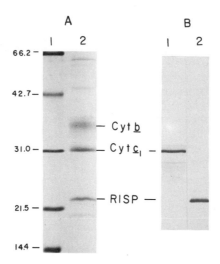

Figure 6. Peptide composition of the R. rubrum cytochrome bc_1 complex. Electrophoresis was performed on a 10-15% gradient polyacrylamide gel in the presence of SDS with 130 pmol of complex per lane. A.Staining for protein with Coomassie Brilliant Blue. B.Western blots after transfer to nitrocellulose paper and treatment with antibody against either Rb. sphaeroides cytochrome c_1(Lane 1) or the Rb. capsulatus Rieske iron-sulfur protein(From Ref. 11).

The peptide composition of the R. rubrum cytochrome bc_1 complex is exteremely simple(11), with only three peptides detected after staining polyacrylamide gel electrophoresis samples run in the presence of SDS with either Coomassie Brilliant Blue (Fig. 6A) or silver(not shown). M_r values estimated for the three components are 35 kDa for cytochrome b, 31 kDa for cytochrome c_1 and 22.4 kDa for the Rieske iron-sulfur protein. The identities of cytochrome c_1(previously established by staining for heme-Ref. 10) and of the Rieske iron-sulfur protein were established by Western blots(Fig. 6B) with antibodies specfic for each of the two proteins(11). Thus, the $M_r=35$ kDa component must be cytochrome b. This component, which appears as a single diffuse, Coomassie Blue-staining band in Fig. 6A, can be resolved into two bands with M_r values of 34 and 36 kDa in some preparations.

As ubiquinol and cytochrome c_2 are, respectively, the electron donating and accepting substrates for the cytochrome bc_1 complex(1-5), it was naturally of interest to explore some details of the binding of these two substrates to the R. rubrum complex. The widely accepted "Q-cycle" model for electron flow through the cytochrome bc_1 complex, in fact, predicts the presence of two sites where quinone interacts with the complex-One where quinol is oxidized by the Rieske iron-sulfur center and the lower potential cytochrome b and another site where the high potential cytochrome b reduces quinone(1,2,4,5). We have utilized photoaffinity labeling with the ubiquinone analog azido-Q(16) to locate the peptide(s) in the R. rubrum cytochrome bc_1 complex that bind ubiquinone(11). UV illumination-induced photolysis of azido-Q in the presence of the R. rubrum complex results in incorporation of [³H]-labeled azido-Q into the cytochrome bc_1 complex, accompanied by inhibition of the ubiquinol:cytochrome c oxidoreductase activity of the complex(11). Approximately 55 to 60% inhibition of the electron transfer activity of the complex could routinely be obtained under these conditions, while no inhibition was observed in control experiments in which either azido-Q was incubated with the R. rubrum complex in the dark or in which the complex was exposed to UV irradiation in the absence of azido-Q(11). Examination of the separated R. rubrum cytochrome bc_1 complex subunits after UV-photolysis in the presence of azido-Q revealed that most of the [³H]label was present on the cytochrome b subunit(11). No significant azido Q-labeling of the Rieske iron-sulfur protein was observed, but some labeling(which we attribute to non-specific binding of azido-Q to the complex) of cytochrome c_1 was seen(11).

Kinetic studies suggest that cytochrome c_1 is the component of the cytochrome bc_1 complexes of photosynthetic bacteria that reduces cytochrome c_2 to complete electron transport through the complex(4). While we have no data that bear directly on this question, we have obtained results supporting the idea that an electrostatically stabilized complex is formed between cytochrome c_2 and the cytochrome bc_1 complex in R. rubrum(12). Furthermore, we have obtained evidence from both binding and kinetic studies that lysine residues located near the exposed heme edge of cytochrome c_2 provide the positive charges involved in stabilizing this complex between the soluble cytochrome c_2 and its membrane-bound electon donor(12,15). The complex between R. rubrum cytochrome c_2 and the purified, detergent-solubilized cytochrome bc_1 complex is sufficiently tight for the two components to co-migrate during gel filtration chromatography at low ionic strength(12). At high ionic strength, where electrostatic forces weaken, no co-migration was observed (12).

We have used the technique of differential chemical modification to identify three specific lysine residues on R. rubrum cytochrome c_2 that are protected against chemical modification by acetic anhydride when the R. rubrum cytochrome bc_1 complex is present(12). In control experiments, no protection by the cytochrome bc_1 complex against acetylation of these cytochrome c_2 lysine residues was observed at high ionic strength. We thus attribute the protection of these residues (Lys 12, 13 and 97, all of which are located on the "front side" of cytochrome c_2 surrounding the exposed heme edge) to their involvement in complex formation between the cytochrome bc_1 complex and cytochrome c_2(12). Although only these three lysines at the "top" of the cytochrome c_2 front side have been implicated in binding to the complex by this method, other front-side lysines may also be involved(See below). A cartoon of R. rubrum cytochrome c_2 shows the location of all the lysine residues on the cytochrome(Fig. 7).

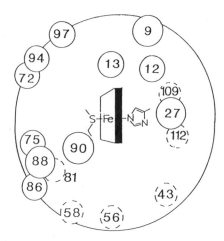

Figure 7. A schematic diagram of R. rubrum cytochrome c_2 as viewed from the front of the heme crevice. The approximate positions of lysine residues are indicated by closed and dashed circles for residues located on the front and back of the molecule, repectively (From Ref. 15).

Further confirmation for the role of "front side" lysines on R. rubrum cytochrome c_2 in positioning the cytochrome for reduction by the cytochrome bc_1 complex comes from kinetic studies using R. rubrum cytochrome c_2 derivatives modified to remove the positive charges on lysine ϵ-amino groups(15). A series of 4-carboxy-2,6-dinitrophenyl(CDNP) derivatives of R. rubrum cytochrome c_2 were prepared which contained one CDNP group per cytochrome c_2(27). The CDNP modifying groups in each derivative were distributed among several different lysines. Four such derivatives(Fractions C1-C4), separable by cation exchange chromatography, had only front side lysine residues modified, while a fifth derivative (Fraction A) had only lysines on the "back side" (situated far from the exposed heme edge) modified by CDNP(27). Eadie-Hofstee plots of the kinetics of electron transfer from $Q_0C_{10}BrH_2$ to R. rubrum cytochrome c_2, equine cytochrome c and the CDNP derivatives, catalyzed by the R. rubrum cytochrome bc_1 complex, indicated that the reaction obeys Michaelis-Menten kinetics for all the acceptors. CDNP modification had little or no effect on v_{max} for the reaction, but modification of front side lysines resulted in a dramatic increase in the K_m for the cytochrome. In contrast, the CDNP derivative of R. rubrum cytochrome c_2 in which only back side lysines are modified had a K_m value much closer to that of the unmodified cytochrome. It should be pointed out that the CDNP derivatives of R. rubrum cytochrome c_2 used in this work had absorbance and circular dichroism spectra indistinguishable from those of native cytochrome c_2 (except for the contributions of the CDNP chromophore) and had E_m values that differed by no more than 10 mV from that of the unmodified cytochrome(27). Thus it appears likely that the effects of CDNP modification of cytochrome c_2 on its reduction by the R. rubrum cytochrome bc_1 complex arise from elimination of positive charges on lysines rather than from some major alteration in cytochrome c_2 conformation. Table 2 summarizes these results and also gives the distribution of the CDNP-modified lysine residues in the four R. rubrum cytochrome c_2 derivatives used in these studies.

Table 2. Kinetic parameters for the reduction of CDNP-cytochrome c_2 derivatives. Lysine positions are shown in bold type(From Ref. 15).

Fraction	% of CDNP-lysine-cyt c_2 derivatives						$(V_{max}/K_m)_{native}$: $(V_{max}/K_m)_{deriv.}$
	9	**13**	**75**	**86**	**88**	**others**	
Fraction C1	36	31	0	24	6	3	22
Fraction C2	5	46	0	25	17	7	23
Fraction C3	4	10	6	6	62	12	22
Fraction C4	3	15	26	9	11	36	23
	58	**81**	**109**	**others**			
Fraction A	19	55	10	16			1.5

As mentioned above, equine cytochrome c can serve as an electron acceptor for the R. rubrum cytochrome bc_1 complex. The kinetic parameters observed for the two cytochrome electron acceptors are very similar despite the large differences in pI's, and thus in net charge, between equine cytochrome c and R. rubrum cytochrome c_2. Examination of the amino acid sequences of these two soluble c cytochromes reveals substantial sequence homologies and, in particular, shows that many of the front side lysines in the vicinity of the exposed heme edge are conserved(28). Modification of these front side lysine residues on equine cytochrome c to eliminate their positive charges produces large increases in the K_m for cytochrome c reduction by the R. rubrum cytochrome bc_1 complex without affecting v_{max}. Modification of back side lysines has little effect on either kinetic parameter(15). Similar effects of lysine modification of R. rubrum cytochrome c_2 or equine cytochrome c on the rate of cytochrome photooxidation by the R. rubrum reaction center had previously been observed, suggesting that the same site on R. rubrum cytochrome c_2 is involved in both its oxidation and reduction(27).

Confirmation of the electrostatic nature of the interaction between the R. rubrum cytochrome bc_1 complex and both R. rubrum cytochrome c_2 and equine cytochrome c came from an investigation of the ionic strength dependence of the rate of cytochrome c_2/c reduction catalyzed by the complex. Increasing the ionic strength had relatively little effect on v_{max} but substantially increased the K_m's for reduction of both cytochromes, as would be expected if electrostaic forces are important in the binding of the soluble cytochromes to the cytochrome bc_1 complex(15). Using a semi-empirical relationship developed earlier(29,30), we were able to estimate that approximately five charged pairs are involved in the R. rubrum cytochrome c_2:cytochrome bc_1 complex interaction and nine pairs are involved in the interaction of equine cytochrome c with the complex(15).

DISCUSSION

A cytochrome bc_1 complex containing only three peptide subunits-cytochrome b, cytochrome c_1 and the Rieske iron-sulfur protein-has been isolated from photosynthetically grown R. rubrum and purified to apparent homogeneity. The bacteriochlorophyll-free complex catalyzes electron flow from quinol to either R. rubrum cytochrome c_2 or equine cytochrome c that is inhibited by antimycin A and myxothiazol. No evidence was found for a small(M_r=8 to 12kDa), quinone-binding subunit, like that reported to be present in the cytochrome bc_1 complexes of other photosynthetic bacteria(1,7,31-34), in the isolated R. rubrum complex. Resonance Raman spectra of the complex have been obtained and have provided information on the environments of the hemes in the complex. ESEEM measurements have established the presence of two nitrogen ligands to iron in the [2Fe-2S] cluster of the R. rubrum Rieske protein.

Evidence has been obtained for the presence of at least one ubiquinone binding site on the cytochrome b subunit of the complex and for electrostatic interaction between the R. rubrum cytochrome bc_1 complex and cytochrome c_2. Because positively charged lysine residues define the binding site on R. rubrum cytochrome c_2 for the R. rubrum cytochrome bc_1 complex, there must exist a complimentary array of negatively charged carboxylate residues on the cytochrome bc_1 complex at the binding site for cytochrome c_2. In mitochondria, cytochrome c(the functional equivalent of cytochrome c_2) binds to cytochrome c_1 and several carboxyl-containing amino acids on cytochrome c_1 have been implicated in this binding(35). These residues are conserved in the Rb. capsulatus cytochrome c_1 sequence(8,9) and one aim of our current gene-sequencing work is to determine whether these residues are also conserved in R. rubrum cytochrome c_1. Another goal of sequencing of the R. rubrum cytochrome bc_1 complex genes will be to calculate the true molecular weight of cytochrome b. The molecular weight of the R. rubrum cytochrome b peptide, estimated from polyacrylamide gel electrophoresis in the presence of SDS, 35 kDa, is considerably smaller than the values near 42 kDa reported for cytochrome b in the cytochrome bc_1 complexes of other photosynthetic bacteria(1,3,7,32,33). While it is possible that the true molecular weight of cytochrome b differs in R. rubrum from that in other photosynthetic bacteria, it is also possible that the R. rubrum cytochrome b either runs anomolously during gel electrophoresis or undergoes some proteolysis during the isolation of the cytochrome bc_1 complex.

ACKNOWLEDGEMENTS

Much of the work done in the author's laboratory was carried out by Dr. R. M. Wynn and Mr. A. Kriauciunas, as part of their of their Ph.D. and M.S. thesis work, respectively. The azido-Q photoaffinity labeling was carried out by Mr. Kriauciunas in collaboration with Professors Chang-An and Linda Yu in the Department of Biochemistry at Oklahoma State University. The differential chemical modification studies were performed at the Biochemical Institute of the University of Zürich in collaboration with Prof. Hans R. Bosshard. The Resonance Raman measurements were performed by Mr. David Hobbs and Prof. Mark Ondrias in the Department of Chemistry at the University of New

Mexico and the ESEEM measurements were performed by Dr. David Britt at the Chemical Biodynamics Laboratory of the University of California, Berkeley. The kinetic studies using lysine-modified derivatives of R. rubrum cytochrome c_2 and equine cytochrome c were performed by Mr. Kriauciunas in collaboration with Dr. Joan Hall and Prof. Francis Millett in the Department of Chemistry and Biochemistry at the University of Arkansas. The identification of the R. rubrum cytochrome bc_1 genes has been carried out by Ms. Savita Shanker, a Ph. D. student in the author's laboratory, in collaboration with Prof. James Harman. Work in the author's laboratory was supported by grants from the U.S. National Science Foundation(PCM-84-08564 and DMB-880609).

REFERENCES

1. Hauska, G., Hurt, E., Gabellini, N. and Lockau, W., Biochim. Biophys. Acta 726:97(1983).
2. Rich, P. R., Biochim Biophys. Acta 768:53(1984).
3. Gabellini, N., J. Bioenerg. Biomembr. 20:58(1988).
4. Crofts, A. R., Meinhardt, S. W., Jones, K. and Snozzi, M., Biochim. Biophys. Acta 723:202(1983).
5. Dutton, P. L., in: "Encyclopedia of Plant Physiology," Vol. 19, Staehlin, A. and Arntzen, C. J., eds., pp. 197-237, Springer-Verlag, W. Berlin (1986).
6. van der Wal, H. N. and van Grondelle, R., Biochim. Biophys. Acta 735:94(1983).
7. Ljungdahl, P. O., Pennoyer, J. B., Roberstson, D. E. and Trumpower, B. L., Biochim. Biophys. Acta 891:227(1987).
8. Gabellini, N. and Sebald, W., Eur. J. Biochem. 154:569(1986).
9. Davidson, E. and Daldal, F., J. Mol. Biology 195:13(1987).
10. Wynn, R. M., Gaul, D. F., Choi, W.-K., Shaw, R. W. and Knaff, D. B., Photosyn. Res. 9:181(1986).
11. Kriauciunas, A., Yu, L., Yu, C.-A., Wynn, R. M. and Knaff, D. B., Biochim. Biophys. Acta, in press.
12. Bosshard, H. R., Wynn, R. M. and Knaff, D. B., Biochem. 6:7688(1987).
13. Hobbs, D., Kriauciunas, A., Güner, S., Knaff, D. B. and Ondrias, M. O., in preparation for submission to Biochim. Biophys. Acta.
14. Britt, D., Sauer, K., Klein, M. P., Knaff, D. B., Kriauciunas, A., Yu, C.-A., Yu, L. and Malkin, R., in preparation for submission to Biochemistry.
15. Hall, J., Kriauciunas, A., Knaff, D. and Millett, F., J. Biol. Chem. 262:14005(1987).
16. Yu, L., Yang, F. and Yu, C.-A., J. Biol. Chem. 260:963(1985).
17. Yu, C.-A. and Yu,L., Biochem. 21:4096(1982).
18. Clayton, R. K., in: "Bacterial Photosynthesis," Gest, H., San Pietro, A. and Vernon, L. P., eds., pp. 495-500, Antioch Press, Yellow Springs Ohio(1963).
19. Haley, P. E., Yu, L., Dong, J. H., Keyser, G. C., Sanborn, M.R. and Yu, C.-A., J. Biol. Chem. 261:14593(1986).
20. Venturoli, G., Fenoll, C. and Zannoni, D., Biochim. Biophys. Acta 892:172(1987).
21. Cramer, W. A., Widger, W. R., Black, M. T. and Girven, M. E., in: "Topics in Photosynthesis," Barber, J., ed., Vol. 8, The Light Reactions, pp. 447-493, Elsevier Medical Press, New York(1987).
22. Fee, J. A., Findling, K.C., Yoshida, T., Hille, R., Tarr, G. E., Hearshen, D. O., Dunham, W. R., Day, E. P., Kent, T. A. and Münck, E., J. Biol. Chem. 259:124(1984).

23. Cline, J. F., Hoffman, B. M., Mims, W. B., LaHaie, E., Ballou, D. P. and Fee, J. A., J. Biol. Chem. 260:3251(1985).

24. Gubriel, R. J., Batie, C. J., Sivaraja, M., True, A.E., Fee, J. A., Hoffman, B. M. and Ballou, D. P., Biochem. 28:4861(1989).

25. Telser, J., Hoffman, B. M., Lo Brutto, R., Ohnishi, T., Tsai, A.-L., Simpkin, D. and Palmer, G., FEBS Lett. 214:117(1987).

26. Mims, W. B. and Peisach, J., in: "Biological Magnetic Resonance," Berliner, C. J. and Reuben, J., eds., Vol. 3, pp. 213-263, Plenum Press, New York(1981).

27. Hall, J., Ayres, M., Zha, X., O'Brien, P., Durham, B., Knaff, D. and Millett, F., J. Biol. Chem. 262:11046(1987).

28. Salemme, F. R., Ann. Rev. Biochem. 46:299(1987).

29. Stonehuerner, J., Williams, J. B. and Millett, F., Biochem. 18:5422(1979).

30. Smith, H. T., Ahmed, A. J. and Millett, F., J. Biol. Chem. 256:4984(1981).

31. Yu, L., Mei, Q.-C. and Yu, C.-A., J. Biol. Chem. 259:5752(1984).

32. Wilson, E., Farley, T. M. and Takemoto, J. Y., J. Biol. Chem. 260:10288(1985).

33. Yu, L. and Yu, C.-A., Biochem. 26:3658(1987).

34. Andrews, K., Ph.D. Thesis, University of Illinois(1988).

35. Stoneheurner, J., O'Brien, P., Geren, L., Millett, F., Steidl, J., Yu, L. and Yu, C.-A., J. Biol. Chem. 260:5392(1985).

NICOTINAMIDE NUCLEOTIDE TRANSHYDROGENASE

IN PHOTOSYNTHETIC BACTERIA

J B Jackson, N P J Cotton, T M Lever, I J Cunningham,
T Palmer and M R Jones

School of Biochemistry
University of Birmingham
P O Box 363
Birmingham B15 2TT, UK

INTRODUCTION

Nicotinamide nucleotide transhydrogenase, found in animal and plant mitochondria and many bacteria, is a membrane protein which catalyses the reversible transfer of hydride equivalents from NADH to $NADP^+$. Biochemically, the enzyme is of particular interest because hydride transfer is coupled to transmembrane proton translocation:

$$NADH + NADP^+ + mH^+ \rightleftharpoons NAD^+ + NADPH + mH^+$$

Hence the rate of the reaction and the ratio $[NAD^+][NADPH]/[NADH][NADP^+]$ are increased by the imposition of a proton electrochemical gradient Δp across the membrane. Transhydrogenase from E.coli has been cloned and sequenced[1]. It has the character of an intrinsic membrane protein with two subunits, $\alpha[M^r 54000]$ and $\beta[M^r 48700]$. The bovine mitochondrial enzyme has only a single subunit $[M^r 109000]$. Its amino acid sequence has been deduced from the corresponding $cDNA^2$ and allowing for 13 extra residues at the N-terminus, 6 at the C-terminus and a frame shift at residue 793, it shows considerable homology with transhydrogenase from E. coli. The hydrophathy plots of the two enzymes are very similar. Three domains can be recognised in each protein. The first 40 residues of the N-terminus of the bovine enzyme (or of the α-subunit of the E. coli enzyme) are relatively hydrophilic but contain one predominantly hydrophobic segment. The central $\sim$ 400 residues of the bovine enzyme (or the last $\sim$ 100 residues of the α-subunit and the first $\sim$ 300 residues of the β of the E. coli enzyme) are hydrophobic and, in principle, could contain up to 14 transmembrane helical

Molecular Biology of Membrane-Bound Complexes in Phototrophic Bacteria
Edited by G. Drews and E. A. Dawes
Plenum Press, New York, 1990

segments. The last $\sim$200 residues at the C-terminus of the mitochondrial protein (or of the E. coli β -subunit) are relatively hydrophilic.

For kinetic studies there are two significant advantages associated with studies of transhydrogenase in chromatophores from photosynthetic bacteria. First, the membranes can be energised for defined periods using light and second, in bacteria such as Rhodobacter capsulatus, the existence of electrochromic absorbance changes makes possible the measurement of membrane potential ($\Delta\psi$) and membrane ionic currents on a rapid timescale. The dependence of the rate of the transhydrogenase reaction in chromatophores on the value of $\Delta\psi$ in steady-state (for Δ pH = 0) is not affected by the manner in which $\Delta\psi$ is varied (with electron transport inhibitors, uncouplers or ionophores[3]). Thus, in this system, there is good reason to believe that $\Delta\psi$ is the sole driving force for transhydrogenase and that local interactions with electron transport components are not significant. In this communication we describe in more detail the relationship between $\Delta\psi$ and the rate of transhydrogenase and discuss mechanisms which could explain the energy-dependence of the reaction.

Structural information on transhydrogenase from photosynthetic bacteria is relatively scarce. However there is evidence that the enzyme in Rhodospirillum rubrum differs fundamentally from that in E. coli and mitochondria. Thus a soluble factor, ostensibly a protein, can be washed off chromatophore membranes from Rhs. rubrum with accompanying loss of transhydrogenase activity[4,5]. Activity can be restored to the membranes after addition of partially purified factor[4]. These observations are confirmed below. The influence of nucleotides and various chemical modifications on reconstitution have been studied in depth[6] but here we show that Mg^{++} concentration is particularly critical. We also show that, like transhydrogenase from Rb. sphaeroides[7], and unlike that from Rhs. rubrum, the enzyme from Rb. capsulatus cannot be separated into soluble and membrane components. A solubilisation and purification procedure for transhydrogenase from Rb. capsulatus is outlined and parallels with the E. coli enzyme are revealed.

MATERIALS AND METHODS

Rb. capsulatus strain 37b4 (from Dr G Drews, Freiburg) and strain N22 (from Dr N G Holmes, Bristol) were grown anaerobically in the light in RCV medium, as described[3]. Rhs. rubrum (from Dr L Slooten, Brussels) was grown under similar conditions in RCV supplemented with biotin.

Chromatophores were prepared as in[3]. Transhydrogenase in chromatophores was measured with NADH and the analogue substrate, thio-

Table 1. Effect of washing on transhydrogenase activity in
chromatophores from Rhs. rubrum and Rb. capsulatus

	Rate of light–driven transhydrogenase (μmol thio-NADPH/μmol BChl/min)	
	No Additions	Plus Fraction of Cell Extract
Rhs.rubrum chromatophores	1.01	
after buffer wash	0.28	1.14
after buffer + 5 mM MgCl$_2$ wash	0.58	1.13
after buffer + 0.5 mM EDTA wash	0.16	0.94
Rb.capsulatus chromatophores	0.53	
after buffer wash	0.65	

Chromatophores were prepared in 10% sucrose, 0.1M Tris–HCl, pH 8.0, but
otherwise as described[3,8]. Washing by centrifugation was in the same
medium or supplemented with MgCl$_2$ or EDTA as shown. The cell extract
from the Rhs. rubrum chromatophore preparation was used to make
"transhydrogenase factor type II" by ammonium sulphate precipitation as
described[6]. Assays were performed in 43 mM Tris–Cl pH 8.0, 125 mM
sucrose, 2.67 mM MgCl$_2$, 32 μM thio-NADP$^+$, 120 μM NADH, 6.7 μM BChl.

NADP$^+$, as described[3]. Transhydrogenase activity in solubilised material
was determined with NADPH and acetylpyridine nicotinamide dinucleotide
(APAD$^+$), see[8]. Measurements of nucleotide concentrations on the approach
to equilibrium were measured by a modification of[9].

RESULTS AND DISCUSSION

Structural aspects of the transhydrogenase enzyme in chromatophores

Fisher and colleagues showed that when chromatophores of Rhs. rubrum
were washed by centrifugation in low ionic strength buffers, there was a
loss of transhydrogenase activity[4-6]. Table 1 confirms this observation
and shows that the presence of Mg^{++} decreased and the presence of
EDTA enhanced the loss of activity during washing. The rate of the
reaction was restored in each case by addition of an ammonium sulphate
precipitate of soluble cell extract obtained during chromatophore
preparation. Under similar conditions the transhydrogenase activity of
Rb. capsulatus was resistant to washing (Table 1).

The transhydrogenase of Rb. capsulatus chromatophores, however,
was solubilised with 0.5% Triton X–100. Generally the recovery of
activity from the membranes was close to 100%. Some inactivation, though
minimised in the presence of dithiothreitol, occurred upon storage at
4° (50% loss of activity in 90 hrs). This was not prevented at 4° by
glycerol, EDTA, phenylmethylsulphonyl fluoride or β–mercaptoethanol but

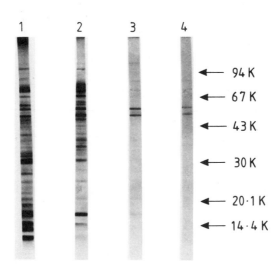

Fig.1 Purification of transhydrogenase from <u>Rb. capsulatus</u> strain 37b4.
Sodium dodecylsulphate polyacrylamide gel electrophoresis of samples
taken during detergent extraction and subsequent chromatography.
Track 1, Triton-X100 extract of membranes. Track 2, after chromatography
on DEAE-Sephacryl. Track 3, after chromatography on Ultrogel
hydroxylapatite. Track 4, after chromatography on AcA-44.

the enzyme was stable at -15° in 25% glycerol for many weeks. The
solubilised material was purified by ion-exchange chromatography in media
containing 0.2% Triton X-100 on DEAE-Sephacryl followed by hydroxylapatite
and gel-exclusion chromatography. The specific activity in the presence of
supplementary phospholipid was typically 3 μmol APAD^{+}/ mg protein/min.
Following polyacrylamide gel electrophoresis in sodium dodecylsulphate there
were two major silver-staining bands at 45k and 50k (Fig.1). The average
molecular weights of bands from a number of preparations were 48k and 53k
which suggested that there may be structural similarities between the
transhydrogenase of <u>E.coli</u> and <u>Rb. capsulatus</u>[10]. The gel also suggests
that transhydrogenase is a prominent component in the membranes of <u>Rb.
capsulatus</u>. Thus two bands attributable to the subunits of transhydrogenase
can be seen to stain heavily in the detergent extract of the chromatophore
membranes. A similar observation was made after staining with Coomassie
(data not shown).

H^{+}/H^{-} ratio of transhydrogenase by an equilibrium method

In agreement with measurements on transhydrogenase from
mitochondria[11] we recently showed using a kinetic approach[8] that
the ratio of H^{+} translocated per H^{-} transferred by transhydrogenase
from <u>Rb. capsulatus</u> could be 1.0 but it is unlikely to be 2.0.

Table 2. The H^+/H^- ratio for transhydrogenase in Rb. capsulatus

Expt.	NADH	NAD$^+$	NADPH	NADP$^+$	K	$\Delta\psi$(v)	H^+/H^-
1	10	178	130	33	70	0.16	0.72
2	13	151	142	15	110		

Chromatophore samples (volume 0.6 ml, 20 μM BChl) were incubated in the light (using a Hero OD–100 GaAlAs IR emitter at 500 mA) in 10% sucrose, 40 mM KCl, 2 mM MgCl$_2$, 50 mM tricine, pH 7.6, 0.26 μg/ml venturicidin, 5 μg/ml rotenone, 0.13 μg/ml nigericin. The initial nucleotide concentrations were 200 μM NADH, 200 μM NADP$^+$. After 20 min, the samples were quenched with either 0.3 ml 14% HClO$_4$ or 0.12 ml 2M KOH, neutralised and assayed enzymically for oxidised or reduced nucleotides, respectively (see[9]). Concentrations are shown as μM. Membrane potentials were measured by electrochromism under similar illumination conditions using the calibration procedure described[12].

Illumination of a suspension of Rb. capsulatus chromatophores in the presence initially of 200 μM NADH and 200 μM NADP$^+$ led to an increase in the ratio [NAD$^+$][NADPH]/[NADH][NADP$^+$] (= K) until a constant value was reached after approx. 15 to 20 min. In the presence of venturicidin and 5 μg/ml rotenone, K typically reached ∼400. In the presence additionally of 0.13 μg/ml nigericin (to selectively reduce ΔpH) or of 2 μM FCCP (to depress Δp), the value of K after 20 min illumination was ∼100 and ∼6, respectively. When illumination was carried out initially with 400 μM NAD$^+$ and 400 μM NADPH (venturicidin and rotenone present) the value of K declined from a high value (> 4500) to ∼110 after 20 min illumination. Rotenone was required in all these experiments to inhibit the NADH dehydro-genase. Separate experiments showed that 5 μg/ml rotenone inhibited chromatophore NADH oxidase activity in the dark by 96%. There was no detectable NADPH oxidase in the presence of rotenone.

Table 2 summarises a series of experiments to estimate the H^+/H^- ratio, assuming that transhydrogenase reaches equilibrium with $\Delta\psi$ (= Δp since ΔpH = 0). $\Delta\psi$ was estimated under similar conditions to those used for nucleotide sampling. The mean H^+/H^- ratio (= $RT\ln K/F\Delta\psi$) was 0.72. The fact that K declined when initially set to a high value (see above) supports the contention that equilibrium was reached. Control experiments in the absence of chromatophores and in the presence of known nucleotide concentrations showed that the quenching and assay procedures were reliable. However, there are two reasons for supposing that the measured value of K (and hence H^+/H^-) is a lower limit. (1) It can not be ruled out that there is a sub-population of membranes with uncoupled transhydrogenase. (2) It was not possible to eliminate completely NADH oxidase activity (see above). Thus, in the presence of

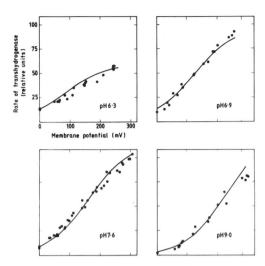

Fig.2 The pH dependence of the relationship between rate of
        transhydrogenase and $\Delta\psi$ in chromatophores of Rb. capsulatus
        strain N22.
Transhydrogenase rates and $\Delta\psi$ (from electrochromic absorbance changes)
were measured in a medium containing 10% sucrose 100 mM KCl, 10 mM
$MgCl_2$, and either K^+-MES (pH 6.3), K^+-Mops (pH 6.9), K^+-Tricine
(pH 7.6) or K^+-Bicine (pH 9.0). The nucleotide concentrations were
200 μM NADH, 300 μM thio-NADP$^+$ (pH 6.3); 800 μM NADH, 300 μM thio-NADP$^+$
(pH 6.9); 800 μM NADH, 200 μM thio-NADP$^+$ (pH 7.6); 600 μM NADH, 500 μM
thio-NADP$^+$ (pH 9.0). FCCP (up to 2.0 μM) was used progressively to
lower $\Delta\psi$. Transhydrogenase rate and $\Delta\psi$ were measured in parallel
cuvettes under similar conditions. The solid lines in each panel were
calculated from equ (iv) using d = 0.4, k_1^o at acid pH = 17.5 and at
alkaline pH = 3.5 with a pK_a of 7.5 and k_2^1 at acid pH = 0 and at
alkaline pH = 120 with a pK_a of 6.3.

5 μg/ml rotenone and 200 μM NADH but in the absence of added NADP$^+$,
low levels of NAD$^+$ (between 0–10 μM) were detected in chromatophore
samples after 20 min illumination. It is not possible even to guess at
how much K is underestimated although it is worth pointing out that for
$\Delta\psi$ = 0.16v, K would be 464 for H^+/H^- = 1 and 215,000 for H^+/H^- = 2.
We have never observed K values in excess of 500.

The dependence of transhydrogenase rate on membrane potential in
chromatophores from Rb. capsulatus

The K_m^{app} values for NADH and thio-NADP$^+$ in the transhydrogenase
reaction were determined at a range of pH values in darkened and
illuminated chromatophore membranes[8]. This enabled us subsequently to
monitor the dependence of the transhydrogenase rate on $\Delta\psi$ (at ΔpH = 0)
as a function of pH under conditions in which the enzyme was essentially
saturated with nucleotide substrates in the forward reaction (nucleotide
concentration in excess of 8 x K^{app}). The results are shown in Fig.2.

The data at pH 7.6 are qualitatively similar to those described[3] at lower nucleotide concentrations. At $\Delta\psi = 0$ there was a low but finite transhydrogenase rate. At low $\Delta\psi$ there was a slightly disproportionate increase in rate with increasing $\Delta\psi$ and thereafter the relation was approximately linear with $\Delta\psi$. When the dependence of transhydrogenase rate on $\Delta\psi$ was determined at other pH values, the following features were observed: (a) At pH 6.9 the data were indistinguishable from those at pH 7.6. (b) At pH 9, the threshold appearance became more pronounced as if the relationship was shifted along the $\Delta\psi$ axis. (c) At pH 6.3 the transhydrogenase rate at $\Delta\psi = 0$ was much larger than at pH 7.6, but the increase in rate was less pronounced when $\Delta\psi$ was increased.

Two models have been considered to explain the data. In both cases it is assumed that the rate of only a single step in the catalytic sequence is dependent on $\Delta\psi$. At high values of $\Delta\psi$ another step in the sequence (with a first order rate constant, k_2) becomes limiting. In steady-state conditions (viz. Fig.2) the sequence of the $\Delta\psi$-dependent and the $\Delta\psi$-independent reactions is not critical. All other reactions are fast (but k_2 can be compounded by two or more first order or pseudo-first order rate constants). The two models lead to clearly different predictions of the pH dependence for the relationship between transhydrogenase rate and $\Delta\psi$.

A. Protons approach the catalytic centre of the transhydrogenase enzyme by way of a cleft or proton well[13] extending from the chromatophore lumen. Across the proton well the $\Delta\psi$ is transformed into ΔpH such that the $[H^+]$ at the bottom of the well is increased according to

$$[H^+]_w = 10^{-(pH_B - \Delta\psi hF/RT)} \qquad \text{equ (i)}$$

where pH_B is the bulk phase pH and h is the fractional depth of the well. Then, assuming that a single proton is involved in the catalytic reaction [see above], the steady-state rate of transhydrogenase is given by

$$v = k_2[E_T]/(1 + k_2/k_1^o[H^+]_w) \qquad \text{equ (ii)}$$

where $[E_T]$ is the total concentration of transhydrogenase and k_1^o is the second order rate constant for proton binding at the bottom of the well.

The curves generated from these equations are shown in Fig.3. In agreement with the data the model predicts (a) a relationship at pH 7.6 which matches the experimental results, (b) a shift of the relationship along the $\Delta\psi$ axis at pH 9.0 and (c) an increased transhydrogenase rate at $\Delta\psi = 0$ at pH 6.3. However quantitative predictions at high and low pH are poor and the divergence in the data points at pH 6.9 and pH 7.6 are not matched by the model.

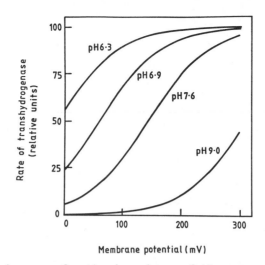

Fig.3   Simulated curves for the dependence of the rate of transhydrogenase on $\Delta\psi$ using a proton well model.
The curves were generated from equ (i) and (ii) using h = 0.5, k_1^o =
2.5 x 10^8, k_2 = 100.

B.   The $\Delta\psi$ –dependent step in transhydrogenase involves the movement of
charge (e.g. H^+) through a region of low dielectric constant (and
therefore large voltage drop) in the membrane protein across an
energy barrier. The rate constant of the $\Delta\psi$ –dependent step is
increased according to[14,15]:

$$k_1 = k_1^o \exp (dF \Delta\psi/RT) \qquad\qquad \text{equ (iii)}$$

where k_1^o is the rate constant at $\Delta\psi = 0$ and d is the fractional
depth of the Eyring barrier for charge translocation through the
transhydrogenase enzyme across the chromatophore membrane. The
steady–state rate of transhydrogenase is then given by

$$v = k_2 [E_T]/\left\{1 + k_2/k_1^o \exp (d F \Delta\psi/RT)\right\} \qquad \text{equ (iv)}$$

With appropriately chosen parameters this model also leads to a
good fit of the experimental data at pH 7.6. Moreover it predicts that
the relationship between transhydrogenase rate and $\Delta\psi$ will be independent
of suspension pH, which is in accordance with the experimental
observations at pH 6.9 and 7.6 (Fig.2). The model fails to predict the
results at extremes of pH (6.3 and 9.0). However it is not unusual, of
course, for rate constants in enzyme catalysed reactions to depend on pH
and the model can be fitted well to the data by ascribing a single acid
dissociation constant to k_1^o and to k_2 (the solid lines in Fig.2).
    Thus, while neither model fits the data precisely without making
secondary assumptions, there are pleasingly predictive aspects of both.
In a sense the two models are complementary. Both encompass the idea of
a proton–conducting path to the catalytic centre from the chromatophore
lumen but in model A it is assumed that the conduction of H^+ is fast
compared with the catalytic step. Thus electrochemical equilibrium of

H^+ is maintained along the length of the proton-conducting path. The driving force for the reaction is the chemical potential difference of H^+ at the bottom of the well and at the H^+ release site on the outside (cytoplasmic side) of the enzyme. However in model B proton conduction along the path limits the rate of reaction (at low $\Delta\psi$) and the driving force is the electric potential difference between the chromatophore lumen and the catalytic centre. Quantitatively both models incorporate implicit assumptions, in particular about the dielectric properties of the protein. Model B assumes a single Eyring barrier and that the electric field is constant within the membrane[16]. The theory can be made more general by including terms in the dielectric constant to calculate the voltage drop across the barrier: effectively to replace d in equ (iii) with the fractional "dielectric depth" of the barrier. Model A assumes that the presence of the proton conducting pathway does not distort the local dielectric constant and therefore the voltage drop across the membrane through the protein[13]. In fact the existence of charged or polar groups in the channel would raise the local dielectric constant so that the dependence of $[H^+]_w$ on $\Delta\psi$ could be less pronounced than equ (i) predicts. Again, this theory could be made more general by adopting the concept of a fractional "dielectric distance" across the membrane for parameter h in equ (i). It may be noted that the two models are not mutually exclusive; it can be envisaged that $\Delta\psi$ drops across the enzyme through a proton well and across an energy barrier operating in series (compare models for ion transporting ATPases, e.g.[17]).

ACKNOWLEDGEMENTS

This work was supported by the Science and Engineering Research Council.

REFERENCES

1. Clarke, D.M., Loo, T.P., Gillam, S. and Bragg, P.D. Nucleotide sequence of the pntA and pntB genes encoding the pyridine nucleotide transhydrogenase of Escherichia coli. Eur. J. Biochem. 158:647-653 (1986)

2. Yamaguchi, M., Hatefi, Y., Trach, K. and Hock, J.A. The primary structure of the mitochondrial energy-linked nicotinamide nucleotide transhydrogenase deduced from the sequence of cDNA clones. J. Biol. Chem. 263:2761-2767 (1988)

3. Cotton, N.P.J., Myatt, J.F. and Jackson, J.B. The dependence of the rate of transhydrogenase on the value of the protonmotive force in chromatophores from photosynthetic bacteria. FEBS Lett. 219:88-92 (1987)

4. Fisher, R.R. and Guillory, R.J. Partial resolution of energy-linked reactions in Rhodospirillum rubrum chromatophores. FEBS Lett. 3:27-30 (1969)

5. Fisher, R.R. and Earle, S.R. in "Pyridine nucleotide coenzymes" Everse et al. eds. Chapter 9. pp.279–324, Academic Press, New York, (1982)

6. Fisher, R.R. and Guillory, R.J. Resolution of enzymes catalysing energy-linked transhydrogenation: Interaction of transhydrogenase factor with the Rhodospirillum rubrum chromatophore membrane. J. Biol. Chem. 246:4679–4686 (1971)

7. Konings, A.W.T. and Guillory, R.J. Specificity of the transhydrogenase factor for chromatophores of Rhodopseudomonas sphaeroides and Rhodospirillum rubrum. Biochim. Biophys. Acta. 283:334–338 (1972)

8. Cotton, N.P.J., Lever, T.M., Nore, B.F., Jones, M.R. and Jackson, J.B. The coupling between the proton motive force and the NAD(P)$^+$ transhydrogenase in chromatophores from photosynthetic bacteria. Eur. J. Biochem. 182:593–603 (1989)

9. Lee, C.-P. and Ernster, L. Equilibrium studies of the energy-dependent and non-energy-dependent pyridine nucleotide transhydrogenase reactions. Biochim. Biophys. Acta. 81:187–190 (1964)

10. Clarke, D.M. and Bragg, P.D. Purification and properties of reconstitutively active nicotinamide nucleotide transhydrogenase of Escherichia coli. Eur. J. Biochem. 149:517–523 (1985)

11. Earle, S.R. and Fisher, R.R. A direct demonstration of proton translocation coupled to transhydrogenation in reconstituted vesicles. J. Biol. Chem. 255:827–830 (1980)

12. Jackson, J.B. and Clark, A.J. Carotenoid absorption band shifts and distribution of butyltriphenylphosphonium ions as membrane potential indicators in intact cells of photosynthetic bacteria, in "Vectorial reactions in electron and ion transport", F. Palmieri et al. eds. pp. 371–379 Elsevier/North Holland Biomedical Press, (1981)

13. Mitchell, P. Chemiosmotic coupling and energy transduction. Theoret. Exp. Biophys. 2:159–215 (1969)

14. Junge, W. The critical electric potential difference for photophosphorylation. Eur. J. Biochem. 14:582–592 (1970)

15. Laüger, P. and Stark, G. Kinetics of carrier mediated ion transport across lipid bilayer membranes. Biochim. Biophys.Acta 211:458–466 (1970)

16. Apell, H.J., Borlinghaus, R. and Laüger, P. Fast charge translocations associated with partial reactions of the Na$^+$, K$^+$ pump. J. Membr. Biol. 97:179–191 (1987)

17. Laüger, P. Thermodynamic and kinetic properties of electrogenic ion pumps. Biochim. Biophys. Acta. 779:307–341 (1984)

THERMODYNAMIC AND KINETIC FEATURES OF THE REDOX CARRIERS OPERATING IN THE
PHOTOSYNTHETIC ELECTRON TRANSPORT OF CHLOROFLEXUS AURANTIACUS

Giovanni Venturoli, Reiner Feick*, Massimo Trotta+
and Davide Zannoni
Department of Biology, Institute of Botany, University
of Bologna Bo (I); +Centro C.N.R., Interazione Luce
Materia Bari(I);*Max-Plank Institut fur Biochemie,
Martinsried (W.G.)

INTRODUCTION

Chloroflexus aurantiacus is a thermophilic green photosynthetic
bacterium containing a reaction center with a pigment composition
different from that of purple bacteria; the reaction center is deficient
in carotenoids and instead of four bacteriochlorophylls (BChls) and two
bacteriopheophytins (BPhs) it contains 3 BChls and 3 BPhs (Blankenship et
al., 1984). Similarly, the protein composition of Chl.aurantiacus RC is
quite peculiar: (a) it is composed of only two protein subunits (L,M);
(b) peptide-mapping of these two polypeptides indicates a high degree of
structural similarity; (c) it is the smallest functionally active RC thus
far and (d) it shows thermal stability (Shiozawa et al. 1987; Schiozawa
et al., 1989; Pierson et al., 1983). Conversely, the photochemical and
early electron-transfer reactions, as determined by fast spectroscopy and
circular dichroism in RCs preparations, suggest many similarities with
the purple bacteria counterpart (Kirmaier and Holten, 1987). In addition,
several aspects of secondary electron transport (Zannoni and Ingledew,
1985) and the light-dependent energy trasducing machinary of
Chloroflexus (Venturoli and Zannoni, 1987) can be brought back to
previous observations in purple non-sulphur bacteria.

Obligate anaerobes such as Chlorobium , Chromatium and
Heliobacterium chlorum contain large amounts of menaquinones (MK).
Interestingly, MK is found associated with ubiquinone (UQ) in species
such as, for examples, Chromatium and R.gelatinosa while in
Chloroflexus , Chlorobium and Heliobacterium is the only quinone
present. In this connection, there are four other groups of Gram
bacteria that do not contain UQ but do grow aerobically. One of these
groups is the thermophilic genus Thermus and another is the gliding
bacteria. Remarkably, thermophilic and gliding properties are found in
Chloroflexus .

A significant difference between Chloroflexus and Chlorobium is

Molecular Biology of Membrane-Bound Complexes in Phototrophic Bacteria
Edited by G. Drews and E. A. Dawes
Plenum Press, New York, 1990

425

that the capacity of aerobic growth is present only in the former. Thus Chloroflexus is the only chemioheterotrophic green photosynthetic bacterium that contains MK as the sole quinone species (Hale et al., 1983).

Here, a brief overview of the most recent acquisitions on the redox carriers involved in both primary and secondary photoactivated electron transport reactions catalyzed by RCs and membrane fragments of Chl.aurantiacus , is presented. Notably, a more detailed analysis of the experiments presented in Fig. 1, will be presented in a subsequent paper.

MATERIALS AND METHODS

Organism cultivation and membrane isolation

The medium used for photoheterotrophic growth of Chloroflexus aurantiacus , strain J-10fl was that described by Pierson and Castenholz (1974). Cells were cultivated in screw-capped bottles at $55^{\circ}C$ with an incident light intensity of 400W m^{-2} for 48h. Membrane fragments were prepared as previously described (Zannoni and Ingledew, 1985).

Isolation of Chl.aurantiacus reaction center
The Chl.aurantiacus RCs were isolated as described by Shiozawa et al. (1987).

Kinetic spectrophotometry

The kinetics of flash induced redox changes of cytochromes and reaction centers were measured using an home-made single beam spectrophotometer with a bandwidth of 1.5nm. Flash excitation was provided by a xenon lamp of 5μs duration at half maximal intensity. Rapid digitisation of the photomultiplier linear amplifier output was done by a Le Croy Transient Recorder (Mod.TR8818A, equipped with a 128K byte memory module) and interfaced to an Olivetti M280 computer. The temperature control was achieved by holding the cuvette in a cylindric massive brass block filled with glycerol and channelled for water circulation. The temperature in the cuvette was measured with a Pt-100 thermo-resistance (R2104, Degussa-Messtechnik F.R.G.) with a tolerance of 0.3 C.

RESULTS AND DUSCUSSION

Chloroflexus aurantiacus complex III

Em values for the Rieske protein in purple bacteria range from +280 to 315mV (pH7.0) similar to those measured for the mitochondrial protein (Prince et al., 1975; Dutton and Leigh, 1973). However, the Rieske proteins from both Chl.aurantiacus (Em=+100mV, Zannoni and Ingledew, 1985) and Chl.limicola (Em6.8=+165mV) are considerably less positive than those from the purple bacteria. The much lower value for the Rieske protein in these green bacteria is probably related to the relatively low Em values of the reaction center and cytochromes c553-554 (Amesz and Knaff, 1988). Among the several b type species detected in

photosynthetically grown Chl.aurantiacus , the 2 hemes with Em7.0 of
+65mV and -70mV might be analogous to those associated with mitochondrial
and bacterial b/c1 complexes (Zannoni and Ingledew, 1985). Cytochromes
with similar mid point potentials have also been detected in aerobically
grown Chloroflexus and functional studies indicated that high
concentrations of antimycin A and/or myxothiazol (20µM each) reduce both
the rate and the extent of the NADH dependent reduction of cytochromes c
(signal at 554-540nm) (Zannoni, 1986). Although this latter evidence
suggests the presence of a putative b/c1 complex, in fast kinetic
experiments where care was taken not to add inhibitors in molar excess
over the estimated content of the cytochrome b/c1 complex, it was found
that photoxidized cyt. c554 is re-reduced in the dark in a reaction
which is sensitive to HQNO (heptyl- hydroxy- quinoline-N-oxide) but
mostly insensitive to antimycin A, mucidin, myxothiazol, and UHDBT
(undecylhydroxy- dioxobenzothiazole) (Zannoni and Venturoli, 1988).
Recently, the effects of HQNO on the light induced redox changes of
cytocromes c of Chl.aurantiacus have been examined at high
temperatures (Fig.1) (Venturoli,Trotta and Zannoni, in preparation).
Similarly to previous results at room temperature (Zannoni and Venuroli,
1988), both the extent and kinetics of the signal at 554-542nm are
dramatically changed so to demonstrate that the affinity of HQNO toward
complex III of Chloroflexus does not seem to be affected at high
temperatures.

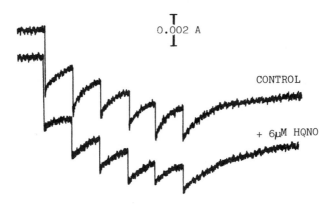

T
0.002 A

CONTROL

+ 6µM HQNO

Fig.1. Flash-induced cytochromes c oxidation, measured at 554-542nm
       (48 °C), in presence or in the absence of HQNO (6µM) in membranes
       from light-grown Chl.aurantiacus . Membranes were suspended in
       20mM MOPS (pH7.5) containing 0.4mM ascorbate. Traces were average
       of 4; sweep, 820ms; filter RC, 1ms.

Membrane-bound cytochromes

    Chromatium vinosum contains two distinct bound c -cytochromes
(Bartsch, 1978), cyt. c555 (Em=+340mV) and cyt. c552 (Em=+10mV). When
cyt. c552 in chromatophores is reduced prior to flash illumination, it
is oxidized by P$^+$ with t1/2=1µs, and no oxidation of cyt. c555 is
observed. Cytochrome c552 appears to be closely associated with the RC

427

since oxidation of such cytochromes by P^+ occurs at a high rate (even at cryogenic temperatures) (Dutton, 1971; Dutton and Prince, 1978). However, the subsequent re-reduction of cyt. c552 is extremely slow excluding a role of this cytochrome in cyclic electron flow. In contrast, the role of the higher potential cyt. c555 in photosynthesis has reasonably been established (Parson, 1968).

Considerably less is known about cytochrome reactions in green, such as Chloroflexus , than in purple bacteria. Nevertless, it is clear that, as in purple bacteria, a membrane bound cytochrome c (cyt. c554) is the immediate electron donor to P^+ (Bruce et al., 1982). Recent evidence, obtained in membranes and in the isolated c554 , suggests that this cytochrome contains multiple hemes and that it is only present in phototrophically grown cells (Zannoni, 1986; Zannoni and Venturoli, 1988; Blankenship et al, 1988). The redox titration of isolated c554 shows a quite complex behaviour since it suggests the presence of at least three distinct redox centers. On the other hand, an estimate of 2 hemes per peptide (Blankenship et al., 1986) seem to be in line with kinetic evidence by Zannoni and Venturoli (1988) indicating that 2 hemes with Em7.0 of +295mV and +140mV are rapidly oxidized (t1/2<15µs) by the RC following a train of flashes. Interestingly, while the extent of the rapid c554 -photooxidation is not affected by HQNO at the 1st flash of light, the extent of the signal at 554-542nm between the 2nd and the 12th flash of light (from 30 to 330ms) is clearly enhanced in a range of ambient redox potential of about 150mV (100mV<Eh7.0<250mV) (see Fig.1, in : Zannoni and Venturoli, 1988). Notably, most of the signal which becomes observable from 30ms to 330ms can be accounted for by the presence of a redox component with an apparent Em of approx +220mV. This latter signal, might be due to a c -type cytochrome (analogous to c1), previously identified in dark equilibrium redox titrations by Zannoni and Ingledew (1985), which goes rapidly oxidized following the 2nd of a series of actinic closely spaced flashes of light.

Soluble electron-transport components

Chloroflexus aurantiacus contains a blue copper protein (auracyanin) with EPR parameters similar to stellacyanin, rusticyanin and plantacyanin (Trost et al., 1988). Auracyanin has a midpoint potential of +240mV, a value close to that reported for stellacyanin (+184mV). Although the function of auracyanin is not certain, preliminary kinetic data suggest that it might act as a redox carrier connecting the b/c1 complex to the RC, as previously shown for plastocyanine or soluble c type cytochromes (McManus et al., 1988). This function would be consistent with the midpoint potential of auracyanin, being 20mV more positive than the photoxidizable c220 (Em7.0=+220mV) and 55mV less positive than c295 (Em7.0=+295), one of the two hemes which are rapidly oxidized by P870 (Zannoni and Venturoli, 1988). On the other hand, it would be also possible for auracyanin to donate electrons directly to the oxidized reaction center (Em7.0=+420mV) (Venturoli and Zannoni, 1988) since it has not been demonstrated unequivocally that cyt. c554 is the obligate electron mediator between b/c1 and P870. Owing to the fact that this latter possibility is rather unlikely, it might be worth to underline

that both the thermophilic nature of <u>Chloroflexus</u> and the lack of
evidence that auracyanin is located in the periplasmic space do not
preclude the possibility of a direct redox interaction between the b/c1
and the "functional-complex" <u>c554-P870</u> .

Primary photochemistry

Previous work indicated that the amount of MK present in light-grown
<u>Chloroflexus</u> is close to the values reported in similar cultures of
<u>Chlorobium</u> and <u>Chromatium</u> (Hale et al., 1983). In these latter genera
a MK/BChl ratio of 1:20 is normally found (Zannoni and Melandri, 1985).
The marked difference between <u>Chloroflexus</u> and other phototrophs is
that MK-10 is the predominant homologue in the former. Recently, a
preliminary estimation of te MK/BChl and MK/RC ratios in membranes from
light grown <u>Chloroflexus</u> has been reported (Venturoli,Trotta and
Zannoni, 1989). These data indicated that <u>Chloroflexus</u> contains a high
amount of MK (MK/RC= 100-120) supporting the concept that most of the
loosely bound quinone might functionn as a pool (Venturoli and
Zannoni,1987). Recent kinetic and thermodynamic studies have also
supported the concept that both Qa and Qb are MK-molecules and also that
the binding of Qb is modulated by the redox and protonation state of Qa
(Venturoli and Zannoni, 1988). These measurements have also indicated
that Qa titrates at -210mV (pH8.2) with a pH dependence of -60mV/pH unit
up to a pK value of 9.3 (pKa), implying that at physiological pHs the
equilibration reduction of Qa is linked to a receipt of a proton.
Analysis of the back reaction at high pHs suggested a pKb close to 10
along with a free energy difference between $Q\bar{a}$ Qb and the $QaQ\bar{b}$ states of
-60meV and -10meV at low and high pH, respectively. Assuming that Qb is
in equilibrium with the MK-pool, the Em7.0 of the loosely bound MK would
be close to -100mV a value which is extremely lower than the optimum
ambient redox potential for light-induced phosphorylation by membrane
fragments from <u>Chloroflexus</u> (Eh8.2= 0mV) (Zannoni and Venturoli, 1988).

In Fig.2, the kinetics of Δ A at 542nm and 550nm induced by a short
flash of light in isolated <u>Chl.aurantiacus</u> RCs in the presence or in
the absence of electron donors and electron acceptors, are shown.

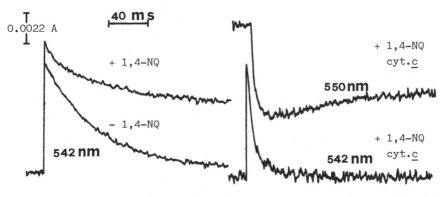

Fig.2. The kinetics of ΔA at 542nm (RC-band) and 550nm (cyt.<u>c</u>-band) in
<u>Chl.aurantiacus</u> RCs (26°C,pH 7.2) (see text for more details).

It is apparent that in the presence of ascorbate (60µM) and 1,4–naphthoquinone (20µM) a slow relaxation (>400ms) of the RC-bleaching takes place due to electron transfer from Qa to 1,4-NQ and stabilization of P^+. Addition of horse-heart cyt. $\underline{c}$ (20µM) induces a fast reduction (t1/2=4ms) of P^+ whith a concomitant oxidation of the soluble $\underline{c}$. These data allowed us to calculate the extinction coefficient of Chl.aurantiacus RC at 542nm. The value obtained, $27m\overline{M}^{1}cm^{1}$, is considerably higher than the correspondent ε ($10.3mM^{-1}cm^{-1}$) of Rhodobacter species. Another interesting point emerging from the kinetics of Fig.2, is that the rate of charge recombination between Qa and P^+ in the isolated RC (t1/2=37ms) is slightly slower than that measured in vivo in the presence of 2mM O-phenantroline (t1/2=30ms, Venturoli and Zannoni, 1988). These results, along with evidence that O-phenantroline abolishes quinone oscillations in RCs with UQ plus diaminodurene acting as electron acceptors (Blankenship et al.1988), demonstrate that the lack of subunit H in Chloroflexus RC, does not preclude the possibility to reconstitute a functionally active reaction center. Interestingly, quite recent data have also demonstrated that the rate of charge recombination in isolated RCs remains unaltered between 8 °C and 55 ℃ (not shown).

ACKNOWLEDGEMENTS

This work was supported by Consiglio Nazionale delle Ricerche. Progetto Finalizzato Biotecnologie e Biostrumentazione (Italy)

REFERENCES

–Amesz,J., and Knaff,D.,1988, Molecular mechanism of bacterial photosynthesis, in :"Biology of Anaerobic Microorganisms", A.J.B.Zehnder,ed., John Wiley and Sons, New York, Chichester, Brisbane, Toronto, Singapore.
–Bartsch,R.G.,1978, Cytochromes, in : "The Photosynthetic Bacteria", R.K.Clayton and W.R.Sistrom, eds.,Plenum Press, New York.
–Blankenship,R.R.,Bruce,D.C.,Freeman,J.H.,King,G.H.,McManus,J.D.,Nozawa,T. Trost,T., and Wittmershaus,B.P.,1988, Energy trapping and electron transfer in Chloroflexus aurantiacus , in : "Green Photosynthetic Bacteria", J.M.Olson, J.G.Ormerod,J.Amesz,E.Stackebrandt, and H.G.Truper, eds.,Plenum Press, New York and London.
–Blankenship,R.E.,Huynh,P.,Gabrielson,H., and Mancino,L.J.,1986, Purification, physical properties and kinetic behaviour of cytochrome c554 from Chloroflexus aurantiacus , Biophys.J. 47:2a.
–Blankenship,R.E.,Mancino,L.J.,Feick,R.,Fuller,R.C.,Machniki,J.,Frank,H.A. Kirkmaier,C., and Holten,D., 1984, Primary photochemistry and pigment composition of reaction centers isolated from the green photosynthetic bacterium Chloroflexus aurantiacus , in :"Advances in Photosynthesis Research", C.Sybesma, ed.,vol.I,p.203, M.Nijhoff-Dr.W.Junk, The Hague.
–Blankenship,R.E.,Trost,J.T., and Mancino,L.J.,1988, Properties of

reaction centers from the green bacterium <u>Chloroflexus aurantiacus</u> , <u>in</u>
: "The Photosynthetic Bacterial Reaction Center. Structure and Dynamics",
J.Breton and A.Vermeglio, eds., vol.149,p.119, Plenum Press New York and
London.

-Bruce,B.D.,Fuller,R.C., and Blankenship,R.E.,1982, Primary
photochemistry in the facultatively aerobic green photosynthetic
bacterium <u>Chloroflexus aurantiacus</u> , <u>Proc.Natl.Acad.Sci.</u> USA 79:6532.
-Dutton,P.L.,1971, Oxidation-reduction potential dependence of the
interactions of cytochromes, bacteriochlorophyll and carotenoids at 77K
in chromatophores of <u>Chromatium vinosum</u> and <u>Rhodopseudomonas
gelatinosa</u> , <u>Biochim. Biophys. Acta</u> 226:63.
-Dutton,P.L., and Leigh,J.S.,1973, Electron spin resonance
characterization of <u>Chromatium D</u> hemes.Non-heme irons and the
components involved in primary photochemistry, <u>Biochim.Biophys.Acta</u> ,
193:93.
-Dutton,P.L., and Prince,R.C.,1978, Reaction center driven cytochrome
interactions in electron and proton translocation and energy coupling,
<u>in</u> : "The Photosynthetic Bacteria", R.K.Clayton and W.R.Sistron, eds.,
Plenum Press, New York.
-Hale,M.B.,Blankenship,R.E., and Fuller,R.C.,1983, Menaquinone is the
sole quinone in the facultatively aerobic green photosynthetic bacterium
<u>Chloroflexus aurantiacus</u> , <u>Biochim.Biophys.Acta</u> 723:376.
-Kirmaier, C., and Holten,D., 1987, Photochemistry of reaction centers
from photosynthetic purple bacteria, <u>Photosynt.Res.</u> 13:225.
-McManus,J.D.,Trost,J.T., and Blankenship,R.E.,1988, Kinetic bahaviour
and N-terminal amino acid sequence of auracyanin, <u>Biophys.J.</u> 53:268a
-Parson,W.W.,1968, The role of P870 in bacterial photosynthesis,
<u>Biochim. Biophys. Acta</u> 153:248.
-Pierson,B.K., and Castenholz,R.W., 1974, A phototrophic gliding
filamentous bacterium of hot springs, <u>Chloroflexus aurantiacus</u> ,
<u>Arch.Microbiol</u> 100:5.
-Pierson,B.,Thornber,J.P., and Seftor,R.E.,1983, Partial purification,
subunit structure and thermal stability of the photochemical reaction
center of the thermophilic green bacterium <u>Chloroflexus aurantiacus</u> ,
<u>Biochim.Biophys.Acta</u> 723:322.
-Prince,R.C.,Lindsay,J.G., and Dutton,P.L.,1975, The Rieske iron sulphur
center in mitochondrial and photosynthetic systems. Em/pH relationship,
<u>FEBS Lett.</u> 51:117
-Trost,J.T.,McManus,J.D.,Freeman,J.C.,Ramarkrishna,B.L., and
Blankenship,R.E., 1988, Auracyanin, a blue copper protein from the green
photosynthetic bacterium <u>Chloroflexus aurantiacus,</u> <u>Biochemistry</u>
27:7858
-Schiozawa,J.A.,Lottspeich,F., and Feick,R., 1987, The photochemical
reaction center of <u>Chloroflexus aurantiacus</u> is composed of two
structurally similar polypeptides, <u>Eur.J.Biochem.</u> 167:595.
-Schiozawa,J.A.,Lottspeich,F.,Oesterhelt,D., and Feick,R., 1989, The prima
ry structure of the <u>Chloroflexus aurantiacus</u> reaction center polypeptides,
<u>Eur.J.Biochem.</u> 180:75.
-Venturoli,G.,Trotta,M., and Zannoni,D.,1989, Comparative aspects of
quinones in bacterial electron transport chains, <u>in</u> :"Highlights in

Ubiquinone Research", G.Lenaz and M.Battino, eds., Taylor and Francis Ltd., in the press

-Venturoli,G., and Zannoni,D.,1988, Oxidation reduction thermodynamics of the acceptor quinone complex in whole-membrane fragments from Chloroflexus aurantiacus, Eur.J.Biochem. 178:503.

-Zannoni,D., and Ingledew,J.W., 1985, A thermodynamic analysis of the plasma membrane electron transport components in photoheterotrophically grown cells of Chloroflexus aurantiacus, FEBS Lett. 193:93.

-Zannoni,D., and Melandri,B.A., 1985, Function of Ubiquinone in Bacteria, in: "Coenzyme Q. Biochemistry,Bioenergetics and Clinical Applications of Ubiquinone", G.Lenaz, ed., John Wiley and Sons, Chichester, New York, Brisbane,Toronto,Singapore.

-Zannoni,D., and Venturoli,G.,1988, The mechanism of photosynthetic electron transport and energy transduction by membrane fragments from Chloroflexus aurantiacus, in: "The Green Photosynthetic Bacteria", J.M.Olson,J.G.Ormerod,J.Amesz,E.Stackebrandt, and G.Truper, eds., Plenum Press, New York and London.

-Zannoni,D.,1986, The branched respiratory chain of heterotrophically dark-grown Chloroflexus aurantiacus, FEBS Lett. 198:119.

THE FUNCTIONS AND COMPONENTS OF THE ANAEROBIC RESPIRATORY

ELECTRON TRANSPORT SYSTEMS IN RHODOBACTER CAPSULATUS

A.G. McEwan[1,2], D.J. Richardson[1,2], M.R. Jones[1],
J.B. Jackson[1] and S.J. Ferguson[2]

[1]School of Biochemistry, University of
Birmingham, P.O. Box 363, Birmingham B15 2TT,
U.K.;
[2]Department of Biochemistry, University of
Oxford, South Parks Road, Oxford OX1 3QU, U.K.

INTRODUCTION

The traditional view of Rhodobacter capsulatus has been of an organism that could grow either phototrophically under anaerobic conditions or aerobically in the absence of illumination. Recognition of anaerobic respiration as a characteristic of R. capsulatus is relatively recent (Ferguson et al., 1987). Nitrate, dimethylsulphoxide (DMSO), trimethylamine-N-oxide (TMAO) and nitrous oxide have been identified as anaerobic electron acceptors, but not all strains can use each of these oxidants. In non-phototrophic bacteria the function of enzymes that allow electron transport to terminate with the reduction of an anaerobic electron acceptor is to permit non-fermentative growth in the absence of oxygen. In some instances this function also applies to the anaerobic respiratory electron transport pathways in R. capsulatus, but, as will be discussed in this paper, phototrophic growth can be facilitated by the possession of the capacity to reduce electron acceptors under anaerobic conditions. The presence in R. capsulatus of certain anaerobic electron transport pathways also provides opportunities to study these pathways because R. capsulatus is much better characterised in terms of its genetics and complement of electron transport proteins than many of the organisms that have been longer recognised to perform anaerobic respiration. R. capsulatus should not be thought of as distinct amongst phototrophs in its possession of anaerobic respiratory pathways; a strain of Rhodobacter sphaeroides catalyses the complete set of denitrification reactions from nitrate to nitrogen gas (Urata and Satoh, 1985; Ito et al., 1989) whilst several other genera of photosynthetic bacteria have been shown to reduce nitrous oxide (McEwan et al., 1985a). Thus it is probable that much of what is discussed here for R. capsulatus will also be applicable to other related organisms.

Molecular Biology of Membrane-Bound Complexes in Phototrophic Bacteria
Edited by G. Drews and E. A. Dawes
Plenum Press, New York, 1990

THE ANAEROBIC ELECTRON ACCEPTORS

    The first recognition of anaerobic respiration in R.
capsulatus occurred when repeated subculturing of a strain of
R. capsulatus N22 on a photosynthetic growth medium with
nitrate as nitrogen source led to the isolation of a mutant
that possessed a respiratory nitrate reductase activity (McEwan
et al., 1982). It emerged that this activity was present in
other strains where reduction of nitrate had been previously
assumed to be exclusively related to the assimilation of
nitrate (McEwan et al., 1984). The discovery of a respiratory
nitrate reductase then led to consideration of the possibility
that reduction of DMSO and TMAO by R. capsulatus, discovered by
Yen and Marrs (1977) and Madigan and Gest (1978) might be a
further example of an anaerobic respiratory process. This
proved to be the case (McEwan et al., 1983). Nitrous oxide
reduction was discovered by testing cells for the capacity to
reduce this gas with a newly introduced electrode system
(McEwan et al., 1985a). Reduction of all these electron
acceptors was shown to be linked to the generation of a
cytoplasmic membrane potential, consistent with the termination
of a proton translocating electron transport chain by each of
the identified electron acceptors. An important finding was
that the enzymes catalysing the terminal reductions were all
water-soluble proteins located in the periplasm (McEwan et al.,
1984, 1985b, 1987). Doubtless other electron acceptors will be
discovered; indeed recently use of a nitric oxide electrode has
shown that at least two strains of R. capsulatus can reduce
this gas (L. Bell et al., unpublished)

ENZYMOLOGY OF ANAEROBIC REDUCTION REACTIONS

Nitrate Reductase

    This water soluble periplasmic enzyme from R. capsulatus
strain N22DNAR[+] has been purified as a single polypeptide chain
of molecular weight 90,000. It possesses the molybdenum
cofactor (McEwan et al., 1987) that is characteristic of
nitrate reductases in general but no other redox centres have
so far been identified in this polypeptide. The enzyme differs
from the membrane-bound respiratory nitrate reductase that is
found in other organisms such as Escherichia coli and
Paracoccus denitrificans, not only in terms of subunit
composition and cellular location, but also in being relatively
insensitive to azide and unable to use chlorate as an
alternative substrate (McEwan et al., 1984). An important
question is that of how electrons reach the periplasmic nitrate
reductase. Several lines of evidence indicate that the
cytochrome bc_1 complex is not involved and the current model is
that electrons destined for nitrate reduction flow from
ubiquinol by an uncharacterised ubiquinol oxidation system that
contains b-type cytochrome(s) and is inhibited by HOQNO and
very low concentrations of cyanide (Richardson 1989; Richardson
et al., 1989). A c-type cytochrome with alpha band at 552 nm
in the reduced form is strongly implicated as a mediator of
electron transport between the step of ubiquinol oxidation and
the nitrate reductase because oxidation of the cytochrome by
nitrate can be observed in periplasmic extracts. Further
support for a role of this cytochrome in electron donation to
nitrate reductase has come from the finding that by

modification of the previously published procedure for purification of this enzyme (McEwan et al., 1984), a preparation containing both the c-type cytochrome and the catalytic subunit can be obtained. This cytochrome has a subunit molecular weight of 13,000 and a redox centre which undergoes a single electron oxidation/reduction with a midpoint potential of +165 mV. The cytochrome in this preparation of the reductase can be reoxidised by nitrate.

It might have been expected that all strains of R. capsulatus that possess respiratory nitrate reductase would contain an identical enzyme. This expectation is not realised because strain BK5 was reported to contain a membrane-bound nitrate reductase some years ago (Wesch and Klemme 1980) and this enzyme has the same functions as the periplasmic enzyme identified in other strains. The nitrate reductase in strain BK5 appears to be very closely related to the respiratory nitrate reductases in P. denitrificans and E. coli (A. Ballard et al., unpublished). The reason for this variation between the type of nitrate reductase in different strains is not clear.

Trimethylamine-N-oxide and Dimethylsulphoxide Reductase

A single periplasmic enzyme is responsible for reduction of both these oxides (Kelly et al., 1988;McEwan et al., 1985b, 1987) which can be conveniently assayed by nuclear magnetic resonance (King et al., 1987). It is responsible for the reduction of chlorate but is inactive towards nitrate. The enzyme has been purified as a single polypeptide chain with a molecular weight of 82,000. It behaves as a monomer on gel filtration and e.p.r. studies have shown it to be a molybdenum protein. A c-type cytochrome with an alpha band maximum in the reduced state at 556 nm (midpoint potential of +105 mV for a one electron oxidation-reduction reaction) co-purifies with the reductase and is oxidised by TMAO or DMSO in the presence of the reductase (McEwan et al., 1989). In common with the nitrate reductase, electrons appear to reach the TMAO/DMSO reductase from ubiquinol independently of the cytochrome bc_1 complex. The working model is that one or more redox proteins, including a component that possesses b-type haem but which is distinct from that associated with the pathway of ubiquinol oxidation by nitrate, catalyse the transfer of electrons from ubiquinol to the cytochrome c_{556} which may be the immediate donor to the reductase. The TMAO/DMSO reductase enzyme is not found in all strains of R. capsulatus; it is, for example, absent from strain BK5 (unpublished observations)

Nitrous Oxide Reductase

The first nitrous oxide reductase to be purified was that from Pseudomonas stutzeri (Zumft and Matsubara 1982). Shortly after the recognition that this enzyme contained a novel type of copper centre, a periplasmic enzyme with very similar spectra and subunit molecular weight was identified in R. capsulatus (McEwan et al., 1985a).

Elucidation of the electron transport pathway to nitrous oxide reductase in R. capsulatus has revealed a hitherto unrecognised complexity. Following the discovery of nitrous oxide respiration in R. capsulatus, it was concluded that the cytochrome bc_1 complex did not participate in the route of electron transfer to this enzyme, principally on the basis that reduction of nitrous oxide was not inhibited by either antimycin or myxothiazol (McEwan et al., 1985a; Richardson et al., 1986). Subsequently it has been recognised that such insensitivity to these two inhibitors is unreliable as evidence against the participation of this electron transport component (Richardson et al., 1989). There are two reasons for this. First, if the catalytic capacity of the cytochrome bc_1 complex is considerably greater than the maximum rate of nitrous oxide reduction then the consequence is that inhibition of a substantial proportion of the total pool of cytochrome bc_1 complexes may be required before any attenuation of the rate of nitrous oxide reduction is noticed (Richardson et al., 1989). Thus only in strains with the highest rates of nitrous oxide reduction is inhibition by myxothiazol clearly observed. The second factor is the presence of an electron transport pathway from ubiquinol to cytochrome c_2 that is independent of the cytochrome bc_1 complex. The evidence for this is: (a) a mutant that is specifically deficient in a functional cytochrome bc_1 complex (Daldal et al., 1987)is able to catalyse nitrous oxide reduction and (b) a mutant (Daldal et al., 1986) that is specifically deficient in cytochrome c_2 is unable to catalyse reduction of nitrous oxide with physiological reductants although the cells possess a functional nitrous oxide reductase (unpublished observations). Together these observations establish that cytochrome c_2 is indispensible for nitrous oxide reduction but that the bc_1 complex can be substituted by at least one other route of electron flow. This second route involves b-type cytochromes as judged by nitrous oxide-oxidised minus reduced spectra of cells of the mutant that lacks the cytochrome bc_1 complex. An important finding was that the nitrous oxide-induced oxidation of cytochromes in wild type cells that had been treated with myxothiazol to block the cytochrome bc_1 complex was indistinguishable from that seen with the mutant deficient in the cytochrome bc_1 complex. Thus the electron transport route that was alternative to the cytochrome bc_1 complex also functioned in the wild type cells. The corollary of this is that inhibition of the cytochrome bc_1 complex by myxothiazol could be compensated by diversion of electron flow to this alternative pathway, thus masking the contribution of bc_1 to nitrous oxide reduction that could occur in the absence of myxothiazol. Despite these findings with nitrous oxide respiration in R. capsulatus several lines of evidence still indicate that the reduction of nitrate, TMAO and DMSO is independent of the cytochrome bc_1 complex (Richardson et al.,1989) and there is no evidence to the contrary. Nitrous oxide reduction that is dependent on the alternative pathway is sensitive to low concentrations of HOQNO which acts on the reducing side of b-type cytochromes, presumably close to the site of ubiquinol oxidation. It is relevant that ubiquinol oxidation linked to nitrate reduction, and also involving a b-type cytochrome, was also sensitive to HOQNO. The possibility that strains expressing both the nitrous oxide and nitrate reductases possess a ubiquinol oxidase that is common to both the route of electron transport to nitrate and the cytochrome bc_1-independent pathway to nitrous oxide in R. capsulatus has been raised by observations that the amount of myxothiazol-insensitive nitrous oxide reduction was increased by growing cells under conditions in which the nitrate reducing pathway was induced to high levels (unpublished observations).

The information obtained about the electron transport pathways to nitrous oxide in R. capsulatus has some wider implications. First, the identification of cytochrome c_2 as an obligatory component naturally implicates a similar role for analogous proteins in other bacteria, for instance P. denitrificans in which cytochrome c_{550} is homologous with cytochrome c_2 of R. capsulatus. Such a role would be fully consistent with the observations of oxidation of a component absorbing at 550 nm in P. denitrificans (Boogerd et al., 1980), a finding which in itself did not definitely identify the molecular species responsible. The observations with R. capsulatus do not, however, establish whether cytochrome c_2 is the direct donor to nitrous oxide reductase; this point is unresolved for any organism in which nitrous oxide reduction occurs. Second, the existence of an electron transfer pathway to cytochrome c_2 that bypasses the bc_1 complex suggests that the latter pathway may not be obligatory for light-driven cyclic electron transport. Indeed a steady-state light-dependent membrane potential which is 30% of the magnitude of that observed in wild type cells has been observed in the mutant that lacks the bc_1 complex. The rate of cyclic electron transport may be approximately 5% of that in the wild type cells. Photosynthetic growth, albeit at a slow rate (generation time 24h) and requiring the presence of DMSO, has also been observed for this mutant (Richardson et al., 1989).

SOME GENERAL COMMENTS ON PERIPLASMIC ELECTRON TRANSPORT

The organisation of the periplasm is still a matter for investigation. As far as electron transport is concerned it is not clear whether electrons are exchanged by a collisional process between independent molecules, as suggested for the inner mitochondrial membrane, or whether there are groups of permanently associated proteins in the periplasm (Ferguson, 1988a; McEwan et al., 1989). The copurification of c-type cytochromes with both the nitrate and TMAO/DMSO reductases suggests that at least these components may be permanently associated in the periplasm. On the other hand, if cytochrome c_2 is required not only for electron transfer to nitrous oxide reductase but also for electron transfer to other components, including the reaction centre, then it would seem probable that a diffusional model for electron transfer may apply in this instance. However, localised pools of cytochrome c_2 associated with different electron transfer functions cannot be discounted, particularly as it has been observed that only some 25% of the total cytochrome c_2 in intact cells or in a periplasmic fraction can be oxidised by nitrous oxide (unpublished observations). An aspect of the periplasm that has to be considered for electron transport reactions is the contention that macromolecular motion is relatively slow owing to a high effective viscosity (Ferguson, 1988a).

FUNCTIONS OF ANAEROBIC ELECTRON TRANSPORT

Anaerobic dark growth of R. capsulatus with malate as carbon source and nitrous oxide as electron acceptor has been observed (McEwan et al., 1985a). Growth with malate and TMAO/DMSO in the dark has been reported although both the rate and extent of growth is very limited (Schulz and Weaver 1982).

A mutant that lacked the TMAO reductase (Kelly et al., 1988) did not grow phototrophically on solid media when supplied with TMAO and malate, whereas the wild type did grow (D.J. Kelly et al. unpublished). TMAO and DMSO are, however, able to support less restricted dark anaerobic growth with glucose or fructose as carbon source (Yen and Marrs, 1977; Madigan and Gest 1978). Nitrate has not been observed to support dark anaerobic growth with any carbon source. These observations can be rationalised in the following way. First, endowment with a reductase for an anaerobic electron acceptor may be advantageous in repects other than facilitating dark anaerobic growth, as will be explained shortly. Second, reduction of nitrate and TMAO/DMSO may generate an inadequate proton motive force to sustain growth on a carbon source from which ATP can be derived solely by respiration. This explanation could account for superior growth of R. capsulatus on TMAO/DMSO with a sugar than with a non-fermentable carbon source. It is worth pointing out that it is not known whether electron flow from ubiquinol to nitrate or TMAO/DMSO is linked to net proton translocation across the cytoplasmic membrane. The observation of a cytoplasmic membrane potential in cells during the reduction of these oxidants could, in principle, be explained solely on the basis of proton translocation by the NADH-ubiquinone oxidoreductase. In the case of nitrous oxide respiration at least some of the electron flow from ubiquinol will be linked to proton translocation, because it is known from other studies that passage of an electron through the cytochrome bc_1 complex leads to net movement of positive charge out of the cell. Thus, even in the absence of any information about the proton translocating properties of the alternative pathway to nitrous oxide reductase, it is possible that nitrous oxide reduction is associated with a larger $H^+/2e$ than electron transport to the other anaerobic acceptors. Clearly these aspects remain to be clarified by further studies. One point is clear, however. The relative redox potentials of the electron acceptors cannot be used as an automatic guide to the relative $H^+/2e$ and thus ATP/2e ratios. The organisation of the electron transport system plays a key role in determining this factor as explained in more detail elsewhere (Ferguson, 1988b). An additional factor that may prevent growth with nitrate as electron acceptor is the accumulation of the toxic nitrite as reaction product.

If a phototrophic organism such as R. capsulatus is to grow on a carbon source which has an average oxidation state that is more reduced that the average oxidation state of the cell biomass then the excess reductant must be redistributed. Carbon sources that fall into this category include propionate and butyrate. The requirement for carbon dioxide in the growth medium with these two carbon sources has been rationalised in this way. The excess reductant in the form of cellular NAD(P)H is believed to be consumed in carbon dioxide fixation via the Calvin Cycle (Lascelles, 1960). An alternative to reduction of carbon dioxide is now known to be reduction of nitrate, TMAO/DMSO or nitrous oxide (Richardson et al., 1988). This is a reason why these compounds are termed auxiliary oxidants. It was demonstrated that the presence in the growth medium of these oxidants facilitated phototrophic growth only if the reductase for the appropriate oxidant was present in the strain of R. capsulatus. In the case of nitrate reduction it has been established that the periplasmic reductase fulfils this function in certain strains and that the membrane-bound enzyme functions in this way in strain BK5 (A. Ballard et al., unpublished). An important finding was that for each molecule of propionate used for the growth of cells one molecule of TMAO was reduced in a two electron reaction, whereas when butyrate was the carbon source the stoichiometry was two molecules of TMAO reduced per molecule of butyrate consumed. If acetate can be regarded as a balanced substrate for growth, in the sense that its state of reduction is probably close to that of the biomass in the culture, then these figures are close to the theoretical values expected for the disposal of the excess reducing power in these two acids (Richardson et al., 1988). These stoichiometries were obtained by nuclear magnetic resonance analysis of the amounts of carbon substrate and oxidants in the growth media at the beginning and end of exponential growth (Richardson et al., 1988).

A second role for the auxiliary oxidants in phototrophic growth can be appreciated from the following considerations. The photosynthetic reaction centre can only function if a supply of an electron acceptor is continuously available. The electron acceptor is ubiquinone and this is regenerated from ubiquinol by the action of the cytochrome bc_1 complex. Under dark anaerobic conditions the $NADH/NAD^+$ ratio in cells is expected to be high and thus, provided there are no kinetic restraints, the ubiquinone pool is expected to be fully reduced. The estimated redox potentials of ubiquinone at binding sites (Q_a and Q_b) on the reaction centre are more negative than that of ubiquinone in the pool. However, when a dark and anaerobic suspension of cells is exposed to a single flash of saturating light, photochemistry and charge separation across the membrane occur, with subsequent oxidation of c-type cytochrome as the reaction centre becomes rereduced. Such photochemical events are abolished if the cells are first treated with phenazine methosulphate which faciltates redox equilibration and presumably therefore reduction of ubiquinone at the Q_a and Q_b sites. If treatment with phenazine methosulphate is followed by addition of one of the auxiliary oxidants, nitrate, nitrous oxide or TMAO, then photochemical events can again take place, thus showing that the availability of the anaerobic respiratory electron transport pathways provides a mechanism for correction of an over reduction of ubiquinone (McEwan et al., 1985c). A similar role can be deduced from a related series of experiments. In these, myxothiazol was present to block the cytochrome bc_1 complex. Under such conditions a first flash of saturating light caused the movement of an electron from the special pair of bacteriochlorophylls to a ubiquinone acceptor, either at the Qa or Qb sites. A second flash resulted in only very limited turnover of the reaction centre whilst a third flash resulted in essentially no photochemical events. The interpretation is as follows. The presence of myxothiazol restricts oxidation of ubiquinol. Hence the restriction to the first flash of substantial photochemistry, with the second flash resulting in a less extensive reaction, must mean that there is no ubiquinone in the membrane pool and that oxidising equivalents must be available only at the Q_a site and to a lesser extent at the Q_b site. If, however, an auxiliary oxidant is added then photochemistry proceeds beyond the second flash even in the presence of myxothiazol because electron flow to the oxidant can remove reducing equivalents from ubiquinol (M.R. Jones et al., unpublished). Thus electrons can continue to pass through the reaction centre from the bacteriochlorophyll special pair to the Q_a and Q_b sites. R. capsulatus is not dependent on the presence of auxiliary oxidants for the provision of sufficient oxidising equivalents in the form of ubiquinone during light-driven cyclic electron transport. Consequently there must exist other mechanisms for preventing over reduction to ubiquinol. Such mechanisms might include reduction of fumarate to succinate. In fact it has been observed that phototrophic growth of cells on malate results in the appearance in the growth medium of fumarate (presumably by the action of fumarase on intracellular malate followed by export of fumarate from the cell) and succinate which is suggested to be formed by intracellular reduction of fumarate with subsequent transport out of the cell (G.F. King et al., unpublished). Further work will be needed to establish whether these observations are related to the problem of avoiding the over reduction of ubiquinone discussed here. It is striking, however, that the organism Erythrobacter sp Och 114 will only grow photosynthetically either microaerobically or in the presence of an auxiliary oxidant (Takamiya et al 1988), presumably for the reasons outlined here.

A final advantage that the possession of anaerobic respiratory pathways may confer on R. capsulatus is supplementation of energy generating reactions at low light intensities. In accordance with this idea enhancement of the growth rates of R. capsulatus has been observed with succinate or malate as carbon source at low light intensities (Richardson et al., 1988).

ACKNOWLEDGEMENTS

We thank the U.K. S.E.R.C. for their support of this work and Dr. F. Daldal for providing specific cytochrome-deficient mutants.

REFERENCES

Boogerd, F.C., van Verseveld, H.W., and Stouthamer, A.H., 1980, Electron transport to nitrous oxide in Paracoccus denitrificans, FEBS Lett. 113: 279.

Daldal, F., Cheng, S., Applebaum, J. Davidson, E. and Prince, R.C., 1986, Cytochrome c_2 is not essential for photosynthetic growth of Rhodopseudomonas capsulata, Proc. Natl. Acad. Sci. U.S. 83: 2012.

Daldal, F., Davidson, E. and Cheng, S.,1987, Isolation of the structural genes for the Rieske Fe-S protein, cytochrome b and cytochrome c_1, all components of the ubiquinol; cytochrome c_2 oxidoreductase complex of Rhodopseudomonas capsulata, J. Mol. Biol. 195: 1.

Ferguson, S.J., 1988a, Periplasmic electron transport reactions, in: "Bacterial Energy Transduction", C. Anthony ed., Academic press, London.

Ferguson, S.J., 1988b, The redox reactions of the nitrogen and sulphur cycles, in: "The Nitrogen and Sulphur Cycles", J.A. Cole and S.J. Ferguson eds., Cambridge University Press, Cambridge.

Ferguson, S.J., Jackson, J.B., and McEwan, A.G., 1987, Anaerobic respiration in the Rhdospirillaceae: characterisation of pathways and evaluation of roles in redox balancing during photosynthesis, FEMS Microbiol. Lett., 46: 117.

Itoh, M., Mizukami, K., and Satoh, T., 1989, Involvement of cytochrome bc_1 complex and cytochrome c_2 in the electron-transfer pathway for NO reduction in a photodenitrifier, Rhodobacter sphaeroides f.s. denitrificans, FEBS Lett. 244: 81.

Kelly, D.J., Richardson, D.J., Ferguson, S.J. and Jackson, J.B., 1988, Isolation of transposon Tn5 insertion mutants of Rhodobacter capsulatus unable to reduce trimthylamine-N-oxide and dimethylsulphoxide, Arch. Microbiol. 150: 138

King, G.F., Richardson, D.J., Jackson, J.B. and Ferguson, S.J., 1987, Dimethylsulphoxide and trimethylamine-N-oxide as bacterial electron accpetors: use of nuclear magnetic resonance to assay and characterise the reductase system in Rhodobacter capsulatus, Arch. Microbiol. 149: 47.

Lascelles, J., 1960, The formation of ribulose-1,5-diphosphate carboxylase by growing cultures of Athiorhodaceae, J. Gen Microbiol. 23: 499.

Madigan, M.T. and Gest, H., 1978, Growth of a photosynthetic bacterium anaerobically in darkness, supported by "oxidant dependent" sugar fermentation, Arch. Microbiol. 117; 119.

McEwan, A.G., George, C.L., Ferguson, S.J. and Jackson, J.B., 1982, A nitrate reductase activity in Rhodopseudomonas capsulatus linked to electron transfer and generation of a membrane potential, FEBS Lett. 150: 277.

McEwan, A.G., Ferguson, S.J., and Jackson, J.B., 1983, Electron flow to dimethylsulphoxide or trimethylamine-N-oxide generates a membrane potential in Rhodopseudomonas capsulata, Arch. Microbiol. 136, 300.

McEwan, A.G., Jackson, J.B., and Ferguson, S.J., 1984, Rationalisation of properties of nitrate reductase in Rhodopseudomonas capsulata, Arch. Microbiol. 137: 344.

McEwan, A.G., Greenfield, A.J., Wetzstein, H.G., Jackson, J.B., and Ferguson, S.J., 1985a, Nitrous oxide reduction by members of the family Rhodospirillaceae and the nitrous oxide reductase of Rhodopseudomonas capsulata, J. Bacteriol. 164: 823.

McEwan, A.G., Wetzstein, H.G., Jackson, J.B., and Ferguson, S.J., 1985b, Periplasmic location of the terminal reductase in trimethylamine-N-oxide and dimethylsulphoxide respiration in the photosynthetic bacterium Rhodopseudomonas capsulata, Biochim. Biophys. Acta 806: 410.

McEwan, A.G. Cotton, N.P.J., Ferguson, S.J., and Jackson, J.B., 1985c, The role of auxiliary oxidants in the maintenance of a balanced redox poise for photosynthesis in bacteria, Biochim. Biophys. Acta, 810: 140

McEwan, A.G., Wetzstein, H.G., Meyer, O., Jackson, J.B., and Ferguson, S.J., 1987, The periplasmic nitrate reductase of Rhodobacter capsulatus; purification, characterisation and distinction from a single reductase for trimethylamine-N-oxide, dimethylsulphoxide and chlorate, Arch. Microbiol., 147: 340.

McEwan, A.G., Richardson, D.J., Hudig, H., Ferguson, S.J., and Jackson, J.B., 1989, Identification of cytochromes involved in electron transport to trimethylamine-N-oxide/dimethylsulphoxide in Rhodobacter capsulatus, Biochim. Biophys. Acta 973: 308.

Richardson, D.J., Kelly, D.J., Jackson, J.B., Ferguson, S.J. and Alef, K., 1986, Inhibitory effects of myxothiazol and 2-n-heptyl-4-hydroxyquinoline-N-oxide on the auxiliary electron transport pathways of Rhodobacter capsulatus, Arch. Microbiol. 146: 159.

Richardson, D.J., King, G.F., Kelly, D.J., McEwan, A.G., Ferguson, S.J., and Jackson, J.B., 1988, The role of auxiliary oxidants in maintaining redox balance during phototrophic growth of Rhodobacter capsulatus on propionate or butyrate, Arch. Microbiol. 150: 138.

Richardson, D.J., 1989, Ph. D Thesis University of Birmingham.

Richardson, D.J., McEwan, A.G., Jackson, J.B. and Ferguson, S.J., 1989, Electron transport pathways to nitrous oxide in Rhodobacter species, Eur. J. Biochem. in press.

Schulz, J.E. and Weaver, P.F., 1982, Fermentation and anaerobic respiration by Rhodospirillum rubrum and Rhodopseudomonas capsulata, J. Bacteriol. 149: 181.

Takamiya, K., Arata, H., Shioi, Y. and Doi, M., 1988, Restoration of the optimal redox state for the photosynthetic electron transfer system by auxiliary oxidants in an aerobic photosynthetic bacterium Erythrobacter sp. OCh 114, Biochim. Biophys. Acta 935:26.

Urata, K. and Satoh, T., 1985, Mechanism of nitrite reduction to nitrous oxide in a photodenitrifier Rhodopseudomonas sphaeroides f. sp. denitrificans, Biochim. Biophys. Acta 841, 201-207.

Wesch, R. and Klemme, J.H., 1980, Catalytic and molecular differences between assimilatory nitrate reductases isolated from two strains of Rhodopseudomonas capsulata, FEMS Microbiol. Lett. 8: 37.

Yen, H.C., and Marrs, B.L., 1977, Growth of Rhodopseudomonas capsulatus under dark anaerobic conditions with dimethylsulfoxide, Arch Biochem Biophys 181: 411.

Zumft, W.G. and Matsubara, T., 1982, A novel kind of multi-copper protein as terminal oxidoreductase of nitrous oxide respiration in Pseudomonas perfectomarinus, FEBS Lett. 148: 107.

OXYGEN-LINKED ELECTRON TRANSFER AND ENERGY

CONVERSION IN *RHODOSPIRILLUM RUBRUM*

Javier Varela and Juan M. Ramírez

Centro de Investigaciones Biológicas, CSIC
Velázquez 144
28006 Madrid, Spain

ABSTRACT

Pigmented *Rhodospirillum rubrum* cells from dark chemotrophic cultures contain several pathways for the transfer of electrons from reduced substrates to O_2. In order of decreasing H^+-translocating efficiency, they are: (i) a cytochrome (oxidase) pathway that is inhibited by low concentrations of KCN and by inhibitors of the cytochrome $b.c_1$ complex, but not by CO; (ii) a CO sensitive or alternative (oxidase) pathway that is partly blocked by inhibitors of the cytochrome $b.c_1$ complex; and (iii) a third pathway that operates in the presence of CO plus antimycin A and that is absent in the presence of CO plus myxothiazol and in a mutant which lacks rhodoquinone. In addition, a significant fraction of the O_2 uptake activity remains when H^+ translocation is completely blocked by inhibitors of electron transfer or mutations. Since the rate of respiratory electron transfer appears to be limited at the substrate level, it is difficult to make a direct estimation of the contribution of each pathway to the final rate of respiration. However, from the relative energy-transducing efficiency of the cytochrome and the CO sensitive pathways (as measured by the H^+/O ratios in O_2 pulses) and the final cell yields of C-limited cultures of wild type and cytochrome-oxidase deficient strains, it seems that the contribution of the cytochrome pathway to the energy-conserving O_2 uptake of pigmented chemotrophic *R. rubrum*, growing on malic and glutamic acids as carbon sources, is close to 80 %.

INTRODUCTION

Many strains of the nonsulfur purple bacteria (*Rhodospirillaceae*) are facultative phototrophs that can also obtain metabolic energy from the aerobic respiration of organic compounds[1]. Their respiratory system includes a membrane-linked chain of electron carriers that is present and functional not only in dark aerobic cultures, but also in cells that are growing phototrophically in the absence of oxygen[2]. However, this does not mean that the levels of the individual respiratory constituents remain fixed during all conditions of growth. Thus, changes of the relative amounts and even of the type of the terminal oxidase have been reported for several strains upon transition from chemotrophic to phototrophic conditions of growth and viceversa[3,4]. Another feature that seems to be common to most respiratory *Rhodospirillaceae* is the simultaneous presence

Molecular Biology of Membrane-Bound Complexes in Phototrophic Bacteria
Edited by G. Drews and E. A. Dawes
Plenum Press, New York, 1990

of two or more aerobic oxidases, usually a ferrocytochrome c_2 oxidase and an ubiquinol oxidase that is usually called the alternative oxidase. Then, it is common for these respiratory chains to be branched at their high potential ends, the branches stemming from the ubiquinone pool[5]. The multiplicity of aerobic oxidases is also frequent among nonphototrophic bacteria[6].

While the terminal oxidases, as well as the substrate and coenzyme dehydrogenases, are specific constituents of the respiratory chain, some other electron carriers of this system participate also in photosynthetic electron transfer. The diffusible pool of ubiquinone, soluble cytochrome c_2 and the cytochrome $b.c_1$ complex that mediates the transfer of electrons from ubiquinol to ferricytochrome c_2 (that are altogether the core segment of the cytochrome chain) have this dual (respiratory and photosynthetic) function[2,5]. Under culture conditions that allow the formation of the photosynthetic constituents, both redox chains coexist in the same membrane and their operation must be co-ordinated by specific regulatory mechanisms that add to those that regulate the traffic of electrons at the intersecting point of the respiratory branches. The early observed light-elicited inhibition of respiration, which appears to be mainly mediated by photochemically generated membrane potential[7], is a likely reflection of this type of mechanisms. The preferential location of each electron transfer system in a different region of the membrane[2,8] may also be of significance for regulation of electron transfer.

Some of these features appear in the current scheme of membrane-linked electron transfer pathways in *Rhodospirillum rubrum* (Fig. 1), that includes both an alternative and a cytochrome oxidase. Although cytochrome oxidase activity had been detected early in this purple bacterium, its participation in respiration under physiological conditions was dubious[9] because its complete inhibition with KCN or NaN$_3$ did not go along with a comparable reduction of the rate of O_2 uptake by intact cells or of the NADH and succinate oxidase activities of isolated membrane vesicles. Nevertheless, it was found that the activity was significantly lower in photoanaerobic than in dark aerobic cultures while other respiratory activities had similar levels in both types of cultures[4]. Later, a mutant strain of *R. rubrum*, that lacked cytochrome oxidase activity while showing unaltered rates of O_2 uptake and of NADH and succinate oxidation by O_2, was isolated[10]. The mutant grew chemotrophically, but at half the rate and with half the final cell yield of the wild type strain. Thus, it was concluded that although cytochrome oxidase participated normally in respiration, the organism had one (or more) additional pathway(s) for the transfer of electrons from the respiratory substrates to O_2. Such an alternative pathway could come into operation (at least) when cytochrome oxidase activity was blocked by chemical inhibitors or by mutation, and was less efficient for energy transduction than the cytocrome oxidase pathway[10]. The alternative oxidase could not mediate the aerobic oxidation of c type cytochromes *in vivo* since these haemoproteins remained reduced in aerobic suspensions of mutant cells, in contrast to what happened in the cytochrome-oxidase containing (wild type) strain[10]. Both terminal oxidases differed also in their sensitivity to classical inhibitors. Thus, whereas cytochrome oxidase was fully inhibited by 0.2 mM KCN and remained uninhibited in the presence of CO, the alternative electron transfer pathway was blocked by CO and required higher concentrations of KCN to exhibit significant inhibition[10].

In later work, an ubiquinol oxidase activity sensitive to mefloquine and CO was demonstrated in isolated membrane vesicles isolated from pigmented chemotrophic cells and was proposed to act as the alternative pathway to oxygen[11]. Two cytochromes of type b, with midpoint redox potentials of 250 and 380 mV, were resolved in these membrane preparations and tenta-

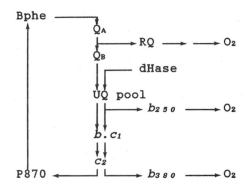

Fig. 1. Light- and O_2-linked electron transfer pathways in *R. rubrum*. The depicted constituents are: the primary photo-reactants (P870, Bphe), the reaction center quinones (Q_A, Q_B), ubiquinone (UQ), rhodoquinone (RQ), ubiquinol cyto-chrome *c* reductase (*b.c₁*), cytochrome c_2, the dehydrogen-ases (dHase) and the respiratory oxidases (b_{250}, b_{380}). See the text for other details.

tively assigned to the alternative and the cytochrome oxidase, respect-ively[11]. Besides, a *c*-type cytochrome and another couple of *b*-type cyto-chromes that participated both in light-dependent and O_2-linked electron flow were attributed to a putative cytochrome $b.c_1$ complex[11]. Such a com-plex was extracted from the *R. rubrum* membrane and purified[12]. It was also shown that the same pool of ferrocytochrome c_2 transferred electrons to O_2 through cytochrome oxidase and to the photooxidized primary donor of the photosynthetic reaction center[13].

Fig 1 also shows a pathway that connects the acceptor side of the cyclic photosynthetic system to O_2. This *photooxidase* pathway includes in *R. rubrum* rhodoquinone[14], an ubiquinone aminoderivative which has been found only in a couple of *Rhodospirillaceae* species[15]. The physiological function of the photooxidase seems to be that of avoiding the photoinhibi-tion caused by the accumulation of reducing equivalents in the low-poten-tial side of the photosynthetic cyclic chain[16]. The pathway is required only for phototrophic growth, since a rhodoquinoneless mutant that lacked photooxidase activity was unable to grow phototrophically in the absence of O_2, but grew normally under chemotrophic conditions in the dark[16]. However, the mutant also lacked a dark NADH-fumarate reductase activity[17], the physiological significance of which remains unclear.

We are interested in understanding the mechanisms that co-ordinate the operation of respiratory and photosynthetic electron transfer in the *R. rubrum* membrane. Since most of the data used to draw the scheme of respiratory pathways in Fig. 1 were obtained using either chemotrophically-grown cells that did not contain the photosynthetic apparatus or isolated membrane vesicles, it seemed convenient to investigate whether that scheme could account for the O_2-linked electron transfer of intact cells in which a functional photosynthetic system was present. Besides, an estimation of the actual contributions of the cytochrome and the alternative pathways to the respiration of cells that contain the photosynthetic constituents seemed also necessary for further understanding of the operation of the branched electron transfer system. The preliminary results of these studies, obtained with chemotrophically-grown cells that had been cultured under limiting O_2 levels to allow the formation of the photosynthetic apparatus, are presented here.

MATERIALS AND METHODS

The *R. rubrum* strains S1 (wild type), F11 (rhodoquinone deficient) and CAF10 (cytochrome-oxidase deficient), used throughout the work, have been described before[10,16]. The culture medium used was that of Lascelles[18], supplemented with 2 g/l yeast extract and malic and glutamic acids, both at 20 mM. The cultures, in flasks filled to 75 % of their total capacity, were subjected to orbital shaking and kept at 30 °C in the dark. Cells, collected before the end of the exponential phase of growth, contained about 10 to 15 nmole bacteriochlorophyll per mg protein.

Cytochrome oxidase activity was assayed in intact cells using the Nadi reaction as described previously[10]. For the assay of O_2 uptake activity, the cells were suspended in fresh culture medium or alternatively, when comparison with transient pH changes was intended, in the low-buffer solution described below. The O_2 level of the cell suspension was determined polarographically at 30 °C, using a Clark type electrode and an O_2 monitor (YSI models 5331 and 53). After replacement of the O_2 electrode by a combined pH electrode, the same setup was used for monitoring the transient pH changes that were elicited by brief (2 s) pulses of white light (70 W.m^{-2}) or by µl additions of air saturated 150 mM KCl. The cell suspension in 50 mM KCl, 1.5 mM glycylglycine and 100 mM KSCN, was allowed to use O_2 up and adjusted to pH 6 before illumination or addition of air saturated KCl.

RESULTS AND DISCUSSION

Effect of respiratory electron-transfer inhibitors on O_2 uptake and proton translocation by intact cells

Fig. 2 shows the effect of the classical oxidase inhibitors, KCN and CO, on the O_2 uptake activity of the wild type *R. rubrum* strain and its cytochrome-oxidase deficient derivative. KCN at 0.2 mM, a concentration that inhibits completely cytochrome oxidase activity but that is too low to have a significant effect on the alternative pathway[10,11], elicited a significant, although incomplete, reduction of the rate of O_2 uptake of wild type cells. The high residual activity may be accounted for by the alternative pathway. In contrast, KCN had little effect on the mutant, as expected from the absence of cytochrome oxidase in this strain. Gassing the cells with CO had the opposite effect: while the respiratory activity of the mutant was almost completely blocked, the wild type strain kept practically unaltered its original rate of O_2 uptake, in accordance with the respective contents of (CO insensitive) cytochrome oxidase activity (Fig. 2). The simultaneous presence of both inhibitors did not suppress completely O_2 uptake in the wild type. This result, that contrasts with the complete inhibition observed in unpigmented chemotrophic cells[10] may indicate the existence of an additional oxidase, resistant to CO plus KCN (cf. Thore et al.[9]). The other possible interpretation, namely that the alternative oxidase had changed its properties to become CO insensitive, is not supported by the effect of this inhibitor on H$^+$ translocation, as it will be shown below.

At least two other points of Fig. 2 deserve some additional comments. First, the added activities of the KCN and the CO resistant pathways exceeded the uninhibited rate. This is in agreement with the previous conclusion where the oxidases of chemotrophically-grown *R. rubrum* are in excess as compared to other constituents of the respiratory chain[4,10]. And second, the loss of KCN sensitive (cytochrome oxidase) activity is compensated in the mutant by an increase of the KCN resistant (alternative) pathway, so that the total capacity for O_2 reduction is maintained at a level similar to that of the cytochrome-oxidase containing strain. This may reflect a requirement for a minimum level of O_2-linked electron transfer activity.

A somewhat unexpected finding was the scarce effect that the inhibitors of the cytochrome $b.c_1$ complex, antimycin A and myxothiazol, had on the O_2 uptake activity of the strain that contained cytochrome oxidase (Fig.3), because *a priori* the scheme of Fig. 1 suggested that the inhibition by either drug should be as high as that elicited by KCN, which is shown here for comparison. In cell suspensions previously gassed with CO, which supposedly were left with the cytochrome oxidase branch as the sole available pathway for respiratory O_2 reduction, the effect of the inhibitors of the cytochrome $b.c_1$ complex was not enhanced (Fig. 3).

Several explanations may be found to account for those results. Thus, for instance, the inhibitors may be unable to reach the appropriate target in the intact cell. Or it is also possible that the sensitive step of the process may be nonlimiting, so that its rate after inhibition can still sustain a large fraction of the original electron transfer from respiratory substrates to O_2. This has been proposed to explain the lack of inhibitory effect of antimycin A on the respiration of phototrophically-grown *Rhodobacter sphaeroides*[19]. A third possibility is that there exists in the cell a bypassing electron transfer pathway that avoids the inhibited cytochrome $b.c_1$ complex.

Therefore, we set to obtain additional data that could help to discern among these interpretations. Although there is not a direct simple assay for the catalytic activity of the cytochrome $b.c_1$ complex in intact cells, the efficiency of its inhibitors could be tested by following their ability to prevent the transient pH changes that are induced by short pulses of light in anerobic, low-buffer suspensions of photosynthetically competent cells. Although complex, such pH changes are largely a direct consequence

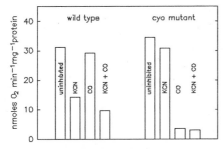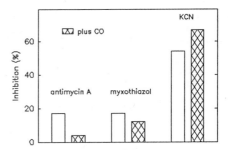

Fig. 2 (left). Effect of oxidase inhibitors on the rate of oxygen uptake by *R. rubrum* cells. Cell suspensions of the wild type strain and the cytochrome oxidase (cyo) mutant contained 1 mg protein/ml. Where indicated, 0.2 mM KCN was added to the suspension 15 min before the assay. Pure CO was bubbled for 1 min through cell suspensions that were diluted with the same volume of aerated medium for the assay. The specific cytochrome oxidase activities, in arbitrary units, were 96 (wild type; 5 in the presence of KCN) and 6 (mutant). Other conditions are described under Methods.

Fig. 3 (right). Effect of cytochrome $b.c_1$ complex inhibitors on the rate of O_2 uptake by *R. rubrum* cells. Wild type or cytochrome oxidase (cyo) mutant cells were suspended in the low-buffer medium described under Methods. Inhibitors were added at 10 µM 5 min prior to the assay. Other conditions as described in Fig. 2 and under Methods.

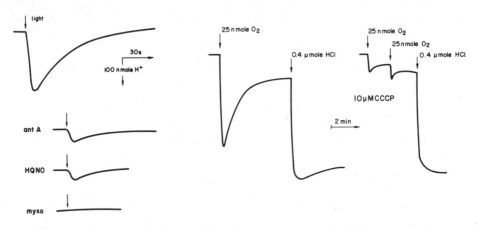

Fig. 4 (left). Effect of cytochrome $b.c_i$ complex inhibitors on the light-elicited transient pH change of wild type cells. Light was turned on for 2 s when indicated by the arrows. Antimycin A (ant A), 2-heptyl-4-hydroxyquinoline-N-oxide (HQNO) and myxothiazol (myxo) were added at 10 μM. Other conditions are described under Methods.

Fig. 5 (right). O_2-induced pH changes in wild type $R.$ $rubrum$ cells. The effect of cyanide m-chlorophenylhydrazone (CCCP), a protonophore, is shown. For other experimental conditions see Methods.

of the electrogenic translocation of protons which is associated to photochemical electron transfer (see Taylor and Jackson[20] for a discussion of these changes), in which the cytochrome $b.c_i$ complex of $R.$ $rubrum$ is involved[11]. As shown in Fig. 4, the inhibitors that failed to elicit a significant reduction of O_2-linked electron transfer (Fig. 3) were able in contrast to reduce between 80 and 100 % the light-dependent pH decrease. Then, it appears that inaccessibility of the inhibitors to the inhibition sites is not a plausible explanation for the high residual rates of O_2 uptake observed in the assays of Fig. 3.

Similarly to the light-induced change, the transient acidification that follows the addition of small amount of O_2 to an anaerobic suspension of cells is due to the electrogenic translocation of protons that is inherent to energy-transducing vectorial electron transfer and, as such, is considerably decreased by protonophores, which increase the proton permeability of the membrane (Fig. 5). Table 1 shows the H^+/O ratios observed in these O_2 pulse assays. They were significantly lower in the cytochrome oxidase mutant, as expected from the decreased efficiency for energy transduction of the alternative pathway that mediates respiration in this strain[10]. The inhibition of this pathway by CO explains both the loss of H^+ translocation in the mutant and the stimulation that the gas elicits in the H^+/O ratio of the wild strain, since it is expected that such inhibition will increase in this strain the fraction of O_2 that is reduced by the more efficient cytochrome pathway. The enhancement of the ratio by CO is consistent, therefore, with the participation of the alternative CO-sensitive oxidase in the respiration of wild type $R.$ $rubrum$.

In order to choose between the two remaining interpretations for the weakness of the inhibition elicited by the cytochrome $b.c_i$ inhibitors on the rate of O_2 uptake, we tested their effect on the transient pH decrease that is induced by an O_2 pulse. If the failure of the inhibitors of the cytochrome $b.c_i$ complex to decrease O_2 uptake were due to their acting on a

Table 1. Oxygen-elicited H^+ translocation in suspensions of *R. rubrum* cells[a]

Strain	H^+/O ratio ± standard deviation	
Wild type	7.7 ± 0.6	(8)[b]
Wild type + CO	8.6 ± 0.3	(8)
Cytochrome oxidase mutant	4.4 ± 0.3	(3)
Cytochrome oxidase mutant + CO	0 ± 0.1	(3)

[a]For experimental conditions see Figs. 2, 5 and Methods.
[b]Number of determinations.

nonlimiting step of the electron transfer process, proton translocation should not be altered. If, on the contrary, inhibition of the complex prevented normal electron transfer and caused the operation of a bypass, it would seem unlikely that such branch could have the same energy transducing efficiency as the original pathway and, consequently, electrogenic proton translocation should be considerably reduced.

The results of such measurements are shown in Fig. 6. The H^+/O ratio of the wild type was clearly reduced by cytochrome $b.c_1$ complex inhibitors, as expected if a significant fraction of the original electron transfer took place along the cytochrome pathway and was sensitive to the drugs. Then, it seems reasonable to conclude that if the effect of the cytochrome $b.c_1$ complex inhibitors on O_2 uptake is small (Fig. 3), it is not due to the nonlimiting nature of the affected electron carrier, but to the operation of bypasses that are less efficient for H^+ translocation. It is remarkable that the presence of CO did not enhance the inhibition caused by antimycin A (Fig. 6). Therefore, it appears that there is still an additional, H^+-translocating pathway that may operate in the presence of CO plus this inhibitor. Surprisingly, antimycin A brought the H^+/O ratio to zero in CO-containing cell suspensions of a mutant that lacks rhodoquinone (Fig. 6). Thus, the residual transient pH change that is induced by O_2 in the presence of CO and antimycin A seems to be associated to electron transfer mediated by that benzoquinone, although the relationship between the lack of rhodoquinone and the absence of the residual H^+ translocation might be indirect. It should be mentioned that double poisoned cells of the

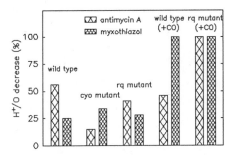

Fig. 6. Effect of cytochrome $b.c_1$ complex inhibitors on H^+ translocation elicited by O_2 pulses. Cells of the wild type strain and the cytochrome oxidase (cyo) and rhodoquinone (rq) deficient mutants were used. Inhibitors were added at 10 μM. For other experimental conditions, see Methods.

mutant showed rates of O_2 uptake similar to those of the wild type strain (data not shown).

Another result of interest among those shown in Fig. 6 is the decrease that the inhibitors of the cytochrome $b.c_1$ complex elicited in the H^+/O ratio of the cytochrome-oxidase deficient mutant. This was unexpected because the alternative pathway proposed for this strain does not include such complex (Venturoli et al.[11]; Fig. 1). The reasons for this effect are not clear at the present moment.

The contribution of the cytochrome pathway to the respiration of chemotrophic cells

The multiplicity of pathways in the respiratory chain of *R. rubrum* raises the question of which of them are actually used during chemotrophic metabolism. One possibility is that respiratory electron transfer takes place normally along a single *main* pathway, while the others are only used when one of the constituents of the main chain becomes inoperative due to the presence of toxic substances in the growth medium. If the sole function of the respiratory chain were H^+ translocation, this possibility would be favoured, being the efficient cytochrome branch the main chain. But the elimination of reducing equivalents for regeneration of the oxidized forms of redox coenzymes is also a physiological function of O_2-linked electron transfer, and the relative requirements for oxidation and H^+ gradient formation depend on factors such as the relative level of reduction of C in the organic growth substrate and in the cell materials. Thus, it seems that at least two respiratory branches of different H^+-translocating efficiencies may operate simultaneously during dark aerobic growth, their actual contributions to total electron transfer being subjected to metabolic control. In fact, the observation that inhibition of the CO sensitive pathway enhances the wild type H^+/O ratio is consistent with the participation of more than one branch in respiratory electron flow.

Since in pigmented chemotrophic cells the joint electron transferring capacity of the oxidases exceeds that of the previous part of the chain, the relative contribution of each oxidase to O_2 uptake cannot be estimated directly from the residual rates in the presence of the specific oxidase inhibitors. However, this parameter and the relative efficiency of each pathway for energy transduction can be estimated from the measured H^+/O ratios of pigmented chemotrophic cells (Table 1) if it is assumed that, in uninhibited mutant cells, the alternative (CO sensitive) pathway accounts for all the activity and that in wild type cells that have been gassed with CO, O_2 consumption is only due to the cytochrome pathway. While the first assumption is supported by the complete suppression by CO of H^+ translocation in the mutant, the second one is sustained by the value of the determined H^+/O ratios, which is similar to those reported for the cytochrome chain in other systems[21]. Then, using the data of Table 1, it results that the cytochrome pathway is about twice (1.96) more effective for H^+ translocation than the alternative one. From this value and the measured H^+/O ratio of uninhibited wild type cells (Table 1), a relative contribution of 77 % to respiration is computed for the cytochrome pathway of wild type *R. rubrum* grown chemotrophically with malic plus glutamic acids as the carbon source (see Material and Methods).

These estimations are based on several other less explicit assumptions. Thus, the possible effects of CO on the cytochrome pathway have not been taken into account. Also, the possible contribution to uninhibited respiration of the pathways that remain in the presence of CO plus an inhibitor of the cytochrome chain has been disregarded. Moreover, since the calculations are based on data obtained from O_2-pulse experiments, it seems that they should not be generalized to include respiration under the quasi steady-state conditions of chemotrophic growth. Then, tests based on other approaches seem necessary. For instance, whether the data obtained from O_2 pulses can be extended to steady-state respiration may be checked by comparing the relative (wild type to cytochrome-oxidase mutant) H^+/O ratios (Table 1) with the relative final cell yields of C-source limited chemotrophic cultures[10]. The resulting values of 1.74 (H^+/O ratios) and 1.88 (yields) are quite close to each other and seem to validate the generalization. This means a participation close to 80 % of the cytochrome pathway in the respiration of *R. rubrum* as cultured in this work.

ACKNOWLEDGEMENTS

This work was supported by a grant from the Plan Nacional de Promoción General del Conocimiento, Spain. The skilful technical assistance of Ms. C. Fernández-Cabrera and Ms. M. Zazo is gratefully acknowledged.

REFERENCES

1. N. Pfennig, General physiology and ecology of photosynthetic bacteria, in: "The Photosynthetic Bacteria", R. K. Clayton and W. R. Sistrom, eds., Plenum Press, New York (1978).

2. L. Smith and P. B. Binder, Oxygen-linked electron transport and energy conservation, in: "The Photosynthetic Bacteria", R. K. Clayton and W. R. Sistrom, eds., Plenum Press, New York (1978).

3. T. Sasaki, Y. Motokawa and G. Kikuchi, Ocurrence of both *a*-type and *b*-type cytochromes as the functional terminal oxidases in *Rhodopseudomonas sphaeroides*, Biochim. Biophys. Acta 197:284 (1977).

4. B. Wakim, B. Georg and J. Oelze, Regulation of respiration and cytochrome c oxidase activities in *Rhodospirillum rubrum* and *Rhodospirillum tenue* during the reversible adaptation from phototrophic to chemotrophic conditions, Arch. Microbiol. 124:97 (1980).

5. D. Zannoni and A. Baccarini-Melandri, Respiratory electron flow in facultative photosynthetic bacteria, in: "Diversity of bacterial respiratory systems", J.C. Knowles, ed., CRC Press, Boca Ratón (1980).

6. R. K. Poole, Bacterial cytochrome oxidases. A structurally and functionally diverse group of electron transfer proteins, Biochim. Biophys. Acta 726:205 (1983).

7. J. Ramírez and L. Smith, Synthesis of ATP in intact cells of *R. rubrum* and *R. spheroides* on oxygenation or illumination, Biochim. Biophys. Acta 153:466 (1968).

8. H. H. Lampe and G. Drews, Differentiation of membranes from *Rhodopseudomonas capsulata* with respect to their photosynthetic and respiratory functions, Arch. Microbiol. 84:1 (1972).

9. A. Thore, D.L. Keister and A. San Pietro, Studies on the respiratory system of aerobically (dark) and anaerobically (light) grown *Rhodospirillum rubrum*, Arch. Mikrobiol. 67:378 (1969).

10. C. Fenoll and J. M. Ramírez, Simultaneous presence of two terminal oxidases in the respiratory system of dark aerobically grown *Rhodospirillum rubrum*, Arch. Microbiol. 137:42 (1984).

11. G. Venturoli, C. Fenoll and D. Zannoni, On the mechanism of respiratory and photosynthetic electron transfer in *R. rubrum*, Biochim. Biophys. Acta 892:172 (1987).

12. R. M. Wynn, F. G. Gaul, W. K. Choi, R. W. Shaw and D. B. Knaff, Isolation of cytochrome bc_1 complexes from the photosynthetic bacteria *Rhodopseudomonas viridis* and *Rhodospirillum rubrum*, Photosynth. Res. Res. 9:181 (1986).

13. C. Fenoll, S. Gómez-Amores, G. Giménez-Gallego and J. M. Ramírez, A single pool of cytochrome c_2 is shared by cytochrome oxidase and photoreaction centers in *Rhodospirillum rubrum*, in: "Advances in

Photosynthesis Research", C. Sybesma, ed., Martinus Nijhoff/Dr W. Junk Publishers, The Hague (1984).

14. M. P. Ramírez-Ponce, G. Giménez-Gallego and J. M. Ramírez, A specific role for rhodoquinone in the photosynthetic electron transfer system of *Rhodospirillum rubrum*, FEBS Lett. 114:319 (1980).

15. J. Imhoff, Quinones of photosynthetic purple bacteria, FEMS Microbiol. Lett. 25:85 (1984).

16. G. Giménez-Gallego, S. del Valle-Tascón and J. M. Ramírez, A possible physiological function of the oxygen-photoreducing system of *Rhodospirillum rubrum*, Arch. Microbiol. 109:119 (1976).

17. M. P. Ramírez-Ponce, J.M. Ramírez and G. Giménez-Gallego, Rhodoquinone as a constituent of the dark electron-transfer system of *Rhodospirillum rubrum*, FEBS Lett. 119:137 (1980).

18. J. Lascelles, The synthesis of porphyrins and bacteriochlorophylls by cell suspensions of *Rhodopseudomonas sphaeroides*, Biochem. J. 62:78 (1956).

19. A. Verméglio and P. Richaud, Effect de l'antimycine A sur la respiration de cellules entières de *Rhodopseudomonas sphaeroides*, Physiol. Vég. 22:581 (1984).

20. M. A. Taylor and J.B. Jackson, Proton translocation in intact cells of the photosynthetic bacterium *Rhodopseudomonas capsulata*, Biochim. Biophys. Acta 810:209 (1985).

21. H. W. van Verseveld and G. Bosma, The respiratory chain and energy conservation in the mitochondrion-like bacterium *Paracoccus denitrificans*, Microbiol. Sci. 4:329 (1987).

PHYSIOLOGY AND GENETICS OF C4-DICARBOXYLATE TRANSPORT IN

Rhodobacter capsulatus

David J. Kelly, Mark J. Hamblin and
Jonathan G.Shaw

Robert Hill Institute
Department of Molecular Biology and
Biotechnology
University of Sheffield, Western Bank
Sheffield S10 2TN, U.K.

INTRODUCTION

Of those carbon sources traditionally used in studies on purple non-sulphur bacteria, the C4-dicarboxylic acids malate and succinate have long been known to be particularly effective in promoting fast growth rates and producing high cell yields under both photo- and chemoheterotrophic growth conditions (Stahl and Sojka, 1973). In Rhodobacter capsulatus, the iron-sulphur centre associated with succinate dehydrogenase is in redox equilibrium with the quinone pool (Zanonni and Ingledew, 1983), so that in addition to providing cell carbon, succinate can also act as a direct electron donor. Alternatively, under different circumstances, the reduction of fumarate to succinate may act as a redox poising mechanism for the removal of excess reducing equivalents (McEwan et al., 1985).

In view of the key role played by C4-dicarboxylates, it is surprising that virtually nothing is known about the transport systems mediating their uptake in photosynthetic bacteria or the nature of the genes and gene products essential for this process. The work of Gibson (1975) did establish, however, that Rb. sphaeroides possessed an inducible transport system for malate, succinate and fumarate and there is evidence for a sodium dependent dicarboxylate symport in Ectothiorhodospira shaposhnikovii (Karzanov and Ivanovsky, 1980). Studies on C4-dicarboxylate transport in enterobacteria and rhizobia have shown that uptake also occurs via a common system that acts on malate, succinate and fumarate (Kay and Kornberg, 1969; Finan et al., 1981). However, the molecular mechanism of transport appears to differ in the two groups. In Rhizobium leguminosarum, the product of the dctA gene is apparently the only protein involved in the actual transport mechanism (Ronson et al., 1984) while in Escherichia coli the

Molecular Biology of Membrane-Bound Complexes in Phototrophic Bacteria
Edited by G. Drews and E. A. Dawes
Plenum Press, New York, 1990

products of at least three genes have been implicated (Lo and Sanwal 1975). A combination of biochemical and genetic evidence strongly suggests that a periplasmic binding protein is involved in the E. coli system but the observation that transport can be mediated efficiently by membrane vesicles alone and that the energy source for uptake may be the proton-motive force (Gutowski and Rosenberg, 1975) is at variance with current views of the operation of binding protein dependent transport systems (Ames, 1988).

In order to study the molecular details of C4-dicarboxylate transport in photosynthetic bacteria, we have investigated the way in which these substrates enter cells of Rhodobacter capsulatus. Here, we describe the isolation by transposon mutagenesis of mutants that are impaired in dicarboxylate uptake, present evidence for the existence of two differentially regulated transport systems in this bacterium and also report the cloning of genes involved in the operation of one of these systems.

RESULTS

Isolation, Growth Characteristics and Enzyme Profiles of C4-Dicarboxylate Transport Mutants

Nine thousand kanamycin resistant transconjugants resulting from five independent matings between E. coli S17-1(pSUP2021; pBR325::mob::Tn5) and Rb. capsulatus 37b4 were selected under aerobic conditions in the dark on minimal pyruvate plates. Each transconjugant was then screened for the inability to grow on minimal malate plates under the same incubation conditions. It was reasoned that this screening procedure would select against the isolation of citric-acid cycle or anaplerotic enzyme mutants, as pyruvate is known to be metabolised through the citric-acid cycle in aerobically grown Rb. capsulatus (Willison, 1988). Five mutants were isolated (MJH 20,22,25,28,40) which showed the desired phenotype. In addition to a lack of aerobic growth on minimal malate plates, none of these mutants grew on media containing either succinate or fumarate as sole carbon source. Plate tests on glucose or acetate, however, showed that all the mutants could utilise these substrates.

Table 1 summarises the growth rates of wild-type and mutant strains on pyruvate and C4-dicarboxylates during cultivation under chemoheterotrophic or photoheterotrophic conditions. The doubling times of all the mutants on pyruvate were similar to the wild-type under both sets of growth conditions. None of the mutants grew on malate, succinate or fumarate under aerobic conditions in the dark, confirming the results of the plate tests described above. Unexpectedly, however, growth did occur under anaerobic conditions in the light when either malate or succinate - but not fumarate - was the carbon source, although the growth rates were significantly slower on succinate compared to malate under these conditions (Table 1). The final optical densities reached by such cultures were similar to those of wild-type cells and it was found possible to repeatedly subculture the mutants under phototrophic conditions on D,L

Table 1. Growth Rates of Rb. capsulatus Wild-Type and
         Mutant Strains on Pyruvate and C4-Dicarboxylates.

Culture Doubling Time (h)

CARBON SOURCE	CHEMOHETEROTROPHIC GROWTH				PHOTOHETEROTROPHIC GROWTH			
		MJH				MJH		
	37b4	22	25	28	37b4	22	25	28
Pyruvate	7	5	10	7	5	4	5	4
D,L-malate	5	/	/	/	4	15	12	15
Succinate	5	/	/	/	5	25	25	18
Fumarate	5	/	/	/	5	/	/	/

/; No Growth. Experiments were performed in RCV minimal
medium (Weaver et al., 1975) and growth was monitored by
regular absorbance readings at 580nm.

malate. However, no growth occured upon subsequent transfer to
malate minimal medium under aerobic conditions in the dark.
The growth characteristics of the mutants isolated are
consistent with the presence of two distinct uptake systems for
C4-dicarboxylates in Rb. capsulatus. Only one of these systems
(designated dct) has been inactivated by Tn5, as the
alternative system (designated pdt; photosynthetic
dicarboxylate transport) must still be expressed to allow
phototrophic growth of the mutants on malate. In order to rule
out the possibility that the mutant phenotype was simply due to
a defect in the aerobic metabolism of C4-dicarboxylates,
measurements were made of the specific activities of the
enzymes of the dicarboxylic acid branch of the citric-acid
cycle in three of the mutants isolated. Malate dehydrogenase,
succinate dehydrogenase and fumarase were present at similar
specific activities in both mutant and wild-type strains grown
aerobically in the dark. In addition, the three C6-branch
enzymes citrate synthase, aconitase and isocitrate
dehydrogenase were also present at wild-type levels in one
mutant, MJH28, which was chosen as a representative strain for
more detailed characterisation.

Southern Hybridisation and Reversion Analysis

In order to establish unequivocally that a single Tn5
insertion was responsible for the unexpected difference in
growth of the mutants under photo- and chemoheterotrophic
conditions, a series of Southern hybridisations were performed
using both a digoxenin labelled Tn5 probe (pRZ104 ; ColE1::Tn5)
and a pSUP202 probe (to rule out the possibility that suicide
vector sequences were also present in the mutant genomes). The
results clearly showed that all five mutants contained a single
band of about 10kb which hybridised to the Tn5 probe in an
EcoR1 digest of chromosomal DNA and that no pSUP202 sequences
were present. As EcoR1 does not cut within Tn5, these results
indicate that each mutant contains only a single copy of the
transposon. In addition, the size of the chromosomal EcoR1

Table 2. Uptake of ^{14}C-L-malate into Rb.capsulatus wild-
type and MJH28 (dct::Tn5) cells.

Growth Conditions	Assay conditions	Initial rate of transport nmol min^{-1}(mg cell protein)$^{-1}$	
		37b4(wild-type)	MJH28
Aerobic/Dark	Aerobic/Dark	22.8	0.01
Anaerobic/ Light	Anaerobic/ Light	8.2	0.01

Transport experiments were performed in RCV medium (minus
malate) at 30^{0}C in an oxygen electrode. Uptake was started
by the addition of radiolabel to 4.4 μM final concentration and 100 μl samples were removed at intervals,
rapidly filtered and scintillation counted.

fragment containing the insertions is about 4.2kb in length,
assuming Tn5 to be 5.8kb (Jorgensen et al., 1979) and that each
insertion has occurred into the same fragment.

A reversion analysis established that the Tn5 insertions
were genetically linked to the phenotype of the mutants
isolated. Revertants of all the mutants were selected on malate
medium and arose at a frequency of between 10^{-8} and 10^{-10}. The
majority of these proved to be kanamycin sensitive and to have
simultaneously regained the ability to grow on succinate and
fumarate as well as malate. These data are consistent with
reversion occuring by the precise excision of a single copy of
Tn5.

Transport of ^{14}C-L-Malate into Wild-Type and Mutant Cells

Cells of MJH28 grown under aerobic conditions in the dark
on pyruvate medium were harvested and then incubated for 2h
under the same conditions in malate medium. The measured rate
of transport of ^{14}C-L-malate into such cells was only 0.04 per
cent that of the wild-type, which was grown and assayed in an
identical manner (Table 2). This result confirms that the
lesion in this mutant is at the level of transport rather than
metabolism.

Transport assays were also carried out with both wild-type
and MJH28 cells grown photoheterotrophically on malate. In
these experiments, the bacteria were maintained initially under
anaerobic conditions in the dark by passing a slow stream of
O_2-free nitrogen gas over the surface of the suspension
contained within the oxygen electrode. No uptake of ^{14}C-L-
malate was observed in either wild-type or MJH28 cells under
such conditions. Upon illumination, however, uptake proceeded
at a linear rate in the wild-type suspension after a reproducible short lag peiod. Surprisingly, only a very low rate of
light-dependent malate transport could be detected using MJH28
cells (Table 2) and this rate was similar to the residual rate

observed with aerobically grown cells. The uptake rate was not
altered significantly if the assays were carried out under
aerobic conditions in the dark. Thus, although MJH28 is capable
of phototrophic but not aerobic growth on malate as sole added
carbon source, this difference was not reflected in transport
assays with radiolabelled malate.

An explanation of this discrepancy was sought from
experiments designed to determine the affinity of cells of
MJH28 for L- and D- malate. Cultures were set up with different
starting concentrations of L- or D- malate as sole added carbon
source and incubated under anaerobic conditions in the light.
The exponential growth rate achieved was plotted as a function
of the initial malate concentration using a double-reciprocal
plot, so that the x-axis intercept could be used to calculate
an approximate K_s value (Pirt, 1980). This method gave a K_s
value of 4.5 mM for L-malate and 6.6 mM for D-malate. Assuming
that these values reflect the K_t value of the pdt system which
allows growth of MJH28 on malate under photosynthetic
conditions, then it is clear that the range of concentrations
of ^{14}C-labelled L-malate that it is practical to use in
transport assays (for economic reasons) is several orders of
magnitude less than the affinity constant of this transport
system. Further work will therefore be needed to devise assay
procedures that do not necessarily rely on the use of
radiolabelled substrates.

Characterisation of Transport via the Dct System

The results described above suggested that in wild-type
cells grown under aerobic conditions in the dark only one
dicarboxylate transport system was present. The affinity
constant (K_t) of this system for L-malate was determined in
intact cells by measuring initial rates of transport at
different concentrations of radiolabelled substrate. The data
conformed to Michaelis-Menten kinetics with a K_t value of 2.9 $\pm$
1.2 uM and V_{max} value of 57.7 $\pm$ 12.3 nmol min^{-1} (mg protein)$^{-1}$,
measured as an average of four separate determinations on
different batches of cells. These data clearly show the dct
system to be of high affinity. The specificity of transport was
studied by measuring the degree of apparent inhibition of L-
malate uptake upon inclusion of unlabelled substrate analogues
in the assay mixture at a concentration of 50 μM. Of those
substrates tested, fumarate and succinate almost completely
prevented malate uptake while D-malate inhibited the rate of
uptake by only 62 %. Malonate, maleate and phthalate inhibited
the rate by about 50 % but aspartate, DL-lactate and pyruvate
did not compete with malate uptake.

The sensitivity of malate transport to cold osmotic shock
is illustrated in Fig 1. In this experiment, wild-type cells
were treated with a low concentration of EDTA in a cold
hypertonic sucrose solution to induce plasmolysis and release
the contents of the periplasm. Such treatment led reproducibly
to the total abolition of transport activity. That this effect
was not due to gross metabolic disturbance was shown in several
ways. Firstly, no significant loss of cell viability resulted
from the EDTA treatment (Fig 1.). Secondly, in separate
experiments, we determined the effect of cold osmotic shock on

457

the ability of photosynthetically grown cells to generate a
light induced membrane potential. This was done using the
electrochromic absorbance change of the endogeneous carotenoid
pigments (Kelly et al., 1988). In one set of experiments,
untreated wild-type cells generated a light induced
electrochromic absorbance change at 528-511nm of 0.009 $\triangle$A.μmol
Bchl^{-1}, while osmotically shocked cells gave a value of 0.007 $\triangle$
A.μmol Bchl^{-1}. This result indicates that shocked cells are
still able to carry out photosynthesis and maintain an intact
and energized cytoplasmic membrane. Importantly, neither
sucrose treatment or EDTA treatment alone led to the complete
abolition of transport activity. Washing the treated cells
several times in fresh growth medium (minus malate) containing
excess divalent cations to eliminate residual EDTA did not
result in restoration of malate transport activity. Taken
together, these data suggest a specific effect of osmotic shock
in removing a periplasmic component involved in dicarboxylate
transport. However, we have thus far been unable to detect
malate binding activity in periplasmic extracts.

Cloning of the dct Locus

A wild-type gene bank of Rb. capsulatus B10 DNA (Colbeau
et al., 1986) consisting of eight pools of recombinant pLAFR1

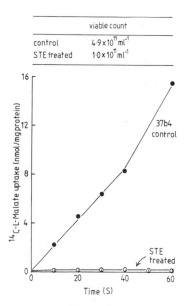

Fig. 1. Effect of Cold Osmotic Shock on L-malate
         Transport into Rb. capsulatus Wild-type
         Cells. The bacteria were harvested, washed
         once in RCV (minus malate) and either re-
         suspended without treatment (filled circles)
         or incubated with STE (0.5M sucrose, 1.3mM
         EDTA in 50mM Tris-HCl, pH8.0) for 10 min on
         ice (open circles). Transport assays were
         performed under aerobic/dark conditions using
         a radiolabelled malate concentration of 4.4 μM.

clones with an average insert size of 20Kb was transferred to
MJH28 at a frequency of approx. 10^{-6} tetracycline resistant
colonies per recipient in separate triparental matings with
pRK2013 as helper plasmid. pLAFR1 clones able to complement the
lesion in MJH28 in trans were selected aerobically on minimal
malate plates containing both kanamycin (to maintain selection
for Tn5) and tetracycline (to select pLAFR1). The frequency of
such mal^+ Km^r Tc^r colonies was approx 10^{-3} per pLAFR1
containing recipient selected on pyruvate. Twelve colonies were
picked, restreaked, purified and then tested for the
restoration of normal photoheterotrophic growth on malate. All
were positive. The pattern of restriction fragments generated
after complete EcoR1 digestion of cosmid mini-preps indicated
that three types of cosmid were responsible for the
complementation of these twelve isolates. One of each type -
designated pDCT100, pDCT200 and pDCT300 - was transformed into
E.coli S17-1, rechecked for complementation of MJH28 and
subjected to detailed restriction mapping to order the EcoR1
fragments within the inserts.

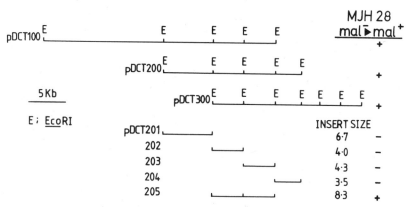

Fig. 2. Physical Maps of Cosmids and Plasmid Subclones
Which Complement the Lesion in dct Mutant MJH28.

The physical maps resulting from this analysis (Fig.2)
showed that the three cosmids contained overlapping insert DNA
with an 8.3kb region in common, consisting of two EcoR1
fragments of 4.0 and 4.3kb. pDCT300 has since proved to be
unstable and to undergo deletions of the insert DNA; it has not
been analysed further. All of the EcoR1 fragments from pDCT200
were subcloned into pRK415 but none of the resulting plasmids
were able to complement MJH28. However, two subclones
(pDCT205/206) which were constructed from a partial EcoR1
digest of pDCT200 and which contained the entire 8.3kb common
region, were able to restore MJH28 to normal aerobic growth on
malate plates.

DISCUSSION

Transposon mutagenesis combined with a simple screening procedure for pyruvate-positive, D,L malate-negative growth under aerobic conditions, has allowed us to isolate mutants of Rb. capsulatus impaired in their ability to transport C4-dicarboxylates. The five mutants isolated in this study appear to have an identical phenotype and each contains a single Tn5 insertion as judged by both Southern blot and reversion analysis. From the present work, it is not possible to determine if the transposon insertions have occured into a structural or regulatory region but evidence from both genetic and physiological experiments suggests that the locus inactivated encodes a transport system that recognises malate, succinate and fumarate. We have cloned this locus and designated it dct by analogy with similar systems identified in both E. coli (Kay and Kornberg, 1969) and Rhizobium (Ronson et al., 1984). The dct genes appear to be located within an 8.3kb region of the Rb. capsulatus B10 chromosome, composed of two contiguous EcoR1 fragments of 4.0 and 4.3kb. The central EcoR1 site appears to be located within the dct gene region itself. Further subcloning and localised Tn5 mutagenesis is now being carried out with pDCT205 to define the number of complementation groups within this region.

The Rb. capsulatus dct system has a high affinity for L-malate, with a K_t value of about 3 μM. It is also highly specific for dicarboxylic acids, as structurally very similar substrates (e.g. aspartate) do not compete with malate for uptake. The sensitivity of transport to osmotic shock is suggestive of the involvement of a periplasmic binding protein but no malate binding activity was detectable in periplasmic extracts. In E. coli, a proportion of the dicarboxylate binding protein is located in the outer membrane. Further work is necessary to investigate this possibility in Rhodobacter.

The dct system is clearly important for the photoheterotrophic assimilation of C4-dicarboxylates, in addition to being essential for chemoheterotrophic growth on these substrates. However, the fact that all of the mutants isolated in this study were still able to grow under photoheterotrophic conditions on malate, yet were not simply deficient in aerobic dicarboxylate metabolism, indicates that a second transport system must be present in both mutant and wild-type cells. Succinate - but not fumarate - also appears to be transported by this route. We have given the designation pdt (photosynthetic dicarboxylate transport) to this second system. The pdt system could not be assayed using the low concentrations of radiolabelled malate which were sufficient to saturate uptake via the dct route. This was apparently due to a low affinity for both D- and L- malate, as judged by the approximate K_s values determined for these substrates in batch cultures.

The physiological role of the pdt system is not clear at present but it is interesting that photosynthetic and not aerobic growth conditions allow its operation. In principle, there are at least two possibilities to explain this differential expression of activity. Firstly, the uptake of malate and succinate by this pathway may require photosynthetic, as opposed to respiratory, energisation. Although the mechanism of energy coupling to the pdt system is not known, this explanation would draw on controversial, but previously advanced, notions of a direct interaction between some component(s) of the electron transport chain and individual solute transport systems (Elferink et al., 1984). A second explanation is that the pdt system is only expressed under phototrophic growth conditions but, when present in the cells,can be energised by a delocalised intermediate (the proton-motive force or ATP) generated by either photosynthetic or respiratory electron transport. This more orthodox view would imply that the pdt genes are subject to regulation by oxygen concentration or light intensity, possibly in a manner similar to the genes encoding the photosynthetic apparatus (i.e. repressed under aerobic conditions).These explanations are not mutually exclusive but can only be tested when appropriate assays and gene fusions for this system become available. To this end, we are developing such assays and constructing pdt mutants which may be useful in cloning the cognate genes and which will help in identifying the function of both dicarboxylate uptake routes in this bacterium.

ACKNOWLEDGEMENTS

We thank the U.K. Science and Engineering Research Council for a research grant to D.J.K. and studentship to J.G.S. We also thank the Society for General Microbiology for financial support.

REFERENCES

Ames, G. F.-L., 1988, Structure and mechanism of bacterial periplasmic transport systems, J. Bioenerg. Biomembr., 20:1.

Colbeau, A., Godfroy, A., and Vignais, P.M., 1986, Cloning of DNA fragments carrying hydrogenase genes of Rhodopseudomonas capsulata, Biochemie., 68:147.

Elferink, M.G.L., Hellingwerf, K.J., van Belkum, F.J., Poolman, B., and Konings, W.N., 1984, Direct interaction between linear electron transfer chains and solute transport systems in bacteria, FEMS Microbiol Letts., 21:293.

Finan, T.M., Wood, J.M., and Jordan., D.C., 1981, Succinate transport in Rhizobium leguminosarum, J. Bacteriol., 148:193.

Gibson, J., 1975, Uptake of C4-dicarboxylates and pyruvate by Rhodopseudomonas sphaeroides, J. Bacteriol., 123:471.

Gutowski, S.J., and Rosenberg, H., 1975, Succinate uptake and related proton movements in Escherichia coli K12, Biochem. J., 152:647.

Karzanov, V.V., and Ivanovsky, R.N., 1980, Sodium dependent succinate uptake in purple bacterium Ectothiorhodospira shaposhnikovii, Biochim. Biophys Acta., 598:91.

Jorgensen, P.A., Rothstein, S.J., and Reznikoff, W.S., 1979, A restriction enzyme cleavage map of Tn5 and location of a region encoding neomycin resistance. Mol. Gen. Genet., 177:65.

Kay, W.W., and Kornberg, H., 1969, Genetic control of the uptake of C4-dicarboxylic acids by Escherichia coli FEBS Letts., 3:93.

Kelly, D.J., Richardson, D.J., Ferguson, S.J. and Jackson J.B., 1988, Isolation of transposon Tn5 insertion mutants of Rhodobacter capsulatus unable to reduce trimethylamine-N-oxide and dimethylsulphoxide, Arch. Microbiol., 150:138.

Lo, T.C.Y., and Sanwal, B.D., 1975, Genetic analysis of mutants of Escherichia coli defective in dicarboxy- late transport, Mol. Gen. Genet., 140:303.

McEwan, A.G., Cotton, N.P.J., Ferguson, S.J., and Jackson, J.B., 1985, The role of auxiliary oxidants in the maintenance of a balanced redox poise for photosynthesis in bacteria, Biochim. Biophys. Acta., 810:140.

Pirt, S.J., 1975, "Principles of Microbe and Cell Cultivation". Blackwell. Oxford.

Ronson, C.W., Astwood, P.M., and Downie, J.A., 1984, Molecular cloning and genetic organisation of C4- dicarboxylate transport genes from Rhizobium leguminosarum, J.Bacteriol., 160:903.

Stahl, C.L., and Sojka, G.A., 1973, Growth of Rhodopseud- omonas capsulata on L- and D- malic acid, Biochim. Biophys. Acta., 299:241.

Weaver, P.F., Wall, J.D., and Gest, H., 1975, Characterisation of Rhodopseudomonas capsulata, Arch. Microbiol., 105:207.

Willison, J.C., 1988, Pyruvate and acetate metabolism in the photosynthetic bacterium Rhodobacter capsulatus J. Gen. Microbiol., 134:2429.

Zannoni, D., and Ingledew, W.J., 1983, A functional character- isation of the membrane bound iron sulphur centres of Rhodopseudomonas capsulata, Arch. Microbiol., 135:176.

SENSORY SIGNALLING IN RHODOBACTER SPHAEROIDES

Judith P. Armitage, Philip S. Poole and Simon Brown

Microbiology Unit
Department of Biochemistry
University of Oxford
Oxford OX1 3QU

INTRODUCTION

Motile bacteria actively swim about their environment, randomly changing direction every few seconds. When faced with a gradient of a chemical they alter the frequency of direction changing to bias their overall direction towards a favourable environment. Bacteria are too small to sense any change in concentration across their body length, and therefore environmental sampling must occur by temporal comparison. A bacterium such as Rhodobacter sphaeroides, living under conditions where any one of many different growth parameters could be limiting, must be able to sense and respond to changes in different metabolites, light intensity and wavelength, oxygen and other terminal electron acceptors and balance these different sensory signals to give an integrated overall response.

Chemotaxis to receptor-dependent chemoeffectors has been intensively studied in enteric bacteria, and the mechanisms are quite well understood[1]. However, bacteria such as Rhodobacter sphaeroides do not show receptor-dependent chemotaxis, and other bacterial species that have receptors for chemoattractants have in addition some receptor-independent responses[2,3]. We have therefore investigated the sensing mechanisms involved in receptor-independent chemotaxis in the hope of elucidating how different types of sensory stimuli can be integrated at the flagellar motor.

Rhodobacter sphaeroides swims using a single medially located flagellum, composed of a polymer of a 54kD protein (see Sockett, Foster and Armitage, this volume). Unlike most other bacterial species the flagellum is only rotated in a clockwise direction[4]. A change in the swimming direction is brought about by stopping flagellar rotation periodically, reorientation being caused by Brownian motion during the stopped period, rather than switching flagellar rotation from counterclockwise to clockwise as occurs in most other bacterial species. This stop-start form of motility occurs even though the driving force for flagellar rotation, the electrochemical proton gradient (pmf), remains high. Within a population the speed of swimming, the stop length and frequency of stopping varies greatly. Speeds can vary from a few μm sec^{-1} to about 100μm sec^{-1}, and stop lengths can vary from less than

Molecular Biology of Membrane-Bound Complexes in Phototrophic Bacteria
Edited by G. Drews and E. A. Dawes
Plenum Press, New York, 1990

100msec to several seconds. The great variation suggests that the control of torque generation comes from transient variations in the structure, and therefore efficiency, of the flagellar motor rather than from changes in the pmf. Bacteria treated with high concentrations of uncouplers to give low measurable membrane potentials showed the same variation in stopping frequency and speed as untreated bacteria, suggesting that the stopping frequency is not directly controlled by the pmf (unpublished observations).

This great variation in swimming pattern made the investigation of tactic responses at anything but a gross descriptive level difficult until the development of computerised motion analysis. It is now possible to follow the swimming patterns of many hundreds of cells and average their behaviour before and after tactic stimulation[5]. Behavioural responses can now be studied not only as gross accumulations of cells in specific environments but also at the level of changes in the behaviour of the flagellar motor that brings about that accumulation.

AEROTAXIS AND PHOTOTAXIS

Rhodobacter sphaeroides can grow as either a photoheterotroph, anaerobically in the light, or as a heterotroph, aerobically in the dark. When photoheterotrophically grown cells are incubated in the light they showed a positive response to increasing light intensity but are repelled by high oxygen concentrations, leaving oxygenated, illuminated areas. On the other hand, when incubated in the dark, cells show a positive response to oxygen. Under both conditions the cells show chemotaxis. One of the interesting questions therefore is how the responses to different stimuli are integrated, and how the incubation conditions can result in a reversed response to oxygen.

There have been two major difficulties involved in measuring phototactic and aerotactic responses. One is the variability in unstimulated swimming behaviour described above, the other that both stimuli can be expected to directly effect the pmf and therefore, possibly, the rate of rotation of the flagellum[6]. Any kinetic response resulting from an increase in pmf must therefore be separated from a true change in behaviour. Many photosynthetic species, e.g. Chromatium, Rhodospirillum show very characteristic reversals when they enter regions of reduced light intensity[6,7,8]. Rhodobacter sphaeroides does not show such a distinct response, however fig.1. shows that a reduction in light intensity causes anaerobic illuminated cells to increase their stopping frequency (the equivalent of increased tumbling, or a repellent response in enteric bacteria). This indicates that there is a true behavioural change in response to changing light quality.

Early work showed the requirement for photosynthetic electron transport and the photosynthetic reaction centre in the phototactic response of R.sphaeroides and other photosynthetic species[9,10], the sensory signal caused by photosynthetic electron transport which results in altered behaviour of the flagellar motor has not yet been identified. Whether the phototactic behavioural changes seen in R.sphaeroides are coincident with the photsynthetic action spectrum is also unknown.

Using the branched respiratory electron transport pathways in R. sphaeroides the requirement for electron transport and/or a change in the pmf in the positive aerotactic response was investigated. By using different combinations of inhibitors it was possible to allow electron transport to the different terminal cytochromes of the branched electron transport chain of R.sphaeroides. Electron transport to some terminal cytochromes allowed an increase in pmf whereas electron transport to others did not. A behavioural response to any oxygen gradient in the dark only occurred if there was not only oxygen binding and electron transport but also an increase in membrane potential[11].

This result suggests that a change in pmf is required for an aerotactic response. However, chemotaxis studies suggest that a change in pmf per se is not enough to cause a behavioural response (see below). It seems probable that the aerotactic signal is not caused directly by a change in pmf affecting the stopping frequency of the motor, but that the signal is an indirect consequence of the pmf change brought about by respiratory electron transport.

If an increase in the size of the pmf can, at least indirectly, result in both a positive phototactic and positive aerotactic response, how are bacteria prevented from swimming up a light gradient into an oxygenated environment? One of the non-coupled terminal oxidases branching from the quinone is present even when R.sphaeroides is grown in strictly anaerobic condition[12]. Using the carotenoid bandshift we showed that if cells were incubated in sub-saturating light and then briefly oxygenated there was a transient fall in the measured membrane potential (unpublished). It seems possible that this high affinity oxidase binds oxygen when the cells move into an aerobic environment changing the redox poise of some of the shared components of the respiratory and photosynthetic electron transport chain, subsequently causing a transient fall in membrane potential. This transient fall in membrane potential, or change in redox poise of an electron transport component would result in a negative behavioural response when the bacterium entered an oxygenated, illuminated environment.

CHEMOTAXIS

When E.coli is chemotactically stimulated it responds in about 200 ms, and adapts to that concentration of chemoeffector in a few minutes[13]. The length of time taken to adapt depends on the concentration of chemoeffector. Many of the attractants act by binding to membrane spanning receptor proteins, also known as methyl-accepting chemotaxis proteins (MCPs). A signal is sent from the MCP to control the flagellar switching via a series of phosphorylated intermediates. Phosphorylation on a slower time scale controls the activity of a specific methyl esterase involved in adaptation of the bacteria to that particular concentration of chemoeffector[14,15].

Intensive investigation at the biochemical, antibody and DNA level has failed to identify any MCPs in R.sphaeroides (although they are present in Rhodospirillum rubrum). R.sphaeroides seems to have only MCP-independent forms of chemotaxis[16]. E.coli also shows a chemotactic response to a range of chemicals not sensed via the MCPs. In addition to oxygen these include sugars transported through the PTS system, and amino-acids such as proline[17,18,19]. The mechanism of sensing non-MCP chemoeffectors and the integration of the signal at the flagellar motor is not understood.

Observation of R.sphaeroides after chemotactic stimulation shows several obvious differences when compared to chemotactically stimulated MCP-containing bacteria. The most obvious differences are (i) not all the cells in the stimulated population respond, although when the overall population behaviour is measured there is an obvious change in the average stopping frequency, (ii) there is little or no adaptation to the chemoeffector over periods of tens of minutes, (iii) the ability to respond to some chemicals appears to depend on the growth environment of the cells (iv) only metabolites and some cations have been shown to cause a behavioural response in R.sphaeroides, and no repellents (other than oxygen in a photosynthetic environment) have been identified.

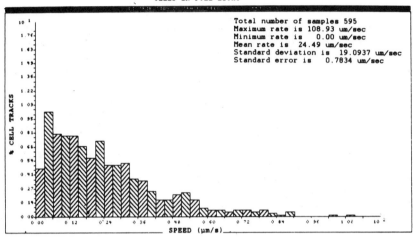

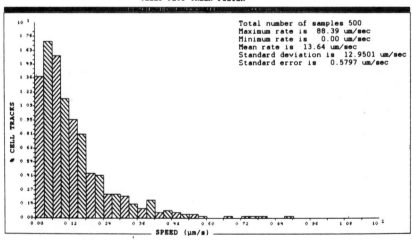

Figure 1. Swimming behaviour of a population of R.sphaeroides incubated in full light (upper trace) and in green light (lower trace). Tracking time was 0.4s per cell

With MCP-dependent chemotaxis transport of the chemoeffector is not
required, only interaction of the chemical with the periplasmic surface
of the MCP molecule. In contrast all the chemoeffectors so far
investigated in R.sphaeroides need to be transported to produce a
chemotactic response[20]. There are many different ways in which the
transported chemoeffectors might alter the behaviour of the flagellar
motor (i) there might be a direct effect on the pmf caused either by the
transport or the metabolism of the chemical, (ii) the intracellular pH
(pHi) might change in response to transport or metabolism (iii) a
metabolic intermediate or the energy state of the cell may effect the
motor (iv) there may not be a single effector, but a range of different
effectors might be able to interact directly with either the flagellar
motor or with a motor control molecule.

An initial problem was to identify the nature of the chemotactic
response in R.sphaeroides, as not all the cells in population showed any
behavioural change, and the responses, in terms of time course of the
response, appeared different for some groups of chemoattractants. Motion
analysis showed that the average speed of swimming of the population
increased after a chemotactic stimulus. The change in average speed
could be caused by either an increase in flagellar rotation rate or a
decrease in both stopping frequency and the length of stops. Experiments
have shown that the latter is certainly true, the stimulated bacteria
stop less frequently, and probably for shorter periods[21]. In addition,
it seems likely that there is a change in flagellar rotation rate in
response to an attractant stimulus. As all chemoattractants identified
for R.sphaeroides are metabolites, and as the flagellum is driven by the
pmf it is therefore possible that both the decrease in stopping frequency
and increase in rotation rate is the result of a change in the pmf caused
by either the transport of the chemoattractant or its metabolism.

The pmf and chemotaxis

The effects of a range of chemicals on cell behaviour, electron
transport rate and membrane potential were examined. There was no
correlation between either the rate of electron transport or change in
membrane potential with the behavioural response of the cells. The
effect of adding chemoattractants on the membrane potential was measured
using the carotenoid bandshift. All chemoattractants were found to
increase the membrane potential in the dark, but only potassium had any
effect in the light. However, all the compounds showed a marked effect
on long term swimming behaviour in the light and all have been shown to
cause accumulation of the bacteria when given as a gradient, irrespective
of the background light intensity. Strong attractants such as propionate
can therefore produce a strong chemoattractant response irrespective of
the baseline membrane potential or the effect of the attractant on the
rate of electron transport.

The lack of requirement for a change in membrane potential during
chemotaxis was confirmed by examining chemoeffector transport, the
chemotactic response and motility as the membrane potential was reduced
by increasing the concentration of the uncoupler CCCP. Chemotaxis was
lost at the membrane potential at which transport of the chemoeffector
was inhibited. At this uncoupler concentration motility continued
normally both in terms of stopping frequency and velocity, until the
membrane potential was reduced to less than 10% of the maximum (fig.2).
These results suggest that in earlier experiments examining the effects

of uncouplers on chemotaxis we were probably examining the effect of transport inhibition rather than the chemotactic signal. The concentration of uncoupler present when chemotaxis was still continuing normally was such that any change in potential would be expected to be clamped by the proton ionophore concentration per cell.

Interestingly if the chemotactic response is not the direct result of a change in the pmf, then the speed increase seen on chemotactic stimulation is not caused by a higher pmf. This result has major implications for the operation of the flagellar motor, suggesting that the speed increase is probably the result of increased efficiency of the coupling of the pmf to torque generation. It also suggests that, at least at high pmf values, the rotation rate of the motor is not directly related to the size of the pmf.

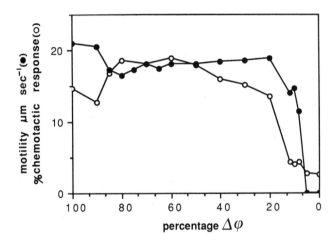

Figure 2. Effect on the chemotactic response and motility of Rhodobacter sphaeroides of reducing the membrane potential with CCCP. Motility continued after the chemotactic reponse was lost. Loss of uptake of the chemoeffector (propionate) exactly paralleled the loss of chemotaxis

The conclusion from the above experiments is therefore that the chemotactic response in R.sphaeroides is not the direct result of any change in the pmf. The role of the pmf in aerotactic and phototactic responses remains to be investigated but it seems probable that the signal is the result not only of the change in pmf directly but requires an additional electron transport dependent event.

pHi and chemotaxis

R.sphaeroides responds to a very wide range of metabolites, including weak acids, ammonia, PTS- and non-PTS sugars, polyols, potassium ions. Ammonia and acetate have the reversed effect on the pHi as measured by NMR, but both cause a positive chemotactic response. In addition benzoate, a non-metabolite, has the same effect on pHi as acetate but causes no chemotactic response[20]. It seems unlikely that a change in pHi causes the motor response.

Metabolism and chemotaxis

Ammonia can be metabolised by R.sphaeroides using either the GDH or GS-GOGAT pathways, depending on whether or not the bacteria are nitrogen limited. If grown on high ammonia there was only a weak response to ammonia as a chemoattractant (see later), but if grown on limiting ammonia to derepress the GS/GOGAT pathway the response was strong. When glutamine synthetase was inhibited by the addition of methionine sulphoximine ammonia transport continued, but both assimilation and chemotaxis towards ammonia were lost [22]. This result suggests that limited metabolism, at least, is required for the response to ammonia.

As mentioned above R.sphaeroides responds to a wide range of chemicals. All the chemoattractants so far identified have been found to be metabolites or have major metabolic consequences. No repellent chemicals have been identifed despite an extensive search. The strongest attractants are organic acids such as acetate, propionate, butyrate, and pyruvate. Amino acids produce weaker responses which may be enhanced by growing the cells on an amino acid such as alanine as the sole carbon-nitrogen source. Some sugars can produce a strong response if grown on them, these include both PTS-sugars such as fructose and non-PTS sugars such as glucose. Interestingly very strong responses are seen to potassium and rubidium. Competition experiments show that e.g. a high background of acetate will reduce the response to propionate but not to fructose, whereas potassium will inhibit only the positive response to rubidium or ammonia. These results suggest that different metabolic pathways may produce independent signals controlling the flagellar behaviour, or that a background chemical may remove the 'limitation' of the related chemoattractant, but not chemoattractants using different metabolic pathways.

The strength of responses to attractants can vary depending on the growth environment of the bacteria. This is particularly obvious for ammonia which is a very strong attractant for cells grown on limiting ammonia but a poor attractant for cells grown on high ammonia. It was originally thought to reflect the pathway used to assimilate the ammonia, but it is now apparent that R.sphaeroides only responds to ammonia if nitrogen is the currently limiting metabolite. Under conditions of nitrogen limitation the cells were found to respond very weakly to an attractant such as pyruvate, although it was a strong attractant for cells grown under high nitrogen. The cells therefore showed the strongest responses to currently growth limiting compounds. If the carbon limitation was overcome by addition of pyruvate, cells would no longer respond to propionate, which earlier caused a response.

When bacteria, resuspended in buffer, were placed in the bottom of a chemotaxis well with a strong attractant such as pyruvate in the upper chamber there was an increase in the number of bacteria swimming into the upper chamber, showing that the bacteria could respond to a gradient of a chemoeffector. If the behaviour was examined under the microscope the stopping frequency of the bacteria decreased and the average swimming speed of the population increased in response to the chemoeffector. Interestingly however if the chemoeffector was added to the cells in the bottom chamber as well as the top chamber there was still an increase in the number swimming into the upper chamber when compared to bacteria in buffer alone, even though it was shown that no effective gradient was formed during the time course of the experiment. The long term change in swimming behaviour in response the chemoattractant, in the absence of any measurable adaptation, therefore resulted in an increased overall mobility of the population, but obviously when presented with a gradient the cells still showed accumulation.

If the cells were resuspended in a saturating concentration of one carbon source, e.g. pyruvate they were unable to respond to a gradient of another carbon source, e.g. propionate. Examination under the microscope showed no additional change in swimming behaviour after addition of the second carbon source There was however a response if the cells were presented with a gradient of a different metabolite e.g. ammonia or potassium. This was seen both in chemotaxis wells as an increase in the number of cells swimming into the upper chamber in response to the gradient and as a further increase in the average swimming speed of the population examined under the microscope. These results suggest that a change in metabolism can directly effect the behaviour of the flagellar motor and that once a metabolic limitation is overcome the cells show no further response to that group, but they do show a change if presented with a new metabolite, until saturation is finally reached. The new increased swimming speed and reduced stopping frequency could continue for tens of minutes (possibly corresponding to metabolic utilisation). These results suggest that at least in the cases of organic acids and cations R. sphaeroides can respond to a gradient in the absence of adaptation. Using this form of response a cell under non-gradient conditions for carbon would disperse from its original position, until an environment with reduced carbon caused a decrease in speed and increase in stopping frequency, reorienting the cell to the high carbon environment. If the cell now encountered a gradient of e.g. ammonia, which was now a limiting metabolite, the direction of movement would be up the nitrogen gradient. Using this response the cells would spread in the environment, the extent of that spreading being dependent on the metabolic state supported by that environment. In addition the cells will respond to gradients of metabolites, as long as the signal detected by the motor is not currently saturated. Current experiments suggest that it is the change in specific intermediary metabolite(s) that is responsible for the change in motor behaviour as mutations in specific pathways can cause the loss of a chemotactic response. The identity of the metabolites responsible, and whether there is one or several is as yet unknown.

SUMMARY

Rhodobacter sphaeroides shows receptor-independent taxis, requiring transport and limited metabolism of the chemoeffectors. The stopping response, and the change in speed, may be controlled by one or more metabolic intermediate(s), this metabolic intermediate may interact directly with a component of the flagellar motor, or may be integrated through a control molecule which can respond to a variety of different metabolic changes. The fact that the speed increase is not the result of an increase in the pmf suggests that the efficiency of the proton driven flagellar motor may change in response to a change in e.g. metabolic rate, perhaps by controlling the extent of interaction between the proteins that make the proton channel and the flagellar motor. We are currently trying to identify the metabolic signal(s) that may alter both the stopping and rotational frequency of the motor. Using transposon mutagenesis we have identified bacteria with metabolic lesions that exhibit altered swimming behaviour and altered chemotactic responses when grown on some compounds, but normal behaviour on other compounds (Havelka and Armitage, unpublished). These mutants should help identify the behavioural control molecule(s).

Acknowledgements

We would like to thank the Wellcome Trust and the SERC for financial support of the work described here.

References

1. Macnab, R. M. 1987. Motility and chemotaxis. In Escherichia coli and Salmonella typhimurium: cellular and molecular biology, vol.1 (F. C. Neidhardt, J. L. Ingraham, K. B. Low, B. Magasanik, M. Schaechter, and H. E. Umbarger, editors). American Society for Microbiology, Washington, D.C. 732-759.
2. Armitage, J. P. 1988. Tactic responses in photosynthetic bacteria. Can.J.Microbiol. 34:475-481.
3. Taylor, B. L. 1983. Role of proton motive force in sensory transduction in bacteria. Ann.Rev.Microbiol. 37: 551-573.
4. Armitage, J. P., and R. M. Macnab. 1987. Unidirectional intermittent rotation of the flagellum of Rhodobacter sphaeroides J.Bacteriol.169:514-518.
5. Poole, P. S., D. R. Sinclair, and J. P. Armitage. 1988. Real time computer tracking of free-swimming and tethered rotating cells. Anal.Biochem. 175:52-58.
6. Hader, D-P. 1987. Photosensory behaviour in prokaryotes. Microbiol.Rev. 51:1-21
7. Clayton, R. K. 1957. Patterns of accumulation resulting from taxes and changes in motility of microorganisms. Arch.Microbiol. 27:311-319
8. Clayton, R. K. 1958. On the interplay of environmental factors affecting taxis and mobility in Rhodospirillum rubrum. Arch.Microbiol. 29: 189-212
9. Harayama, S. 1977. Phototaxis and membrane potential in the photosynthetic bacterium Rhodospirillum rubrum. J.Bacteriol.131:34-41.
10. Armitage, J. P., and M. C. W. Evans 1981. The reaction centre in the phototactic and chemotactic responses of Rhodopseudomonas sphaeroides. FEMS Microbiol.Letts.11:89-92.
11. Armitage, J. P., C. Ingham, and M. C. W. Evans. 1985. Role of the proton motive force in phototactic and aerotactic responses of Rhodopseudomonas sphaeroides. J.Bacteriol. 163:967-972
12. Ferguson, S. J., J. B. Jackson, and A. G. McEwan. 1987. Anaerobic respiration in the Rhodospirillaceae: characterisation of pathways and evaluation of roles in redox balancing during photosynthesis. FEMS.Microbiol.Rev. 46:117-143.

13   Segall, J. E., A. Ishihara, and H. C.Berg 1985.   Chemotactic
      signalling in filamentous cells of Escherichia coli.J.Bacteriol.
      161:51-59.

14.   Hess, J. F., K. Oosawa, N. Kaplan, and M. I. Simon. 1988.
      Phosphorylation of three proteins in the signalling pathway of
      bacterial chemotaxis. Cell 53:79-87

15.   Parkinson, J. S. 1988.   Protein phosphorylation in bacterial
      chemotaxis. Cell 53:1-2.

16.   Sockett, R. E., J. P. Armitage, and M. C. W. Evans. 1987.
      Methylation-independent and methylation-dependent chemotaxis in
      Rhodobacter sphaeroides and Rhodospirillum rubrum. J.Bacteriol.
      169:5808-5814.

17.   Shioi, J., C. V. Dang, and B. L. Taylor. 1987.   Oxygen as
      attractant and repellent in bacterial chemotaxis. J.Bacteriol.
      169:3118-3123.

18.   Taylor, B. L. 1983. How do bacteria find the optimum concentration
      of oxygen? Trends Biochem.Sci. 8:438-441

19.   Postma, P. W., and J. W. Lengeler 1985. Phosphoenol-
      pyruvate:carbohydrate phosphotransferase system of
      bacteria. Microbiol.Rev. 49:232-269

20.   Ingham, C. J., and J. P. Armitage. 1987. Involvement of transport
      in Rhodobacter sphaeroides. J. Bacteriol. 169:5801-5807.

21.   Poole, P. S., and J. P. Armitage. 1988. Motility response of
      Rhodobacter sphaeroides chemotaxis to chemotactic stimulation. J.
      Bacteriol. 170:5673-5679.

22.   Poole, P. S. and J. P. Armitage. 1989.   Role of metabolism in the
      chemotactic response of Rhodobacter sphaeroides to ammonia. J.
      Bacteriol. 171:2900-2902.

MOLECULAR BIOLOGY OF THE RHODOBACTER SPHAEROIDES FLAGELLUM

R. Elizabeth Sockett, Jocelyn C. A. Foster
and Judith P. Armitage

Microbiology Unit, Department of Biochemistry
University of Oxford, Oxford OX1 3QU

BACKGROUND

The first experiments on the motility of bacteria were conducted by the German scientist Theodore Engelmann in the 1880's. He cultured purple photosynthetic bacteria (probably *Chromatium* and *Rhodospirillum* species) from the River Rhine and noted that they swam rapidly around their illuminated environment in a series of stops, turns, and starts[1]. Since those times there has been considerable interest in the structure and function of flagella from non-photosynthetic bacteria but little interest in the flagella of photosynthetic species.

Flagella are a feature of nearly all photosynthetic, purple, non-sulphur bacteria. In species of the genus *Rhodospirillum* the flagella originate from the poles of the cell in bundles. These bundles rotate in clockwise or counterclockwise directions, a change in the direction of rotation of the flagellar bundle results in a reversal in the direction of swimming of the organism. In the case of members of the genus *Rhodobacter*, there is a single flagellum which rotates and transiently stops rotating[2]. This action powers the bacteria in a random stop-start motility at speeds of up to 100μm per second.

FLAGELLAR ROTATION IN *RHODOBACTER SPHAEROIDES*

The action of the single flagellum of *Rhodobacter sphaeroides* WS8 has been studied bioenergetically in wild type cells using metabolic inhibitors to characterise the factors governing its rotation[2,3,4]. *R. sphaeroides* is an ideal model organism with which to investigate how flagella rotate as it has been found that the flagellum of this organism rotates only in the clockwise direction and stops periodically presumably with a full membrane potential in operation [2]. It is also known that, although *R. sphaeroides* exhibits chemotaxis, it lacks the complex system of methylated sensory transducers and phosphorylated signal compounds that modulate flagellar action in tactic responses in other bacteria[5,6,7] (for a review of the taxis system in this organism see Armitage, Poole and Brown this volume). The absence of these proteins, some of which bind to flagellar components, facilitates a study of the bare flagellar

Molecular Biology of Membrane-Bound Complexes in Phototrophic Bacteria
Edited by G. Drews and E. A. Dawes
Plenum Press, New York, 1990

motor in *R. sphaeroides* without the complication of a highly developed taxis system.

There are several other features of the *R. sphaeroides* flagellum which make it an interesting system to study with regard to membrane differentiation, protein targeting and assembly, in addition to the lure of the age old question of how, mechanistically a gradient of protons is transformed into flagellar rotation.

SITE OF INSERTION OF THE FLAGELLUM

The cytoplasmic membrane of purple non-sulphur bacteria becomes highly differentiated and invaginated to house the photosynthetic apparatus when the bacteria are shifted to low oxygen tensions from highly oxygenated conditions[8,9]. The bacteria are motile under both these conditions and the flagellum is always located in the middle of the long side of the bacterial cell whether the bacteria are growing under high or low oxygen tensions. This raises the questions :
(i) Is there a specially differentiated area of the cytoplasmic membrane which houses the flagellar motor?
(ii) Is the flagellum shed during initiation of photosynthetic membrane synthesis, and is a new flagellum subsequently synthesised?
(iii) How is the medial positioning of the flagellum achieved at cell division?
We have found evidence that there is cessation of motility when *R. sphaeroides* cells are shifted from growth under high light intensity (100 W/m^2) to low light intensity (5 W/m^2). Whether this reflects a shedding of the flagellum, a rearrangement of membrane lipids or proteins around the motor, turnover of a specific motor component, or a temporary depletion in cellular energy is currently under investigation.

Isolation of flagellated vesicles[10] using Protein A-Sepharose columns and anti-flagellar antiserum mixed with vesicles derived from Sarkosyl treated sphaeroplasts photosynthetically grown *R. sphaeroides*, yielded a population of pigmentless, flagellate vesicles which were retarded by the column. The population of vesicles that were not retarded by the column were highly pigmented and did not contain flagella. This suggests that flagella are inserted into pigmentless regions of the membrane devoid of photosynthetic pigment-protein complexes. At present yields from this procedure are not high enough to permit a detailed analysis of the lipid and protein components of the flagellated vesicles. Work is continuing to scale up the procedure and to characterise this apparently non-pigmented, differentiated region of the membrane surrounding the flagellar motor, and also to identify any specific components present that are associated with flagellar rotation. There has been a previous report of non-pigmented populations of membranes being isolated from photosynthetically grown *R. sphaeroides*[8], and this is an intriguing aspect of prokaryotic membrane organisation to be investigated further.

WHY STUDY FLAGELLAR GENE EXPRESSION IN *R. SPHAEROIDES* ?

A lot of work has been carried out characterising the synthesis and assembly of the pigment-protein complexes of purple non-sulphur bacteria into the intracytoplasmic membrane (ICM), and how expression of the genes encoding these products is regulated[9]. The *R. sphaeroides* flagellum provides an unique opportunity to study the expression of genes probably not subject to regulation by oxygen tension or light intensity. It must be remembered that the flagellum is a single multi-enzyme complex which is

present at one copy per cell. Studying the swimming behaviour of a single
R. sphaeroides cell gives a measure of the activity of that single multi-
enzyme complex, not, as in the case of photosynthetic reaction centres, a
value which is the mean activity of all complexes, be they in the process of
assembly, newly synthesised, or fully assembled and inherited from a
parental cell upon cell division. Studying the level of expression of
flagellar genes during the lifetime of R. sphaeroides cells in synchronous
cultures will enable us to investigate whether there is turnover of
flagellar components during the functional life of these complexes. It is
also interesting to elucidate the mechanisms controlling the assembly
of flagellar components outside the bacterial cell such as the flagellar
filament and hook. We have found that if flagella are mechanically sheared
from R. sphaeroides cells motility stops and that it is regained after
periods in excess of sixty minutes. It is interesting to speculate how the
bacterium recognises the removal of its flagellum, and how synthesis and
assembly of new flagellin is induced. It is possible that flagellin is
continually synthesised in R. sphaeroides but the time delay before
motility is regained after shearing suggests that if this is the case,
either flagellar assembly is a very slow process or that the induction of
synthesis of other transport, chaperonin, or scaffolding proteins is
required for flagellar assembly to take place. It seems unlikely that R.
sphaeroides recognises the removal of the flagellum by binding flagellin on
its outer surface as this would not allow discrimination between the
detachment of an individual bacterium's flagellum and shearing of flagella
from other members of the population. A more intriguing possibility is that
the change in torque resulting from the removal of the flagellum is sensed
by the bacterium and causes the induction of flagellin synthesis.

We have undertaken to characterise the R. sphaeroides flagellum at the
molecular level, isolating and characterising flagellar and motility genes
with the hope of understanding:
(i) How protein structures can be rotated at great speeds within a
cytoplasmic membrane.
(ii) How the targeting of flagellar components to a specific location in the
cytoplasmic membrane is achieved.
(iii) Whether flagellar components turnover during operation of the
flagellum and how flagellar gene expression is regulated.

APPROACHES TAKEN TO IDENTIFY FLAGELLAR GENES

Protein Purification

Flagella can be easily isolated and purified from R. sphaeroides cells by
passage of bacteria through fine gauge cannula tubing followed by
differential centrifugation to harvest the bacterial cells and then to
harvest the flagella from the sheared suspension. Flagella prepared in this
way are seen to consist solely of the filaments without any portions of the
hook structure. The flagellar filament has an obviously different helical
array of subunits when negatively stained compared to that of the hook.
SDS polyacrylamide gel electrophoresis of this filament preparation shows
that the filament is a polymer of 54 kilodalton subunits. Preparations of
flagellin have been used to raise a polyclonal antiserum in rabbits, this
antiserum recognises a single 54 kilodalton protein in Western blots of R.
sphaeroides proteins. This antiserum is being used to screen an R.
sphaeroides Lambda gt11 gene library for the presence of the flagellin
gene. The N terminal sequence of the flagellin has been determined and used
to construct a mixed oligonucleotide probe. This probe hybridises with
unique 6kb Stu I, and 2kb Bgl II restriction endonuclease fragments from
chromosomal digests of R. sphaeroides DNA on "unblots". This
oligonucleotide will be used to further analyse positive lambda gt11 clones
isolated from

the library with the anti-flagellar antiserum. There is some amino acid sequence homology between the N terminal sequence of the *R. sphaeroides* flagellum and that of *Escherichia coli*, but the anti-flagellar antiserum does not react with any protein on a Western blot of total *E. coli* proteins. This suggests that there is little overall homology along the length of the flagellins and that only the ends are conserved, as has been found for other bacterial flagellins.

Electron microscopic analysis of isolated "intact" flagella from *R. sphaeroides* has shown that the basal body of the flagellum consists of a number of ring-like structures threaded onto a rod and connected to the filament by a straight, narrow hook structure . We hope to identify the flagellar gene products making up these structures using antisera raised to these gene products made in an over-expression vector system.

Mutagenesis and Screening

The main approach that we have taken to identify flagellar genes is one of transposon mutagenesis coupled with a plate assay for swimming. Mutagenesis was achieved by diparental mating between *E. coli* S17-1 carrying the transposon TnphoA on the plasmid pSup202[11] , and a nalidixic acid resistant derivative of the wild type *R. sphaeroides* strain WS8. Individual kanamycin and nalidixic acid resistant colonies containing the transposon were picked into soft agar swarm plates[12] containing 1/20th the normal level of growth substrates. Wild type *R. sphaeroides* form large circular swarms on these plates as nutrients are depleted at the point of inoculation and motile bacteria swim out responding tactically to attractant concentration gradients. Mutants defective in motility or taxis exhibit small swarming phenotypes on these plates. Strains showing small swarms were picked into liquid culture, grown to mid-log phase and examined microscopically for motility.

Using this procedure we have isolated some 20 individual mutants which are non motile, have no detectable flagellar structure, and show no cross reactivity to an antiserum raised against flagellar filament protein. The absence of flagellin protein in all these mutants suggests that flagellin synthesis is tightly coupled to the synthesis of other flagellar components, this phenomenon has been observed in other bacteria and we wish to investigate the mechanism of this regulation in *R. sphaeroides* It is known that flagellar genes in *E. coli* have upstream regulatory sequences which are only recognised by forms of RNA polymerase including an alternate sigma factor to that usually found[13]. There are no reports of alternate sigma factors being found in association with the *R. sphaeroides* RNA polymerase for recognition of differentially regulated operons, such as those encoding proteins only found under photosynthetic growth conditions. It is possible that this mode of regulation of gene expression may not be present in *R. sphaeroides*, considered an evolutionarily "ancient" bacterium. It will be interesting to determine the organisation of flagellar operons of *R. sphaeroides* and examine regulatory sequences to see if there is any evidence for alternate sigma factor mediated recognition in this system.

Two paralysed mutants, PARA1 and PARA2 were also isolated using the transposon mutagenesis and screening approach. Although these strains are non-motile, they do cross react with anti-flagellar antiserum, and one of them, PARA1 does make seemingly intact flagellar structures complete with basal bodies, visible in the electron microscope.

Chromosomal DNA was isolated from all these mutants and subject to restriction mapping and Southern blotting using transposon DNA as a probe. This established that all thesemutants had independent transposon insertions. DNA flanking the transposon insertion was cloned from three representative mutants, by a shotgun cloning approach. Recombinant puc18 derivatives were subject to selection for resistance to a high level of kanamycin in addition to resistance to ampicillin due to expression of vector sequences, this selected for the presence of the transposon kanamycin resistance gene, and flanking sequences on the multi-copy vector. Clones containing the kanamycin resistance gene of the transposon, along with 5' flanking *R. sphaeroides* chromosomal DNA were obtained and fragments of *R. sphaeroides* DNA without the transposon sequences were isolated from these clones. These DNA fragments were used to probe a pLA2917 [14] broad host range conjugal cosmid library of *R. sphaeroides* WS8 DNA which we have constructed. Small scale preparations of cosmid DNA from the library were made in groups of eight and subject to restriction endonuclease digestion, agarose gel electrophoresis and Southern analysis. Southern blots were hybridised with radioactively labelled probes, and positive cosmid clones were identified and isolated.

Positive cosmid clones were grown up and used in conjugation experiments with the original mutants from whose DNA the probes had been derived. Exconjugants were plated out onto solid agar plates containing tetracycline to select for the presence of the cosmid, and kanamycin to maintain the transposon insertion in the chromosome. After growth under anaerobic, illuminated conditions for 2-3 days at 30°C, exconjugant colonies were cultured in liquid medium containing kanamycin and tetracycline and were examined by phase contrast light microscopy for motility. This procedure has resulted in the isolation of three cosmids:

(i) One which restores flagellar synthesis and motility to a non flagellate mutant.
(ii) One which hybridises to a fragment flanking the transposon insert in another non flagellate but which does not restore motility and flagella formation in trans.
(iii) The third cosmid restores swimming to the mutant PARA1.

Complementation studies are in progress to determine whether these cosmids complement any of the other motility mutants and so examine the clustering of motility and flagellar genes on the chromosome. Genes restoring motility to mutants with paralysed flagella will be sequenced to gain an idea of the nature of the proteins they encode. It is interesting to note that there is no hybridization even at low stringency conditions between motility genes[15] (*mot* A and B) from *E. coli*, an organism with bidirectionally rotating flagella and an MCP chemotaxis system, and the DNA of *R. sphaeroides*. This absence of hybridization may be due solely to the differing G+C contents of the DNA of the two bacteria or it may suggest that there are unique gene products required for the rotation of the unidirectional flagellum of *R. sphaeroides* which are not found in *E. coli*. It may turn out that there is homology between *E.coli* and *R. sphaeroides* motility gene products at the protein level, which was masked in Southern hybridization experiments by differing codon usage. If this is the case then an analysis of protein domains conserved between *R. sphaeroides* and *E.coli* motility genes may give an insight into their functional mechanism which has not been gained by an examination of *E. coli mot* gene sequences alone.

CONCLUSION

A framework has been established within which to analyse the action of the *R. sphaeroides* flagellum at the molecular level. Gene probes have been isolated to use in studies on the regulation of synthesis and assembly of this complex organelle which spans all layers of the bacterial cell, is targeted to a specific location in the membrane, and which is always present as a single copy . An analysis of flagellar motor gene sequences from *R. sphaeroides* will hopefully shed light on the mechanism which causes a proteinaeceous structure to rotate at up to 100 revolutions per second within the membrane of a living cell. *R. sphaeroides* has been consistently one of the best characterised organisms in terms of the mechanisms of energy production and the time has now come to investigate the uses to which this energy isput, including flagellar motility.

ACKNOWLEDGEMENT

This work was supported by grants from the S.E.R.C. and the M.R.C., U.K..

REFERENCES

1. Engelmann, T.W. 1883. *Bacterium photometricum* : An article on the comparitive physiology of the sense for light and colour. Pfluger's Arch. Ges. Physiol. 30: 95-124.
2. Armitage, J.P. and R.M. Macnab. 1987. Unidirectional, intermittent rotation of the flagellum of *R. sphaeroides*. J. Bacteriol. 169: 514-518.
3. Armitage, J.P. and M.C.W. Evans. 1985. Control of the protonmotive force in *Rhodopseudomonas sphaeroides* in the light and dark and its effect on the initiation of flagellar rotation. Bioch. Biophys. Acta. 806: 42-55.
4. Evans, M.C.W. and J.P. Armitage. 1985. Initiation of flagellar rotation in *Rhodopseudomonas sphaeroides*. FEBS Lett. 186: 93-97.
5. Sockett, R.E., J.P. Armitage and M.C.W.Evans.1987. Methylation -independent and methylation-dependent chemotaxis in *Rhodobacter sphaeroides* and *Rhodospirillum rubrum*. J. Bacteriol. 169: 5808-5814.
6. Hazelbauer, G.L. 1988. The bacterial chemosensory system. Can. J. Microbiol. 34: 466-474.
7. Hess, J.F., K. Oosawa, N. Kaplan and M.I. Simon. 1988. Phosphorylation of three proteins in the signalling pathway of bacterial chemotaxis. Cell 53 :79-87.
8. Inamine, G.S., P.A. Reilly and R.A. Neiderman. 1983. Differential protein insertion into developing photosynthetic membrane regions of *Rhodopseudomonas sphaeroides*. J. Cell. Biochem. 24: 69-77.
9. Kiley, P.J. and S. Kaplan. 1988. Molecular genetics of photosynthetic membrane biosynthesis in *Rhodobacter sphaeroides* Micro. Revs. 52: 50 -69.
10. Huguenel, E. and A. Newton. Isolation of flagellated membrane vesicles from *Caulobacter crescentus* cells. Proc. Natl Acad. Sci. USA. 81:3409 -3413.
11. Simon, R., U. Priefer, and A. Puhler. 1983. Vector plasmids for *in vivo* and *in vitro* manipulations of Gram negative bacteria. p 98-106. In A. Puhler (ed) "Molecular genetics of bacterial-plant interactions". Springer- Verlag KG, Berlin.
12. Armstrong, J.B., J. Adler, and M.M. Dahl. 1967. Non chemotactic mutants of *Escherichia coli*. J. Bacteriol. 93: 390-398.
13. Helmann, J.D. and M.J. Chamberlin. 1987. DNA sequence analysis suggests that expression of the flagellar and chemotaxis genes in *Escherichia coli* and *Salmonella typhimurium* is controlled by an alternate sigma factor. Proc. Natl. Acad. Sci. USA 84: 6422-6424.
14. Allen, L.N., and Hanson, R.S. 1985. Construction of broad-host-range cosmid cloning vectors: identification of genes necessary for growth of *Methylobacterium organophilum* on methanol. J. Bacteriol. 161: 955-962.
15. Silverman, M., P. Matsumura, and M. Simon. 1976. The identification of the *mot* gene product with *Escherichia coli*-lambda hybrids. Proc. Natl. Acad. Sci. USA 73: 3126-3130.

THE STRUCTURE OF PORIN FROM *Rhodobacter capsulatus* AT 6 Å

W.Welte[1], T.Wacker[1], U.Nestel,[1]
D.Woitzik[2], J.Weckesser[2,]
M.S.Weiss[3], and G.E.Schulz[3]

[1] Institut für Biophysik und Strahlenbiologie, Albertstr. 23,
[2] Institut für Biologie II, Mikrobiologie, Schänzlestr. 1,
[3] Institut für Organische Chemie und Biochemie, Albertstr. 21,
7800 Freiburg i.Br., FRG

INTRODUCTION

Porins are a class of integral membrane proteins which confer well-defined permeability properties to the outer membrane of gram-negative bacteria. They form aqueous channels across the outer membrane which allow the passage of small polar molecules up to an exclusion size of, typically, 600 daltons. For recent reviews see [1,2,3]. The geometry of the channel and the distribution of the amino acid side chains which form its surface, determine the individual characteristics of porins : the exclusion limit as well as the selectivity as e.g. for anions, cations, phosphates or sugars[3,4]. Except for this selectivity the solutes seem to pass the channel by free diffusion. The exclusion limit is large enough to allow for the passage of antibiotics[5].

The studding of the outer membrane with different types of porins is changed in response to environmental factors. While the cation selective OmpF (matrix porin) and OmpC are present under usual laboratory conditions in *E.coli* (some 100'000 copies per cell[6]), the synthesis of an anion selective pore, PhoE, is induced under conditions of phosphate limitation. Some porins act as phage receptors, as e.g. maltoporin(λ-receptor protein or lamB), T6, phoE and OmpA[7-11]. LamB is involved in the uptake of maltose in *E.coli* . It has an unusually high molecular weight of 47400, while porins usually have molecular weights between 35 kDa and 45 kDa. T6 is believed to facilitate nucleoside permeation across the outer membrane of *E.coli* [12]. OmpA resembles the other porins in several respects [3], a role as a pore could , however, not be demonstrated.

Porins from *E.coli* in vivo [13] and after mild solubilization with detergents [14] , form trimers. These trimers can be dissociated into monomers by heating [14,15] or treatment with guanidinium hydrochloride [16] in the presence of SDS , but the process is accompanied by a profound conformational change as monitored by CD spectroscopy. Porin

Molecular Biology of Membrane-Bound Complexes in Phototrophic Bacteria
Edited by G. Drews and E. A. Dawes
Plenum Press, New York, 1990

479

oligomers from phototrophic purple bacteria, however, could be dissociated into monomers by a mild treatment with EDTA which conserves the pore activity [17].

Two-dimensional crystals of porins have been found in outer membrane preparations[13,18] as well as by reconstitution of detergent solubilized porin with lipids[19-22]. Several different types of these crystals were found : "large hexagonal crystals" (a=93Å) [18,19], "small hexagonal crystals" (a=79Å) [19,20,22], each with one trimer per unit cell, a "rectangular" (a=79Å, b=139Å) crystal form [19,20] with 2 trimers per unit cell and another rectangular crystal form[21] (a=150Å, b=129Å) with 4 trimers per unit cell. These crystals were used for electron microscopic studies using image filtering and three-dimensional reconstruction techniques from both negatively stained and unstained crystals. Engel et al.[23] calculated a three-dimensional reconstruction of negatively stained OmpF crystals at 23Å resolution. According to the map, the trimer possesses three independent aqueous channels near the external side of the membrane, which merge into a single channel near the periplasmic surface. Jap [21] calculated a similar map of negatively stained PhoE crystals at a resolution of 18Å and an electron density projection of unstained crystals at 6.5Å. His map shows three independent aqueous channels with no indications of a fusion into a single channel.

The secondary structure of porins differs from the majority of integral membrane proteins, whose membrane embedded parts aré believed to consist of a bundle of transmembranous hydrophobic α-helices, connected by hydrophilic loops. Porins, in contrast, are known to possess a considerable proportion of β-sheet structure (some 60% of the polypeptide) with their strands being orientated roughly perpendicular to the membrane plane [24,25]. Similar conclusions were drawn from electron density projections of unstained two-dimensional porin crystals at 3.5 Å resolution [26]. The ratio of hydrophilic to hydrophobic amino acids also differs from what is found in the majority of integral membrane proteins. Porins, typically, possess some 50% of polar amino acids [6] and, in this respect, bear more resemblance to water soluble proteins. Primary sequences show a conspicuous pattern of alternation of residues of high and low polarity along stretches of 10 to 12 amino acids. It is tempting to arrange them as amphiphilic β-sheets with every second residue pointing either to the polar or to the apolar side[10]. In this arrangement, they could run repeatedly through the membrane connected by hydrophilic loops and forming amphiphilic β-sheets which are polar on one surface and apolar on the other . In this hypothetical model, the hydrophilic loops are more extended on the external than on the periplasmic surface.

This secondary structure seems to confer to them an unusual stability against attack by proteases, heat and detergents. Solubilization by sodiumdodecylsulfate even at temperatures above room temperature does not cause denaturation. Porins are thus sturdy proteins designed to protect the cytoplasmic membrane of the cell against the attack by detergents and proteases and to function in this hostile environment by allowing, with some degree of specificity, the influx of vitally important small polar compounds and ions. Porins interact electrostatically with the peptidoglycan layer located beneath the outer membrane.

The extraordinary stability may be one reason why OmpF porin from *E.coli* was the first integral membrane protein of which three-dimensional crystals could be obtained which diffracted to high resolution[26,28]. However, the structure analysis of these crystals was tied up by difficulties with the determination of the phases by the heavy metal atom isomorphous replacement method. Garavito also reported the crystallization of lamB and OmpA[29]. As mentioned above, porin trimers from *R.sphaeroides* can be dissociated into monomers by mild treatment with EDTA. It, therefore, appeared interesting to see if different crystal forms or crystals of different quality could be obtained from these monomeric solutions as compared to trimeric solutions.

We have recently crystallized porin from the purple bacterium *Rhodobacter capsulatus* [30]. More recently, we have solved the structure to 6 Å resolution using the multiple isomorphous heavy metal atom replacement method [31] . Here we give a summary of the crystallization method and the structural results.

PURIFICATION AND CRYSTALLIZATION OF PORIN FROM *R.CAPSULATUS*

R.capsulatus 37b4 was grown semiaerobically or aerobically at 32°C in R8ÄH medium [32] containing 0.3% yeast extract. Freshly harvested cells were disrupted mechanically with a cooled vibrogen shaker and cell envelopes were prepared by differential centrifugation as described in detail in [30].

The cell envelopes were first extracted in a buffer (20mM Tris-HCl, pH 8) containing 2% SDS, 10% glycerol at 50°C for 30 min. After centrifugation at 113000 x g, 20°C, 1 h, the pellet was resuspended in the same buffer plus 0.5 M NaCl at 37°C for 30 min. After repeating the aforementioned centrifugation step, the supernatant was enriched with porin. In order to exchange the nonionic detergent octyltetraoxyethylene for SDS, a dialysis procedure was applied. Porin solutions were dialysed against bidestilled water with 3 mM sodium azide at room temperature for several days until the dialysis buffer contained no SDS, using the assay of Hayashi [33]. Subsequently the protein solution was dialysed against the buffer containing 300 mM LiCl, 0.6% w/v octyltetraoxyethylene and either 0 or 20 mM EDTA, for 2 days at room temperature. Finally the protein was subjected to gel filtration chromatography with Sephacryl S200 (Pharmacia Fine Chemicals, Freiburg, FRG , with a gelbed volume of 32 ml).

When no EDTA was present in the dialysis buffer, the elution diagram showed two bands, a prominent one within the void volume at 11 ml and a small one at 16 ml (see **Fig. 1**).

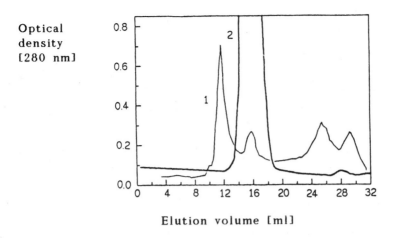

Fig. 1   Elution diagram of porin from gelfiltration with Sephacryl S200
1: without EDTA, 2: after dialysis against 20 mM EDTA .

Calibration with proteins of known molecular weight allowed to estimate a molecular mass of 90000 for the band eluting at 16 ml. Assuming
the mass of a micelle to be roughly 50000, a protein mass of some
40000 can be estimated. When 20 mM EDTA was present in the dialysis
buffer, only one band eluted at 16 ml (see **Fig.1**).

An SDS-PAGE analysis of the latter material showed the presence of
one protein at an apparent molecular mass of 40000 [30]. These results
indicate, that porin from *Rhodobacter capsulatus* forms trimers after the
solubilization from the outer membrane which dissociate to monomers
upon dialysis against EDTA, similarly as reported for porin from *Rhodobacter sphaeroides* [17]. For crystallization, the mixed porin-octyltetra-
oxyethylene micelles were concentrated to 13 mg protein/ml with an
Amicon ultrafiltration cell. Alternatively, lauryldimethylamineoxide was
exchanged for octyltetraoxyethylene . The protein solution was loaded
to a column with DEAE-Cellulose (Whatman, DE52), equilibrated with
25mM Tris-HCl, pH8 and 0.08 % w/v lauryldimethylemineoxide (Fluka
GmbH, Neu-Ulm, FRG, ≥98% pure) , washed with 1 column volume of the
equilibration buffer and eluted by a gradient of 0 to 300 mM sodium
chloride in the aforementioned equilibration buffer. For crystallization
the vapor diffusion technique was used . Polyethyleneglycol(PEG) 400,
600 and 1000 as well as ammonium sulfate were successfully used for
obtaining crystals. For crystallization, 5 µl of the porin solution were
diluted with an equal amount of PEG or ammonium sulfate solution in
25 mM Tris-HCl, pH 7.2, 300 mM LiCl, 5 mM EDTA and either 0.6%
octyltetraoxyethylene or 0.08% lauryldimethylemineoxide. When using
PEG 400, phase separation phenomena were observed together with

crystal growth. With PEG 1000 , amorphous precipitation and crystal growth was observed. With PEG 600 crystal growth predominated . The best results with PEG 600 were obtained with a presaturation of 12% w/v and a reservoir solution of 21% w/v. When using ammonium sulfate, crystals were obtained with a presaturation of 400 mM and a reservoir concentration of 700 mM.

Most crystals had a rhombohedral habit. The largest crystals grew to a size of approximately 0.5 mm cubed. Crystals were mounted in 1 mm quartz capillaries and precession photographs were made with a rotating anode X-ray source (Siemens AG, FRG).

X-RAY STRUCTURE ANALYSIS

The space group was found to be R3 which can be described with hexagonal or rhombohedral axes. The lattice constants are a = b = 95 Å , c = 147 Å or a = 73 Å (α = 82 Å), respectively. The symmetry of the space group with 9 parallel three-fold axes per unit cell, revealed the arrangement of the porin molecules in the unit cell. The three-fold axes arrange the porins into trimers. These trimers arrange into a hexagonally packed two-dimensional sheet with parallel trimer axes and a lattice constant of 95 Å. Finally these sheets packed ontop of each other with a vertical displacement of 49 Å and a displacement of 32 Å along both of the hexagonal sheet axes. In still photographs, reflexions were observed to 2.8 Å , the crystals were stable in the X-ray beam to ~50 h.

Native data have been collected on a four circle diffractometer. Four data sets up to a resolution of 6 Å have been collected and averaged with an internal R-factor 1) of 2.9 % while R-factors between 2 full native data sets were around 5-10 %. In order to obtain heavy atom derivatives of the native molecule, attempts were made to soak crystals in solutions of heavy atom salts and to cocrystallize porin in the presence of heavy atom compounds. The quality of the derivative data as checked by an internal R-factor 2) was about as good as that of the native data, the R-factors between native and derivative data sets were around 20 %.

$$1) \quad R = 2 \cdot \frac{\sum_{hkl} |F^+ - F^-|}{\sum_{hkl} |F^+ + F^-|} \quad ; \quad F^+ = |\underline{F}(0kl)| \ , \ F^- = |\underline{F}(k\bar{k}l)|$$

$$2) \quad R = 2 \cdot \frac{\sum_{hkl} |F_1 - F_2|}{\sum_{hkl} |F_1 + F_2|} \quad ; \quad F_1 = |\underline{F}_1(hkl)| \ , \ F_2 = |\underline{F}_2(hkl)|$$

To obtain the heavy atom binding sites, difference Patterson maps of the derivatives were calculated at 6 and 9 Å resolution. The difference Patterson map of one platinum derivative could be interpreted in terms of three sites. All other sites were then derived from difference Fourier maps. With one platinum and one uranium derivative an electron density

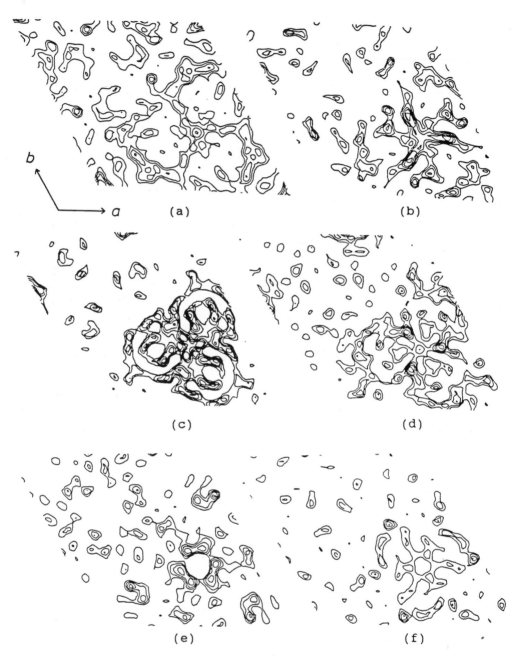

Fig. 2   Sections of the electron density map cut parallel to the trimer
sheets. The displayed sections are at z=11/54, 14/54, 17/54,
20/54, 23/54 and 26/54. The density contours are at 12%, 24%
etc. of the maximum density in the map.

map could be calculated using the multiple isomorphous replacement (MIR) method. **Fig.2** shows six z-sections of the electron density map with a vertical distance of 8.2 Å between two sections and a horizontal scale of 1 cm = 14.9 Å.

The trimeric structure of the pore-forming complex is clearly visible. The central region of the trimer covers an area of about 8000 Å^2 in the xy-plane, with a vertical length of about 50 Å.

Because of the density shapes the secondary structure seems to be mainly β-pleated sheet structure. Only few α-helices, if any, seem to be possible. This is consistent with earlier spectroscopic investigations on OmpF porin from *E.coli* [24,25]. It is also in accord with projection maps of the electron density of OmpF [26] and of PhoE [21], calculated by image filtering of electron micrographs from unstained two dimensional porin crystals. These maps show the cylindrical structure of the mono- mers bordered by high density which accounts for the projected β-sheets and a low density area in the center of each cylinder which accounts for the possible pore region. It is therefore tempting, to describe the whole trimeric complex as three merged cylinders whose walls are built up by β-sheets forming a barrel-like structure. The β-strands run roughly perpendicular to the membrane plane and are connected by hydrophilic loops, which extend both into the exterior and into the periplasmic space. The exact size of the pore and the geometry are determined by side chains protruding into the interior. The electron density map shows, that the whole molecule is remarkable porous with several regions of low electron density in the interior of the molecule. On certain z-sec- tions, there exists some density suggesting a lateral domain which could act as a spacer.

The analysis is continued to higher resolution.

References

[1]   Nikaido,H. and Vaara, M., Outer membrane, in: *E.coli* and
       S. typhimurium ,Cellular and Molecular Biology (Neidhard, F.C.,
       ed.), American Society for Microbiol., Washington D.C. (1987), 7-22.

[2]   Benz,R. and Bauer,K.,Permeation of hydrophilic molecules through
       the outer membrane of gram-negative bacteria,
       Eur.J.Biochem. 176 :1-19 (1988).

[3]   Benz,R., Porin from bacterial and mitochondrial outer membranes,
       in: CRC Critical Reviews in Biochem. 19 (1985) , 145-190.

[4]   Benz,R., Schmid,A., van der Ley,P. and Tommassen,P.,Molecular
       basis of porin selectivity : membrane experiments with
       OmpC-PhoE and OmpF-PhoE hybrid proteins of *Escherischia
       Coli* K-12, *Biochim.Biophys.Acta 981* : 8-14 (1989)

[5]   Nikaido,H., Rosenberg,E. and Foulds,J., Porin Channels
       in *Escherischia Coli* : Studies with β-lactams in intact cells,
       J. Bacteriol. 153 ,232-240 (1983).

[6]   Rosenbusch, J.P., Characterization of the major envelope protein
       from *Escherischia coli*, *J. Biol. Chem.* 249 :8019-8029 (1974).

[7]   Thirion,J.P. and Hofnung,M., On some genetic aspects of phage
       λ resistance in *E.coli* K12, *Genetics 71* , 207-216 (1972).

[8] Randall-Hazelbauer,L. and Schwartz,M., Isolation of the Bacteriophage Lambda Receptor from *Escherischia Coli*, *J. Bacteriol. 116* :1436-1446. (1973).

[9] Neuhaus,J.M. ,The receptor protein of phage λ: purification. caracterization and preliminary electrical studies in planar lipid bilayers, *Ann. Microbiol. (Inst.Pasteur) 133A* ,27-32 (1982).

[10] van der Ley,P. and Tomassen,J. ,PhoE protein structure and function, in : Phosphate Metabolism and Cellular Regulation in Microorganisms (Torriani-Corini,T. ed) American Society for Microbiol. , Washington D.C. (1987), 159-163.

[11] Morona,R., Klose,M. and Henning,U., *Escherischia coli* K-12 outer membrane protein (OmpA) as a bacteriophage receptor : analysis of mutant genes expressing altered proteins, *J. Bacteriol. 159* :570-578 (1984).

[12] Munch-Petersen,A., Mygind,B., Nicolaisen,A. and Pihl,N.J., Nucleoside Transport in Cells and Membrane Vesicles from *Escherischia Coli* K12, *J.Biol.Chem. 254* :3730-3737 (1979).

[13] Steven,A.C., tenHeggeler,B., Müller,R., Kistler,J. and Rosenbusch,J.P., Ultrastructure of a Periodic Protein Layer in the Outer membrane of *Escherischia Coli*, *J.Cell Biol. 72* :292-301 (1977).

[14] Nakae,T., Ishii,J. and Tokunaga,M., Subunit structure of functional porin oligomers that form permeability channels in the outer membrane of *Escherischia coli*, *J.Biol.Chem. 254* :1457-1461 (1979).

[15] Touaga,M., Tokunaga,H., Okajima,Y. and Nakae,T., Characterization of porins from the outer membrane of *Salmonalla typhimurium*, *Eur. J. Bioch. 95* :441-448 (1979).

[16] Nakae,T. and Ishii,J.N., Molecular weight and subunit structure of LamB proteins, *Ann. Microbiol.(Inst.Pasteur) 133A* :21-25 (1982).

[17] Weckesser,J., Zalman,L.S. and Nikaido,H., Porin from *Rhodopseudomonas sphaeroides*, *J.Bacteriol. 159* :199-205 (1984).

[18] Chalcroft,J.P., Engelhardt,H. and Baumeister,W., Structure of the porin from a bacterial stalk, *FEBS Lett. 211* :53-58 (1987).

[19] Dorset,D.L., Engel, A., Häner,M., Massalski,A. and Rosenbusch,J.P., Two-dimensional crystal packing of matrix porin, *J.Mol.Biol. 165* :701-710 (1983).

[20] Lepault,J., Dargent,B., Tichelaar,W., Rosenbusch,J.P., Leonard,K. and Pattus,F., Three-dimensional reconstruction of maltoporin from electron microscopy and image processing, *EMBO J. 7* :261-268 (1988).

[21] Jap,B.K., Molecular design of PhoE porin and its functional consequences, *J.Mol.Biol. 205* :407-419 (1989).

[22] Chang,C., Mizushima,S. and Glaeser,R., Projected Structure of the Pore-Forming PmpC Protein from *Escherischia Coli* Outer Membrane, *Biophys. J. 47* : 629-639 (1985).

[23] Engel,A., Massalski,A., Schindler,H., Dorset,D.L. and Rosenbusch,J.P., Porin channel triplets merge into single outlets in *Escherischia coli* outer membranes, *Nature 317* :643-645 (1985).

[24]   Kleffel,B., Garavito,R.M., Baumeister,W. and Rosenbusch, J.P., Secondary structures of a channel-forming protein : porin from *E.coli* outer membranes, *EMBO J. 4*:1589-1592 (1985).

[25]   Nabedryk, E., Garavito, R.M. and Breton, J., The orientation of β-sheets in porin. A polarized Fourier transform infrared spectroscopic investigation, *Biophys. J. 53* :671-676 (1988).

[26]   Sass,H.J., Büldt,G., Beckmann,E., Zenlin,F., vanHeel,M., Zeitler,E., Rosenbusch, J.P., Dorset,D.L. and Massalski,A., Densely packed β-structure at the Protein-Lipid Interface of Porin is revealed by High Resolution Cryo-Electron Microscopy , *J.Mol.Biol. 209* : 171-175 (1989).

[27]   Garavito,R.M. and Rosenbusch,J.P. , Three dimensional crystals of an integral membrane protein, *J.Cell Biol. 86* :327-329 (1980)

[28]   Garavito,R.M., Jenkins,J., Jansonius,J.N., Karlsson,R. and Rosenbusch,J.P., X-ray diffraction analysis of matrix porin, an integral membrane protein from *Escherischia coli* outer membranes, *J.Mol.Biol. 164* :313-327 (1983).

[29]   Garavito,R.M., Hinz,U. and Neuhaus,J.M., The crystallization of outer membrane proteins from *Escherischia coli* , *J.Biol.Chem. 259* :4254-4257 (1984)

[30]   Nestel,U., Wacker,T., Woitzik,D., Weckesser,J., Kreutz,W. and Welte,W., Crystallization and preliminary X-ray analysis of porin from *Rhodobacter capsulatus. FEBS Lett. 242* :405-408 (1989).

[31]   Weiss,M.S., Wacker,T., Nestel,U., Woitzik,D., Weckesser,J., Kreutz,W., Welte,W. and Schulz,G.E. , The structure of porin from *Rhodobacter capsulatus* at 0.6 nm resolution, *FEBS Lett.* in press (1989).

[32]   Flamann,H.T. and Weckesser,J., Characterization of the Cell Wall and Outer Membrane of *Rhodopseudomonas capsulata* ,*J.Bacteriol. 159* : 191-198 (1984).

[33]   Hayashi,K., A rapid determination of sodium dodecyl sulfate with methylene blue, *Anal. Biochem. 67* : 503-506 (1975).

DATE DUE

DEC 0 8 1997	OCT 2 6 2011		